Modern Statistical Methods for Astronomy

Modern astronomical research is beset with a vast range of statistical challenges, ranging from reducing data from megadatasets to characterizing an amazing variety of variable celestial objects or testing astrophysical theory. Linking astronomy to the world of modern statistics, this volume is a unique resource, introducing astronomers to advanced statistics through ready-to-use code in the public-domain **R** statistical software environment.

The book presents fundamental results of probability theory and statistical inference, before exploring several fields of applied statistics, such as data smoothing, regression, multivariate analysis and classification, treatment of nondetections, time series analysis, and spatial point processes. It applies the methods discussed to contemporary astronomical research datasets using the **R** statistical software, making it an invaluable resource for graduate students and researchers facing complex data analysis tasks.

A link to the authors' website for this book can be found at www.cambridge.org/msma. Material available on their website includes datasets, **R** code and errata.

Eric D. Feigelson is a Professor in the Department of Astronomy and Astrophysics at Pennsylvania State University. He is a leading observational astronomer and has worked with statisticians for 25 years to bring advanced methodology to problems in astronomical research.

G. Jogesh Babu is Professor of Statistics and Director of the Center for Astrostatistics at Pennsylvania State University. He has made extensive contributions to probabilistic number theory, resampling methods, nonparametric methods, asymptotic theory, and applications to biomedical research, genetics, astronomy, and astrophysics.

Modern Statistical Methods for Astronomy

With R Applications

ERIC D. FEIGELSON

Pennsylvania State University

G. JOGESH BABU

Pennsylvania State University

CAMBRIDGE
UNIVERSITY PRESS

CAMBRIDGE
UNIVERSITY PRESS

University Printing House, Cambridge CB2 8BS, United Kingdom

One Liberty Plaza, 20th Floor, New York, NY 10006, USA

477 Williamstown Road, Port Melbourne, VIC 3207, Australia

4843/24, 2nd Floor, Ansari Road, Daryaganj, Delhi - 110002, India

79 Anson Road, #06-04/06, Singapore 079906

Cambridge University Press is part of the University of Cambridge.

It furthers the University's mission by disseminating knowledge in the pursuit of
education, learning and research at the highest international levels of excellence.

www.cambridge.org
Information on this title: www.cambridge.org/9780521767279

© E. D. Feigelson and G. J. Babu 2012

First published 2012
4th printing 2015

A catalogue record for this publication is available from the British Library

Library of Congress Cataloging in Publication data
Feigelson, Eric D.
Modern statistical methods for astronomy : with R applications /
Eric D. Feigelson, G. Jogesh Babu.
p.   cm.
ISBN 978-0-521-76727-9 (hardback)
1. Statistical astronomy.   I. Babu, Gutti Jogesh, 1949–   II. Title.
QB149.F45   2012
520.72'7 – dc23   2012009113

ISBN 978-0-521-76727-9 Hardback

Additional resources for this publication: www.cambridge.org/msma

For Zoë, Clara and Micah

In memory of my parents,
Nagarathnam and Mallayya

Contents

The color plates are to be found between pages 398 and 399.

Preface

Motivation and goals

For many years, astronomers have struggled with the application of sophisticated statistical methodologies to analyze their rich datasets and address complex astrophysical problems. On one hand, at least in the United States, astronomers receive little or no formal training in statistics. The traditional method of education has been informal exposure to a few familiar methods during early research experiences. On the other hand, astronomers correctly perceive that a vast world of applied mathematical and statistical methodologies has emerged in recent decades. But systematic, broad training in modern statistical methods has not been available to most astronomers.

This volume seeks to address this problem at three levels. First, we present fundamental principles and results of broad fields of statistics applicable to astronomical research. The material is roughly at a level of advanced undergraduate courses in statistics. We also outline some recent advanced techniques that may be useful for astronomical research to give a flavor of the breadth of modern methodology. It is important to recognize that we give only incomplete introductions to the fields, and we guide the astronomer towards more complete and authoritative treatments.

Second, we present tutorials on the application of both simple and more advanced methods applied to contemporary astronomical research datasets using the **R** statistical software package. **R** has emerged in recent years as the most versatile public-domain statistical software environment for researchers in many fields. In addition to a coherent language for data analysis and common statistical tools, over 3000 packages have been added for advanced analyses in the **CRAN** archive. We have culled these packages for functionalities that may be useful to astronomers. **R** can also be linked to other analysis systems and languages such as C, FORTRAN and Python, so that legacy codes can be included in an **R**-based analysis and *vice versa*.

Third, we hope the book communicates to astronomers our enthusiasm for statistics as a substantial and fascinating intellectual enterprise. Just as astronomers use the latest engineering to build their telescopes and apply advanced physics to interpret cosmic phenomena, they can benefit from exploring the many roads of analyzing and interpreting data through modern statistical analysis.

Another important purpose of this volume is to give astronomers and other physical scientists a bridge to the vast library of specialized texts and monographs in statistics and allied fields. We strongly encourage researchers who are engaged in statistical data analysis to read more detailed treatments in the 'Recommended reading' at the end of each chapter; they are carefully chosen from many available volumes. Most of the material in the book which is not specifically referenced in the text is presented in more detail in these recommended readings. To further this goal, the present book does not shy away from technical language that, though unfamiliar in the astronomical community, is critical for further learning from the statistical literature. For example, the astronomers' "upper limits" are "left-censored data points", a "power-law distribution" is a "Pareto distribution", and "$1/f$ noise" is a "long-memory process". The text make these connections between the languages of astronomy and statistics, and the comprehensive index can assist the reader in finding material in both languages.

The reader may find the appendices useful. An introduction to **R** is given in Appendix B. It includes an overview of the programming language and an outline of its statistical functionalities, including the many **CRAN** packages. **R** applications to astronomical datasets are given at the end of each chapter which implement methods discussed in the text. Appendix C presents 18 astronomical datasets illustrating the range of statistical challenges that arise in contemporary research. The full datasets and **R scripts** are available online at http://astrostatistics.psu.edu/MSMA. Readers can thus easily reproduce the **R** results in the book.

In this volume, we do not present mathematical proofs underlying statistical results, and we give only brief outlines of a few computational algorithms. We do not review research at the frontiers of astrostatistics, except for a few topics where astronomers have contributed critically important methodology (such as the treatment of truncated data and irregularly spaced time series). Only a small fraction of the many methodological studies in the recent astronomical literature are mentioned. Some fields of applied statistics useful for astronomy (such as wavelet analysis and image processing) are covered only briefly. Finally, only $\sim$2500 **CRAN** packages were examined for possible inclusion in the book; roughly one new package is added every day and many others are extended.

Audience

The main audience envisioned for this volume is graduate students and researchers in observational astronomy. We hope it serves both as a textbook in a course on data analysis or astrostatistics, and as a reference book to be consulted as specific research problems are encountered. Researchers in allied fields of physical science, such as high-energy physics and Earth sciences, may also find portions of the volume helpful. Statisticians can see how existing methods relate to questions in astronomy, providing background for astrostatistical research initiatives.

Our presentation assumes that the reader has a background in basic linear algebra and calculus. Familiarity of elementary statistical methods commonly used in the physical sciences is also useful; this preparatory material is covered in volumes such as Bevington & Robinson (2002) and Cowan (1998).

Outline and classroom use

The introduction (Chapter 1) reviews the long historical relationship between astronomy and statistics and philosophical discussions of the relationship between statistical and scientific inference. We then start with probability theory and proceed to lay foundations of statistical inference: hypothesis testing, estimation, modeling, resampling and Bayesian inference

(Chapters 2 and 3). Probability distributions are discussed in Chapter 4 and nonparametric statistics are covered in Chapter 5.

The volume proceeds to various fields of applied statistics that rest on these foundations. Data smoothing is covered in Chapters 5 and 6. Regression is discussed in Chapter 7, followed by analysis and classification of multivariate data (Chapters 8 and 9). Treatments of nondetections are covered in Chapter 10, followed by the analysis of time-variable astronomical phenomena in Chapter 11. Chapter 12 considers spatial point processes. The book ends with appendices introducing the **R** software environment and providing astronomical datasets illustrative of a variety of statistical problems.

We can make some recommendation regarding classroom use. The first part of a semester course in astrostatistics for astronomy students would be devoted to the principles of statistical inference in Chapters 1–4 and learning the basics of **R** in Appendix B. The second part of the semester would be topics of applied statistical methodology selected from Chapters 5–12. We do not provide predefined student exercises with definitive answers, but rather encourage both instructors and students to develop open-ended explorations of the contemporary astronomical datasets based on the **R** tutorials distributed throughout the volume. Suggestions for both simple and advanced problems are given in the dataset presentations (Appendix C).

Astronomical datasets and R scripts

The datasets and **R** scripts in the book can be downloaded from Penn State's Center for Astrostatistics at http://astrostatistics.psu.edu/MSMA. The **R** scripts are self-contained; simple cut-and-paste will ingest the datasets, perform the statistical operations, and produce tabular and graphical results.

Extensive resources to pursue issues discussed in the book are available on-line. The **R** system can be downloaded from http://www.r-project.org and **CRAN** packages are installed on-the-fly within an **R** session. The primary astronomy research literature, including full-text articles, is available through the NASA–Smithsonian *Astrophysics Data System* (http://adswww.harvard.edu). Thousands of astronomical datasets are available from the Vizier service at the Centre des Données Stellaires (http://vizier.u-strasbg.fr) and the emerging *International Virtual Observatory Alliance* (http://ivo.net). The primary statistical literature can be accessed through MathSciNet (http://www.ams.org/mathscinet/) provided by the American Mathematical Society. Considerable statistical information is available on Wikipedia (http://en.wikipedia.org/wiki/Index_of_statistics_articles). Astronomers should note, however, that the best way to learn statistics is often through textbooks and monographs written by statisticians, such as those in the recommended reading.

Acknowledgements

This book emerged from 25 years of discussion and collaboration between astronomers and statisticians at Penn State under the auspices of the Center for Astrostatistics. The volume particularly benefited from the lectures and tutorials developed for the *Summer Schools in Statistics for Astronomers* since 2005 and taught at Penn State and Bangalore's Indian Institute of Astrophysics. We are grateful to our dozens of statistician colleagues who have taught at the *Summer Schools in Statistics for Astronomers* for generously sharing their knowledge and perspectives. David Hunter and Arnab Chakraborty developed **R** tutorials for astronomers. Donald Percival generously gave detailed comments on the time series analysis chapter. We are grateful to Nancy Butkovich and her colleagues for providing excellent library services. Finally, we acknowledge the National Science Foundation, National Aeronautics and Space Administration, and the Eberly College of Science for supporting astrostatistics at Penn State over many years.

Eric D. Feigelson
G. Jogesh Babu
Center for Astrostatistics
Pennsylvania State University
University Park, PA, U.S.A.

1 Introduction

1.1 The role of statistics in astronomy

1.1.1 Astronomy and astrophysics

Today, the term "astronomy" is best understood as shorthand for "astronomy and astrophysics". Astronomy (*astro* = star and *nomen* = name in ancient Greek) is the observational study of matter beyond Earth: planets and bodies in the Solar System, stars in the Milky Way Galaxy, galaxies in the Universe, and diffuse matter between these concentrations of mass. The perspective is rooted in our viewpoint on or near Earth, typically using telescopes on mountaintops or robotic satellites to enhance the limited capabilities of our eyes. Astrophysics (*astro* = star and *physis* = nature) is the study of the intrinsic nature of astronomical bodies and the processes by which they interact and evolve. This is an indirect, inferential intellectual effort based on the (apparently valid) assumption that physical processes established to rule terrestrial phenomena – gravity, thermodynamics, electromagnetism, quantum mechanics, plasma physics, chemistry, and so forth – also apply to distant cosmic phenomena. Figure 1.1 gives a broad-stroke outline of the major fields and themes of modern astronomy.

The fields of astronomy are often distinguished by the structures under study. There are planetary astronomers (who study our Solar System and extra-solar planetary systems), solar physicists (who study our Sun), stellar astronomers (who study other stars), Galactic astronomers (who study our Milky Way Galaxy), extragalactic astronomers (who study other galaxies), and cosmologists (who study the Universe as a whole). Astronomers can also be distinguished by the type of telescope used: there are radio astronomers, infrared astronomers, visible-light astronomers, X-ray astronomers, gamma-ray astronomers, and physicists studying cosmic rays, neutrinos and the elusive gravitational waves. Astrophysicists are sometimes classified by the processes they study: astrochemists, atomic and nuclear astrophysicists, general relativists (studying gravity) and cosmologists.

The astronomer might proceed to investigate stellar processes by measuring an ordinary main-sequence star with spectrographs at different wavelengths of light, examining its spectral energy distribution with thousands of absorption lines. The astrophysicist interprets that the emission of a star is produced by a sphere of 10^{57} atoms with a specific mixture of elemental abundances, powered by hydrogen fusion to helium in the core, revealing itself to the Universe as a blackbody surface at several thousand degrees temperature. The development of the observations of normal stars started in the late-nineteenth century,

Overview of modern astronomy and astrophysics

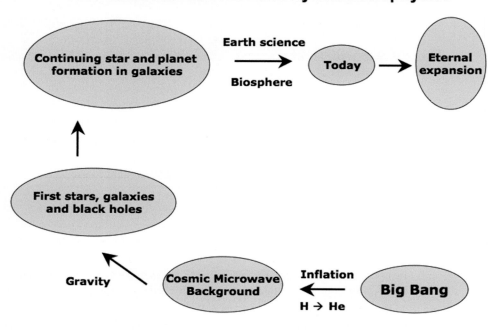

Lifecycle of the stars

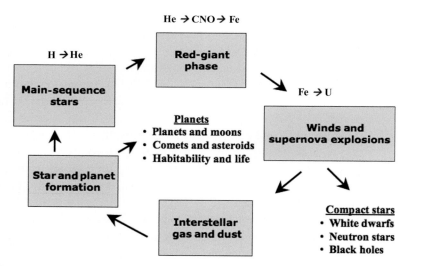

Fig. 1.1 Diagrams summarizing some important fields and themes of modern astronomy. *Top*: the history and growth of structures of the expanding Universe; *bottom*: the evolution of stars with generation of heavy elements and production of long-lived structures.

and the successful astrophysical interpretation emerged gradually throughout the twentieth century. The vibrant interwoven progress of astronomy and astrophysics continues today as many other cosmic phenomena, from molecular clouds to black holes, are investigated. Cosmology, in particular, has emerged with the remarkable inference that the familiar atoms around us comprise only a small fraction of the "stuff" in the Universe which is dominated by mysterious dark matter and dark energy inaccessible to normal telescopes or laboratory instruments.

1.1.2 Probability and statistics

While there is little debate about the meaning and goals of astronomy and astrophysics as intellectual enterprises, the meaning and goals of probability and statistics has been widely debated. In his volume *Statistics and Truth*, C. R. Rao (1997) discusses how the term "statistics" has changed meaning over the centuries. It originally referred to the collection and compilation of data. In the nineteenth century, it accrued the goal of the mathematical interpretation of data, often to assist in making real-world decisions. Rao views contemporary statistics as an amalgam of a science (techniques derived from mathematics), a technology (techniques useful for decision-making in the presence of uncertainty), and an art (incompletely codified techniques based on inductive reasoning).

Barnett (1999) considers various viewpoints on the meaning of statistics. The first group of quotes see statistics as a very useful, but essentially mechanical, technology for treating data. In this sense, it plays a role similar to astronomy's role *vis á vis* astrophysics.

1. "The first task of a statistician is cross-examination of data." (Sir R. A. Fisher, quoted by Rao 1997)
2. "[S]tatistics refers to the methodology for the collection, presentation, and analysis of data, and for the uses of such data." (Neter *et al.* 1978)
3. "Broadly defined, statistics encompasses the theory and methods of collecting, organizing, presenting, analyzing, and interpreting data sets so as to determine their essential characteristics." (Panik 2005)

The following interpretations of statistics emphasize its role in reducing random variations in observations to reveal important effects in the underlying phenomenon under study.

4. "A statistical inference carries us from observations to conclusions about the populations sampled." (Cox 1958)
5. "Uncertain knowledge + Knowledge of the amount of uncertainty in it = Usable knowledge." (Rao 1997)
6. "My favourite definition [of statistics] is bipartite: statistics is both the science of uncertainty and the technology of extracting information from data." (Hand 2010)
7. "In statistical inference experimental or observational data are modeled as the observed values of random variables, to provide a framework from which inductive conclusions may be drawn about the mechanism giving rise to the data." (Young & Smith 2005)

1.1.3 Statistics and science

Opinions differ widely when considering the relationship between statistical analysis of empirical data and the underlying real phenomena. A group of prominent twentieth-century statisticians express considerable pessimism that statistical models are anything but useful fictions, much as Renaissance Europe debated the meaning of Copernicus' heliocentric cosmological model. These scholars view statistical models as useful but often trivial or even misleading representations of a complex world. Sir D. R. Cox, towards the end of a long career, perceives a barrier between statistical findings and the development or validation of scientific theories.

8. "There is no need for these hypotheses to be true, or even to be at all like the truth; rather one thing is sufficient for them – that they should yield calculations which agree with the observations." (Osiander's preface to Copernicus' *De Revolutionibus*, quoted by Rao 1997)
9. "Essentially, all models are wrong, but some are useful." (Box & Draper 1987)
10. "[Statistical] models can provide us with ideas which we test against data, and about which we build up experience. They can guide our thinking, lead us to propose courses of action, and so on, and if used sensibly, and with an open mind, and if checked frequently with reality, might help us learn something that is true. Some statistical models are helpful in a given context, and some are not. . . . What we do works (when it does) because it can be seen to work, not because it is based on true or even good models of reality." (Speed 1992, addressing a meeting of astronomers)
11. "It is not always convenient to remember that the right model for a population can fit a sample of data worse than a wrong model, even a wrong model with fewer parameters. We cannot rely on statistical diagnostics to save us, especially with small samples. We must think about what our models mean, regardless of fit, or we will promulgate nonsense." (Wilkinson 2005).
12. "The object [of *statistical* inference] is to provide ideas and methods for the critical analysis and, as far as feasible, the interpretation of empirical data . . . The extremely challenging issues of *scientific* inference may be regarded as those of synthesising very different kinds of conclusions if possible into a coherent whole or theory . . . The use, if any, in the process of simple *quantitative* notions of probability and their numerical assessment is unclear . . . " (Cox 2006)

Other scholars quoted below are more optimistic. The older Sir R. A. Fisher bemoans a mechanistic view of statistics without meaning in the world. G. Young and R. Smith imply that statistical modeling can lead to an understanding of the causative mechanisms of variations in the underlying population. I. Hacking, a philosopher, believes statistics can improve our scientific inferences but not lead to new discovery. B. Efron, in an address as President of the American Statistical Association, feels that statistics can propel many sciences towards important results and insights.

13. "To one brought up in the free intellectual atmosphere of an earlier time there is something rather horrifying in the ideological movement represented by the doctrine that

reasoning, properly speaking, cannot be applied to empirical data to lead to inferences valid in the real world." (Fisher 1973)

14. "The quiet statisticians have changed our world, not by discovering new facts or technical developments, but by changing the ways we reason, experiment, and form our opinions." (Hacking 1990)

15. "Statistics has become the primary mode of quantitative thinking in literally dozens of fields, from economics to biomedical research. The statistical tide continues to roll in, now lapping at the previously unreachable shores of the hard sciences. ... Yes, confidence intervals apply as well to neutrino masses as to disease rates, and raise the same interpretive questions, too." (Efron 2004)

16. "The goal of science is to unlock nature's secrets. ... Our understanding comes through the development of theoretical models which are capable of explaining the existing observations as well as making testable predictions. ... Fortunately, a variety of sophisticated mathematical and computational approaches have been developed to help us through this interface, these go under the general heading of statistical inference." (Gregory 2005)

Leading statisticians are thus often more cautious, or at least less self-confident, about the value of their labors for understanding phenomena than are astronomers. Most astronomers believe implicitly that their observations provide a clear window into the physical Universe, and that simple quantitative statistical interpretations of their observations represent an improvement over qualitative examination of the data.

We generally share the optimistic view of statistical methodology in the service of astronomy and astrophysics, as expressed by P. C. Gregory (2005). In the language of the philosophy of science, we are positivists who believe that underlying causal relationships can be discovered through the detection and study of regular patterns of observable phenomena. While quantitative interpretation and models of complex biological and human affairs attempted by many statisticians may be more useful for prediction or decision-making than understanding the underlying behaviors, we feel that quantitative models of many astrophysical phenomena can be very valuable. A social scientist might interview a sample of voters to accurately predict the outcome of an election, yet never understand the beliefs underlying these votes. But an astrostatistician may largely succeed in understanding the orbits of binary stars, or the behavior of an accretion disk around a black hole or the growth of structure in an expanding Universe, that must obey deterministic mathematical laws of physics.

However, we wish to convey throughout this volume that the process of linking statistical analysis to reality is not simple and challenges must be faced at all stages. In setting up the calculation, there are often several related questions that might be asked in a given scientific enterprise, and their statistical evaluation may lead to apparently different conclusions. In performing the calculation, there are often several statistical approaches to a given question asked about a dataset, each mathematically valid under certain conditions, yet again leading to different scientific inferences. In interpreting the result, even a clear statistical finding may give an erroneous scientific interpretation if the mathematical model is mismatched to physical reality.

Astronomers should be flexible and sophisticated in their statistical treatments, and adopt a more cautious view of the results. A "3-sigma" result does not necessarily represent astrophysical reality. Astronomers might first seek consensus about the exact question to be addressed, apply a suite of reasonable statistical approaches to the dataset with clearly stated assumptions, and recognize that the link between the statistical results and the underlying astrophysical truth may not be straightforward.

1.2 History of statistics in astronomy

1.2.1 Antiquity through the Renaissance

Astronomy is the oldest observational science. The effort to understand the mysterious luminous objects in the sky has been an important element of human culture for tens of thousands of years. Quantitative measurements of celestial phenomena were carried out by many ancient civilizations. The classical Greeks were not active observers but were unusually creative in the applications of mathematical principles to astronomy. The geometric models of the Platonists with crystalline spheres spinning around the static Earth were elaborated in detail, and this model endured in Europe for fifteen centuries.

The Greek natural philosopher Hipparchus made one of the first applications of mathematical principles in the realm of statistics, and started a millennium-long discussion on procedures for combining inconsistent measurements of a physical phenomenon (Sheynin 1973, Hald 2003). Finding scatter in Babylonian measurements of the length of a year, defined as the time between solstices, he took the middle of the range – rather than the mean or median – for the best value. Today, this is known as the midrange estimator of location, and is generally not favored due to its sensitivity to erroneous observations. Ptolemy and the eleventh-century Persian astronomer Abu Rayhan Biruni (al-Biruni) similarly recommended the average of extremes. Some medieval scholars advised against the acquisition of repeated measurements, fearing that errors would compound uncertainty rather than compensate for each other. The utility of the mean of discrepant observations to increase precision was promoted in the sixteenth century by Tycho Brahe and Galileo Galilei. Johannes Kepler appears to have inconsistently used arithmetic means, geometric means and middle values in his work. The supremacy of the mean was not settled in astronomy until the eighteenth century (Simpson 1756).

Ancient astronomers were concerned with observational errors, discussing dangers of propagating errors from inaccurate instruments and inattentive observers. In a study of the corrections to astronomical positions from observers in different cities, al-Biruni alludes to three types of errors: ". . . the use of sines engenders errors which become appreciable if they are added to errors caused by the use of small instruments, and errors made by human observers" (quoted by Sheynin 1973). In his 1609 *Dialogue on the Two Great World Views, Ptolemaic and Copernican*, Galileo also gave an early discussion of observational errors concerning the distance to the supernova of 1572. Here he outlined in nonmathematical language many of the properties of errors later incorporated by Gauss into his quantitative theory of errors.

1.2.2 Foundations of statistics in celestial mechanics

Celestial mechanics in the eighteenth century, in which Newton's law of gravity was found to explain even the subtlest motions of heavenly bodies, required the quantification of a few interesting physical quantities from numerous inaccurate observations. Isaac Newton himself had little interest in quantitative probabilistic arguments. In 1726, he wrote concerning discrepant observations of the Comet of 1680 that, "From all this it is plain that these observations agree with theory, in so far as they agree with one another" (quoted by Stigler 1986).

Others tackled the problem of combining observations and estimating physical quantities through celestial mechanics more earnestly. In 1750 while analyzing the libration of the Moon as head of the observatory at Göttingen, Tobias Mayer developed a "method of averages" for parameter estimation involving multiple linear equations. In 1767, British astronomer John Michell similarly used a significance test based on the uniform distribution (though with some technical errors) to show that the Pleiades is a physical, rather than chance, grouping of stars. Johann Lambert presented an elaborate theory of errors, often in astronomical contexts, during the 1760s. Bernouilli and Lambert laid the foundations of the concept of maximum likelihood later developed more thoroughly by Fisher in the early twentieth century.

The Marquis Pierre-Simon de Laplace (1749−1827), the most distinguished French scientist of his time, and his competitor Adrien-Marie Legendre, made seminal contributions both to celestial mechanics and to probability theory, often intertwined. Their generalizations of Mayer's methods for treating multiple parametric equations constrained by many discrepant observations had great impact. In astronomical and geodetical studies during the 1780s and in his huge 1799–1825 opus *Mécanique Céleste*, Laplace proposed parameter estimation for linear models by minimizing the largest absolute residual. In an 1805 appendix to a paper on cometary orbits, Legendre proposed minimizing the sum of the squares of residuals, or the method of least squares. He concluded "that the method of least squares reveals, in a manner of speaking, the center around which the results of observations arrange themselves, so that the deviations from that center are as small as possible" (quoted by Stigler 1986).

Both Carl Friedrich Gauss, also director of the observatory at Göttingen, and Laplace later placed the method of least squares onto a solid mathematical probabilistic foundation. While the method of least squares had been adopted as a practical convenience by Gauss and Legendre, Laplace first treated it as a problem in probabilities in his *Théorie Analytique des Probabilités*. He proved by an intricate and difficult course of reasoning that it was the most advantageous method for finding parameters in orbital models from astronomical observations, the mean of the probabilities of error in the determination of the elements being thereby reduced to a minimum. Least-squares computations rapidly became the principal interpretive tool for astronomical observations and their links to celestial mechanics. These and other approaches to statistical inference are discussed in Chapter 3.

In another portion of the *Théorie*, Laplace rescued from obscurity the postulation of the Central Limit Theorem by the mathematician Abraham De Moivre who, in a remarkable article published in 1733, used the normal distribution to approximate the distribution of the number of heads resulting from many tosses of a fair coin. Laplace expanded De Moivre's finding by approximating the binomial distribution with the normal distribution. Laplace's proof was flawed, and improvements were developed by Siméon-Denis Poisson, an astronomer at Paris' Bureau des Longitudes, and Friedrich Bessel, director of the observatory in Königsberg. Today, the Central Limit Theorem is considered to be one of the foundations of probability theory (Section 2.10).

Gauss established his famous error distribution and related it to Laplace's method of least squares in 1809. Astronomer Friedrich Bessel introduced the concept of "probable error" in a 1816 study of comets, and demonstrated the applicability of Gauss' distribution to empirical stellar astrometric errors in 1818. Gauss also introduced some treatments for observations with different (heteroscedastic) measurement errors and developed the theory for unbiased minimum variance estimation. Throughout the nineteenth century, Gauss' distribution was widely known as the "astronomical error function".

Although the fundamental theory was developed by Laplace and Gauss, other astronomers published important contributions to the theory, accuracy and range of applicability of the normal distribution and least-squares estimation during the latter part of the nineteenth century (Hald 1998). They include Ernst Abbe at the Jena Observatory and the optics firm of Carl Zeiss, Auguste Bravais of the Univerity of Lyons, Johann Encke of the Berlin Observatory, Britain's Sir John Herschel, Simon Newcomb of the U.S. Naval Observatory, Giovanni Schiaparelli of Brera Observatory, and Denmark's Thorvald Thiele. Sir George B. Airy, British Royal Astronomer, wrote an 1865 text on least-squares methods and observational error.

Adolphe Quetelet, founder of the Belgian Royal Observatory, and Francis Galton, director of Britain's Kew Observatory, did little to advance astronomy but were distinguished pioneers extending statistical analysis from astronomy into the human sciences. They particularly laid the groundwork for regression between correlated variables. The application of least-squares techniques to multivariate linear regression emerged in biometrical contexts by Karl Pearson and his colleagues in the early 1900s (Chapter 7).

The intertwined history of astronomy and statistics during the eighteenth and nineteenth centuries is detailed in the monographs by Stigler (1986), Porter (1986) and Hald (1998).

1.2.3 Statistics in twentieth-century astronomy

The connections between astronomy and statistics considerably weakened during the first decades of the twentieth century as statistics turned its attention principally to biological sciences, human attributes, social behavior and statistical methods for industries such as life insurance, agriculture and manufacturing. Advances in astronomy similarly moved away from the problem of evaluating errors in measurements of deterministic processes of celestial mechanics. Major efforts on the equilibrium structure of stars, the geometry of the Galaxy, the discovery of the interstellar medium, the composition of stellar atmospheres, the study of solar magnetic activity and the discovery of extragalactic nebulae generally did not involve statistical theory or application. Two distinguished statisticians wrote series of papers in the astronomical literature – Karl Pearson on correlations between stellar properties around 1907–11, and Jerzy Neyman with Elizabeth Scott on clustering of galaxies around 1952–64 – but neither had a strong influence on further astronomical developments.

The least-squares method was used in many astronomical applications during the first half of the twentieth century, but not in all cases. Schlesinger (1916) admonished astronomers estimating elements of binary-star orbits to use least-squares rather than trial-and-error techniques. The stellar luminosity function derived by Jacobus Kapteyn, and thereby the inferred structure of the Galaxy, were based on subjective curve fitting (Kapteyn & van Rhijn 1920), although Kapteyn had made some controversial contributions to the mathematics of skewed distributions and correlation. An important study on dark matter in the Coma Cluster fits the radial distribution of galaxies by eye and does not quantify its similarity to an isothermal sphere (Zwicky 1937). In contrast, Edwin Hubble's seminal studies on galaxies were often based on least-squares fits (e.g. the redshift–magnitude relationship in Hubble & Humason 1931), although an early study reports a nonstandard symmetrical average of two regression lines (Hubble 1926, Section 7.3.2). Applications of statistical methods based on the normal error law were particularly strong in studies involving positional astronomy and star counts (Trumpler & Weaver 1953). Astronomical applications of least-squares estimation were strongly promoted by the advent of computers and Bevington's (1969) useful volume with FORTRAN code. Fourier analysis was also commonly used for time series analysis in the latter part of the twentieth century.

Despite its formulation by Fisher in the 1920s, maximum likelihood estimation emerged only slowly in astronomy. Early applications included studies of stellar cluster convergent points (Brown 1950), statistical parallaxes from the Hertzsprung–Russell diagram (Jung 1970), and some early work in radio and X-ray astronomy. Crawford *et al.* (1970) advocated use of maximum likelihood for estimating power-law slopes, a message we reiterate in this volume (Section 4.4). Maximum likelihood studies with truly broad impact did not emerge until the 1970s. Innovative and widely accepted methods include Lynden-Bell's (1971) luminosity function estimator for flux-limited samples, Lucy's (1974) algorithm for restoring blurry images, and Cash's (1979) algorithm for parameter estimation involving photon counting data. Maximum likelihood estimators became increasingly important in extragalactic astronomy; they were crucial for the discovery of galaxy streaming towards the Great Attractor (Lynden-Bell *et al.* 1988) and calculating the galaxy luminosity function from flux-limited surveys (Efstathiou *et al.* 1988). The 1970s also witnessed the first use and rapid acceptance of the nonparametric Kolmogorov–Smirnov statistic for two-sample and goodness-of-fit tests.

The development of inverse probability and Bayes' theorem by Thomas Bayes and Laplace in the late eighteenth century took place largely without applications to astronomy. Despite the prominence of the leading Bayesian proponent Sir Harold Jeffreys, who won the Gold Medal of the Royal Astronomical Society in 1937 and served as Society President in the 1950s, Bayesian methods did not emerge in astronomy until the latter part of the twentieth century. Bayesian classifiers for discriminating stars and galaxies (based on the 2001 text written for engineers by Duda *et al.*) were used to construct large automated sky survey catalogs (Valdes 1982), and maximum entropy image restoration gained some interest (Narayan & Nityananda 1986). But it was not until the 1990s that Bayesian methods became widespread in important studies, particularly in extragalactic astronomy and cosmology.

The modern field of astrostatistics grew suddenly and rapidly starting in the late 1990s. This was stimulated in part by monographs on statistical aspects of astronomical image processing (Starck *et al.* 1998, Starck & Murtagh 2006), galaxy clustering (Martínez & Saar 2001), Bayesian data analyses (Gregory 2005) and Bayesian cosmology (Hobson *et al.* 2010). Babu & Feigelson (1996) wrote a brief overview of astrostatistics. The continuing conference series *Statistical Challenges in Modern Astronomy* organized by us since 1991 brought together astronomers and statisticians interested in forefront methodological issues (Feigelson & Babu 2012). Collaborations between astronomers and statisticians emerged, such as the California–Harvard Astro-Statistical Collaboration (http://hea-www.harvard.edu/AstroStat), the International Computational Astrostatistics Group centered in Pittsburgh (http://www.incagroup.org), and the Center for Astrostatistics at Penn State (http://astrostatistics.psu.edu). However, the education of astronomers in statistical methodology remains weak. Penn State's Center and other institutes operate week-long summer schools in statistics for young astronomers to partially address this problem.

1.3   Recommended reading

We offer here a number of volumes with broad coverage in statistics. Stigler's monograph reviews the history of statistics and astronomy. Rice, Hogg & Tanis, and Hogg *et al.* are well-respected textbooks in statistical inference at undergraduate and graduate levels, and Wasserman gives a modern viewpoint. Lupton, James, and Wall & Jenkins are written by and for physical scientists. Ghosh *et al.* and Gregory introduce Bayesian inference.

Ghosh, J. K., Delampady, M. & Samanta, T. (2006) *An Introduction to Bayesian Analysis: Theory and Methods*, Springer, Berlin

A graduate-level textbook in Bayesian inference with coverage of the Bayesian approach, objective and reference priors, convergence and large-sample approximations, model selection and testing criteria, Markov chain Monte Carlo computations, hierarchical Bayesian models, empirical Bayesian models and applications to regression and high-dimensional problems.

Gregory, P. (2005) *Bayesian Logical Data Analysis for the Physical Sciences*, Cambridge University Press

This monograph treats probability theory and sciences, practical Bayesian inference, frequentists approaches, maximum entropy, linear and nonlinear model fitting, Markov chain Monte Carlo, harmonic time series analysis, and Poisson problems. Examples are illustrated using *Mathematica*.

Hogg, R., McKean, J. & Craig, A. (2005) *Introduction to Mathematical Statistics*, 6th ed., Prentice Hall, Englewood Cliffs
A slim text aimed at graduate students in statistics that includes Bayesian methods and decision theory, hypothesis testing, sufficiency, confidence sets, likelihood theory, prediction, bootstrap methods, computational techniques (e.g. bootstrap, MCMC), and other topics (e.g. pseudo-likelihoods, Edgeworth expansion, Bayesian asymptotics).

Hogg, R. V. & Tanis, E. (2009) *Probability and Statistical Inference*, 8th ed., Prentice-Hall, Englewood Cliffs
A widely used undergraduate text covering random variables, discrete and continuous distributions, estimation, hypothesis tests, linear models, multivariate distributions, non-parametric methods, Bayesian methods and inference theory.

James, F. (2006) *Statistical Methods in Experimental Physics*, 2nd ed., World Scientific, Singapore
This excellent volume covers concepts in probability, distributions, convergence theorems, likelihoods, decision theory, Bayesian inference, point and interval estimation, hypothesis tests and goodness-of-fit.

Lupton, R. (1993) *Statistics in Theory and Practice*, Princeton University Press
This slim monograph explains probability distributions, sampling statistics, confidence intervals, hypothesis tests, maximum likelihood estimation, goodness-of-fit and nonparametric rank tests.

Rice, J. A. (2007) *Mathematical Statistics and Data Analysis*, 3rd ed., Duxbury Press
An undergraduate-level text with broad coverage of modern statistics with both theory and applications. Topics covered include probability, statistical distributions, Central Limit Theorem, survey sampling, parameter estimation, hypothesis tests, goodness-of-fit, data visualization, two-sample comparisons, bootstrap, analysis of variance, categorical data, linear least squares, Bayesian inference and decision theory.

Stigler, S. M. (1986) *The History of Statistics: The Measurement of Uncertainty before 1900*, Harvard University Press
This readable monograph presents the intellectual history of the intertwined developments in astronomy and statistics during the eighteenth and nineteenth centuries. Topics include combining observations, least squares, inverse probability (Bayesian inference), correlation, regression and applications in astronomy and biology.

Wall, J. V. & Jenkins, C. R. (2003) *Practical Statistics for Astronomers*, Cambridge University Press
A useful volume on statistical methods for physical scientists. Coverage includes concepts of probability and inference, correlation, hypothesis tests, modeling by least squares and maximum likelihood estimation, bootstrap and jackknife, nondetections and survival analysis, time series analysis and spatial point processes.

Wasserman, L. (2004) *All of Statistics: A Concise Course in Statistical Inference*, Springer, Berlin
A short text intended for graduate students in allied fields presenting a wide range of topics with emphasis on mathematical foundations. Topics include random variables, expectations, empirical distribution functions, bootstrap, maximum likelihood, hypothesis testing, Bayesian inference, linear and loglinear models, multivariate models, graphs, density estimation, classification, stochastic processes and simulation methods. An associated Web site provides **R** code and datasets.

2 Probability

2.1 Uncertainty in observational science

Probability theory models uncertainty. Observational scientists often come across events whose outcome is uncertain. It may be physically impossible, too expensive or even counterproductive to observe all the inputs. The astronomer might want to measure the location and motions of all stars in a globular cluster to understand its dynamical state. But even with the best telescopes, only a fraction of the stars can be located in the two dimensions of sky coordinates with the third distance dimension unobtainable. Only one component (the radial velocity) of the three-dimensional velocity vector can be measured, and this may be accessible for only a few cluster members. Furthermore, limitations of the spectrograph and observing conditions lead to uncertainty in the measured radial velocities. Thus, our knowledge of the structure and dynamics of globular clusters is subject to considerable restrictions and uncertainty.

In developing the basic principles of uncertainty, we will consider both astronomical systems and simple familiar systems such as a tossed coin. The outcome of a toss, heads or tails, is completely determined by the forces on the coin and Newton's laws of motion. But we would need to measure too many parameters of the coin's trajectory and rotations to predict with acceptable reliability which face of the coin will be up. The outcomes of coin tosses are thus considered to be uncertain even though they are regulated by deterministic physical processes. Similarly, the observed properties of a quasar have considerable uncertainty, even though the physics of accretion disks and their radiation are based on deterministic physical processes.

The uncertainty in our knowledge could be due to the current level of understanding of the phenomenon, and might be reduced in the future. Consider, for example, the prediction of solar eclipses. In ancient societies, the motions of Solar System bodies were not understood and the occurrence of a solar eclipse would have been modeled as a random event (or attributed to divine intervention). However, an astronomer noticing that solar eclipses occur only on a new moon day could have revised the model with a monthly cycle of probabilities. Further quantitative prediction would follow from the Babylonian astronomers' discovery of the 18-year saros eclipse cycle. Finally, with Newtonian celestial mechanics, the phenomenon became essentially completely understood and the model changed from a random to a deterministic model subject to direct prediction with known accuracy.

The uncertainty of our knowledge could be due to future choices or events. We cannot predict with certainty the outcome of an election yet to be held, although polls of the voting public will constrain the prediction. We cannot accurately predict the radial velocity of a globular star prior to its measurement, although our prior knowledge of the cluster's velocity dispersion will constrain the prediction. But when the election results are tabulated, or the astronomical spectrum is analyzed, our level of uncertainty is suddenly reduced.

When the outcome of a situation is uncertain, why do we think that it is possible to model it mathematically? In many physical situations, the events that are uncertain at the micro-level appear to be deterministic at the macro-level. While the outcome of a single toss of a coin is uncertain, the proportion of heads in a large number of tosses is stable. While the radial velocity of a single globular cluster star is uncertain, we can make predictions with some confidence based on a prior measurement of the global cluster velocity and our knowledge of cluster dynamics from previous studies. Probability theory attempts to capture and quantify this phenomenon; the Law of Large Numbers directly addresses the relationship between micro-level uncertainty and macro-level deterministic behavior.

2.2 Outcome spaces and events

An **experiment** is any action that can have a set of possible results where the actually occurring result cannot be predicted with certainty prior to the action. Experiments such as tossing a coin, rolling a die, or counting of photons registered at a telescope, all result in sets of outcomes. Tossing a coin results in a set Ω of two outcomes $\Omega = \{H, T\}$; rolling a die results in a set of six outcomes $\Omega = \{1, 2, 3, 4, 5, 6\}$; while counting photons results in an infinite set of outcomes $\Omega = \{0, 1, 2, \dots\}$. The number of neutron stars within 1 kpc of the Sun is a discrete and finite sample space. The set of all outcomes Ω of an experiment is known as the **outcome space** or sample space.

An **event** is a subset of a sample space. For example, consider now the sample space Ω of all exoplanets, where the event E describes all exoplanets with eccentricity in the range 0.5–0.6, and the event F describes that the host star is a binary system. There are essentially two aspects to probability theory: first, assigning probabilities to simple outcomes; and second, manipulating probabilities or simple events to derive probabilities of complicated events.

In the simplest cases, such as a well-balanced coin toss or die roll, the inherent symmetries of the experiment lead to equally likely outcomes. For the coin toss, $\Omega = \{H, T\}$ with probabilities $P(H) = 0.5$ and $P(T) = 0.5$. For the die roll, $\Omega = \{1, 2, 3, 4, 5, 6\}$ with $P(i) = \frac{1}{6}$ for $i = 1, 2, \dots, 6$. Now consider the more complicated case where a quarter, a dime and a nickel are tossed together. The outcome space is

$$\Omega = \{HHH, HHT, HTH, HTT, THH, THT, TTH, TTT\}, \qquad (2.1)$$

where the first letter is the outcome of the quarter, the second of the dime and the third of the nickel. Again, it is reasonable to model all the outcomes as equally likely with probabilities

$\frac{1}{8}$. Thus, when an experiment results in m equally likely outcomes, $\{e_1, e_2, \ldots, e_m\}$, then the probability of any event A is simply

$$P(A) = \frac{\#A}{m}, \tag{2.2}$$

where $\#$ is read "the number of". That is, $P(A)$ is the ratio of the number of outcomes favorable to A and the total number of outcomes.

Even when the outcomes are not equally likely, in some cases it is possible to identify the outcomes as combinations of equally likely outcomes of another experiment and thus obtain a model for the probabilities. Consider the three-coin toss where we only note the number of heads. The sample space is $\Omega = \{0, 1, 2, 3\}$. These outcomes cannot be modeled as equally likely. In fact, if we toss three coins 100 times, then we would observe that $\{1, 2\}$ occur far more frequently than $\{0, 3\}$. The following simple argument will lead to a logical assignment of probabilities. The outcome $\omega \in \Omega$ in this experiment is related to the outcomes in (2.1):

$\omega = 0$ when TTT occurs
$\omega = 1$ when HTT, THT or TTH occurs
$\omega = 2$ when HHT, HTH or THH occurs
$\omega = 3$ when HHH occurs.

Thus $P(0) = P(3) = 0.125$ and $P(1) = P(2) = 0.375$.

For finite (or countably infinite) sample spaces $\Omega = \{e_1, e_2, \ldots\}$, a probability model assigns a nonnegative weight p_i to the outcome e_i for every i in such a way that the p_i's add up to 1. A finite (or countably infinite) sample space is sometimes called a **discrete sample space**. For example, when exploring the number of exoplanets orbiting stars within 10 pc of the Sun, we consider a discrete sample space. In the case of countable sample spaces, we define the probability $P(A)$ of an event A as

$$P(A) = \sum_{i\,:\,e_i \in A} p_i. \tag{2.3}$$

In words, this says that the probability of an event A is equal to the sum of the individual probabilities of outcomes e_i belonging to A.

If the sample space Ω is uncountable, then not all subsets are allowed to be called events for mathematical and technical reasons. Astronomers deal with both countable spaces — such as the number of stars in the Galaxy, or the set of photons from a quasar arriving at a detector — and uncountable spaces — such as the variability characteristics of a quasar, or the background noise in an image constructed from interferometry observations.

2.3 Axioms of probability

A **probability space** consists of the triplet $(\Omega, \mathcal{F}, P)$, with sample space Ω, a class $\mathcal{F}$ of events, and a function P that assigns a probability to each event in $\mathcal{F}$ that obey three axioms of probability:

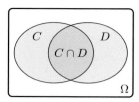 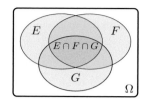

Fig. 2.1 Union and intersection of events.

Axiom 1 $0 \leq P(A) \leq 1$, for all events A

Axiom 2 $P(\Omega) = 1$

Axiom 3 For mutually exclusive (pairwise disjoint) events $A_1, A_2, \ldots$,

$$P(A_1 \cup A_2 \cup A_3 \cdots) = P(A_1) + P(A_2) + P(A_3) + \cdots,$$

that is, if for all $i \neq j$, $A_i \cap A_j = \emptyset$ ($\emptyset$ denotes the empty set or null event), then

$$P\left(\bigcup_{i=1}^{\infty} A_i\right) = \sum_{i=1}^{\infty} P(A_i).$$

Here, $\cup$ represents the union of sets while $\cap$ represents their intersection. Axiom 3 states that the probability that at least one of the mutually exclusive events A_i occurs is the same as the sum of the probabilities of the events A_i, and this should hold for infinitely many events. This is known as the **countable additivity** property. This axiom, in particular, implies that the finite additivity property holds; that is, for mutually exclusive (or disjoint) events A, B (i.e. $A \cap B = \emptyset$),

$$P(A \cup B) = P(A) + P(B). \tag{2.4}$$

This in particular implies that for any event A, the probability of its complement $A^c = \{\omega \in \Omega : \omega \notin A\}$, the set of points in the sample space that are not in A, is given by

$$P(A^c) = 1 - P(A). \tag{2.5}$$

(A technical comment can be made here: in the case of an uncountable sample space Ω, it is impossible to define a probability function P that assigns zero weight to singleton sets and satisfying these axioms for all subsets of Ω.)

Using the above axioms, it is easy to establish that for any two events C, D

$$P(C \cup D) = P(C) + P(D) - P(C \cap D); \tag{2.6}$$

that is, the probability of the union of the two events is equal to the sum of the event probabilities minus the probability of the intersection of the two events. This is illustrated in the left-hand panel of Figure 2.1.

For three events E, F, G,

$$\begin{aligned} P(E \cup F \cup G) = {} & P(E) + P(F) + P(G) - P(E \cap F) - P(F \cap G) \\ & - P(E \cap G) + P(E \cap F \cap G) \end{aligned} \tag{2.7}$$

as shown in the right-hand panel of Figure 2.1. The generalization to n events, $E_1, \ldots, E_n$ is called the **inclusion–exclusion formula**;

$$P(E_1 \cup E_2 \cup \cdots \cup E_n) = \sum_{i=1}^{\infty} P(E_i) - \sum_{i_1 < i_2} P(E_{i_1} \cap E_{i_2}) + \cdots$$
$$\times (-1)^{r+1} \sum_{i_1 < i_2 < \cdots < i_r} P(E_{i_1} \cap E_{i_2} \cap \cdots \cap E_{i_r})$$
$$+ \cdots + (-1)^{n+1} P(E_1 \cap E_2 \cap \cdots \cap E_n), \qquad (2.8)$$

where the summation

$$\sum_{i_1 < i_2 < \cdots < i_r} P(E_{i_1} \cap E_{i_2} \cap \cdots \cap E_{i_r}) \qquad (2.9)$$

is taken over all of the $\binom{n}{r}$ possible subsets of size r of the set $\{1, 2, \ldots, n\}$.

2.4 Conditional probabilities

Conditional probability is one of the most important concepts in probability theory and can be tricky to understand. It often helps in computing desired probabilities, particularly when only partial information regarding a result of an experiment is available. Bayes' theorem at the foundation of Bayesian statistics uses conditional probabilities.

Consider the following simple example. When a die is rolled, the probability that it turns up one of the numbers $\{1, 2, 3\}$ is $1/2$, as each of the six outcomes is equally likely. Now consider that someone took a brief glimpse at the die and found that it turned up an even number. How does this additional information influence the assignment of probability to $A = \{1, 2, 3\}$? In this case, the weights assigned to the points are reassessed by giving equal weights, $1/3$, to each of the three even integers in $B = \{2, 4, 6\}$ and zero weights to the odd integers. Since it is already known that B occurred, it is intuitive to assign the scale and probability 1 to B and probability 0 to the complementary event B^c. Now as all the points in B are equally likely, it follows that the required probability is the ratio of the number of points of A that are in B to the total number of points in B. Since 2 is the only number from A in B, the required probability is $\#(A \cap B)/\#B = 1/3$.

Generalizing this, let us consider an experiment with m equally likely outcomes and let A and B be two events. If we are given the information that B has happened, what is the probability that A has happened in light of the new knowledge? Let $\#A = k$, $\#B = n$ and $\#(A \cap B) = i$. Then, as in the rolled die example above, given that B has happened, the new probability allocation assigns probability $1/n$ to all the outcomes in B. Out of these n, $\#(A \cap B) = i$ outcomes belong to A. Noting that $P(A \cap B) = i/m$ and $P(B) = n/m$, it leads to the conditional probability, $P(A \mid B) = i/n$, of A given B,

$$P(A \mid B) = \frac{P(A \cap B)}{P(B)}. \qquad (2.10)$$

Equation (2.10) can be considered a formal definition of conditional probabilities providing $P(B) > 0$, even in the more general case where outcomes may not be equally likely. As a consequence, the multiplicative rule of probability for two events,

$$P(A \cap B) = P(A \mid B)P(B) \tag{2.11}$$

holds. The **multiplication rule** easily extends to n events:

$$P(A_1 \cap A_2 \cap \ldots A_n) = P(A_1)\, P(A_2 \mid A_1) \ldots P(A_{n-1} \mid A_1, \ldots A_{n-2})$$
$$\times\ P(A_n \mid A_1, \ldots A_{n-1}). \tag{2.12}$$

These concepts are very relevant to observational sciences such as astronomy. Except for the rare circumstance when an entirely new phenomenon is discovered, astronomers are measuring properties of celestial bodies or populations for which some distinctive properties are already available. Consider, for example, a subpopulation of galaxies found to exhibit Seyfert-like spectra in the optical band (property A) that have already been examined for nonthermal lobes in the radio band (property B). Then the conditional probability that a galaxy has a Seyfert nucleus given that it also has radio lobes is given by Equation (2.10), and this probability can be estimated from careful study of galaxy samples. The composition of a Solar System minor body can be predominately ices or rock. Icy bodies are more common at large orbital distances and show spectral signatures of water (or other) ice rather than the spectral signatures of silicates. The probability that a given asteroid, comet or Kuiper Belt Object is mostly icy is then conditioned on its semi-major axis and spectral characteristics.

2.4.1 Bayes' theorem

We are now ready to derive the famous Bayes' theorem, also known as Bayes' formula or Bayes' rule. It is named for the mid-eighteenth-century British mathematician and Presbyterian minister Thomas Bayes, although it was recognized earlier by James Bernoulli and Adrian de Moivre, and was later fully explicated by Pierre Simon Laplace. Let $B_1, \ldots, B_k$ be a partition of the sample space Ω. A partition of Ω is a collection of mutually exclusive (pairwise disjoint) sets whose union is Ω; that is, $B_i \cap B_j = 0$ for $i \neq j$. If A is any event in Ω, then to compute $P(A)$, one can use probabilities of pieces of A on each of the sets B_i and add them together to obtain

$$P(A) = P(A \mid B_1)P(B_1) + \cdots + P(A \mid B_k)P(B_k). \tag{2.13}$$

This is called the **law of total probability** and follows from the observation

$$P(A) = P(A \cap B_1) + \cdots + P(A \cap B_k), \tag{2.14}$$

and the multiplicative rule of probability, $P(A \cap B_i) = P(A \mid B_i)P(B_i)$.

Now consider the following example, a bit more complicated than those treated above. Suppose a box contains five quarters, of which one is a trick coin that has heads on both sides. A coin is picked at random and tossed three times. It was observed that all three tosses turned up heads.

If the type of the coin chosen is known, then one can easily compute the probability of the event H that all three tosses yield heads. If it is the two-headed coin, then the probability is 1, otherwise it is $1/8$. That is, $P(H \mid M) = 1$ and $P(H \mid M^c) = 1/8$, where M denotes the event that the two-headed coin is chosen and M^c is the complimentary event that a regular quarter is chosen. After observing three heads, what is the probability that the chosen coin has both sides heads? Bayes' theorem helps to answer this question. Here, by using the law of total probability and the multiplication rule, one obtains,

$$
P(M \mid H) = \frac{P(M \cap H)}{P(H)} = \frac{P(H \mid M)P(M)}{P(H \mid M)P(M) + P(H \mid M^c)P(M^c)}
$$
$$
= \frac{1/5}{(1/5) + (1/8) \times (4/5)} = \frac{2}{3}. \tag{2.15}
$$

For a partition $B_1, \ldots, B_k$ of Ω, Bayes' theorem generalizes the above expression to obtain $P(B_i \mid A)$ in terms of $P(A \mid B_j)$ and $P(B_j)$, for $j = 1, \ldots, k$. The result is very easy to prove, and is the basis of Bayesian inference discussed in Section 3.8.

Theorem 2.1 (Bayes' theorem) *If $B_1, B_2, \ldots, B_k$ is a partition of the sample space, then for $i = 1, \ldots, k$,*

$$
P(B_i \mid A) = \frac{P(A \mid B_i)P(B_i)}{P(A \mid B_1)P(B_1) + \cdots + P(A \mid B_k)P(B_k)}. \tag{2.16}
$$

Bayes' theorem thus arises directly from logical inference based on the three axioms of probability. While it applies to any form of probabilities and events, modern Bayesian statistics adopts a particular interpretation of these probabilities, which we will present in Section 3.8.

2.4.2 Independent events

The examples above show that, for any two events A and B, the conditional probability of A given B, $P(A \mid B)$, is not necessarily equal to the unconditional probability of A, $P(A)$. Knowledge of B generally changes the probability of A. In the special situation where $P(A \mid B) = P(A)$ where the knowledge that B has occurred has not altered the probability of A, A and B are said to be **independent events**. As the conditional probability $P(A \mid B)$ is not defined when $P(B) = 0$, the multiplication rule $P(A \cap B) = P(A \mid B)P(B)$ will be used to formally define independence:

Definition 2.2 Two events A and B are defined to be independent if

$$
P(A \cap B) = P(A)P(B).
$$

This shows that if A is independent of B, then B is independent of A. It is not difficult to show that if A and B are independent, then A and B^c are independent, A^c and B are independent and also A^c and B^c are independent.

Note that three events E, F, G satisfying $P(E \cap F \cap G) = P(E)P(F)P(G)$ cannot be called independent, as it does not guarantee independence of E, F or independence of F, G or independence of E, G. This can be illustrated with a simple example of a sample space $\Omega = \{1, 2, 3, 4, 5, 6, 7, 8\}$, where all the points are equally likely. Consider the events $E = \{1, 2, 3, 4\}$, $F = G = \{4, 5, 6, 7\}$. Clearly $P(E \cap F \cap G) = P(E)P(F)P(G)$. But neither E and F, F and G, nor E and G are independent.

Similarly, we note that independence of A and B, B and C, and A and C together does not imply $P(A \cap B \cap C) = P(A)P(B)P(C)$. If we consider the events $A = \{1, 2, 3, 4\}$, $B = \{1, 2, 5, 6\}$, $C = \{1, 2, 7, 8\}$ then clearly A and B are independent, B and C are independent, and also A and C are independent, as A, B, C each contain exactly four numbers,

$$P(A) = P(B) = P(C) = \frac{4}{8} = \frac{1}{2}, \qquad (2.17)$$

but $A \cap B = B \cap C = A \cap C = A \cap B \cap C = \{1, 2\}$ and

$$P(A \cap B) = P(B \cap C) = P(A \cap C) = P(\{1, 2\}) = \frac{2}{8} = \frac{1}{4}. \qquad (2.18)$$

However,

$$P(A \cap B \cap C) = P(\{1, 2\}) = \frac{1}{4} \neq \frac{1}{8} = P(A)P(B)P(C). \qquad (2.19)$$

Though A and B are independent, and A and C are independent, $P(A \mid B \cap C) = 1$. So A is not independent of $B \cap C$. This leads to the following definition:

Definition 2.3 (Independent events) A set of $A_1, \ldots, A_n$ events is said to be independent if, for every subcollection $A_{I_1}, \ldots, A_{I_r}, r \leq n$,

$$P(A_{I_1} \cap A_{I_2} \cdots \cap A_{I_r}) = P(A_{I_1})P(A_{I_2}) \cdots P(A_{I_r}). \qquad (2.20)$$

An infinite set of events is defined to be independent if every finite subcollection of these events is independent. It is worth noting that for the case of three events, A, B, C are independent if all the following four conditions are satisfied:

$$\begin{aligned}
P(A \cap B \cap C) &= P(A)P(B)P(C), \\
P(A \cap B) &= P(A)P(B), \\
P(B \cap C) &= P(B)P(C), \\
P(A \cap C) &= P(A)P(C). \qquad (2.21)
\end{aligned}$$

2.5 Random variables

Often, instead of focusing on the entire outcome space, it may be sufficient to concentrate on a summary of outcomes relevant to the problem at hand, say a function of the outcomes. In tossing a coin four times, it may be sufficient to look at the number of heads instead of the order in which they are obtained. In observing photons from an astronomical source, it may be sufficient to look at the mean number of photons in a spectral band over some time interval, or the ratio of photons in two spectral bands, rather than examining each photon individually.

These real-valued functions on the outcome space or sample space are called **random variables**. Data are realizations of random variables. Typically a random variable X is a function on the sample space Ω. In the case of countable sample spaces Ω, this definition always works. But in the case of uncountable Ω, one should be careful. As mentioned earlier, not all subsets of an uncountable space can be called an event, or a probability assigned to them. A random variable is a function such that $\{\omega \in \Omega : X(\omega) \leq a\}$, is an event for all real numbers a. In practical situations, the collection of events can be defined to be inclusive enough that the set of events follows certain mathematical conditions (closure under complementation, countable unions and intersections). So in practice, the technical aspects can be ignored.

Note that in casual usage, some people label a phenomenon as "random" to mean that the events have equal chances of possible outcomes. This concept is correctly called **uniformity**. The concept of **randomness** does not require uniformity. Indeed, the following sections and Chapter 4 are largely devoted to phenomena that follow nonuniform distributions.

2.5.1 Density and distribution functions

A random variable is called a **discrete** random variable if it maps a sample space to a countable set (e.g. the integers) with each value in the range having probability greater than or equal to zero.

Definition 2.4 (Cumulative distribution function) The **cumulative distribution function (c.d.f.)** or simply the **distribution function** F of a random variable X is defined as

$$F(x) = P(X \leq x) = P(\omega \in \Omega : X(\omega) \leq x), \qquad (2.22)$$

for all real numbers x. In the discrete case when X takes values $x_1, x_2, \ldots$, then

$$F(x) = P(X \leq x) = \sum_{x_i \leq x} P(X = x_i). \qquad (2.23)$$

The c.d.f. F is a nondecreasing, right-continuous function satisfying

$$\lim_{x \to -\infty} F(x) = 0 \text{ and } \lim_{x \to \infty} F(x) = 1. \qquad (2.24)$$

The c.d.f. of a discrete random variable is called a **discrete distribution**. A random variable with a continuous distribution function is referred to as a **continuous** random variable. A continuous random variable maps the sample space to an uncountable set (e.g. the real numbers). While the probability that a continuous random variable takes any specific value is zero, the probability that it belongs to an infinite set of values such as an interval may be positive. It should be understood clearly that the requirement that X is a continuous random variable does not mean that $X(\omega)$ is a continuous function; in fact, continuity does not make sense in the case of an arbitrary sample space, Ω.

Often some continuous distributions are described through the **probability density function** (p.d.f.). A nonnegative function f is called the probability density function of a distribution function if for all x

$$F(x) = \int_{-\infty}^{x} f(y)dy. \qquad (2.25)$$

We warn astronomers that the statistician's term "density" has no relationship to "density" in physics where it measures mass per unit volume. If the probability density exists, then the distribution function F is continuous. The converse of this statement is not always true, as there exist continuous distributions that do not have corresponding density functions.

Instead of a single function on a sample space, we might consider several functions at the same time. For example, consider the sample space (or outcome space) where Ω consists of exoplanets within 50 pc of the Sun, and the random variables $\{X_1, X_2, X_3, X_4\}$ denote the exoplanet masses, radii, surface temperatures, and orbital semi-major axes. There may be important relationships among these variables of scientific interest, so it is crucial to study these variables together rather than individually. Such a vector $(X_1, \ldots, X_k)$ of random variables is called a **random vector**. The distribution of random vector F is given by

$$F(x_1, \ldots, x_k) = P(X_1 \leq x_1, \ldots, X_k \leq x_k)$$
$$= P(\omega \in \Omega : X_1(\omega) \leq x_1, \ldots, X_k(\omega) \leq x_k), \qquad (2.26)$$

where $x_1, \ldots, x_k$ are real numbers. Note that F is a nondecreasing right-continuous function in each coordinate. The one-dimensional distributions F_i of individual variables X_i are called **marginal distributions** given by

$$F(x_i) = P(X_i \leq x_i), \quad i = 1, \ldots, k. \qquad (2.27)$$

Definition 2.5 (Independent random variables) The random variables $X_1, \ldots, X_n$ are said to be independent if the joint distribution is the product of the marginal distributions. That is

$$P(X_1 \leq a_1, \ldots, X_n \leq a_n) = P(X_1 \leq a_1) \ldots P(X_n \leq a_n),$$

for all real numbers $a_1, \ldots, a_n$.

An infinite collection $\{X_i\}$ of random variables are said to be independent if every finite subcollection of random variables is independent.

An important indicator of the central location of a random variable's distribution is the first moment or **mean** of the random variable. The mean of a random variable is defined as the weighted average where the weight is obtained from the associated probabilities. The terms **mathematical expectation** or the **expected value** of a random variable are often used interchangeably with the more familiar term "mean". For a random variable X taking values $x_1, x_2, \ldots, x_n$, the expected value of X is denoted by $E[X]$ defined by

$$E[X] = \sum_i x_i P(X = x_i). \qquad (2.28)$$

If h is a real-valued function, then $Y = h(X)$ is again a random variable. The expectation of this Y can be computed without deriving the distribution of Y using only the distribution of X:

$$E[Y] = E[h(X)] = \sum_i h(x_i)P(X = x_i). \qquad (2.29)$$

The notation $E[X]$ is equivalent to the notations $\langle X \rangle$ or $\bar{X}$ familiar to physical scientists.

The same definition of expectation as in (2.28) and (2.29) can be used for any discrete random variable X taking infinitely many nonnegative values. However, difficulties may be encountered in defining the expectation of a random variable taking infinitely many positive and negative values. Consider the case where W is a random variable satisfying

$$P(W = 2^j) = P(W = -2^j) = 2^{-j-1}, \text{ for } j = 1, 2, \ldots \tag{2.30}$$

In this case, the expectation $E[W]$ cannot be defined, as both the positive part $W^+ = \max(0, W)$ and the negative part $W^- = \max(0, -W)$ have infinite expectations. This would make $E[X]$ to be $\infty - \infty$, which is meaningless. However, for a general discrete random variable, $E[X]$ can be defined as in (2.28) provided

$$\sum_i |x_i| P(X = x_i) < \infty. \tag{2.31}$$

In case the distribution F of a random variable X has density f as in (2.25), then the **expectation** is defined as

$$E[X] = \int_{-\infty}^{\infty} y f(y) dy, \text{ provided } \int_{-\infty}^{\infty} |y| f(y) dy < \infty. \tag{2.32}$$

The expectation of a function h of a random variable X can be defined similarly as in (2.29), provided $\sum_i |h(x_i)| P(X = x_i) < \infty$ in the discrete case, and

$$E[h(X)] = \int h(y) f(y) dy \quad \text{provided} \quad \int |h(y)| f(y) dy < \infty \tag{2.33}$$

in case the distribution of X has density f.

Another important and commonly used function of a distribution function that quantifies the spread is the second moment centered on the mean, known as the **variance** and often denoted by σ^2. The variance is defined by

$$\sigma^2 = Var(X) = E\left[(X - \mu)^2\right] = E[X^2] - \mu^2, \tag{2.34}$$

where $\mu = E[X]$.

The mean and variance need not be closely related, as seen in the following simple example. Let X be a random variable taking values 1 and -1 with probability 0.5 each, and let Y be a random variable taking values 1000 and -1000 with probability 0.5. Both X and Y have the same mean ($= 0$), but $Var(X) = 1$ and $Var(Y) = 10^6$.

It is helpful to derive the variance of the sum of random variables. If $X_1, X_2, \ldots, X_n$ are n random variables, we find that

$$E\left[\sum_{i=1}^n X_i\right] = \sum_{i=1}^n E[X_i] \tag{2.35}$$

and the variance of the sum $\sum_{i=1}^n X_i$ can be expressed as

$$Var\left(\sum_{i=1}^n X_i\right) = \sum_{i=1}^n Var(X_i) + \sum_{i=1}^n \sum_{\substack{j=1 \\ i \neq j}}^n Cov(X_i, X_j), \quad \text{where}$$

$$Cov(X, Y) = E[(X - E[X])(Y - E[Y])]. \tag{2.36}$$

The *Cov* quantity is the **covariance** measuring the relation between the scatter in two random variables. If X and Y are independent random variables, then $Cov(X, Y) = 0$ and $Var(\sum_{i-1}^{n} X_i) = \sum_{i=1}^{n} Var(X_i)$, while the converse is not true; some dependent variables may have zero covariance.

If all the X_i variables have the same variance σ^2, the situation is called **homoscedastic**. If $X_1, X_2, \ldots, X_n$ are independent, the variance of the sample mean $\bar{X} = (1/n) \sum_{i=1}^{n} X_i$ is given by

$$Var(\bar{X}) = \frac{1}{n^2} \sum_{i=1}^{n} Var(X_i) = \frac{\sigma^2}{n}. \tag{2.37}$$

The variance essentially measures the mean square deviation from the mean of the distribution. The square root of the variance, σ, is called the **standard deviation**. The mean μ and the standard deviation σ of a random variable X are often used to convert X to a **standardized form**

$$X_{std} = \frac{X - \mu}{\sigma} \tag{2.38}$$

with mean zero and variance unity. This important transformation also removes the units of the original variable. Other transformations also reduce scale and render a variable free from units, such as the logarithmic transformation often used by astronomers. It should be recognized that the logarithmic transformation is only one of many optional variable transformations. Standardization is often preferred by statisticians with mathematical properties useful in statistical inference.

The third central moment $E[(X - E[X])^3]$ provides information about the **skewness** of the distribution of a random variable; that is, whether the distribution of X leans more towards right or left. Higher order moments like the k-th order moment $E[X^k]$ also provide some additional information about the distribution of the random variable.

2.5.2 Independent and identically distributed random variables

When repeated observations are made, or when an experiment is repeated several times, the successive observations lead to independent random variables. If the data are generated from the same population, then the resultant values can be considered as random variables with a common distribution. These are a sequence of **independent and identically distributed** or i.i.d. random variables. In the i.i.d. case, the random variables have a common mean and variance (if these moments exist).

Some observational studies in astronomy produce i.i.d. random variables. The redshifts of galaxies in an Abell cluster, the equivalent widths of absorption lines in a quasar spectrum, the ultraviolet photometry of a cataclysmic variable accretion disk, and the proper motions of a sample of Kuiper Belt bodies will all be i.i.d. if the observational conditions are unchanged. But the i.i.d. conditions are often violated. The sample may be heterogeneous with objects drawn from different underlying distributions. The observations may have been taken under different conditions such that the measurement errors differ. This leads to a condition called **heteroscedasticity** that violates the i.i.d. assumption. Heteroscedasticity means that different data points have different variances.

Since a great many methods of statistics, both classical and modern, depend on the i.i.d. assumption, it is crucial that astronomers understand the concept and its relationship to the datasets under study. Incorrect use of statistics that require i.i.d. will lead to incorrect quantitative results, and thereby increase the risk of incorrect or unsupported scientific inferences.

2.6 Quantile function

The cumulative distribution function $F(x)$ estimates the value of the population distribution function at a chosen value of x. But the astronomer often asks the inverse question: "What value of x corresponds to a specified value of $F(x)$?" This answers questions like "What fraction of galaxies have luminosities above L^*?" or "At what age have 95% of stars lost their protoplanetary disks?" This requires estimation of the **quantile function** of a random variable X, the inverse of F, defined as

$$Q(u) = F^{-1}(u) = \inf\{y : F(y) \geq u\} \tag{2.39}$$

where $0 < u < 1$. Here inf (infimum) refers to the smallest value of y with the property specified in the brackets.

When large samples are considered, the quantile function is often convenient for scientific analysis as the large number of data points are reduced to a smaller controlled number of interesting quantiles such as the 5%, 25%, 50%, 75% and 95% quantiles. A quantile function for an astronomical dataset is compared to the more familiar histogram in Figure 6.1 of Chapter 6. Quantile-quantile (Q-Q) plots are often used in visualization to compare two samples or one sample with a probability distribution. Q-Q plots are illustrated in Figures 5.4, 7.2, 8.2 and 8.6.

But when small samples are considered, the quantile function can be quite unstable. This is readily understood: for a sample of $n = 8$ points, the 25% and 75% quartiles are simply the values of the second and sixth data points, but for $n = 9$ interpolation is needed based on very little information about the underlying distribution of $Q(u)$. Also, the asymptotic normality of the quantile function breaks down when the distribution function has regions with low density.

For many datasets, computation of the quantile function is straightforward: the data are sorted in order of increasing X, and the values of chosen quantiles are estimated by local interpolation. But for extremely large datasets, sorting the entire vector is computationally infeasible and the total number of data points may be ill-defined. In these cases, quantiles can be estimated from a continuous data stream if the sampling is random using the **data skeleton algorithm** (McDermott *et al.* 2007).

2.7   Discrete distributions

We have encountered the discrete uniform distribution in some examples above, such as the probability $P(i) = 1/6$ for $i = 1, 2, \ldots, 6$ for a single roll of a balanced die. This is only the simplest of a range of probability distributions that take on discrete values which are frequently encountered. We outline other discrete probability distributions here; more mathematical properties of some with particular importance in astronomy are presented in Chapter 4.

Bernoulli distribution: Suppose that an experiment results in a success or failure (True or False) and we define $X = 1$ if the outcome is a success and $X = 0$ if the outcome is a failure. X is called a **binary variable** or **dichotomous variable**, and the distribution $P(X = 1) = p = 1 - P(X = 0)$ is called the **Bernoulli distribution**, where $0 < p < 1$.

Binomial distribution: Suppose this experiment is repeated n times independently and the number of successes are denoted by X. Then the distribution of X is given by

$$P(X = i) = \binom{n}{i} p^i (1 - p)^{n-i}, \text{ where } \binom{n}{i} = \frac{n!}{i!(n-i)!}, \tag{2.40}$$

for $i = 0, 1, \ldots, n$. In this case, the random variable X is said to have the **binomial distribution** and is denoted by $X \sim Bin(n, p)$. The mean and variance of X are given by

$$E[X] = np, \text{ and } Var[X] = np(1 - p). \tag{2.41}$$

The binomial probabilities $P(X = i)$ first increase monotonically and then decrease. Its highest value is reached when i is the largest integer less than or equal to $(n + 1)p$.

If the number of trials n in a binomial distribution is large, it is practically impossible to compute the probabilities. A good approximation is needed. In 1773, De Moivre established the special case for a binomial random variable $X \sim Bin(n, p)$,

$$P\left(a \leq \frac{X - np}{\sqrt{np(1 - p)}} \leq b\right) \to \Phi(b) - \Phi(a), \tag{2.42}$$

as $n \to \infty$, where Φ is the cumulative normal distribution function. This is a special case of the Central Limit Theorem (CLT) (Section 2.10). Note that the CLT applies when the underlying probability p is any fixed number between 0 and 1 and n approaches infinity.

Poisson distribution: A random variable X is said to have a **Poisson distribution**, denoted $X \sim Poi(\lambda)$, with rate $\lambda > 0$ if

$$P(X = i) = \frac{\lambda^i}{i!} e^{-\lambda}, \text{ for } i = 0, 1, \ldots \tag{2.43}$$

In this case

$$E[X] = Var(X) = \lambda. \tag{2.44}$$

The binomial and Poisson distributions are closely related. Suppose $X \sim Bin(n, p_n)$. If n is large and i is close to $n/2$, it is extremely difficult to compute the binomial coefficients $\binom{n}{i}$ in Equation (2.40). However, the binomial probabilities can be approximated by the Poisson probabilities when p_n is small, n is large, and $\lambda = np_n$ is not too large or too small. That is,

$$P(X = i) = \binom{n}{i} p_n^i (1 - p_n)^{n-i} \approx e^{-\lambda} \frac{\lambda^i}{i!}, \tag{2.45}$$

for $i = 0, 1, \ldots$. Poisson distributed random variables appear in many astronomical studies and it is thus very important to understand them well.

Negative binomial distribution: Let the random variable X_r denote the number of trials until a total of r successes is accumulated. Then

$$P(X_r = i) = \binom{n-1}{r-1} (1 - p)^{n-r} p^r, \text{ for } n = r, r+1, \ldots,$$

$$E[X_r] = \frac{r}{p}, Var(X_r) = \frac{qr}{p^2}, \tag{2.46}$$

and X_r is said to have a **negative binomial distribution** with parameters (r, p). When $r = 1$ the distribution is known as the **geometric distribution**.

2.8 Continuous distributions

We introduce here four important continuous distributions: uniform, exponential, normal or Gaussian, and lognormal distributions. Figure 2.2 at the end of this chapter plots some of these distributions for a few typical parameter values, showing both the p.d.f. f and the c.d.f. F defined in Equations (2.23) and (2.25). Further properties of these and other continuous distributions are discussed in Chapter 4.

Uniform distribution: The distribution function F is called **uniform** if its p.d.f. f is given by

$$f(x) = \frac{1}{b - a} \quad \text{for} \quad a < x < b, \tag{2.47}$$

and zero otherwise. A random variable X is uniformly distributed on (a, b) if its distribution is uniform on (a, b). In this case it is denoted by $X \sim U(a, b)$. For such a random variable, $E[X] = (a + b)/2$ and $Var(X) = \frac{1}{12}(b - a)^2$. If Y has a continuous distribution F, then $F(Y) \sim U(0, 1)$.

Uniformly distributed random variables play an important role in simulations. To simulate a uniform random variable, flip a coin repeatedly, and define $X_n = 1$ or $X_n = 0$ according as head or tail turned up on the n-th flip. Then it can be shown that

$$V = \sum_{n=1}^{\infty} \frac{X_n}{2^n} \sim U(0, 1). \tag{2.48}$$

If F is any c.d.f., continuous or otherwise, the random variable

$$W = \inf\{x : F(x) \geq V\}, \tag{2.49}$$

the smallest x such that $V \leq F(x)$, has distribution F. This provides a method for generating realizations of a random variable with any given distribution. A random variable with any distribution can be generated this way by flipping a coin.

Exponential distribution: A random variable with the distribution function

$$F(x) = 1 - e^{-\lambda x} \tag{2.50}$$

for $x > 0$ and $F(x) = 0$ for $x \leq 0$ is called **exponential with rate** $\lambda > 0$. F has p.d.f. f given by

$$f(x) = \lambda \exp(-\lambda x), \quad x \geq 0, \tag{2.51}$$

and zero otherwise. A random variable with the exponential distribution has the so-called **memoryless property**

$$P(X > t + s \mid X > s) = P(X > t) \quad \text{for all } s, t \geq 0. \tag{2.52}$$

This property is essential in modeling the waiting times in Poisson processes. The mean and variance of an exponential random variable X are given by

$$E[X] = \frac{1}{\lambda} \quad \text{and} \quad Var(X) = \frac{1}{\lambda^2}. \tag{2.53}$$

The top panels of Figure 2.2 show the exponential density and distribution for three different values for λ.

Normal or Gaussian distribution: In 1733, French mathematician Abraham De Moivre introduced the probability density function (p.d.f.)

$$\phi(x; \mu, \sigma^2) = \frac{1}{\sqrt{2\pi}\sigma} \exp\left\{-\frac{(x-\mu)^2}{2\sigma^2}\right\} \quad \text{where} \quad -\infty < \mu < \infty, \ \sigma > 0, \tag{2.54}$$

to approximate the binomial distribution $Bin(n, p)$ when n is large. He named Equation (2.54) the "exponential bell-shaped curve". Its full potential was realized only in 1809, when the German mathematician Karl Friedrich Gauss used it in his astronomical studies of celestial mechanics. During the following century, statisticians found that many datasets representing many types of random variables have histograms closely resembling the Gaussian density. This is explained by the Central Limit Theorem (Section 2.10). The curve (2.54) has come to be known both as the Gaussian and the normal probability density. The probability distribution function obtained from the bell-shaped curve (2.54) is called the **Gaussian** or **normal distribution function**,

$$\Phi(x; \mu, \sigma^2) = \int_{-\infty}^{x} \phi(y; \mu, \sigma^2)dy = \int_{-\infty}^{x} \frac{1}{\sqrt{2\pi}\sigma} \exp\left\{-\frac{(y-\mu)^2}{2\sigma^2}\right\}dy. \qquad (2.55)$$

A random variable with a p.d.f. (2.54) is called a **normal random variable**. A normal random variable X with parameters μ and σ^2 is denoted by $X \sim N(\mu, \sigma^2)$. Its mean and variance are given by

$$E[X] = \mu \quad \text{and} \quad Var(X) = \sigma^2. \qquad (2.56)$$

The middle panels of Figure 2.2 show the normal density and distribution for four different combinations of mean μ and variance σ^2.

Lognormal distribution: For a random variable $X \sim N(\mu, \sigma^2)$, $Y = e^X$ has a **lognormal distribution** with p.d.f. f given by

$$f(x) = \frac{1}{\sigma x \sqrt{2\pi}} \exp\left\{-\frac{(\ln(x)-\mu)^2}{2\sigma^2}\right\}, \quad \text{for } x > 0, \qquad (2.57)$$

and $f(x) = 0$ for $x \leq 0$. A lognormal random variable Y with parameters μ and σ^2 is denoted by $Y \sim \ln N(\mu, \sigma^2)$. The mean and variance of Y are given by

$$E[Y] = e^{\mu + (1/2)\sigma^2} \quad \text{and} \quad Var(Y) = \left(e^{\sigma^2} - 1\right)e^{2\mu + \sigma^2}. \qquad (2.58)$$

The bottom panels of Figure 2.2 show the lognormal density and distribution for $\mu = 0$ and four values of variance σ^2.

2.9 Distributions that are neither discrete nor continuous

Of course, there are distributions that are neither discrete nor continuous. Suppose X is uniformly distributed on $(-1, 1)$, that is

$$P(a < X < b) = \frac{1}{2}(b-a) \text{ for all } -1 < a < b < 1. \qquad (2.59)$$

If $Y = \max(0, X)$, then clearly $P(Y = 0) = P(X \leq 0) = 1/2$ and

$$P(a < X < b) = \frac{1}{2}(b-a) \text{ for all } 0 < a < b < 1. \qquad (2.60)$$

So the distribution F of Y is continuous except at 0. In the most general case, where the random variable X is neither discrete nor has a density, as in this example, $E[X]$ is defined as

$$E[X] = \int_{-\infty}^{\infty} y \, dF(y), \text{ provided } \int_{-\infty}^{\infty} |y| dF(y) < \infty, \qquad (2.61)$$

where the integral is the Riemann–Steiljes integral with respect to the distribution function F, where F is the c.d.f. of X.

2.10 Limit theorems

Probability theory has many powerful mathematical results which establish or constrain properties of random variables. One profound phenomenon, mentioned in Section 2.1, is that uncertainty at the micro-level (e.g. a single measurement or event) leads to deterministic behavior at the macro-level. This is a consequence of the Law of Large Numbers:

Theorem 2.6 (Law of Large Numbers) *Let $X_1, X_2, \ldots$ be a sequence of independent random variables with a common distribution and $E[|X_1|] < \infty$. Then*

$$\bar{X}_n = \frac{1}{n} \sum_{i=1}^{n} X_i \to \mu = E[X_1], \tag{2.62}$$

as $n \to \infty$, i.e. *for all $\epsilon > 0$, $P(|\bar{X}_n - \mu| > \epsilon) \to 0$, as $n \to \infty$.*

The theorem states that the **sample mean** $\bar{X}_n$ gives a good approximation to the **population mean** $\mu = E[X_1]$ when n is large. There is a crucial distinction between the sample mean, which is a random quantity, and the population mean, which is a population parameter and is a fixed number. It is important to note that this result is valid for discrete random variables, continuous random variables, and for general random variables.

An even more powerful result is the Central Limit Theorem (CLT). This states that, for a sequence of independent random variables $X_1, X_2, \ldots$ with a common distribution, the distribution of $\bar{X}_n$ can be approximated by a Gaussian (normal) distribution provided $E[|X_1|^2] < \infty$. A formal statement of the theorem follows.

Theorem 2.7 (Central Limit Theorem) *Let $X_1, X_2, \ldots$ be a sequence of i.i.d. random variables with mean $\mu = E[X_1]$ and finite variance $\sigma^2 = E[(X_1 - \mu)^2] > 0$. Then*

$$P\left(\sqrt{n}(\bar{X}_n - \mu) \leq x\sigma\right) \to \Phi(x), \tag{2.63}$$

for all x, where

$$\Phi(x) = \int_{-\infty}^{x} \phi(t)dt \quad and \quad \phi(x) = \frac{1}{\sqrt{2\pi}} e^{-\frac{1}{2}x^2}.$$

The CLT is an extremely useful result for approximating probability distributions of sums of large numbers of independent random variables. In later chapters, we will repeatedly refer to a statistic or estimator exhibiting **asymptotic normality**. This arises directly from the CLT where the distribution approaches the Gaussian for large sample sizes.

2.11 Recommended reading

Ross, S. (2010) *Introduction to Probability Models* 10th ed., Academic Press, New York
    An excellent undergraduate textbook laying the foundations of probability theory. Coverage includes conditional probability, Bayes' theorem, random variables, elementary

probability distributions, renewal theory, queueing theory, reliability, stationary stochastic processes and simulation techniques.

2.12   R applications

We give here the first **R** scripts of the volume with associated graphics. The reader is encouraged to first read the introduction to **R** in Appendix B.

R provides automatic computation of about 20 common probability distribution functions, and many more are available in **CRAN** packages. In the script below, we make the upper left panel of Figure 2.2 showing the p.d.f. of the exponential distribution for three values of the rate λ in Equation (2.51). The script starts with the construction of a closely spaced vector of x-axis values using **R**'s *seq* function. Here we make a vector of 250 elements ranging from 0 to 5. The function *dexp* gives the density (p.d.f.) of the exponential distribution to be evaluated at these values.

A bivariate graphics frame is created by the generic *plot* function. We first calculate it for the rate $\lambda = 0.5$ and then add curves for $\lambda = 1.0$ and 1.5. Note how the function *dexp* can be compactly embedded within the *plot* function call. Several commonly used parameters of *plot* are used: plot type (points, lines, histogram, steps), axis limits, axis labels, font size scaling, line width, and line type (solid, dashed, dotted). After the first curve is plotted, we superpose other curves using the *lines* command using the 'add=TRUE' option to place new curves on the same plot. These and other options are used very often, and the **R** practitioner is encouraged to learn them from the documentation given by *help*(par). The *legend* function permits annotations inside the plotting window. Another generic option for annotating graphs is the *text* function.

```
# Set up 6 panel figure

par(mfrow=c(3,2))

# Plot upper left panel with three illustrative exponential p.d.f. distributions

xdens <- seq(0,5,0.02)
plot(xdens,dexp(xdens,rate=0.5), type='l', ylim=c(0,1.5), xlab='',
   ylab='Exponential p.d.f.',lty=1)

lines(xdens,dexp(xdens,rate=1), type='l', lty=2)
lines(xdens,dexp(xdens,rate=1.5), type='l', lty=3)
legend(2, 1.45, lty=1, substitute(lambda==0.5), box.lty=0)
legend(2, 1.30, lty=2, substitute(lambda==1.0), box.lty=0)
legend(2, 1.15, lty=3, substitute(lambda==1.5), box.lty=0)

# Help files to learn these functions

help(seq) ; help(plot) ; help(par) ; help(lines) ; help(legend)
```

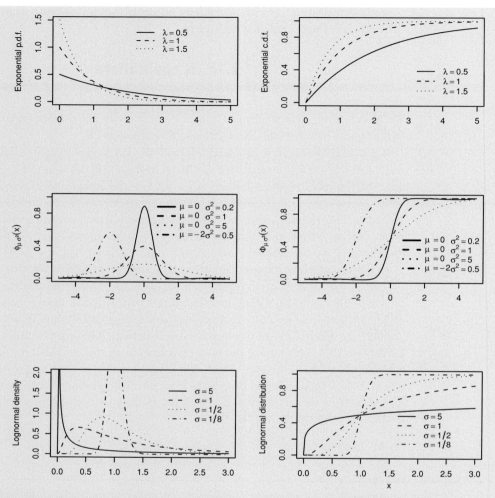

Fig. 2.2 Examples of three continuous statistical distributions: exponential (*top*), normal (*middle*), and lognormal (*bottom*). Left panels are (differential) probability density functions (p.d.f.) and right panels are cumulative distribution functions (c.d.f).

The other five panels of Figure 2.2 illustrating other distribution shapes are made with very similar scripts shown below. The top right panel substitutes *dexp* giving the exponential p.d.f. with the *pexp* function giving the exponential c.d.f. The middle panels similarly use *dnorm* and *pnorm* for the p.d.f. and c.d.f. of the normal (Gaussian) distribution, and the bottom panels use the *dlnorm* and *plnorm* functions for the lognormal distribution p.d.f. and c.d.f.

The labels and annotations of these graphs have options commonly used by astronomers such as Greek letters, superscripts and subscripts. One option shown above uses the *substitute* function to show Greek letters, while another more flexible option below uses the *expression* function. The "$==$" operator is used to assign specific values to parameters within functions.

```
# Plot upper right panel with three illustrative exponential c.d.f. distributions

plot(xdens, pexp(xdens,rate=0.5), type='l', ylim=c(0,1.0), xlab='',
   ylab='Exponential c.d.f.', lty=1)
lines(xdens, pexp(xdens,rate=1), type='l', lty=2)
lines(xdens, pexp(xdens,rate=1.5), type='l', lty=3)
legend(3, 0.50, lty=1, substitute(lambda==0.5), box.lty=0)
legend(3, 0.38, lty=2, substitute(lambda==1.0), box.lty=0)
legend(3, 0.26, lty=3, substitute(lambda==1.5), box.lty=0)

# Plot middle panels with illustrative normal p.d.f. and c.d.f.

xdens <- seq(-5, 5, 0.02)
ylabdnorm <- expression(phi[mu~sigma^2] (x))
plot(xdens, dnorm(xdens, sd=sqrt(0.2)), type='l', ylim=c(0,1.0), xlab='',
   ylab=ylabdnorm, lty=1)
lines(xdens, dnorm(xdens, sd=sqrt(1.0)), type='l', lty=2)
lines(xdens, dnorm(xdens, sd=sqrt(5.0)), type='l', lty=3)
lines(xdens, dnorm(xdens, mean=-2.0, sd=sqrt(0.5)), type='l', lty=4)
leg1 <- expression(mu^' '==0, mu^' '==0, mu^' '==0, mu^' '==-2)
leg2 <- expression(sigma^2==0.2, sigma^2==1.0, sigma^2==5.0, sigma^2==0.5)
legend(0.5, 1.0, lty=1:4, leg1, lwd=2, box.lty=0)
legend(3.0, 1.01, leg2, box.lty=0)

ylabpnorm <- expression(Phi[mu~sigma^2] (x))
plot(xdens, pnorm(xdens, sd=sqrt(0.2)), type='l', ylim=c(0,1.0), xlab='',
   ylab=ylabpnorm, lty=1)
lines(xdens, pnorm(xdens, sd=sqrt(1.0)), type='l', lty=2)
lines(xdens, pnorm(xdens, sd=sqrt(5.0)), type='l', lty=3)
lines(xdens, pnorm(xdens, mean=-2.0, sd=sqrt(0.5)), type='l', lty=4)
leg1 <- expression(mu^' '==0, mu^' '==0, mu^' '==0, mu^' '==-2)
leg2 <- expression(sigma^2==0.2, sigma^2==1.0, sigma^2==5.0, sigma^2==0.5)
legend(0.5, 0.6, lty=1:4, leg1, lwd=2, box.lty=0)
legend(3.0, 0.61, leg2, box.lty=0)

# Plot bottom panels with illustrative lognormal p.d.f. and c.d.f.

xdens <- seq(0,3, 0.02)
plot(xdens, dlnorm(xdens, meanlog=0, sdlog=5), type='l', ylim=c(0,2), xlab='',
   ylab='Lognormal density', lty=1)
```

```
lines(xdens, dlnorm(xdens, meanlog=0, sdlog=1), type='l', lty=2)
lines(xdens, dlnorm(xdens, meanlog=0, sdlog=1/2), type='l', lty=3)
lines(xdens, dlnorm(xdens, meanlog=0, sdlog=1/8), type='l', lty=4)
leg1 <- expression(sigma==5, sigma==1, sigma==1/2, sigma==1/8)
legend(1.8,1.8,lty=1:4,leg1,box.lty=0)

plot(xdens, plnorm(xdens, meanlog=0, sdlog=5), type='l', ylim=c(0,1), xlab='x',
    ylab='Lognormal distribution',lty=1)
lines(xdens, plnorm(xdens, meanlog=0, sdlog=1), type='l', lty=2)
lines(xdens, plnorm(xdens, meanlog=0, sdlog=1/2), type='l', lty=3)
lines(xdens, plnorm(xdens, meanlog=0, sdlog=1/8), type='l', lty=4)
leg1 <- expression(sigma==5, sigma==1, sigma==1/2, sigma==1/8)
legend(1.5, 0.6, lty=1:4, leg1, box.lty=0)

# Return plot to single-panel format

par(mfrow=c(1,1))
```

The **CRAN** package *prob* provides capabilities to define random variables based on experiments and to examine their behaviors. The package has built-in simple experiments such as coin tosses and urn draws, but allows the user to define new calculated associated probabilities, including marginal and conditional distributions. This could be useful to astronomers considering the statistical outcomes of hierarchical experiments; for instance, the selection of a sample from an underlying population using a sequence of observational criteria. The code requires that the full sample space be specified.

3 Statistical inference

3.1 The astronomical context

Statistical inference helps the scientist to reach conclusions that extend beyond the obvious and immediate characterization of individual datasets. In some cases, the astronomer measures the properties of a limited sample of objects (often chosen to be brighter or closer than others) in order to learn about the properties of the vast underlying population of similar objects in the Universe. Inference is often based on a **statistic**, a function of random variables. At the early stages of an investigation, the astronomer might seek simple statistics of the data such as the average value or the slope of a heuristic linear relation. At later stages, the astronomer might measure in great detail the properties of one or a few objects to test the applicability, or to estimate the parameters, of an astrophysical theory thought to underly the observed phenomenon.

Statistical inference is so pervasive throughout these astronomical and astrophysical investigations that we are hardly aware of its ubiquitous role. It arises when the astronomer:

– smooths over discrete observations to understand the underlying continuous phenomenon
– seeks to quantify relationships between observed properties
– tests whether an observation agrees with an assumed astrophysical theory
– divides a sample into subsamples with distinct properties
– tries to compensate for flux limits and nondetections
– investigates the temporal behavior of variable sources
– infers the evolution of cosmic bodies from studies of objects at different stages
– characterizes and models patterns in wavelength, images or space

and many other situations. These problems are discussed in later chapters of this volume: nonparametric statistics (Chapter 5), density estimation (Chapter 6), regression (Chapter 7), multivariate analysis (Chapter 8) and classification (Chapter 9), censoring and truncation (Chapter 10), time series analysis (Chapter 11) and spatial analysis (Chapter 12). In this chapter, we lay some of the foundations of statistical inference.

Consider the effort to understand the dynamical state and evolution of a globular cluster of stars. Images give two-dimensional locations of perhaps 1% of the stars, with individual stellar masses estimated from color–magnitude diagrams. Radial velocity measurements from spectra are available for even fewer cluster members. Thus, only three of the six dimensions of phase space (three spatial and three velocity dimensions) are accessible. From this limited information, we have gleaned significant insights with the use of astrophysical models. The three-dimensional structure was originally modeled as a truncated isothermal sphere with two-body dynamical interactions causing the more massive stars to settle towards the core. Modern models involve computationally intensive N-body dynamical simulations to study the effects of binary star systems and stellar collisions on cluster evolution. Statistical inference allows quantitative evaluation of parameters within the context of astrophysical models, giving insights into the structure and dynamics of the globular star cluster. Statistical modeling is thus a crucial component of the scientific inferential process by which astrophysical understanding is developed from incomplete observational information.

3.2 Concepts of statistical inference

Statistical inference helps in making judgments regarding the likelihood that a hypothesized effect in data arises by chance or represents a real effect. It is particularly designed to draw conclusions about the underlying population when the observed samples are subject to uncertainties.

The term **statistical inference** is very broad. Two main aspects of inference are **estimation** and the **testing of hypotheses**. Regression, goodness-of-fit, classification and many other statistical procedures fall under its framework. Statistical inference can be parametric, nonparametric and semi-parametric. Parametric inference requires that the scientist makes some assumptions regarding the mathematical structure of the underlying population, and this structure has parameters to be estimated from the data at hand. Linear regression is an example of parametric inference. Nonparametric procedures make no assumption about the model structure or the distribution of the population. The Kolmogorov–Smirnov hypothesis test and the rank-based Kendall's τ correlation coefficient are examples of nonparametric procedures. Semi-parametric methods combine nonparametric and parametric procedures; local regression models are examples of semi-parametric procedures.

A classic development of the theory of statistical inference is presented by Erich Lehmann in his volumes *Theory of Point Estimation* (Lehmann & Casella 1998) and *Testing Statistical Hypotheses* (Lehmann & Romano 2005). The undergraduate-level texts by Rice (1995) and Hogg *et al.* (2005), and the graduate-level text by Wasserman (2004), are recommended for modern treatments. We outline here several central concepts that these volumes cover in more detail.

Point estimation If the shape of the probability distribution, or relationship between variables, of the underlying population is well-understood, then it remains to find the parameters of the distribution or relationship. For a dataset drawn from a Poisson distribution (Sections 2.7 and 4.2), for example, we seek the value of the rate λ, while for a normal distribution, we want to estimate the mean μ and variance σ^2. Typically a probability distribution or relationship is characterized by a p-dimensional vector of model parameters $\theta = (\theta_1, \theta_2, \ldots, \theta_p)$. For example, the model of a planet in a Keplerian orbit around a star has a vector of six parameters: semi-major axis, eccentricity, inclination, ascending node longitude, argument of periastron and true anomaly. Analogous parameter lists can be made for a tidally truncated isothermal sphere of stars (King model), a turbulent viscous accretion disk (Shakura–Sunyaev α-disk model), the consensus model of cosmology with dark matter and dark energy (ΛCDM model), and many other well-developed astrophysical theories. The goal of estimating plausible or "best" values of θ based on observations is called **point estimation** (Section 3.3).

Likelihood methods   One of the most popular methods of point estimation is **maximum likelihood estimation** (MLE). Likelihood is the hypothetical probability that a past event would yield a specific outcome. The concept differs from that of a probability in that a probability refers to the occurrence of future events, while a likelihood refers to past events with known outcomes. MLE is an enormously popular statistical method for fitting a mathematical model to data. Modeling real-world data by maximizing the likelihood offers a way to tune the free parameters of the model to provide a good fit. Developed in the 1920s by R. A. Fisher, MLE is a conceptual alternative to the least-squares method of the early nineteenth century (Section 1.2.3), but is equivalent to least squares under Gaussian assumptions. The Cramér–Rao inequality, which sets a lower bound on the variance of a parameter, is an important mathematical result of MLE theory (Section 3.4.4). MLE can be used for nontrivial problems such as mixture models for multimodal distributions or nonlinear models arising from astrophysical theory, and is readily applied to multivariate problems. While the likelihood function can be maximized using a variety of computational procedures, the EM algorithm developed in the 1970s is particularly effective (Section 3.4.5).

Confidence intervals   Point estimates cannot be perfectly accurate as different datasets drawn from the same population will give rise to different inferred parameter values. To account for this, we estimate a range of values for the unknown parameter that is usually consistent with the data. This is the parameter's **confidence interval** or confidence set around the best-fit value (Section 3.4.4). The confidence intervals will vary from sample to sample. A confidence interval associated with a particular confidence level, say 95%, is a random interval that is likely to contain a one-dimensional parameter with probability 95%. Confidence intervals can be estimated for different methods of point estimation including least squares and MLE.

Resampling methods   Understanding the variability of a point estimation is essential to obtaining a confidence interval, or to assess the accuracy of an estimator. In many situations encountered in the physical sciences, the variance may not have a closed-form expression. Resampling methods developed in the 1970s and 1980s come to the rescue in such cases. Powerful theorems demonstrated that they provide inference on a wide range of statistics under very general conditions. Methods such as the "bootstrap" involved constructing hypothetical populations from the observations, each of which can be analyzed in the same way to see how the statistics of interest depend on plausible random variations in the observations. Resampling the original data preserves whatever structures are truly present in the underlying population, including non-Gaussianity and multimodality. Although they typically involve random numbers, resampling is not an arbitrary Monte Carlo simulation; it is simulation from the observed data.

Testing hypotheses As the name implies, the goal here is not to estimate parameters of a function based on the data, but to test whether a dataset is consistent with a stated hypothesis. The scientist formulates a null hypothesis and an alternative hypothesis. The result of the test is to either reject or not reject the null hypothesis at a chosen significance level. Note that failure to reject the null hypothesis does not mean that the null hypothesis is correct. Statistical testing of a hypothesis leads to two types of error: wrongly rejecting the null hypothesis (**Type 1 errors** or **false positives**) and wrongly failing to reject the null hypothesis (**Type 2 errors** or **false negatives**). It is impossible to bring these two errors simultaneously to negligible values. Classical hypothesis testing is not symmetric; interchanging the null and alternative hypotheses gives different results. An important astronomical application is the detection of weak signals in noise. Hypothesis tests can also address questions like: Is the mean, or other statistic, of the data equal to a preset value (perhaps zero)? Is the current dataset consistent with the same underlying population as a previous dataset (two-sample test)? Are a group of datasets consistent with each other (k-sample test)? Is the dataset consistent with an underlying population that follows a specified functional relationship (goodness-of-fit test)?

Bayesian inference A growing approach to statistical inference is based on an interpretation of Bayes' theorem (Section 2.4.1) where observational evidence is used to infer (or update) inferences. As evidence accumulates, the degree of belief in a hypothesis or model ought to change. Bayesian inference is based not only on the likelihood that the data follow the model but also on a **prior** distribution. The quality of a Bayesian analysis depends on how one can best convert the prior information into a mathematical prior probability. Various Bayesian methods for parameter estimation, model assessment and other inference problems are outlined in Section 3.8.

3.3 Principles of point estimation

In parametric point estimation, the astronomer must be very careful in setting up the statistical calculation. Two decisions must be made. First, the functional model and its parameters must be specified. If the model is not well-matched to the astronomical population or astrophysical process under study, then the best fit obtained by the inferential process may be meaningless. This problem is called **model misspecification**. Statistical procedures are available to assist the scientist in **model validation** (or **goodness-of-fit**) and **model selection**.

Second, the method by which best-fit parameters are estimated must be chosen. The method of moments, least squares (LS) and maximum likelihood estimation (MLE) are important and commonly used procedures for constructing estimates of the parameters. The choice of estimation method is not obvious, but can be guided by the scientific goal. Statistical measures are available to assist the scientist in choosing a method: some may give best-fit parameters closest to the true value, with the greatest accuracy (smallest variance), or with the highest probability (maximum likelihood). Fortunately, for many situations we can find single best-fit parameter values that are simultaneously unbiased, have minimum variance and have maximum likelihood.

In classical parametric estimation, the observations are assumed to be independently and identically distributed (i.i.d.) random variables with known probability distributions. The dataset $x_1, x_2, \ldots, x_n$ is assumed to be a realization of independent random variables $X_1, X_2, \ldots, X_n$ having a common probability distribution function (p.d.f.) f. We now consider distribution functions characterized by a small number of parameters, $\theta = (\theta_1, \theta_2, \ldots, \theta_p)$. The point estimator of the vector of true parameter values θ is designated $\hat{\theta}$, pronounced "theta-hat". The estimator $\hat{\theta}$ of θ is a function of the random variables $(X_1, X_2, \ldots, X_n)$ under study,

$$\hat{\theta} = g(X_1, X_2, \ldots, X_n). \tag{3.1}$$

The point estimator is thus a function of random variables of the underlying population that is computed from a realization of the population in a particular data sample.

Providing one validates that the data are consistent with the model, the result of parameter estimation can be a great simplification of a collection of data into a few easily interpretable parameters. Often the astronomer believes the data instantiate a deterministic astrophysical theory where the functional relationship is established by physical processes such as gravity, quantum mechanics and thermodynamics. Here the functions can be relatively simple, such as an elliptical orbit of two orbiting bodies, or extremely complex, such as the spectrum from a stellar atmosphere with different atomic species in different excited states subject to radiative transfer.

A great deal of mathematics and discussion lies behind the simple goal of obtaining the "best" estimates of the parameters θ. One aspect relates to the method of obtaining estimators. During the nineteenth century, the method of moments and method of least squares were developed. Astronomy played a crucial role in least-squares theory, as outlined in Section 1.2.2. As a young man in the 1910s and 1920s, R. A. Fisher formulated the "likelihood" that a dataset fits a model, and inaugurated the powerful methods of maximum likelihood estimation (MLE). Minimum variance unbiased estimators (MVUEs) later rose to prominence. As computers became more capable, numerically intensive methods with fewer limitations than previous methods became feasible. The most important, developed in the 1970s and 1980s, is the bootstrap method. With the advent of numerical methods like Markov chain Monte Carlo simulations, nontrivial Bayesian inferential computations became feasible during the 1990s. Bayesian computational methods are being actively developed today.

It is thus common that different point estimators, often derived with different methods, are available to achieve the same data-analytic or scientific goal. A great deal of effort has been exerted to interpret possible meanings of the word "best" in "best-fit parameter" because statistical point estimators have several important properties that often cannot be simultaneously optimized. Statisticians take into consideration several important criteria of a point estimator:

Unbiasedness The bias of an estimator $\hat{\theta}$ is defined to be the difference between the mean of estimated parameter and its true value,

$$B(\hat{\theta}) = E[\hat{\theta}] - \theta. \tag{3.2}$$

This is not the error of a particular instantiation of $\hat{\theta}$ from a particular dataset. This is an intrinsic offset in the estimator.

An estimator $\hat{\theta}$ of a parameter θ is called **unbiased** if $B(\hat{\theta}) = 0$; that is, the expected value of $\hat{\theta}$ is $E[\hat{\theta}] = \theta$. For some biased estimators $\hat{\theta}$, $B(\hat{\theta})$ approaches zero as the data size approaches infinity. In such cases, $\hat{\theta}$ is called **asymptotically unbiased**. Heuristically, $\hat{\theta}$ is an unbiased if its long-term average value is equal to θ. If $\hat{\theta}$ is an unbiased estimator of θ, then the variance of the estimator $\hat{\theta}$ is given by $E[(\hat{\theta} - \theta)^2]$. The smaller the variance of the estimator, the better the estimation procedure. However, if the estimator $\hat{\theta}$ is biased, then $E[(\hat{\theta} - \theta)^2]$ is not the variance of $\hat{\theta}$. In this case,

$$E[(\hat{\theta} - \theta)^2] = \text{Var}(\hat{\theta}) + (E[\hat{\theta} - \theta])^2 \tag{3.3}$$
$$MSE = \text{Variance of } \hat{\theta} + (\text{Bias})^2. \tag{3.4}$$

This quantity, the sum of the variance and the square of the bias, is called the **mean square error** (MSE) and is very important in evaluating estimated parameters.

Minimum variance unbiased estimator (MVUE) Among a collection of unbiased estimators, the most desirable one has the smallest variance, $Var(\hat{\theta})$.

Consistency This criterion states that a **consistent estimator** will approach the true population parameter value as the sample size increases. More precisely, an estimator $\hat{\theta}$ for a parameter θ is **weakly consistent** if for any small $\epsilon > 0$

$$P[|\hat{\theta} - \theta| \geq \epsilon] \longrightarrow 0 \tag{3.5}$$

as $n \to \infty$. The estimator is **strongly consistent** if

$$P[\hat{\theta} \longrightarrow \theta \text{ as } n \to \infty] = 1. \tag{3.6}$$

Asymptotic normality This criterion requires that an ensemble of consistent estimators $\hat{\theta}(n)$ has a distribution around the true population value θ that approaches a normal (Gaussian) distribution with variance decreasing as $1/n$.

3.4 Techniques of point estimation

Parameter estimation is motivated by the problem of fitting models from probability distributions or astrophysical theory to data. Many commonly used probability distributions (such as Gaussian, Poisson, Pareto or power-law) or astrophysical models (such as the temperature and pressure of a uniform gas, or masses and eccentricity in a planetary orbit) depend only on a few parameters. Once these parameters are known, the shape and scale of the curve, and the corresponding properties of the underlying population, are completely determined. For example, the one-dimensional Gaussian distribution depends only on two parameters, the mean μ and the standard deviation σ, as the probability density function ϕ of the Gaussian distribution is given by

$$\phi(x; \mu, \sigma^2) = \frac{1}{\sqrt{2\pi}\sigma} \exp\left\{-\frac{(x-\mu)^2}{2\sigma^2}\right\}. \tag{3.7}$$

Point estimates $\hat{\mu}$ and $\hat{\sigma}$ can be obtained using a variety of techniques. The most common methods for constructing estimates are: the method of moments, least squares and maximum likelihood estimation (MLE). The methods are assessed by the scientific goal of the estimation effort, and by criteria that are important to the scientist.

We will illustrate the common methods of estimation using the two parameters of a population that satisfies a normal (Gaussian) density, the mean μ and standard deviation σ. In this simple and familiar case, the different methods often — but not always — give the same estimators. But in more complicated situations, as we discuss in later chapters, the estimators will often differ.

In the Gaussian case (Equation 3.7), the sample mean and sample variance

$$\hat{\mu} = \bar{X} = \frac{1}{n}\sum_{i=1}^{n} X_i$$

$$\hat{\sigma}^2 = S_X^2 = \frac{1}{n-1}\sum_{i=1}^{n}(X_i - \bar{X})^2 \tag{3.8}$$

are estimators of μ and σ^2, respectively. The factor $n-1$ instead of n in the denominator of the estimator of σ^2 is required for unbiasedness. S_X is not an unbiased estimator of the standard deviation σ and $E[S_X] < \sigma$. In the Gaussian case, $\bar{X}$ and S_X^2 are unbiased estimators of μ and σ^2. This is because $E[X_i] = \mu$ and $E[(X_i - \mu)^2] = \sigma^2$ for each i. A simple calculation indicates that the estimator S_X^2 is not an unbiased estimator of σ^2 if $n-1$ in the denominator is replaced by n.

3.4.1 Method of moments

The **method of moments** for parameter estimation dates to the nineteenth century. The moments are quantitative measures of the parameters of a distribution: the first moment describes its central location; the second moment its width; and the third and higher moments describe asymmetries. As presented in Section 2.5.1, the distribution function F of a random variable is defined as $F(x) = P(X \leq a)$ for all a. The k-th moment of a random variable X with distribution function F is given by

$$\mu_k(X) = E[X^k] = \int x^k dF(x). \tag{3.9}$$

For the random sample X_i, the k-th sample moment is

$$\hat{\mu}_k = \frac{1}{n}\sum_{i=1}^{n} X_i^k. \tag{3.10}$$

Various parameters of a distribution can be estimated by the method of moments if one can first express the parameters as simple functions of the first few moments. Replacing the population moments in the functions with the corresponding sample moments gives the estimator.

To illustrate some moment estimators, first consider the exponential distribution with p.d.f. $f(x) = \lambda \exp(-\lambda x)$ introduced in Equation (2.51). The moment estimator for the rate λ is $\hat{\lambda} = 1/\bar{X}$ as $E[X] = 1/\lambda$. This result is not immediately obvious from a casual examination of the distribution.

Second, consider the normal distribution with the mean μ and variance σ^2. The population moments and their sample estimators are

$$\mu = E[X] \quad \text{and} \quad \hat{\mu} = \frac{1}{n}\sum_{i=1}^{n} X_i$$

$$\sigma^2 = E[X^2] - \mu^2 \quad \text{and} \quad \hat{\sigma}^2 = \frac{1}{n}\sum_{i=1}^{n} X_i^2 - \hat{\mu}^2. \tag{3.11}$$

Note that the variance is the central second moment. The moment-based variance estimator is not unbiased, as the unbiased variance has a factor $1/(n-1)$ rather than $1/n$ before the summation.

3.4.2   Method of least squares

As discussed in our historical overview in Section 1.2.2, parameter estimation using least squares was developed in the early nineteenth century to solve problems in celestial mechanics, and has since been very widely used in astronomy and other fields. We discuss least-squares estimation extensively for regression problems in Section 7.3, and give only a brief introduction here.

Consider estimation of the population mean μ. The least-squares estimator $\hat{\mu}$ is obtained by minimizing the sum of the squares of the differences $(X_i - \mu)$,

$$\hat{\mu}_{LS} = \arg\min_{\mu} \sum_{i=1}^{n}(X_i - \mu)^2; \tag{3.12}$$

that is, $\hat{\mu}_{LS}$ is the value of μ that minimizes $\sum_{i=1}^{n}(X_i - \mu)^2$. We can derive this as follows:

$$\sum_{i=1}^{n}(X_i - \mu)^2 = \sum_{i=1}^{n}(X_i - \bar{X})^2 + n(\bar{X} - \mu)^2 + 2(\bar{X} - \mu)\sum_{i=1}^{n}(X_i - \bar{X})$$

$$= \sum_{i=1}^{n}(X_i - \bar{X})^2 + n(\bar{X} - \mu)^2$$

To show the unbiasedness of S_x^2, note that

$$E\left[\sum_{i=1}^{n}(X_i - \mu)^2\right] = \sum_{i=1}^{n}E[(X_i - \mu)^2] = n\sigma^2$$

$$E[\bar{X} - \mu]^2 = Var(\bar{X}) = \frac{1}{n}\sigma^2$$

$$E\left[\sum_{i=1}^{n}(X_i - \bar{X})^2\right] = n\sigma^2 - \sigma^2 = (n-1)\sigma^2. \tag{3.13}$$

We thus find that $S_X^2 = 1/(n-1) \sum_{i=1}^{n} (X_i - \bar{X})^2$ satisfies $E[S_X^2] = \sigma^2$, demonstrating that the sample variance S_X^2 is an unbiased estimator of the population variance σ^2. If we use instead $T^2 = 1/n \sum_{i=1}^{n} (X_i - \bar{X})^2$ as an estimator of σ^2, then $E[T^2] = [(n-1)/n]\sigma^2 \neq \sigma^2$ showing that T^2 is a biased estimator of the variance.

In this simple case, $\hat{\mu} = \bar{X}$ which is the intuitive solution. But in more complex estimation problems, particularly in the context of regression with a functional relationship between two or more variables, this method provides solutions that are not intuitively obvious. Consider the linear regression $Y_i = \xi_i + \epsilon_i$ where ϵ_i are random variables with mean zero and $\xi_i = \sum_{j=1}^{k} a_{ij}\beta_j$ is a known linear combination of parameters $(\beta_1, \beta_2, \ldots, \beta_k)$. The estimators of β_j can be obtained by minimizing the sum of squares of $(Y_i - \xi_i)$ provided all the ϵ_i variables have the same variance (homoscedastic errors). If the error variances σ_i^2 of X_i are also different (heteroscadastic), then one can minimize the weighted sum of squares

$$\sum_{i=1}^{n} \frac{1}{\sigma_i^2} \left(X_i - \sum_{j=1}^{k} a_{ij}\beta_j \right)^2, \tag{3.14}$$

over $\beta_1, \ldots, \beta_k$. This is called the weighted least-squares method and is related to some procedures astronomers call **minimum χ^2 regression**. Least-squares regression weighted by measurement error is discussed in detail in Section 7.4.

3.4.3 Maximum likelihood method

British mathematicians were actively discussing theoretical and practical approaches to estimation during the early twentieth century. The most brilliant was R. A. Fisher, starting with his critique of least squares as an undergraduate in 1912. He advocated that one should instead calculate the "chance of a given set of observations occurring" as the product of individual probabilities given the data and the model, stating that "the most probable set of values for the θ's [model parameters] will make P [the product of probabilities, later called the likelihood] a maximum". During the next decade, he actively criticized Pearson's χ^2 procedure, as well as the classical method of moments and least squares, promoting instead methods that give the most probable outcome.

In a crucial paper, Fisher (1922) clearly formulated the principle underlying maximum likelihood: "The likelihood that any parameter (or set of parameters) should have any assigned value (or set of values) is proportional to the probability that if this were so, the totality of observation should be that observed." This paper introduced the concepts of consistency, sufficiency, efficiency and information. Aldrich (1997) gives an interesting historical account of this decade in the history of statistics. Even today, Fisher's 1922 arguments play a fundamental role in much of the conceptual and operational methodology for statistical inference.

The method is based on the "likelihood", the probability density (or for discrete distributions, the probability mass) function viewed as a function of the data given particular values of the model parameters. Here we use the notation $f(\cdot; \theta)$ for a probability density with parameter θ. For example, for an exponential random variable, $f(\cdot; \theta)$ is given by $f(x; \theta) = \theta \exp(-\theta x)$ for $x > 0$, and $f(x; \theta) = 0$ for $x \leq 0$. For i.i.d. random variables

$X_1, X_2, \ldots, X_n$ with a common density function $f(\cdot; \theta)$, the likelihood L and loglikelihood ℓ are given by

$$L(\theta) = \prod_{i=1}^{n} f(X_i; \theta)$$

$$\ell(\theta) = \ln L(\theta) = \sum_{i=1}^{n} \ln f(X_i; \theta) \tag{3.15}$$

where "ln" represents the natural logarithm. The likelihood at parameter θ is thus the product of densities evaluated at all the random variables. The sample realization of the likelihood is

$$L(\theta) = \prod_{i=1}^{n} f(x_i; \theta) \tag{3.16}$$

when $(x_1, x_2, \ldots, x_n)$ are the observed data. It is usually computationally easier to work with ℓ than L.

To illustrate maximum likelihood estimation, let us consider a situation where the data follow the geometric distribution. This is a discrete analog of the exponential distribution. Let $X_1, X_2, \ldots, X_n$ be an i.i.d. sequence of random variables with probability mass function

$$f(x; p) = (1 - p)^{x-1} p, \quad \text{where } x = 1, 2, 3, \ldots \tag{3.17}$$

and $0 < p < 1$. The likelihood function is given by

$$L(p) = (1 - p)^{X_1 - 1} p \ldots (1 - p)^{X_n - 1} p$$
$$= p^n (1 - p)^{\sum X_i - n}. \tag{3.18}$$

Defining $\ell(p) = \ln L(p)$, the maximum likelihood estimator for the parameter p is obtained by equating the derivative of $\ell(p)$ to zero,

$$\frac{d\ell(p)}{dp} = \frac{n}{p} - \frac{(\sum_{i=1}^{n} X_i) - n}{1 - p} = 0. \tag{3.19}$$

This leads to the solution

$$\hat{p}_{MLE} = \frac{n}{\sum_{i=1}^{n} X_i} = \frac{1}{\bar{X}}. \tag{3.20}$$

If the actual data were 3, 4, 7, 2, 9 with $n = 5$, then the MLE of p would be $p_{MLE} = 5/(3 + 4 + 7 + 2 + 9) = 1/5 = 0.2$. If all of the observed values are equal to 1, then there is no MLE.

MLEs have many strong mathematical properties. For most probability structures considered in astronomy, the MLE exists and is unique. As we saw for the normal variance, the MLE $\hat{\theta}$ is usually consistent but may not be unbiased. But this can often be overcome by multiplying $\hat{\theta}$ by a constant. Another property is that, for many nice functions g of the parameter, $g(\hat{\theta})$ is the MLE of $g(\theta)$, whenever $\hat{\theta}$ is the MLE of θ. A crucial property is that, for many commonly occurring situations, maximum likelihood parameter estimators $\hat{\theta}$ have an approximate normal distribution when n is large. This asymptotic normality is very useful for estimating confidence intervals for MLE parameters. In most cases, the MLE estimator satisfies

$$\hat{\theta} \doteq \theta$$
$$Var(\hat{\theta}) \doteq \frac{1}{I(\theta)} \quad \text{where}$$
$$I(\theta) = nE\left(\frac{\partial}{\partial \theta} \log f(\mathbf{X}; \theta)\right)^2, \tag{3.21}$$

where the symbol $\doteq$ means "approximately". $I(\theta)$ is called the **Fisher information**. When θ is a vector of parameters, this is the **Fisher information matrix** with off-diagonal terms of the form $\partial^2 f/\partial \theta_i \partial \theta_j$.

3.4.4 Confidence intervals

The confidence interval of a parameter θ, a statistic derived from a dataset X, is defined by the range of lower and upper values $[l(X), u(X)]$ that depend on the variable(s) X defined such that

$$P[l(X) < \theta < u(X)] = 1 - \alpha \tag{3.22}$$

where $0 < \alpha < 1$ is usually a small value like $\alpha = 0.05$ or 0.01. That is, if θ is the true parameter, then the **coverage probability** that the interval $[l(X), u(X)]$ contains θ is at least $1 - \alpha$. The quality of confidence intervals is judged using criteria including validity of the coverage probability, optimality (the smallest interval possible for the sample size), and invariance with respect to variable transformations. For $\alpha = 0.05$, the estimated 95% confidence interval of an estimator of some parameter θ is an interval (l, μ) such that $P(l < \hat{\theta} < u) = 0.95$. If the experiment were repeated 100 times, an average of 95 intervals obtained will contain the parameter value θ.

The idea is illustrated with a simple example for the random variables $X_1, X_2, \ldots, X_n$ drawn from a normal distribution with mean μ and variance 1. Then $\bar{X}$ is a good estimator of μ, its variance is $1/n$ and $\sqrt{n}(\bar{X} - \mu)$ has exactly the normal distribution with mean zero and unit variance. This can be expressed in two ways,

$$P(-1.96 < \sqrt{n}(\bar{X} - \mu) < 1.96) = 0.95,$$
$$P(\bar{X} - 1.96/\sqrt{n} < \mu < \bar{X} + 1.96/\sqrt{n}) = 0.95 \tag{3.23}$$

for all values of μ. Thus, $(\bar{X} - 1.96/\sqrt{n}, \bar{X} + 1.96/\sqrt{n})$ is the 95% confidence interval for μ. A confidence interval for the variance can be similarly derived based on the χ^2 distribution which applies to the variables that are squares of normally distributed variables.

If there are two or more unbiased estimators, the one with the smaller variance is often preferred. Under some regularity conditions, the **Cramér–Rao inequality** gives a lower bound on the minimum possible variance for an unbiased estimator. It states that if $\hat{\theta}$ is an unbiased estimator based on i.i.d. random variables $X_1, X_2, \ldots, X_n$ with a common density function $f(\cdot; \theta)$ where θ is a parameter, then the smallest possible value that $Var(\hat{\theta})$ can attain is $1/I(\theta)$ where I is the Fisher information in Equation (3.21).

Consider again the situation where X is exponentially distributed with density f that is,

$$f(x; \theta) = \theta^{-1} \exp(-x/\theta), \quad x > 0. \tag{3.24}$$

A simple application of the MLE shows that $\hat{\theta} = \bar{X}$ and the Fisher information is $I(\theta) = n\theta^{-2}$. From the Cramér–Rao inequality, the smallest possible value of the variance of an estimator of θ is θ^2/n. This is attained by $\bar{X}$; hence, $\bar{X}$ is the best possible unbiased estimator of θ. It is the minimum variance unbiased estimator (MVUE).

A subtlety, often missed by astronomers, is that the resulting MLE confidence intervals on the mean depend on a precise statement of what is known in advance about other parameters of the problem, the variance σ^2 and the sample size n. The $100(1-\alpha)\%$ confidence interval for μ is

$$\left[\bar{X} - \frac{1}{\sqrt{n}} c_{\alpha/2}, \bar{X} + \frac{1}{\sqrt{n}} c_{\alpha/2} \right],$$

where

$$c_b = \begin{cases} z_b \, \sigma & \text{if } \sigma \text{ is known} \\ z_b \, S_X & \text{if } \sigma \text{ is unknown and } n \text{ is large} \\ t_b(n-1) \, S_X & \text{if } \sigma \text{ is unknown and } n \text{ is small.} \end{cases} \tag{3.25}$$

Here $S_X = 1/(n-1) \sum_{i=1}^{n} (X_i - \bar{X})^2$, and z_b, t_b denote respectively the numbers such that $P(Z > z_b) = P(T_m > t_b(m)) = \alpha$, where Z has a standard normal distribution and T_m has a t-distribution with m degrees of freedom. The middle of the three solutions for c_b is most commonly used, but is not always the appropriate solution.

Similarly, the confidence intervals for differences of two means are not always the simple case of quadratic error propagation commonly used by astronomers (Bevington 1969). Let $X_1, X_2, \ldots, X_n$ and $Y_1, Y_2, \ldots, Y_m$ be two independent samples from two normal populations with means μ and ν and variances σ^2 and τ^2, respectively. Let $\bar{Y}$ denote the sample mean of the Y's, and define

$$S_Y^2 = \frac{1}{m-1} \sum_{i=1}^{m} (Y_i - \bar{Y})^2, \quad S_p^2 = \frac{1}{n+m-2} \left[\{ (n-1)S_X^2 + (m-1)S_Y^2 \} \right].$$

The maximum likelihood estimator for the difference between the two means $\mu - \nu$ is $\bar{X} - \bar{Y}$, as expected, but the $100(1-\alpha)\%$ confidence interval for the difference depends on the situation with the other parameters:

$$\left[\bar{X} - \bar{Y} - \frac{1}{\sqrt{n}} d_b(\alpha/2, m, n), \bar{X} - \bar{Y} + \frac{1}{\sqrt{n}} d_b(\alpha/2, m, n) \right],$$

where

$$d_b(m, n) = \begin{cases} z_b \sqrt{\frac{\sigma^2}{n} + \frac{\tau^2}{m}} & \text{if } \sigma, \tau \text{ are known} \\ z_b \sqrt{\frac{S_X^2}{n} + \frac{S_Y^2}{m}} & \text{if } \sigma, \tau \text{ are unknown and } n, m \text{ are large} \\ t_b(n+m-2) S_p \sqrt{\frac{1}{n} + \frac{1}{m}} & \text{if } \sigma = \tau \text{ but the common value is} \\ & \text{unknown and } n, m \text{ are small.} \end{cases}$$

$$\tag{3.26}$$

The confidence limits above are **two-sided** where the scientific question permits values either higher or lower than the best-fit parameter value. But sometimes the scientific question involves an asymmetrical **one-sided** confidence limit. The one-sided lower $100(1 - \alpha)\%$ confidence interval for μ is given by $[\bar{X} - \frac{1}{\sqrt{n}}c_\alpha, \infty)$. Similarly, the one-sided lower $100(1 - \alpha)\%$ confidence interval for $\mu - \nu$ is $[\bar{X} - \bar{Y} - d_\alpha(m, n), \infty)$.

Finally, we consider the problem of estimating a proportion, or ratio, in a binary trial experiment. While typically phrased in terms of the proportion of heads and tails from flipping a coin, the issue arises in astronomy when we seek the fraction of quasars that are radio-loud, the ratio of brown dwarfs to stars, or the "hardness" ratio of photons above and below some critical energy. Let y be the number of successes in n Bernoulli trials with probability $0 < p < 1$ of success on each trial. The best-fit value (e.g. by MLE) for the unknown fraction p is easily found to be $p = y/n$, the intuitive value.

But it is less easy to obtain the confidence intervals for such a proportion. When n is large and asymptotic normality applies, the $100(1 - \alpha)\%$ approximate confidence interval for p is

$$\left[\frac{y}{n} - z_{\alpha/2}\sqrt{\frac{1}{n}\left(\frac{y}{n}\left(1 - \frac{y}{n}\right)\right)}, \frac{y}{n} + z_{\alpha/2}\sqrt{\frac{1}{n}\left(\frac{y}{n}\left(1 - \frac{y}{n}\right)\right)} \right]. \tag{3.27}$$

The case of the difference, rather than the ratio, between two Bernoulli trial experiments can also be considered. Let y_1 and y_2 be two independent binomial random variables with parameters p_1 and p_2, and the corresponding number of trials n_1 and n_2. Then the estimate of the difference is

$$\widehat{p_1 - p_2} = y_1/n_1 - y_2/n_2 \tag{3.28}$$

and the $100(1 - \alpha)\%$ approximate confidence interval is given by

$$\left[\frac{y_1}{n_1} - \frac{y_2}{n_2} - z_{\alpha/2}\, a(n_1, n_2, y_1, y_2), \frac{y_1}{n_1} - \frac{y_2}{n_2} + z_{\alpha/2}\, a(n_1, n_2, y_1, y_2) \right],$$

where

$$a(n_1, n_2, y_1, y_2) = \sqrt{\frac{1}{n_1}\left(\frac{y_1}{n_1}\left(1 - \frac{y_1}{n_1}\right)\right) + \frac{1}{n_2}\left(\frac{y_2}{n_2}\left(1 - \frac{y_2}{n_2}\right)\right)}. \tag{3.29}$$

We will return to this and similar problems involving ratios of discrete distributions in Sections 3.8 and 4.1.1 to illustrate the subtleties and options that can arise in estimation even for easily stated problems that often appear in astronomy.

3.4.5 Calculating MLEs with the EM algorithm

Likelihoods can be maximized by any numerical optimization method. During the mid-twentieth century before computers, simple analytical statistical models were emphasized where the maximum of the likelihood function could be obtained by differential calculus. For a model with p parameters $\theta_1, \theta_2, \ldots, \theta_p$, the equations

$$\frac{\partial L(\theta_i)}{\partial \theta_i} = 0 \tag{3.30}$$

often gave a system of p equations in p unknowns that could be solved using algebraic techniques.

Alternatively, the maximum of the likelihood could be found numerically using iterative numerical techniques like the Newton–Raphson and gradient descent methods and their modern variants (e.g. Levenberg–Marquardt, Davidon–Fletcher–Powell, and Broyden–Fletcher–Goldfarb–Shanno methods; Nocedal & Wright 2006). These techniques may converge slowly, encountering problems when the derivative is locally unstable and when the likelihood function has multiple maxima. Other techniques, such as simulated annealing and genetic algorithms, are designed to assist in finding the global maximum in likelihood functions with complex structure.

One particular numerical procedure, the **EM algorithm**, has been enormously influential in promoting maximum likelihood estimation since the seminal papers of Dempster *et al.* (1977) and Wu (1983). In accord with the statisticians' concern with modeling uncertain and incomplete data to describe a hidden phenomenon, the EM algorithm considers the mapping of a set of datasets to an unknown complete dataset. The method was independently developed in astronomy for image deconvolution by Richardson (1972) and Lucy (1974). Here the data are the observed image of the sky blurred by the telescope's optics, the model is the telescope point spread function, and the missing dataset is the true sky image.

One begins the EM algorithm with initial values of the model parameter values θ and the dataset. These might be estimated by least squares, or represent guesses by the scientist. The algorithm proceeds by iteration of two steps. The **expectation step** (E) calculates the likelihood for the current values of the parameter vector θ. The **maximization step** (M) updates the missing dataset values with the criterion that the likelihood of the values with respect to the current model is maximized. This updated dataset then takes the place of the original dataset, and the algorithm is iterated until convergence. In many situations, maximization of the complete data is easier than the original incomplete data.

The algorithm is successful for many MLE problems because each iteration is guaranteed to increase the likelihood over the previous iteration. Local minima are ignored and convergence is usually rapid. However, there is still no guarantee that the achieved maximum is global over the full parameter space. Research on accelerating the EM algorithm and improving its convergence for maximum likelihood estimation is still actively pursued (McLachlan & Krishnan 2008).

The EM algorithm has been fruitful for many types of MLE calculations even when missing data are not obviously present. These include and linear and nonlinear regression (Chapter 7), normal (Gaussian) mixture models, pattern recognition and image deconvolution, modeling binned or categorical data, multivariate analysis and classification (Chapters 8 and 9), modeling censored and truncated data (Chapter 10) and time series anlaysis (Chapter 11).

3.5 Hypothesis testing techniques

We now turn to statistical testing of hypotheses. This is formulated in terms of deciding between two competing statements, H_0 and H_a, called the **null hypothesis** and **alternative**

hypothesis, based on the data. There are two possible errors in adjudicating between these hypotheses:

Type 1 error Here one wrongly rejects the null hypothesis H_0 giving a **false positive** decision. For example, when searching for a faint signal in noise, this occurs when we incorrectly infer that a signal is present when it truly is not.

Type 2 error Here one fails to reject the null hypothesis when the alternative is true, giving a **false negative** decision. In our example, we would incorrectly infer that a signal is absent when it truly is present.

Ideally, one likes to minimize these two errors to negligible levels, but it is not possible to achieve this. So the scientist must decide what errors are more important for the goals of the test.

A traditional choice is to construct the critical regions to keep Type 1 errors under control at the 5% level, allowing Type 2 errors to be uncontrolled. This choice of 5% is called the **significance level** of the hypothesis test, and represents the probability of generating false positives; that is, incorrectly rejecting the null hypothesis. The **power** of a test is the probability of correctly rejecting the null when the alternative hypothesis is true. The power is $1 - \beta$ where β here is the Type 2 error or false negative rate. The **uniformly most powerful** (UMP) test is the test statistic that give the highest power for all parameter values for a chosen significance level. This is often the preferred test statistic.

A result of a hypothesis test is called **statistically significant** if it is unlikely to have occurred by chance. That is, the hypothesis is significant at level α if the test rejects the null hypothesis at the prescribed significance level α. Typical significance levels used in many fields are $\alpha = 0.05$ or 0.01. Note that the common standard in astronomy of 3σ, where σ is the standard deviation corresponding to $\alpha = 0.003$ for the normal distribution.

Along with the binary "Yes/No" results of a statistical test, the so-called ***p*-value** is often reported. The *p*-value of a hypothesis test is the probability, assuming the null hypothesis is true, of observing a result at least as extreme as the value of the test statistic. It is important to note that the null hypothesis and the alternative hypothesis are not treated symmetrically. We can only reject the null hypothesis at a given level of significance; we can never accept a null hypothesis. In the case of signal detection, rejecting the null hypothesis leads to detecting a signal. So it is often the case that the alternative hypothesis is chosen as the statement for which there is likely to be supporting evidence.

A common difficulty is that significance levels must be adjusted when many hypothesis tests are conducted on the same dataset. This situation often occurs in astronomical image analysis. A large image or data cube may be searched at millions of locations for faint sources, so that one must seek a balance between many false positives and sensitivity. A new procedure for combining multiple hypothesis tests called the **false detection rate** provides a valuable way to control for false positives (Benjamini & Hochberg 1995).

For a two-sided hypothesis test $H_0 : \theta = \theta_0$ vs. $H_a : \theta \neq \theta_0$, the set theoretic complement of the rejection region at level of significance α serves as the $(1 - \alpha)100\%$ confidence region for θ. Recall that z_α and $t_\alpha(m)$ denote, respectively, the numbers such that $P(Z > z_\alpha) = P(T_m > t_\alpha(m)) = \alpha$, where Z has the standard normal distribution and T_m has the t distribution with m degrees of freedom.

Table 3.1 Hypotheses for one proportion ($H_0\colon p = p_0$).

	H_1	Critical region
$x = \dfrac{(y/n) - p_0}{\sqrt{p_0(1 - p_0)/n}},$	$p > p_0$	$x \geq z_\alpha$
	$p < p_0$	$x \leq -z_\alpha$
	$p \neq p_0$	$\lvert x \rvert \geq z_{\alpha/2}$

Table 3.2 Hypotheses for two proportions ($H_0\colon p_1 = p_2$).
$\hat{p}_1 = \frac{y_1}{n_1}$, $\hat{p}_2 = \frac{y_2}{n_2}$, $\hat{p} = \frac{y_1 + y_2}{n_1 + n_2}$

	H_1	Critical region
$x = \dfrac{\hat{p}_1 - \hat{p}_2}{\sqrt{\hat{p}(1 - \hat{p})(1/n_1 + 1/n_2)}},$	$p_1 > p_2$	$x \geq z_\alpha$
	$p_1 < p_2$	$x \leq -z_\alpha$
	$p_1 \neq p_2$	$\lvert x \rvert \geq z_{\alpha/2}$

Table 3.3 Hypotheses for one mean for normal data ($H_0\colon \mu = \mu_0$).

H_1	Critical region, σ known	Critical region, variance unknown
$\mu > \mu_0$	$\bar{X} \geq \mu_0 + z_\alpha \sigma/\sqrt{n}$	$\bar{X} \geq \mu_0 + t_\alpha(n-1)S_X/\sqrt{n}$
$\mu < \mu_0$	$\bar{X} \leq \mu_0 - z_\alpha \sigma/\sqrt{n}$	$\bar{X} \leq \mu_0 - t_\alpha(n-1)S_X/\sqrt{n}$
$\mu \neq \mu_0$	$\lvert \bar{X} - \mu_0 \rvert \geq z_{\alpha/2}\sigma/\sqrt{n}$	$\lvert \bar{X} - \mu_0 \rvert \geq t_{\alpha/2}(n-1)S_X/\sqrt{n}$

Table 3.4 Hypotheses for one mean for nonnormal data when n is large.

H_0	H_1	Critical region
$\mu = \mu_0$	$\mu > \mu_0$	$\bar{X} \geq \mu_0 + z_\alpha S_X/\sqrt{n}$
$\mu = \mu_0$	$\mu < \mu_0$	$\bar{X} \leq \mu_0 - z_\alpha S_X/\sqrt{n}$
$\mu = \mu_0$	$\mu \neq \mu_0$	$\lvert \bar{X} - \mu_0 \rvert \geq z_{\alpha/2}S_X/\sqrt{n}$

Critical regions (also called **rejection regions**) for some commonly used tests of hypotheses involving proportions, means and variances, are given in Tables 3.1–3.8. The tables provide inequalities involving values of statistics computed from the data. Some of the critical regions are based on the Central Limit Theorem. In these tables, $\bar{X}$ is the sample mean of n i.i.d random variables from a population with mean μ_X and variance σ^2, and $\bar{Y}$ is the sample mean of m i.i.d random variables, independent of the X's, from a population with mean μ_Y and variance τ^2. S_X^2 denotes the sample variance.

In the testing for a single proportion (Table 3.1), the denominator $\sqrt{p_0(1 - p_0)/n}$ can be replaced by $\sqrt{(y/n)(1 - (y/n))/n}$. Similarly in testing for equality of two proportions (Table 3.2), the denominator of z can be replaced by $\sqrt{\hat{p}_1(1 - \hat{p}_1)/n_1 + \hat{p}_2(1 - \hat{p}_2)/n_2}$. Here the y, y_1 and y_2 values are observations from the binomial distribution with population proportions p, p_1 and p_2 with n, n_1 and n_2 trials, respectively.

If both the X's and Y's are from normal populations and m and n are not large, then asymptotic theory does not apply and different critical regions of the hypothesis tests are

Table 3.5 Hypotheses for the equality of two means when n and m are large.

H_0	H_1	Critical region
$\mu_X = \mu_Y$	$\mu_X > \mu_Y$	$\bar{X} - \bar{Y} \geq z_\alpha \sqrt{\frac{S_X^2}{n} + \frac{S_Y^2}{m}}$
$\mu_X = \mu_Y$	$\mu_X < \mu_Y$	$\bar{X} - \bar{Y} \leq -z_\alpha \sqrt{\frac{S_X^2}{n} + \frac{S_Y^2}{m}}$
$\mu_X = \mu_Y$	$\mu_X \neq \mu_Y$	$|\bar{X} - \bar{Y}| \geq z_{\alpha/2} \sqrt{\frac{S_X^2}{n} + \frac{S_Y^2}{m}}$

Table 3.6 Confidence regions for common small-sample hypothesis tests. Hypotheses for the equality of two means for normal populations.

H_0	H_1	Critical region
$\mu_X = \mu_Y$	$\mu_X > \mu_Y$	$\bar{x} - \bar{y} \geq d_\alpha(m, n)$
$\mu_X = \mu_Y$	$\mu_X < \mu_Y$	$\bar{x} - \bar{y} \leq -d_\alpha(m, n)$
$\mu_X = \mu_Y$	$\mu_X \neq \mu_Y$	$|\bar{x} - \bar{y}| \geq d_{\alpha/2}(m, n)$

Table 3.7 Hypotheses for one variance $(H_0 : \sigma^2 = \sigma_0^2)$.

H_1	Critical region
$\sigma^2 > \sigma_0^2$	$(n-1)S_X^2 \geq \sigma_0^2 \chi_\alpha^2(n-1)$
$\sigma^2 < \sigma_0^2$	$(n-1)S_X^2 \leq \sigma_0^2 \chi_{1-\alpha}^2(n-1)$
$\sigma^2 \neq \sigma_0^2$	$(n-1)S_x^2 \geq \sigma_0^2 \chi_{\alpha/2}^2(n-1)$
	or $(n-1)S_x^2 \leq \sigma_0^2 \chi_{1-\alpha/2}^2(n-1)$

Table 3.8 Hypotheses for the equality of two variances $(H_0 : \sigma_X^2 = \sigma_Y^2)$.

H_1	Critical region
$\sigma_X^2 > \sigma_Y^2$	$S_X^2 / S_Y^2 \geq F_\alpha(n-1, m-1)$
$\sigma_X^2 < \sigma_Y^2$	$S_Y^2 / S_X^2 \geq F_\alpha(m-1, n-1)$
$\sigma_X^2 \neq \sigma_Y^2$	$S_X^2 / S_Y^2 \geq F_{\alpha/2}(n-1, m-1)$
	or $S_Y^2 / S_X^2 \geq F_{\alpha/2}(m-1, n-1)$

used as given in Table 3.6. Here the quantity $d_b(m, n)$ is defined in Equation (3.26). Recall that if $X_1, X_2, \ldots, X_n$ are i.i.d. normal random variables with mean μ and variance σ^2, then $(n-1)S_X^2 / \sigma^2$ has a χ^2 distribution with $n-1$ degrees of freedom. This fact is used to construct tests concerning variances (Tables 3.7–3.8). Here we let $\chi_\beta^2(m)$ denote the number such that the probability that a χ^2 random variable with m degrees of freedom exceeds $\chi_\beta^2(m)$ is β. Finally, for the test of equality of variances of two populations, we rely on the property that the ratio of two independent χ^2 random variables, normalized by their degrees of freedom, follows an F distribution. Thus, if $\sigma_X^2 = \sigma_Y^2$, then S_X^2/S_Y^2 has an F distribution with $(n-1, m-1)$ degrees of freedom. Let $F_\beta(n-1, m-1)$ be such that $P(F \geq F_\beta(n-1, m-1)) = \beta$.

3.6   Resampling methods

The classical statistical methods of earlier sections concentrated mainly on the statistical properties of the estimators that have a simple closed form and which can be analyzed mathematically. Except for a few important but simple statistics, these methods often involve unrealistic model assumptions. While it is often relatively easy to devise a statistic that measures a property of scientific interest, it is almost always difficult or impossible to determine the distribution of that statistic.

These limitations have been overcome in the last two decades of the twentieth century with a class of computationally intensive procedures known as **resampling methods** that provide inferences on a wide range of statistics under very general conditions. Resampling methods involve constructing hypothetical populations derived from the observations, each of which can be analyzed in the same way to see how the statistics depend on plausible random variations in the observations. Resampling the original data preserves whatever distributions are truly present, including selection effects such as truncation and censoring.

The **half-sample method** may be the oldest resampling method, where one repeatedly chooses at random half of the data points, and estimates the statistic for each resample. The inference on the parameter can be based on the histogram of the resampled statistics. It was used by P. C. Mahalanobis in 1946 under the name **interpenetrating samples**. An important variant is the Quenouille–Tukey **jackknife method**. For a dataset with n data points, one constructs exactly n hypothetical datasets each with $n-1$ points, each one omitting a different point. The most important of the resampling methods proved to be the **bootstrap** or resampling with replacement. B. Efron introduced the bootstrap method in 1979. Here one generates a large number of datasets, each with n data points randomly drawn from the original data. The constraint is that each drawing is made from the entire dataset, so a simulated dataset will miss some points and have duplicates or triplicates of others. Thus, the bootstrap can be viewed as a Monte Carlo method to simulate from existing data, without any assumption about the underlying population.

3.6.1   Jackknife

The jackknife method was introduced by M. Quenouille (1949) to estimate the bias of an estimator. The method was later shown to be useful in reducing the bias as well as in estimating the variance of an estimator. Let $\hat{\theta}_n$ be an estimator of θ based on n i.i.d. random vectors $X_1, \ldots, X_n$. That is, $\hat{\theta}_n = f_n(X_1, \ldots, X_n)$, for some function f_n. Let

$$\hat{\theta}_{n,-i} = f_{n-1}(X_1, \ldots, X_{i-1}, X_{i+1}, \ldots, X_n) \tag{3.31}$$

be the corresponding recomputed statistic based on all but the i-th observation. The jack-knife estimator of bias $E(\hat{\theta}_n) - \theta$ is given by

$$bias_J = \frac{(n-1)}{n} \sum_{i=1}^{n} \left(\hat{\theta}_{n,-i} - \hat{\theta}_n\right). \tag{3.32}$$

The jackknife estimator θ_J of θ is given by

$$\theta_J = \hat{\theta}_n - bias_J = \frac{1}{n} \sum_{i=1}^{n} \left(n\hat{\theta}_n - (n-1)\hat{\theta}_{n,-i}\right). \tag{3.33}$$

Such a bias-corrected estimator hopefully reduces the overall bias of the estimator. The summands above

$$\theta_{n,i} = n\hat{\theta}_n - (n-1)\hat{\theta}_{n,-i}, \ i = 1, \ldots, n \tag{3.34}$$

are called **pseudo-values**.

In the case of the sample mean $\hat{\theta}_n = \bar{X}_n$, it is easy to check that the pseudo-values are simply

$$\theta_{n,i} = n\hat{\theta}_n - (n-1)\hat{\theta}_{n,-i} = X_i, \ i = 1, \ldots, n. \tag{3.35}$$

This provides motivation for the jackknife estimator of the variance of $\hat{\theta}_n$,

$$var_J(\hat{\theta}_n) = \frac{1}{n(n-1)} \sum_{i=1}^{n} (\theta_{n,i} - \theta_J)(\theta_{n,i} - \theta_J)'$$
$$= \frac{n-1}{n} \sum_{i=1}^{n} (\hat{\theta}_{n,-i} - \bar{\theta}_n)(\hat{\theta}_{n,-i} - \bar{\theta}_n)', \tag{3.36}$$

where $\bar{\theta}_n = \frac{1}{n} \sum_{i=1}^{n} \hat{\theta}_{n,-i}$. For most statistics, the jackknife estimator of the variance is consistent; that is

$$Var_J(\hat{\theta}_n)/Var(\hat{\theta}_n) \to 1, \tag{3.37}$$

as $n \to \infty$ almost surely. In particular, this holds for a **smooth functional model**. To describe this, let the statistic of interest, $\hat{\theta}_n$, based on n data points be defined by $H(\bar{Z}_n)$, where $\bar{Z}_n$ is the sample mean of the random vectors $Z_1, \ldots, Z_n$ and H is continuously differentiable in a neighborhood of $E(\bar{Z}_n)$. Many commonly occurring statistics fall under this model, including: sample means, sample variances, central and noncentral t statistics (with possibly nonnormal populations), sample coefficient of variation, least-squares estimators, maximum likelihood estimators, correlation coefficients, regression coefficients and smooth transforms of these statistics.

However, consistency does not always hold; for example, the jackknife method fails for nonsmooth statistics such as the sample median. If $\hat{\theta}_n$ denotes the sample median in the univariate case, then in general,

$$Var_J(\hat{\theta}_n)/Var(\hat{\theta}_n) \to \left(\frac{1}{2}\chi_2^2\right)^2 \tag{3.38}$$

in distributions where χ_2^2 denotes a *chi-square* random variable with two degrees of freedom (Efron 1982, Section 3.4). So in this case, the jackknife method does not lead to a consistent estimator of the variance. However, the bootstrap resampling method does lead to a consistent estimator for this case.

3.6.2 Bootstrap

Bootstrap resampling constructs datasets with n points, rather than $n-1$ for the jackknife, where each point was selected from the full dataset; that is, resampling with replacement. The importance of the bootstrap emerged during the 1980s when mathematical study demonstrated that it gives a nearly optimal estimate of the distribution of many statistics under a wide range of circumstances, including the smooth function models listed above for the jackknife. In several cases, the method yields better results than those obtained by the classical normal approximation theory. However, one should caution that bootstrap is not the solution for all problems. The theory developed in the 1980s and 1990s shows that bootstrap fails in some nonsmooth situations. Hence, caution should be used and one should resist the temptation to use the method inappropriately.

We first describe an application of the bootstrap to estimate the variance of the sample mean. Let $\mathbf{X} = (X_1, \ldots, X_n)$ be data drawn from an unknown population distribution F. Suppose $\hat{\theta}_n$, based on the data $\mathbf{X}$, is a good estimator of θ, a parameter of interest. The interest lies in assessing its accuracy in estimation. Determining the confidence intervals for θ requires knowledge of the sampling distribution G_n of $\hat{\theta}_n - \theta$; that is, $G_n(x) = P(\hat{\theta}_n - \theta \leq x)$, for all x. The sample mean $\bar{X}_n = n^{-1} \sum_{i=1}^{n} X_i$ is a good estimator of the population mean μ. To get the confidence interval for μ, we must find the sampling distribution of $\bar{X}_n - \mu$ which depends on the shape and other characteristics of the unknown distribution F.

Classical statistical theory applies the normal approximation obtained from the Central Limit Theorem to the sampling distribution. The problem is that, even if the sampling distribution is not symmetric, the CLT approximates using a normal distribution, which is symmetric. This can be seen from the following example. If $(X_1, Y_1), \ldots, (X_n, Y_n)$ denote observations from a bivariate normal population, then the maximum likelihood estimator of the correlation coefficient ρ is given by **Pearson's linear correlation coefficient**,

$$\hat{\rho}_n = \frac{\sum_{i=1}^{n}(X_i Y_i - \bar{X}_n \bar{Y}_n)}{\sqrt{\left(\sum_{i=1}^{n}(X_i - \bar{X}_n)^2\right)\left(\sum_{i=1}^{n}(Y_i - \bar{Y}_n)^2\right)}}. \tag{3.39}$$

For statistics with asymmetrical distributions, such as that of $\hat{\rho}_n$, the classical theory suggests variable transformations. In this case, **Fisher's Z transformation Z** given by

$$Z = \frac{\sqrt{(n-3)}}{2}\left(\ln\left(\frac{1 + \hat{\rho}_n}{1 - \hat{\rho}_n}\right) - \ln\left(\frac{1 + \rho}{1 - \rho}\right)\right) \tag{3.40}$$

gives a better normal approximation. This approximation corrects skewness and is better than the normal approximation of $\sqrt{n}(\hat{\rho}_n - \rho)$. The bootstrap method, when properly used, avoids such individual transformations by taking into account the skewness of the sampling distribution. It automatically corrects for skewness.

The bootstrap method avoids such clumsy transformations for statistics with asymmetrical (or unknown) distributions. The bootstrap presumes that if $\hat{F}_n$ is a good approximation to the unknown population distribution F, then the behavior of the samples from $\hat{F}_n$ closely resemble that of the original data. Here $\hat{F}_n$ can be the empirical distribution function (e.d.f.) or a smoothed e.d.f. of the data $X_1, \ldots, X_n$, or a parametric estimator of the function F. The e.d.f. will be further discussed in Section 5.3.1. Once $\hat{F}_n$ is provided, datasets $\mathbf{X}^* = (X_1^*, \ldots, X_n^*)$ are resampled from $\hat{F}_n$ and the statistic θ^* based on $\mathbf{X}^*$ is computed for each resample. Under very general conditions Babu & Singh (1984) have shown that the difference between the sampling distribution G_n of $\hat{\theta}_n - \theta$ and the **bootstrap distribution** G_b (that is, the distribution of $\theta^* - \hat{\theta}_n$) is negligible. G_b can thus be used to draw inferences about the parameter θ in place of the unknown G_n.

In principle, the bootstrap distribution G_b, which is a histogram, is completely known, as it is constructed entirely from the original data. However, to get the complete bootstrap distribution, one needs to compute the statistics for nearly all of the $M = n^n$ possible bootstrap samples. For the simple example of the sample mean, presumably one needs to compute

$$
\begin{aligned}
X_1^{*(1)}, \ldots, X_n^{*(1)}, \quad & r_1 = \bar{X}^{*(1)} - \bar{X} \\
X_1^{*(2)}, \ldots, X_n^{*(2)}, \quad & r_2 = \bar{X}^{*(2)} - \bar{X} \\
\ddots \qquad \ddots \ddots \qquad & \ddots \\
X_1^{*(M)}, \ldots, X_n^{*(M)}, \quad & r_M = \bar{X}^{*(M)} - \bar{X}.
\end{aligned}
$$

The bootstrap distribution is given by the histogram of $r_1, \ldots, r_M$. Even for $n = 10$ data points, M turns out to be ten billion.

In practice, the statistic of interest, $\theta^* - \hat{\theta}_n$, can be accurately estimated from the histogram of a much smaller number of resamples. Asymptotic theory shows that the sampling distribution of $\theta^* - \hat{\theta}_n$ can be well-approximated by generating $N \simeq n(\ln n)^2$ bootstrap resamples (Babu & Singh 1983). Thus, only $N \sim 50$ simulations are needed for $n = 10$ and $N \sim 50,000$ for $n = 1000$. Thus, we can estimate the distribution of the statistic of interest for the original dataset from the histogram of the statistic obtained from the bootstrapped samples.

So far, we have described the most popular and simple bootstrap – the **nonparametric bootstrap** where the resampling with replacement is based on the e.d.f. of the original data. This gives equal weights to each of the original data points. Table 3.9 gives bootstrap versions of some common statistics. In the case of the ratio estimator and the correlation coefficient, the data pairs are resampled from the original data pairs (X_i, Y_i).

Bootstrap resampling is also widely used for deriving confidence intervals for parameters. However, one can only invert the limiting distribution to get confidence intervals when the limiting distribution of the point estimator is free from the unknown parameters. Such

Table 3.9 Statistics and their bootstrap versions.

Statistic	Bootstrap version
Mean, $\bar{X}_n$	$\bar{X}_n^*$
Ratio estimator, $\bar{X}_n/\bar{Y}_n$	$\bar{X}_n^*/\bar{Y}_n^*$
Variance, $\frac{1}{n}\sum_{i=1}^n (X_i - \bar{X}_n)^2$	$\frac{1}{n}\sum_{i=1}^n (X_i^* - \bar{X}_n^*)^2$
Correlation coefficient, $\frac{\sum_{i=1}^n (X_iY_i-\bar{X}_n\bar{Y}_n)}{\sqrt{\left(\sum_{i=1}^n (X_i-\bar{X}_n)^2\right)\left(\sum_{i=1}^n (Y_i-\bar{Y}_n)^2\right)}}$	$\frac{\sum_{i=1}^n (X_i^*Y_i^*-\bar{X}_n^*\bar{Y}_n^*)}{\sqrt{\left(\sum_{i=1}^n (X_i^*-\bar{X}_n^*)^2\right)\left(\sum_{i=1}^n (Y_i^*-\bar{Y}_n^*)^2\right)}}$

quantities are called **pivotal statistics**. It is thus important to focus on pivotal or approximately pivotal quantities in order to get reliable confidence intervals for the parameter of interest.

Consider the confidence interval of the sample mean. If the data are normally distributed, $X_i \sim N(\mu, \sigma^2)$, then $\sqrt{n}(\bar{X} - \mu)/S_n$ has a t distribution with $n - 1$ degrees of freedom, and hence it is pivotal. In the nonnormal case, it is approximately pivotal where $S_n^2 = \frac{1}{n}\sum_{i=1}^n (X_i-\bar{X})^2$. To obtain the bootstrap confidence interval for μ, we compute $\sqrt{n}(\bar{X}^{*(j)} - \bar{X})/S_n$ for N bootstrap samples, and arrange the values in increasing order, $h_1 < h_2 < \cdots < h_N$. One can then read from the histogram (say) the 90% confidence interval of the parameter. That is, the 90% confidence interval for μ is given by

$$\bar{X} - h_m \frac{S_n}{\sqrt{n}} \leq \mu < \bar{X} - h_k \frac{S_n}{\sqrt{n}}, \tag{3.41}$$

where $k = [0.05N]$ and $m = [0.95N]$.

It is important to note that even when σ is known, the bootstrap version of $\sqrt{n}(\bar{X} - \mu)/\sigma$ is $\sqrt{n}(\bar{X}^* - \bar{X})/S_n$. One should not replace $\sqrt{n}(\bar{X}^* - \bar{X})/S_n$ by $\sqrt{n}(\bar{X}^* - \bar{X})/\sigma$.

For datasets of realistic size, the sampling distributions of several commonly occurring statistics are closer to the corresponding bootstrap distribution than the normal distribution given by the CLT. If pivotal statistics are used, then the confidence intervals are similarly reliable under very general conditions. The discussion here is applicable to a very wide range of functions that includes sample means and variances, least squares and maximum likelihood estimators, correlation and regression coefficients, and smooth transforms of these statistics.

A good overview of the bootstrap is presented by Efron and Tibshirani (1993), and Zoubir & Iskander (2004) provide a practical handbook for scientists and engineers. Bootstrap methodology and limiting theory are reviewed by Babu & Rao (1993). The bootstrap method has found many applications in business, engineering, biometrics, environmental statistics, image and signal processing, and other fields. Astronomers have started using the bootstrap also. In spite of their many capabilities, one should recognize that bootstrap methods fail under certain circumstances. These include nonsmooth statistics (such as the maximum value of a dataset), heavy-tailed distributions, distributions with infinite variances, and some nonlinear statistics.

3.7 Model selection and goodness-of-fit

The aim of model fitting and parameter estimation is to provide the most parsimonious "best fit" of a mathematical model to data. The model might be a simple, heuristic function to approximate phenomenological relationships between observed properties in a sample; for example, a linear or power-law (Pareto) function. The same procedures of mathematical statistics can be used to link data to complicated astrophysical models. The relevant methods fall under the rubrics of regression, goodness-of-fit, and model selection.

The common procedure has four steps:

1. choose a model family based on astrophysical knowledge or a heuristic model based on exploratory data analysis;
2. obtain best-fit parameters for the model using the methods described in Section 3.3;
3. apply a goodness-of-fit hypothesis test to see whether the best-fit model agrees with the data at a selected significance level; and
4. repeat with alternative models to select the best model according to some quantitative criterion. In this last step, a dataset may be found to be compatible with many models, or with no model at all.

The coupled problems of goodness-of-fit and model selection, steps 3 and 4 above, are among the most common in modern statistical inference. The principles are clear. After a model has been specified and best-fit parameters estimated, a reliable and broadly applicable test for the fit's validity given the dataset is needed. Once a satisfactory model has been found, exploration of alternative models is needed to find the optimal model. A final principle is parsimony: a good statistical model should be among the simplest consistent with the data. This idea dates to the Middle Ages, when the fourteenth-century English philosopher William of Ockham proposed that the simplest solution is usually the correct one. In statistical modeling, **Ockham's Razor** suggests that we leave off extraneous ideas better to reveal the truth. We thus seek a model that neither underfits, excluding key variables or features, nor overfits, incorporating unnecessary complexity. As we will discuss further in Section 6.4.1 on data smoothing, underfitting induces bias and overfitting induces high variability. A model selection criterion should balance the competing objectives of conformity to the data and parsimony.

The model selection problem is more tractable when **nested** models are compared, where the simpler model is a subset of the more elaborated model. In interpreting a continuum astronomical spectrum, for example, modeling might start with a single-temperature blackbody and proceed to multiple temperatures as needed to fit the spectrum. Perhaps the underlying physics is synchrotron rather than blackbody, and the spectrum should be fitted with synchrotron models. Comparison of **nonnested** models, such as thermal and nonthermal, is more difficult than comparison of nested models.

The most common goodness-of-fit procedure in astronomy involves the **reduced χ^2 statistic** (Bevington 1969; Press *et al.* 1997; see Chapter 7). Here the model is first compared to grouped (binned) data and best-fit parameters are obtained by weighted least squares where the weights are obtained from the dataset (e.g. $\sigma_i = \sqrt{N_i}$ where N_i is the number of objects placed into the i-th bin) or from ancillary measurements (e.g. the noise level in featureless regions of the image, spectrum or time series). Mathematically, this procedure can be expressed as

$$\hat{\theta} = \arg\min_{\theta} X^2(\theta) = \arg\min_{\theta} \sum_{i=1}^{N} \left(\frac{y_i - M_i(\theta)}{\sigma_i} \right)^2 . \tag{3.42}$$

After the **minimum χ^2 parameters** are found, goodness-of-fit is evaluated using a criterion $X_{\nu}^2 \simeq 1$ where ν represents the degrees of freedom in the problem. If X_{ν}^2 is acceptably close to unity based on a chosen significance level and the assumption of an asymptotic χ^2 distribution, then the model is deemed acceptable. Model selection can proceed informally, with the $\chi_{\nu}^2 \simeq 1$ test applied for each model examined. Note that we use the designation X^2 rather than χ^2 in Equation (3.42) because, under many conditions, it may not asymptotically follow the χ^2 distribution. Indeed, as discussed in Section 7.4, the astronomers' common use of minimum χ^2 procedures has a number of problems and is not recommended as a regression procedure under many circumstances. Similar difficulties arise when it is used as a general statistic for goodness-of-fit or model selection.

Both goodness-of-fit and model selection involve statistical hypothesis testing (Section 3.5), first to test model validity assuming a specified significance level, and second to compare two models as the null (H_0) and alternative (H_a) hypotheses. Classical hypothesis testing methods are generally used for nested models. However, they do not treat models symmetrically.

Informative reviews of the discussions on goodness-of-fit methods and model validation can be found in the multi-authored volume edited by Huber-Carol *et al.* (2002). Model selection techniques are discussed in the monograph by Burnham & Anderson (2002) and the review articles in the volume edited by Lahiri (2001). Information criteria are discussed in detail by Konishi & Kitagawa (2008).

3.7.1 Nonparametric methods for goodness-of-fit

The three well-known nonparametric hypothesis tests — based on the Kolmogorov–Smirnov (K-S), Cramér–von Mises, and Anderson–Darling statistics — can be used to compare the empirical distribution function of univariate data to the hypothesized distribution function of the model. These statistics and their corresponding tests are described in Section 5.3.1. They measure different distances between the data and the model, and have powerful theory establishing that their distribution functions are independent of the underlying distributions. We note that e.d.f.-based goodness-of-fit tests are not applicable when the problem has two or more dimensions.

However, for the purpose of the goodness-of-fit, it is important to recognize the distinction between the K-S-type **statistic** and the K-S-type **test** with tabulated probabilities. These widely available probabilities are usually not correct when applied to model-fitting situations where some or all of the parameters of the model are estimated from the same dataset. The standard probabilities of the K-S and related tests are only valid if the parameters are estimated independently of the dataset at hand, perhaps from some previous datasets or prior astrophysical considerations. The inapplicability of these probabilities was demonstrated by Lilliefors (1969) and was developed further by others (e.g. Babu & Rao 2004). Astronomers are often not aware of this limitation, and are obtaining unreliable goodness-of-fit probabilities from the e.d.f.-based tests.

Fortunately, bootstrap resampling of the data, and repeated calculation of the K-S-type statistic with respect to a single model, gives a histogram of the statistic from which a valid goodness-of-fit probability can be evaluated (Babu & Rao 1993). We outline the methodology underlying these goodness-of-fit bootstrap calculations. Let $\{F(.; \theta) : \theta \in \Theta\}$ be a family of continuous distributions parametrized by θ. We want to test whether the dataset $X_1, \ldots, X_n$ comes from $F = F(.; \theta)$ for some $\theta = \theta_0$. The K-S-type statistics (and a few other goodness-of-fit tests) are continuous functionals of the process,

$$Y_n(x; \hat{\theta}_n) = \sqrt{n}\big(F_n(x) - F(x; \hat{\theta}_n)\big). \tag{3.43}$$

Here F_n denotes the e.d.f. of $X_1, \ldots, X_n$, $\hat{\theta}_n = \theta_n(X_1, \ldots, X_n)$ is an estimator of θ derived from the dataset, and $F(x; \hat{\theta}_n)$ is the model being tested. For a simple example, if $\{F(.; \theta) : \theta \in \Theta\}$ denotes the Gaussian family with $\theta = (\mu, \sigma^2)$, then $\hat{\theta}_n$ can be taken as $(\bar{X}_n, S_n^2)$ where $\bar{X}_n$ is the sample mean and S_n^2 is the sample variance based on the data $X_1, \ldots, X_n$. In modeling an astronomical spectrum, F may be the family of blackbody or synchrotron models.

The bootstrap can be computed in two different ways. The **parametric bootstrap** consists of simulated datasets obtained from the best-fit model $F(x; \hat{\theta}_n)$. Techniques for obtaining Monte Carlo realizations of a specified function are well-known (Press *et al.* 1997). The **nonparametric bootstrap**, discussed in Section 3.6.2, gives Monte Carlo realizations of the observed e.d.f. using a random-selection-with-replacement procedure.

In the parametric bootstrap, $\hat{F}_n = F(.; \hat{\theta}_n)$; that is, we generate data $X_1^*, \ldots, X_n^*$ from the model assuming the estimated parameter values $\hat{\theta}_n$. The process based on the bootstrap simulations,

$$Y_n^P(x) = \sqrt{n}\big(F_n^*(x) - F(x; \hat{\theta}_n^*)\big), \tag{3.44}$$

and the sample process,

$$Y_n(x) = \sqrt{n}\big(F_n(x) - F(x; \hat{\theta}_n)\big), \tag{3.45}$$

converge to the same Gaussian process Y. Consequently, for the K-S test,

$$L_n = \sqrt{n} \sup_x |F_n(x) - F(x; \hat{\theta}_n)| \text{ and}$$

$$L_n^* = \sqrt{n} \sup_x |F_n^*(x) - F(x; \hat{\theta}_n^*)| \tag{3.46}$$

have the same limiting distribution. The critical values of L_n for the K-S statistic can be derived by constructing B resamples based on the parametric model ($B \sim 1000$ usually suffices), and arrange the resulting L_n^* values in increasing order to obtain 90 or 99 percentile points for getting 90% or 99% critical values. This procedure replaces the often incorrect use of the standard probability tabulation of the K-S test.

The nonparametric bootstrap involving resamples from the e.d.f.,

$$Y_n^N(x) = \sqrt{n}\big(F_n^*(x) - F(x; \hat{\theta}_n^*)\big) - B_n(x)$$
$$= \sqrt{n}\big(F_n^*(x) - F_n(x) + F(x; \hat{\theta}_n) - F(x; \hat{\theta}_n^*)\big), \tag{3.47}$$

is operationally easy to perform but requires an additional step of bias correction

$$B_n(x) = \sqrt{n}(F_n(x) - F(x; \hat{\theta}_n)). \tag{3.48}$$

The sample process Y_n and the bias-corrected nonparametric process Y_n^N converge to the same Gaussian process Y. That is,

$$L_n = \sqrt{n} \sup_x |F_n(x) - F(x; \hat{\theta}_n)| \text{ and}$$
$$J_n^* = \sup_x |\sqrt{n}\big(F_n^*(x) - F(x; \hat{\theta}_n^*)\big) - B_n(x)| \tag{3.49}$$

have the same limiting distribution. The critical values of the distribution of L_n can then be derived as in the case of the parametric bootstrap. The regularity conditions under which these results hold are detailed by Babu & Rao (2004).

3.7.2 Likelihood-based methods for model selection

To set up a general framework for model selection, let D denote the observed data and let $M_1, \ldots, M_k$ denote the models for D under consideration. For each model M_j, let $L(D|\theta_j; M_j)$ and $\ell(\theta_j) = \ln f(D|\theta_j; M_j)$ denote the likelihood and loglikelihood respectively, where θ_j is a p_j-dimensional parameter vector. Here $L(D|\theta_j; M_j)$ denotes the probability density function (in the continuous case) or the probability mass function (in the discrete case) evaluated at the data D. Most of the methodology can be framed as a comparison between two models, M_1 and M_2.

The model M_1 is said to be nested in M_2 if some elements of the parameter vector θ_1 are fixed (and possibly set to zero). That is, $\theta_2 = (\alpha, \gamma)$ and $\theta_1 = (\alpha, \gamma_0)$, where γ_0 is some known fixed constant vector. Comparison of M_1 and M_2 can then be considered as a classical hypothesis testing problem where $H_0 : \gamma = \gamma_0$. Nested models of this type occur frequently in astronomical modeling. In astrophysical modeling, stellar photometry might be modeled as a blackbody (M_1) with absorption (M_2), the structure of a dwarf elliptical galaxy might be modeled as an isothermal sphere (M_1) with a tidal cutoff (M_2), or a hot plasma might be modeled as an isothermal gas (M_1) with nonsolar elemental abundances (M_2).

A simple example in the statistics of nested models might be a normal model with mean μ and variance σ^2 (M_2) compared to a normal with mean 0 and variance σ^2 (M_1). This model selection problem can be framed as a hypothesis test of $H_0 : \mu = 0$ with free parameter σ. There are some objections to using hypothesis testing to decide between the two models M_1 and M_2, as they are not treated symmetrically by the test in which the null hypothesis is M_1. We cannot accept H_0 and show it is true. We can only reject, or fail to reject, H_0. Note that with very large samples, even very small discrepancies lead to rejection of the null hypothesis, while for small samples, even large discrepancies may not lead to rejection.

We now look at three classical methods for testing H_0 based on maximum likelihood estimators that were developed during the 1940s: the Wald test, the likelihood ratio test, and Rao's score test. The three tests are equivalent to each other to the first order of asymptotics, but differ in the second-order properties. No single test among these three is uniformly better than the others. Here we restrict consideration to a one-dimensional vector of model parameters θ where θ_0 is a preselected value of the parameter.

To test the null hypothesis $H_0 : \theta = \theta_0$, the **Wald test** uses the statistic

$$W_n = \frac{(\hat{\theta}_n - \theta_0)^2}{Var(\hat{\theta}_n)}, \tag{3.50}$$

the standardized distance between θ_0 and the maximum likelihood estimator $\hat{\theta}_n$ based on data of size n. A. Wald showed that the distribution of W_n is approximately the χ^2 distribution with one degree of freedom. We saw in Section 3.4.4 that, although in general the variance of $\hat{\theta}_n$ is not known, a close approximation is $Var(\hat{\theta}_n) = 1/I(\hat{\theta}_n)$ where $I(\theta)$ is Fisher's information. Thus $I(\hat{\theta}_n)(\hat{\theta}_n - \theta_0)^2$ has a χ^2 distribution in the limit, and the Wald test rejects the null hypothesis H_0, when $I(\hat{\theta}_n)(\hat{\theta}_n - \theta_0)^2$ is large.

The **likelihood ratio test**, as its name implies, uses the ratio of likelihoods as a model selection statistic, or in logarithmic form,

$$LRT = \ell(\hat{\theta}_n) - \ell(\theta_0), \tag{3.51}$$

where $\ell(\theta)$ denotes the loglikelihood at θ. The likelihood ratio test is widely used in astronomy but not always correctly, as explained by Protassov *et al.* (2002).

The **score test** developed by C. R. Rao, also known as the Lagrangian multiplier test, uses the statistic

$$S(\theta_0) = \frac{\ell'(\theta_0))^2}{nI(\theta_0)}, \tag{3.52}$$

where ℓ' denotes the derivative of ℓ, and I again is Fisher's information.

3.7.3 Information criteria for model selection

While the classical Wald, likelihood ratio, and score tests are still widely used, an alternative approach based on **penalized likelihoods** has dominated model selection since the 1980s. If the model M_1 is nested within model M_2, then the largest likelihood achievable by M_2 will always be larger than that achievable by M_1 simply because there are more parameters that can adjust to more detailed variations. If a **penalty** is applied to compensate for the obligatory difference in likelihoods due to the different number of parameters in M_1 and M_2, the desired balance between overfitting and underfitting might be found. Many model selection procedures based on information criteria use penalty terms.

The traditional maximum likelihood paradigm, as applied to statistical modeling, provides a mechanism for estimating the unknown parameters of a model having a specified dimension and structure. H. Akaike (1973) extended this paradigm by considering a framework in which the model dimension is also unknown. He proposed a framework where both model estimation and selection could be simultaneously accomplished. Grounding in the concept of entropy, Akaike proposed **an information criterion** now popularly known as the **Akaike information criterion** (AIC) defined for model M_j, as

$$AIC = 2\ell(\hat{\theta}_j) - 2p_j. \tag{3.53}$$

where p_j is the number of parameters in the j-th model. The term $2\ell(\hat{\theta}_j)$ is the goodness-of-fit term, and $2p_j$ is the penalty term. The penalty term increases as the complexity of the model grows and thus compensates for the necessary increase in the likelihood. The AIC selects the model M_i if $i = \arg\max_j 2\ell(\hat{\theta}_j) - 2p_j$. That is, the maximum AIC attempts to find the model that best explains the data with a minimum of free parameters.

Unlike the classical Wald, likelihood ratio, and score tests, the AIC treats all the models symmetrically, not requiring an assumption that one of the candidate models is the "correct" model. The AIC can be used to compare nonnested as well as nested models. The AIC can also be used to compare models based on different families of probability distributions. One of the disadvantages of AIC is the requirement of large samples, especially in complex modeling frameworks. In addition, it is not a consistent statistic: if p_0 is the correct number of parameters, and $\hat{p} = p_i$ ($i = \arg\max_j [2\ell(\hat{\theta}_j) - 2p_j]$), then $\lim_{n\to\infty} P(\hat{p} > p_0) > 0$. That is, even if we have a very large number of observations, $\hat{p}$ does not approach the true value.

The Schwarz information criterion, more commonly called the **Bayesian information criterion** (BIC), is the other widely used choice for penalized likelihood model selection. The BIC defined as

$$BIC = 2\ell(\hat{\theta}_j) - p_j \ln n, \tag{3.54}$$

where n is the number of data points, is consistent. It is derived by giving all the models under consideration equal weights; that is, equal prior probabilities to all the models under consideration. The BIC selects the model with highest marginal likelihood that, expressed as an integral, is approximated using Laplace's method. This in turn leads to the expression (3.54). Another model evaluation criterion based on the concept of **minimum description length** (MDL) is used in transmitting a set of data by coding using a family of probability models. The MDL model selection criterion is $-(1/2)BIC$. Like AIC, the models compared by the BIC need not be nested.

Conditions under which these two criteria are mathematically justified are often ignored in practice. AIC penalizes free parameters less strongly than does the BIC. The BIC has a greater penalty for larger datasets, whereas the penalty term of the AIC is independent of the sample size. Different user communities prefer one over the other. The user should also beware that sometimes these criteria are expressed with a minus sign so the goal changes to minimizing rather than maximizing the information criterion.

3.7.4 Comparing different model families

We briefly address the more difficult problem of comparing different model family fits to the dataset. Rather than asking "Which model within a chosen model family best fit the observed dataset?" we need to address "How far away is the unknown distribution underlying the observed dataset from the hypothesized family of models?". For example, an astronomer might ask whether the continuum emission of a quasar is better fit by a power-law model associated with a nonthermal relativistic jet or by a multiple-temperature blackbody model associated with a thermal accretion disk. The functional form and parameters of the nonthermal and thermal models are completely unrelated to each other. If we assume a nonthermal model family and the quasar is actually emitting due to thermal processes, then we have **misspecified** the model. Least squares, maximum likelihood estimation and other inferential methods assume that the probability model is "correctly specified" for the problem under study. When misspecification is present, the classical model selection tests (Wald, likelihood ratio, and score tests) and the widely used Akaike information criterion are not valid for comparing fits from different model families. The treatment of this problem has been addressed by a number of statisticians (e.g. Huber 1967, White 1982, Babu & Rao 2003) and the findings rely on somewhat advanced mathematics.

We outline here one approach to the problem (Babu & Rao 2003). Let the original dataset $X_1, \ldots, X_n$ come from an unknown distribution H that may or may not belong to the family $\{F(.; \theta) : \theta \in \Theta\}$. Let $F(., \theta_0)$ be the specific model in the family that is "closest" to H where proximity is based on the **Kullback–Leibler information**,

$$\int \ln\left(\frac{h(x)}{f(x; \theta)}\right) dH(x) \geq 0, \tag{3.55}$$

which arises naturally due to maximum likelihood arguments and has advantageous properties. Here h and f are the densities (i.e. derivatives) of H and F. If the maximum likelihood estimator $\hat{\theta}_n \to \theta_0$, then for any $0 \leq \alpha \leq 1$,

$$P\left(\sqrt{n} \sup_x |F_n(x) - F(x; \hat{\theta}_n) - (H(x) - F(x; \theta_0))| \leq C_\alpha^*\right) - \alpha \to 0 \tag{3.56}$$

where C_α^* is the α-th quantile of $\sup_x |\sqrt{n}\left(F_n^*(x) - F(x; \hat{\theta}_n^*)\right) - \sqrt{n}(F_n(x) - F(x; \hat{\theta}_n))|$. This provides an estimate of the distance between the true distribution and the family of distributions under consideration.

3.8 Bayesian statistical inference

In Section 2.4.1, we showed how Bayes' theorem is a logical consequence of the axioms of probability. Bayesian inference is founded on a particular interpretation of Bayes' theorem. Starting with Equation (2.16), we let A represent an observable X, and B represent the space of models M that depend on a vector of parameters θ. Bayes' theorem can then be rewritten as

$$P(M_i(\theta) \mid X) = \frac{P(X \mid M_i(\theta))P(M_i(\theta))}{P(X \mid M_1(\theta))P(M_1(\theta)) + \cdots + P(X \mid M_k(\theta))P(M_k(\theta))}. \tag{3.57}$$

In this context, the factors in Bayes' theorem have the following interpretations and designations:

$P(M_i(\theta) \mid X)$ is the conditional probability of the i-th model given the data X. This is called the **posterior probability**.

$P(X \mid M_i(\theta))$ is the conditional probability of the observable for the i-th model. This is called the **likelihood function**.

$P(M_i(\theta))$ is the marginal probability of the i-th model. The observable has no input here. This is called the **prior information**.

In this framework, Bayes' theorem states that the posterior distribution of a chosen set of models given the observable is equal to the product of two factors — the likelihood of the data for those models and the prior probability on the models — with a normalization constant.

More generally, we can write Bayes' theorem in terms of the parameters θ of the models M and the likelihood $L(X \mid \theta)$,

$$P(\theta \mid X) = \begin{cases} \dfrac{P(\theta)L(X|\theta)}{\sum_j P(\theta_j)L(X|\theta_j)} & \text{if } \theta \text{ is discrete;} \\[2ex] \dfrac{P(\theta)L(X|\theta)}{\int P(u)L(X|u)\,du} & \text{if } \theta \text{ is continuous.} \end{cases} \qquad (3.58)$$

Bayesian inference applies Bayes' theorem to assess the degree to which a given dataset is consistent with a given set of hypotheses. As evidence accumulates, the scientists' belief in a hypothesis ought to change; with enough evidence, it should become very high or very low. Thus, proponents of Bayesian inference say that it can be used to discriminate between conflicting hypotheses: hypotheses with very high support should be accepted as true and those with very low support should be rejected as false. However, detractors say that this inference method may be biased because the results depend on the prior distribution based on the initial beliefs that one holds before any evidence is ever collected. This is a form of inductive bias; it is generally larger for small datasets than large datasets, and larger for problems without much prior study than well-developed problems with extensive prior knowledge.

3.8.1 Inference for the binomial proportion

We examine here the Bayesian approach to statistical inference through applications to a simple problem: the estimation of the success ratio of a simple binary ("Yes/No") variable using Equation (3.57) with $k = 2$. This is called the binomial proportion problem.

Example 3.1 Consider the astronomer who has in hand a sample of optically selected active galactic nuclei (AGN such as Seyfert galaxies and quasi-stellar objects) and surveys them with a radio telescope to estimate the fraction of AGN that produce nonthermal radio emission from a relativistic jet. Let X denote the random variable giving the result of the radio-loudness test, where $X = 1$ indicates a positive and $X = 0$ indicates a negative result. Let θ_1 denote that radio-loudness is present and θ_2 denote that the AGN is radio-quiet. $P(\theta = \theta_1)$ denotes the probability that a randomly chosen AGN is radio-loud; this is the prevalence of radio-loudness in the optically selected AGN population.

From previous surveys, the astronomer expects the prevalence is around 0.1, but this is not based on the observational data under study. This is **prior** information that can be incorporated in the Bayesian analysis. Our goal is to use the new radio data to estimate posterior information in the form of $P(\theta|X = x)$, where as described above $x = 1$ denotes the existence of radio-loudness and $x = 0$ denotes radio-quietness. We start by adopting a prior of $P(\theta = \theta_1) = 0.1$. Based on our knowledge of the sensitivity of the radio telescope and the AGN redshift distribution, we establish that our radio survey is sufficiently sensitive to measure the radio emission 80% of the time in radio-loud AGN, $P(X = 1|\theta_1) = 0.8$. The failures are due to AGN at very high redshifts. However, the sample also includes some AGN at small redshifts where the telescope may detect thermal radio emission from star formation in the host galaxy. Let us say that this irrelevant positive detection of radio emission occurs 30% of the time, $P(X = 1|\theta_2) = 0.3$.

We are now ready to calculate the chances that an AGN is truly radio-loud when a positive radio detection is obtained, $P(\theta = \theta_1|X = 1)$. Bayes' theorem gives

$$P(\theta = \theta_1 \mid X = 1) = \frac{0.8 \times 0.1}{0.8 \times 0.1 + 0.3 \times 0.9} = \frac{0.08}{0.35} = 0.23. \tag{3.59}$$

This shows that the observational situation outlined here does not answer the scientific question very effectively: only 23% of the true radio-loud AGN are clearly identified, and 77% are either false negatives or false positives. The result is moderately sensitive to the assumed prior; for example, the fraction of true radio-loud AGN discoveries rises from 23% to 40% if the prior fraction is assumed to be 20% rather than 10%. But the major benefit would arise if we had additional data that could reduce the detection of irrelevant thermal radio emission. This might arise from ancillary data in the radio band (such as the radio polarization and spectral index) or data from other bands (such as optical spectral line-widths and ratios). If this fraction of erroneous detection is reduced from 30% to 5%, then the discovery fraction of radio-loud AGN increases from 23% to a very successful 95%. Significant astronomical efforts are devoted to addressing this problem (e.g. Ivecić *et al.* 2002).

This example shows that the measured conditional probabilities $P(X = x|\theta)$ are "inverted" to estimate the conditional probabilities of interest, $P(\theta|X = x)$. Inference based on Bayes' theorem is thus often called the theory of **inverse probability**. It originated with Thomas Bayes and Pierre Simon Laplace in the eighteenth and nineteenth centuries, and was revived by Sir Harold Jeffreys (who held the Plumian Professorship of Astronomy and Experimental Philosophy at the University of Cambridge), E. Jaynes, and other scholars in the mid-twentieth century.

3.8.2 Prior distributions

Bayesian inference is best performed when prior information is available to guide the choice of the prior distribution. This information can arise either from earlier empirical studies or from astrophysical theory thought to determine the observed phenomena. Even when prior knowledge is available, it may not be obvious how to convert that knowledge into a workable prior probability distribution. The choice of prior distribution influences the final result. The numerator of Bayes' theorem can be viewed as an average of the likelihood function weighted by the prior, whereas the maximum likelihood estimator gives the mode of the likelihood function without any weighting. There is often a relationship between the type of prior distribution and the type of posterior distribution.

Consider, for example, that we have some prior knowledge about θ that can be summarized in the form of a **beta distribution** p.d.f.

$$P(x; \alpha, \gamma) = \frac{x^{\alpha-1}(1-x)^{\gamma-1}}{B(\alpha, \gamma)}, \quad 0 < x < 1, \quad \text{where}$$

$$B(\alpha, \gamma) = \frac{\Gamma(\alpha)\Gamma(\gamma)}{\Gamma(\alpha + \gamma)} \quad \text{and} \quad \Gamma(\alpha) = \int_0^\infty t^{\alpha-1}e^{-t}dt \qquad (3.60)$$

with two shape parameters, $\theta = (\alpha, \gamma)$. Note that special cases of the beta include the uniform distribution. In the case when the likelihood is binomial, the posterior will be distributed as beta with parameters $x + \alpha$ and $n - x + \gamma$ where n is the sample size. Such priors which result in posteriors from the same family are called **natural conjugate priors**.

Example 3.2   Consider the problem in Example 3.1 but restricted to the situation where we have no prior information available on the probability of radio-loudness in AGN; that is, the distribution of θ is unconstrained. For simplicity, we might assume that θ is uniformly distributed on the interval $(0, 1)$. Its prior density is then $p(\theta) = 1$, $0 < \theta < 1$. This choice is sometimes called the **non-informative prior**. Often, Bayesian inference from such a prior coincides with classical inference. The likelihood for a binary variable is given by the binomial distributions discussed in Section 4.1. The posterior density of θ given X is now

$$P(\theta|X) = \frac{p(\theta)L(X|\theta)}{\int P(u)L(X|u)\,du}$$

$$= \frac{(n+1)!}{x!(n-x)!}\theta^x(1-\theta)^{n-x}, \quad 0 < \theta < 1.$$

In this example, the resulting function of θ is the same as the likelihood function $L(x|\theta) \propto \theta^X(1-\theta)^{n-X}$. Thus, maximizing the posterior probability density will give the same estimate as the maximum likelihood estimate,

$$\hat{\theta}_B = \hat{\theta}_{MLE} = \frac{X}{n}, \qquad (3.61)$$

where the subscript B denotes Bayesian. However, the interpretation of $\hat{\theta}_B$ as an estimate of θ is quite different from the interpretation of $\hat{\theta}_{MLE}$. The Bayesian estimator is the most probable value of the unknown parameter θ conditional on the sample data x. This is called the **maximum a posteriori** (MAP) estimate or the **highest posterior density** (HPD) estimate; it is the maximum likelihood estimator of the posterior distribution. Note that if we chose a different prior distribution, the Bayesian maximum posterior result would differ from classical results.

The uncertainty in the most probable posterior value of θ can also be estimated because the $P(\theta|X)$ is now described in terms of a genuine probability distribution concentrated around $\hat{\theta}_B = X/n$ to quantify our post-experimental knowledge about θ. The classical Bayes estimate $\hat{\theta}_B$ minimizes the posterior mean square error,

$$E[(\theta - \hat{\theta}_B)^2 | X] = \min_a E[(\theta - a)^2 | X]. \tag{3.62}$$

Hence $\hat{\theta}_B = E(\theta | X)$ is the mean of the posterior distribution. If $\hat{\theta}_B$ is chosen as the estimate of θ, the natural measure of variability of this estimate is obtained in the form of the posterior variance, $E[(\theta - E(\theta | X))^2 | X]$, with the posterior standard deviation serving as a natural measure of the estimation of error. The resulting estimated confidence interval of the form

$$\hat{\theta}_B \pm c \sqrt{E[(\theta - E(\theta | x))^2 | X]}. \tag{3.63}$$

We can compute the posterior probability of any interval containing the parameter θ. A statement such as $P(\hat{\theta}_B - k_1 \leq \theta \leq \hat{\theta}_B + k_2 | X) = 0.95$ is perfectly meaningful, conditioned on the given dataset.

In Example 3.2, if the prior is a beta distribution with parameters $\theta = (\alpha, \gamma)$, then the posterior distribution of $\theta | X$ will be a beta distribution with parameters $X + \alpha, n - X + \gamma$. So the Bayes estimate of θ will be

$$\hat{\theta}_B = \frac{(X + \alpha)}{(n + \alpha + \gamma)} = \frac{n}{n + \alpha + \gamma} \frac{X}{n} + \frac{\alpha + \gamma}{n + \alpha + \gamma} \frac{\alpha}{\alpha + \gamma}. \tag{3.64}$$

This is a convex combination of the sample mean and prior mean with weights depending upon the sample size and the strength of the prior information as measured by the values of α and γ.

Bayesian inference relies on the concept of conditional probability to revise one's knowledge. In the above example, prior to the collection of sample data, the astronomer had some (perhaps vague) information constraining the distribution of θ. Then measurements of the new sample were made. Combining the model density of the new data with the prior density gives the posterior density, the conditional density of θ given the data.

3.8.3 Inference for Gaussian distributions

We now consider Bayesian analysis of a problem of univariate signal estimation where the data are thought to follow a normal distribution due either to intrinsic scatter in the population or to error in the measurement process. Let $\mathbf{X} = (X_1, X_2, \ldots, X_n)$ be n independent and identically distributed (i.i.d.) random variables drawn from a normal distribution with mean μ and variance σ^2. We can express the data as

$$X_i = \mu + \sigma \eta_i \tag{3.65}$$

where μ is the signal strength (the parameter of scientific interest) and η_i are i.i.d. standard normal random variables. The likelihood is given by

$$L(\mathbf{X}; \mu, \sigma^2) = (2\pi\sigma^2)^{-n/2} \exp\left\{ -\frac{1}{2\sigma^2} \left(\sum_{i=1}^{n} (X_i - \bar{X}_n)^2 + n (\bar{X}_n - \mu)^2 \right) \right\} \tag{3.66}$$

where $\bar{X}_n$ is the mean of **X**. This likelihood is combined with an assumed prior in the numerator of Bayes' theorem. We consider two cases representing different levels of prior knowledge.

Case 1: σ is known Here the likelihood simplifies to

$$L(\mathbf{X}; \mu, \sigma^2) \propto \exp\left[-\frac{n}{2\sigma^2}(\bar{X}_n - \mu)^2\right]. \tag{3.67}$$

If we assume an informative prior π on μ and that it is normally distributed with mean μ_0 and variance τ^2, then the posterior is $P(\mu|\mathbf{X}) \propto L(\mathbf{X}; \mu, \sigma^2)\Pi(\mu)$. Here Π is the normal density with mean μ_0 and variance τ^2 evaluated at the parameter μ. After some algebra, this leads to the result that

$$P(\mu|\mathbf{X}) \propto \exp\left\{-\frac{1}{\gamma^2}(\mu - \hat{\mu})^2\right\} \quad \text{where}$$

$$\gamma^2 = \frac{\sigma^2}{n}\frac{\tau^2}{\tau^2 + \sigma^2/n} \quad \text{and}$$

$$\hat{\mu} = \gamma^2\left(\frac{\mu_0}{\tau^2} + \frac{n\bar{X}_n}{\sigma^2}\right) = \left(\frac{\gamma^2}{\tau^2}\right)\mu_0 + \left(1 - \frac{\gamma^2}{\tau^2}\right)\bar{X}_n. \tag{3.68}$$

That is, the posterior distribution of μ, given the data, is normally distributed with mean $\hat{\mu}$ and variance γ^2. Consequently the 95% HPD "credible interval" for μ is given by $\hat{\mu} \pm 1.96\gamma$. Bayesian versions of confidence intervals are referred to as **credible intervals**.

Note that the HPD Bayesian estimator for the signal strength μ_B is not the simple MLE value $\bar{X}_n$, but rather is a convex combination of the sample mean $\bar{X}_n$ and the prior mean μ_0. If $\tau \to \infty$, the prior becomes nearly flat and $\hat{\mu}_B \to \hat{\mu}_{MLE}$ and $\gamma^2 \to \sigma^2/n$, producing the frequentist results. The Bayesian estimator of the signal strength could be either larger or smaller than the sample mean, depending on the chosen values of μ_0 and τ^2 of the assumed prior distribution.

Case 2: σ is unknown In this case, the variance σ^2 is a nuisance parameter that should be removed, if possible. In the Bayesian context, we **marginalize** over the nuisance parameters by integrating σ^2 out of the joint posterior distribution. That is, if $P(\mu, \sigma^2)$ denotes the prior, then the corresponding joint posterior is

$$P(\mu, \sigma^2|\mathbf{X}) \propto L(\mathbf{X}; \mu, \sigma^2)P(\mu, \sigma^2) \quad \text{and}$$

$$P(\mu|\mathbf{X}) = \int_0^\infty P(\mu, \sigma^2|\mathbf{X})d\sigma^2. \tag{3.69}$$

If we use the uninformative Jeffreys prior, then $P(\mu, \sigma^2) \propto 1/\sigma^2$. For the Jeffreys prior, algebraic manipulation shows that the posterior estimator μ_B follows the Student's t distribution with $n-1$ degrees of freedom. Thus, the Jeffreys prior reproduces frequentist results.

While our examples here are restricted to rather simple parameter estimation problems, Bayes' theorem can be applied to any problem where the likelihood and the prior can be clearly specified. Complex real-life modeling problems often require a **hierarchical Bayes** approach, where the statistical model requires two or more interrelated likelihoods. In astronomy, this arises in regression models incorporating measurement errors (Kelly 2007; Mandel *et al.* 2009; Hogg *et al.* 2010), instrumental responses (van Dyk *et al.* 2001), systematic errors (Shkedy *et al.* 2007) and many other problems. Time series and other parameter estimation situations with fixed variables can be formulated in a Bayesian framework. Bayesian applications have grown into many branches of applied statistics and, in some cases, have become the preferred approach.

3.8.4 Hypotheses testing and the Bayes factor

Bayesian principles can be effectively applied to estimate probabilities for hypothesis tests where one seeks probabilities associated with binary (e.g. Yes vs. No, H_0 vs. H_1) questions as discussed in Section 3.5. These include two-sample tests, treatments of categorical variables, composite and multiple hypothesis testing.

We start testing alternative hypotheses based on the random vector **X**,

$$H_0 : \theta \in \Theta_0 \quad \text{against} \quad H_1 : \theta \in \Theta_1. \tag{3.70}$$

If Θ_0 and Θ_1 are of the same dimension (for example, $H_0 : \theta \leq 0$ and $H_1 : \theta > 0$), we may choose a prior density that assigns positive prior probability to Θ_0 and Θ_1.

Next, one calculates the posterior probabilities $P\{\Theta_0|\mathbf{X}\}$ and $P\{\Theta_1|\mathbf{X}\}$ as well as the **posterior odds ratio**, $P\{\Theta_0|X\}/P\{\Theta_1|X\}$. Here one might set a threshold like 1/9 for a 90% significance level or 1/19 for a 95% significance level to decide what constitutes evidence against H_0. Unlike classical tests of hypotheses, Bayesian testing treats H_0 and H_1 symmetrically.

Another way to formulate the problem is to let $\pi_0 = P(\Theta_0)$ and $1 - \pi_0 = P(\Theta_1)$ be the prior probabilities of Θ_0 and Θ_1. Further, let $g_i(\theta)$ be the prior p.d.f. of θ under Θ_i (or H_i), so that $\int_{\Theta_i} g_i(\theta) d\theta = 1$. This leads to $\pi(\theta) = \pi_0 g_0(\theta) I\{\theta \in \Theta_0\} + (1 - \pi_0) g_1(\theta) I\{\theta \in \Theta_1\}$ where I is the indicator function. This formulation is quite general, allowing Θ_0 and Θ_1 to have different dimensions.

We can now calculate posterior probabilities or posterior odds. Letting $f(.|\theta)$ be a probability density, the marginal density of X under the prior π can be expressed as

$$m_\pi(x) = \int_\Theta f(x|\theta)\pi(\theta)\, d\theta = \pi_0 \int_{\Theta_0} f(x|\theta)g_0(\theta)\, d\theta + (1 - \pi_0) \int_{\Theta_1} f(x|\theta)g_1(\theta)\, d\theta. \tag{3.71}$$

The posterior density of θ given the data $X = x$ is therefore $\pi_0 f(x|\theta)g_0(\theta)/m_\pi(x)$ or $(1 - \pi_0)f(x|\theta)g_1(\theta)/m_\pi(x)$ depending on whether $\theta \in \Theta_0$ or $\theta \in \Theta_1$. The posterior distributions of $\Theta_0|x$ and $\Theta_1|x$ are given respectively by

$$P^\pi(\Theta_0|x) = \frac{\pi_0}{m_\pi(x)} \int_{\Theta_0} f(x|\theta)g_0(\theta)\, d\theta \quad \text{and} \quad P^\pi(\Theta_1|x) = \frac{(1 - \pi_0)}{m_\pi(x)} \int_{\Theta_1} f(x|\theta)g_1(\theta)\, d\theta. \tag{3.72}$$

This result can also be reported as the **Bayes factor** of H_0 relative to H_1,

$$BF_{01} = \frac{P(\Theta_0|x)}{P(\Theta_1|x)} \bigg/ \frac{P(\Theta_0)}{P(\Theta_1)} = \frac{\int_{\Theta_0} f(x|\theta)g_0(\theta)\, d\theta}{\int_{\Theta_1} f(x|\theta)g_1(\theta)\, d\theta}. \tag{3.73}$$

Clearly, $BF_{10} = 1/BF_{01}$, and the posterior odds ratio of H_0 relative to H_1 is

$$\frac{P(\Theta_0|x)}{P(\Theta_1|x)} = \left(\frac{\pi_0}{1-\pi_0}\right) BF_{01}. \tag{3.74}$$

Thus, the posterior odds ratio of H_0 relative to H_1 is BF_{01} whenever $\pi_0 = \frac{1}{2}$. The smaller the value of BF_{01}, the stronger the evidence against H_0. This can easily be extended to any number of hypotheses.

3.8.5 Model selection and averaging

We discussed in Section 3.7 the challenging task of evaluating the relative merits of different models for the same dataset, both nested and nonnested. The Bayesian approach to these issues is often attractive. We illustrate Bayesian model selection with a model of a univariate dataset with an unknown number of normal (Gaussian) components. This is the **normal mixture model**.

Let $X_1, X_2, \ldots, X_n$ be i.i.d. random variables with common density f given by a mixture of normals

$$f(x|\theta) = \sum_{j=1}^{k} p_j \phi(x|\mu_j, \sigma_j^2), \tag{3.75}$$

where $\phi(.|\mu_j, \sigma_j^2)$ denotes the normal density with mean μ_j and variance σ_j^2, k is the number of mixture components, and p_j is the weight given to the j-th normal component. Let M_k denote a k-component normal mixture model. Bayesian model selection procedures involve computing

$$m(x|M_k) = \int \pi(\theta_k) f(x|\theta_k)\, d\theta_k, \tag{3.76}$$

for each k of interest. We estimate the number of components $\hat{k}$ that gives the largest value.

From the Bayesian point of view, a natural approach to model uncertainty is to include all models M_k under consideration for later decisions. This is suitable for prediction, eliminating the underestimation of uncertainty that results from choosing a single best model $M_{\hat{k}}$. This leads to Bayesian **model averaging**. Consider the parameter space encompassing all models under consideration, $\Theta = \cup_k \Theta_k$. Let the likelihood h and prior π be given by

$$h(.|\theta) = f_k(.|\theta_k) \text{ and } \pi(\theta) = p_k g_k(\theta_k) \text{ if } \theta \in \Theta_k, \tag{3.77}$$

where $p_k = P_\pi(M_k)$ is the prior probability of M_k and g_k integrates to 1 over Θ_k. Given X, the posterior is the weighted average of the model posteriors,

$$\begin{aligned} \pi(\theta|X) &= \frac{h(X|\theta)\pi(\theta)}{m(X)} = \sum_k \frac{p_k}{m(X)} f_k(X|\theta_k) g_k(\theta_k) I_{\Theta_k}(\theta_k) \\ &= \sum_k P(M_k|X) g_k(\theta_k|X) I_{\Theta_k}(\theta_k), \end{aligned} \tag{3.78}$$

where the normalizing constant based on the data is given by

$$m(X) = \int_{\Theta} h(X|\theta)\pi(\theta)\,d\theta, \quad P(M_k|X) = p_k m_k(X)/m(X),$$

$$g_k(\theta_k|X) = f_k(X|\theta_k)g_k(\theta_k)/m_k(X), \quad \text{and } m_k(X) = \int_{\theta_k} f_k(X|\theta_k)g_k(\theta_k)\,d\theta_k.$$

$$(3.79)$$

The needed predictive density $\ell(y|X)$ of **future** y values given X is obtained by integrating out the **nuisance** parameter θ. That is,

$$\begin{aligned} \ell(y|X) &= \int_{\Theta} h(y|\theta)\pi(\theta|X)\,d\theta \\ &= \sum_k P(M_k|X) \int_{\Theta_k} f_k(y|\theta_k)g_k(\theta_k|X)\,d\theta_k \\ &= \sum_k P(M_k|X)\ell_k(y|X), \end{aligned}$$

$$(3.80)$$

averages of predictive densities where $\ell_k(y|X) = \int_{\Theta_k} f_k(y|\theta_k)g_k(\theta_k|X)\,d\theta_k$. This is obtained by averaging over all models.

3.8.6 Bayesian computation

Least-squares estimation can usually be calculated as a solution to a system of linear equations and maximum likelihood estimation can be accomplished with an optimization algorithm for nonlinear equations such as the EM algorithm. But characterizing the multivariate posterior distribution, finding the highest posterior density (HPD), or the maximum a priori (MAP) model in Bayesian estimation requires integrating over the full space of possible models. While this is not demanding for simple models such as estimation of the binomial proportion or the mean and variance of a univariate normal function, it can be quite difficult in realistic models of scientific interest. For example, a ΛCDM cosmological model fit to unpolarized and polarized fluctuations of the Cosmic Microwave Background radiation and to the galaxy clustering correlation function can have $\sim$15 parameters with nonlinear effects on different statistics derived from the datasets (Trotta 2008; see also the interactive model simulations of M. Tegmark at http://space.mit.edu/home/tegmark/movies.html). Complete coverage of the $\sim$15-dimensional model space is not computationally feasible even if the likelihood is not difficult to calculate.

The practical solution to this problem is to obtain a limited number of independent draws from model parameter space, $\theta \in \Omega$, so that the desired results are estimated with reasonable reliability and accuracy. A commonly desired product is the distribution of the posterior density, $P(\theta|x)$, for a range of model parameter values. For inferences around the peak of low-dimensional unimodal models, a small number of simulations (say, $\sim$100−1000) may suffice. But for models with complex likelihood functions, hierarchical model structures, high dimensionality, and/or interest in the tails of the model distributions, efficient strategies are needed to sample the model space to address the scientific question.

A variety of computational tools is available for computing posterior densities for a model space. Under some conditions where the bounds of the probability density can be estimated, **rejection sampling** and **importance sampling** can be applied (Press *et al.* 2007). More generally, **Markov chain Monte Carlo (MCMC)** simulations are used. The calculation begins with initial draws from the probability distribution from random or selected locations in the parameter space. Sequential samples are drawn; simple Markov chains are random walks that depend only on the previous location in the space. The goal is to design the Markov chain to converge efficiently to the desired posterior distribution.

The **Gibbs sampler** and the **Metropolis–Hastings algorithm** are often found to be effective in MCMC simulations in multidimensional parameter spaces. First, one divides the full parameter vector θ into lower dimensional subvectors. At a given iteration of the Markov chain, draws from these subvectors are made with other subvectors kept fixed. When each subvector is updated separately, a new iteration is begun. This technique can be combined with the Metropolis–Hastings algorithm that uses a rule to accept or reject the next location in the Markov chain depending on the ratio of the probability density function at the two locations. Steps forward are accepted only if they increase the posterior density. A jumping distribution is constructed to map the parameter space in an efficient fashion. Jumps can be made either within or between subspaces defined by the Gibbs sampler subvectors.

Many strategies have been developed to improve the convergence and accuracy of these methods. Initial iterations of the MCMC that depend strongly on the initial starting points in parameter space can be ignored; this is the **burn-in** fraction. Convergence can be assessed by comparing chains based on different starting points. Jumping distributions can be weighted by local gradients in the posterior distribution, and designed to have covariance structure similar to the posterior yet be more easily computed. Complex posterior distributions can be approximated by simpler functions, such as normal mixture models.

The technologies of Bayesian computation are rapidly developing, and specialized study is needed to be well-informed. The monographs by Gelman *et al.* (2003) and Gemerman & Lopes (2006) give thorough reviews of these issues. Recent developments are described by Brooks *et al.* (2011). Astrophysical applications of these methods for modeling planetary orbits from sparse, irregularly spaced radial velocity time series are described by Clyde *et al.* (2007) and Ford & Gregory (2007). Cosmological applications are described by Feroz *et al.* (2009).

3.9 Remarks

Mathematical statisticians have worked industriously for decades to establish the properties (biasedness, consistency, asymptotical normality and so forth) of many statistics derived using various procedures under various conditions. Those statistics that perform well are widely promulgated for practical use.

However, the logical inverse of this situation is also true: when mathematical statisticians have not established the properties of a statistic, then the properties of that statistic cannot be assumed to be known. Indeed, even subtle changes in the assumptions (e.g. a sample is i.i.d. but not normally distributed) can invalidate a critical theorem. There is no guarantee that a new statistic, or a standard statistic examined under nonstandard conditions, will have desirable properties such as unbiasedness and asymptotic normality. Astronomers are acting in a valid manner scientifically when they construct a new functionality (i.e. statistic) of their data that measures a characteristic of scientific interest. But they may not then assume that the mathematical behavior of that statistic is known, even if it superficially resembles a standard and well-studied statistic. As a result, many statements of estimated values (which assume unbiasedness) or confidence intervals in the astronomical literature have uncertain reliability. When novel statistics are considered, and even with standard statistics applied to small or non-Gaussian samples, the bootstrap approach to estimation and confidence intervals is recommended, especially when there is no closed-form expression for the estimator.

The twentieth century has witnessed long arguments on the relative value of Fisher's MLE and Bayesian approaches to statistical modeling. Today, these debates are muted, and most statisticians agree that neither frequentist nor Bayesian approaches are self-evidently perfect for all situations. If prior information is available and can be effectively stated as a mathematical prior distribution, then Bayesian statistics may be more appropriate. In many situations the frequentist and Bayesian solutions are similar or identical. Essays on this issue particularly oriented towards astrophysicists, some of which frankly advocate a Bayesian orientation, include Loredo (1992), Cousins (1995) and Efron (2004). Instructive and balanced discussions of different approaches to statistical inference appear in the monographs by Barnett (1999), Cox (2006) and Geisser (2006).

The past decade has witnessed a surge of Bayesian applications in astronomy. When prior information is available from astrophysical constraints and can be readily incorporated into a prior distribution for the model, this is an excellent approach. Bayesian inference is also powerfully applied when the astronomer seeks more than simple estimators and confidence intervals. Astrophysical insights can emerge from examination of complicated multimodal posterior distributions or from marginalization of ancillary variables. However, Bayesian methods can be overused. Specification of priors is often tricky, and computation in multidimensional parameter space can require millions of iterations without guarantee of convergence.

Inference problems encountered in astronomy are incredibly varied, so no single approach can be recommended for all cases. When a parametric model, either heuristic or astrophysical, can be reasonably applied, then point estimation can be pursued. For many problems, least-squares and maximum likelihood estimation provide good solutions. Bayesian inference can be pursued when the situation is known to have external constraints.

3.10 Recommended reading

Gelman, A., Carlin, J. B., Stern, H. S. & Rubin, D. B. (2003) *Bayesian Data Analysis*, 2nd ed., Chapman & Hall, London
A comprehensive and practical volume on Bayesian methodology. Topics include fundamentals of Bayesian inference, estimation of single- and multiple-parameter modes,

hierarchical models, model comparison, sample inadequacies, Markov chain simulation techniques, regression models, hierarchical and generalized linear models, robust inference, mixture models, multivariate models and nonlinear models.

Hogg, R. V., McKean, J. W. & Craig, A. T. (2005) *Introduction to Mathematical Statistics*, 6th ed., Prentice Hall, Englewood Cliffs
A widely used textbook for graduate students in statistics with broad coverage on the theory and practice of inference. Topics include probability distributions, unbiasedness and consistency, hypothesis testing and optimality, χ^2 tests, bootstrap procedures, maximum likelihood methods, sufficiency, nonparametric statistics, Bayesian statistics and linear modeling.

James, F. (2006) *Statistical Methods in Experimental Physics*, 2nd ed., World Scientific, Singapore
This excellent volume for physical scientists covers many of the topics on statistical inference discussed here: likelihood functions, information and sufficiency, bias and consistency, asymptotic normality, point estimation methods, confidence intervals, hypothesis tests, goodness-of-fit tests and various maximum likelihood methods. Additional topics include concepts of probability, probability distributions, information theory, decision theory and Bayesian inference.

Rice, J. A. (2007) *Mathematical Statistics and Data Analysis*, 2nd ed., Duxbury Press
A high-quality text with broad coverage at an upper-level undergraduate level. Topics on inference include the method of moments, maximum likelihood, sufficiency, hypothesis testing and goodness-of-fit.

Wasserman, L. (2004) *All of Statistics: A Concise Course in Statistical Inference*, Springer, Berlin
A slim, sophisticated volume with broad scope intended for computer science graduate students needing a background in statistics, it is also valuable for physical scientists. Topics include theory of probability and random variables, bootstrap resampling, methods of parametric inference, hypothesis testing, Bayesian inference, decision theory, linear and loglinear models, multivariate models, graph theory, nonparametric density estimation, multivariate classification, stochastic processes and Bayesian computational methods.

3.11 R applications

We do not provide **R** scripts for inference here because most of the examples presented in later chapters implement some type of statistical inference. A few particular applications can be highlighted. In Section 4.7.1, we compare the performance of several point estimation methods — method of moments, least squares, weighted least squares (minimum χ^2), MLE and MVUE — on simulated datasets following the Pareto (power-law) distribution.

Estimation of parameters for a real dataset using a normal model follows, with hypothesis tests of normality. Chapter 5 implements a variety of nonparametric hypothesis tests including Wilcoxon tests, two-sample tests, and e.d.f.-based tests such as the Kolgomorov–Smirnov test. Chapter 6 includes some nonparametric and semi-parametric model-fitting methods including spline fitting, kernel density estimators and local regression. Chapter 7 on regression covers important inferential procedures including least-squares and weighted least-squares estimation, maximum likelihood estimation, robust estimation and quantile regression. Multivariate models are treated in Chapter 8, and some multivariate classifiers that can be considered to be examples of inferential modeling are treated in Chapter 9. A variety of specialized inferential modeling tasks is shown for censored, time series and spatial data in Chapters 10–12.

Elementary Bayesian computations using **R** are presented in the volumes by Albert (2009) and Kruschke (2011). Scripts often call **CRAN** packages such as *rbugs* that provide links to independent codes specializing in MCMC and other computational methods used in Bayesian inference.

4 Probability distribution functions

Certain probability distribution functions have proved to be useful for scientific problems. Some are discrete with countable jumps while others are continuous. In Sections 2.7 and 2.8, we introduced several probability distribution functions that emerged from elementary experiments. These included the discrete uniform, Bernoulli, binomial, Poisson, and negative binomial distributions and the continuous uniform, exponential, normal, and lognormal distributions.

In this chapter, we deepen our discussion of a few probability distribution functions particularly relevant to astronomical research. The mathematical properties of these distributions have been extensively studied, sometimes for over two centuries. These include moments, methods for parameter estimation, order statistics, interval and extrema estimation, two-sample tests, and computational algorithms. We pay particular attention to the Poisson and Pareto (or power-law) distributions which are especially important in astronomy.

Table 4.1 summarizes basic properties of the distributions discussed here. The formulae give the probability mass function (p.m.f.) for discrete distributions and probability density function (p.d.f.) for continuous distributions. The p.m.f.'s and p.d.f.'s have normalizations that may be unfamiliar to astronomers because their sum or total integral needs to be equal to unity, as they represent probability measures. For example, the astronomer's familiar power-law distribution $f(x) = bx^{-\alpha}$ is not a p.d.f. for arbitrary choices of α and b, but the Pareto distribution, $f(x) = \alpha b^{\alpha}/x^{\alpha+1}$ for $x \geq b > 0$ and $f(x) = 0$ for $x < b$, has the correct normalization for a p.d.f. The domain and range in Table 4.1 give the permitted span of parameters and variable x, respectively. The mean and variance are the first and second central moments of the distribution.

The results given in Table 4.1 and other mathematical properties of these p.m.f.'s and p.d.f.'s (such as their cumulative distribution function, inverse distribution function, survival function, hazard function, probability generating function, characteristic function, median, mode, moments about zero, skewness, and kurtosis) can be found in Evans *et al.* (2000), Krishnamoorthy (2006) and other references listed at the end of the chapter. Many properties are tabulated online in Wikipedia. We also discuss different approaches to estimation of probability distribution parameters, both for simple p.d.f.'s and for their common extensions.

4.1 Binomial and multinomial

The binomial is a discrete distribution providing probabilities of outcomes from independent binary trials, experiments with success or failure. The distribution was first discussed by

Table 4.1 Some univariate probability distribution functions for astronomy.

Distribution	Designation	Probability function	Domain	Range	Mean	Variance
		Discrete distributions (x integer, p.m.f.s)				
Binomial	$Bin(n,p)$	$\binom{n}{x}p^x(1-p)^{n-x}$	$0<p<1$	$x=0,1,2,\ldots,n$	np	$np(1-p)$
Poisson	$Pois(\lambda)$	$\lambda^x e^{-\lambda}/x!$	$\lambda>0$	$x=0,1,2,\ldots$	λ	λ
		Continuous distributions (x real, p.d.f.s)				
Normal	$N(\mu,\sigma^2)$	$(1/\sigma\sqrt{2\pi})e^{-(x-\mu)^2/(2\sigma^2)}$	$-\infty<\mu<\infty,\sigma>0$	$-\infty<x<\infty$	μ	σ^2
Student's	T_ν	$\frac{\Gamma\left(\frac{\nu+1}{2}\right)}{\sqrt{\nu\pi}\Gamma\left(\frac{\nu}{2}\right)}\left(1+\frac{t^2}{\nu}\right)^{\frac{-\nu+1}{2}}$	$\nu>0$, integer	$-\infty<t<\infty$	0	$\frac{\nu}{\nu-2}$ for $\nu>2$
Lognormal	$LN(\mu,\sigma^2)$	$(1/x\sigma\sqrt{2\pi})e^{-(\log x-\mu)^2/(2\sigma^2)}$	$-\infty<\mu<\infty,\sigma>0$	$0<x<\infty$	$e^\mu e^{\sigma^2/2}$	$e^{2\mu}e^{\sigma^2}(e^{\sigma^2}-1)$
Chi-square	χ^2_ν	$x^{(\nu-2)/2}e^{-x/2}/[2^{\nu/2}\Gamma(\nu/2)]$	$\nu>0$, integer	$0<x<\infty$	ν	2ν
Pareto	$P(\alpha,b)$	$\alpha b^\alpha/x^{\alpha+1}$	$\alpha>0,b>0$	$b<x<\infty$	$\alpha b/(\alpha-1)$ for $\alpha>1$	$\alpha b/[(\alpha-1)^2(\alpha-2)]$ for $\alpha>2$
Gamma	$Gamma(\alpha,b)$	$(x/b)^{\alpha-1}e^{-x/b}/b\Gamma(\alpha)$	$\alpha>0,b>0$	$0<x<\infty$	αb	αb^2

B. Pascal, J. Bernouilli and T. Bayes in the eighteenth century. This situation is quite relevant for astronomy, as we often classify objects into two categories: radio-quiet and radio-loud active galaxies; metal-rich and metal-poor globular clusters; red-sequence and blue-sequence galaxies; signal and background photons; and so forth. Providing all objects fall into one or another of these categories, the situation is equivalent to tossing a coin.

The multinomial distribution is a k-dimensional generalization of the binomial that provides probabilities for trials that result in k possible outcomes. Here again, many applications to astronomy are evident: Type Ia, Ib, Ic and II supernovae; Class 0, I, II and III pre-main-sequence stars; Flora, Themis, Vesta and other asteroid families; and so forth.

As presented in Equation (2.40), the binomial probability distribution has one parameter, the Bernouilli probability p, which gives the chance of a positive outcome. If the n objects are sampled from an underlying population, then the sample mean will be an unbiased estimate of the population p. Recall that in the binomial p.m.f.

$$P(i) = \binom{n}{i} p^i (1-p)^{n-i} \tag{4.1}$$

the notation $\binom{n}{i}$ represents $n!/[i!(n-i)!]$.

The distribution is asymmetrical with skewness $(1-2p)/\sqrt{np(1-p)}$, but for sufficiently large n, it is approximately normal with mean np and standard deviation $\sqrt{np(1-p)}$. This is based on the De Moivre–Laplace theorem. A standardized variable

$$X_{std} = \frac{X + 0.5 - np}{\sqrt{np(1-p)}} \tag{4.2}$$

can be constructed that is asymptotically normally distributed, where the 0.5 is a continuity correction that improves accuracy for small n. Another transformation of a binomial to an approximately normal variable is by Anscombe,

$$X_{Ans} = \arcsin\sqrt{\frac{X + 3/8}{n + 3/4}} \tag{4.3}$$

with mean $\arcsin(\sqrt{p})$ and variance $1/(4n)$. A common rule of thumb for using the normal approximation is $np(1-p) > 9$. In the particular limit where $n \to \infty$ and $p \to 0$ such that their product $np = \lambda$ is constant, then the binomial distribution is approximately the Poisson distribution. This approximation is recommended when $n^{0.31} p < 0.47$.

Parameter estimation for binomial experiments is simple. If n is known, then the method of moments, maximum likelihood estimation (MLE), minimum χ^2, and minimum variance unbiased estimator (MVUE) all give

$$\hat{p} = \frac{X}{n} \tag{4.4}$$

where X is the number of successes and p is the probability of success occurring in the population. This is the intuitive value: if a sample of 72 quasars has 12 radio-loud and 60 radio-quiet objects, the estimate for the population ratio of radio-loud to radio-quiet objects is $\hat{p}_{MLE} = 12/72 = 0.17$. For large samples, one sigma confidence intervals around p can be estimated using the Nayatani–Kurahara approximation

$$\frac{x}{n} \pm \left[\frac{k}{n} \left(1 - \frac{k}{n} \right) \right] \tag{4.5}$$

which can be applied when $np > 5$ and $0 < p \le 0.5$. The MLE is unstable if n as well as p is unknown.

Bayesian analysis does not give Equation (4.4) for any simple prior assumed for p (Chew 1971, Irony 1992). P. Laplace first derived the Bayesian estimator for a uniform prior, $\hat{p} = (X + 1)(n + 2)$, and H. Jeffreys recommended $\hat{p} = (X + 0.5)(n + 1)$ based on a beta distribution prior.

Treatments of two or more binomial samples have been studied. The sum of independent binomial random variables with the same p is also distributed as a binomial, so that samples can be directly added to give a merged sample and improved estimate of p. For k samples with X_i successes and n_j objects, the MLE for the combined sample is

$$\hat{p}(k)_{MLE} = \frac{\sum_{i=1}^{k} X_i}{\sum_{i=1}^{k} n_i}. \tag{4.6}$$

The sum of binomial samples from populations with different p values is more complicated.

A random vector $(n_1, n_2, \ldots, n_k)$ follows a multinomial distribution when the experiment has $k > 2$ classes. Here the parameters p_i satisfy $\sum_{i=1}^{k} p_i = 1$ and the number of trials occurring in each class n_i satisfy $\sum_{i=1}^{k} n_i = N$. The maximum likelihood estimators for the class probabilities are the same as with the binomial parameter, n_i/N, with variances $Np_i(1 - p_i)$. For example, an optical photometric survey may obtain a sample of 43 supernovae consisting of 16 Type 1a, 3 Type 1b and 24 Type II supernovae. The sample estimator of the Type Ia fraction is then $\hat{p}_1 = 16/43$.

Using the multivariate Central Limit Theorem (Section 2.10), it can be established that the random variable

$$X^2 = \sum_{i=1}^{k} \frac{(n_i - Np_i)^2}{Np_i} \tag{4.7}$$

has approximately a χ^2 distribution with $k - 1$ degrees of freedom. The X^2 quantity is the sum of the ratio $(O_i - E_i)^2/E_i$ where O_i is the observed frequency of the i-th class and E_i is the expected frequency given its probability p_i (Section 7.4). The accuracy of the χ^2 approximation can be poor for large k or if any p_i is small.

4.1.1 Ratio of binomial random variables

Estimation of the ratio of binomial random variables has proved to be a tricky problem. In astronomy, such ratios include the **signal-to-noise ratio** based on photon numbers, the **hardness ratio** if the photons have measured energies, or the amplitude of variability for a time series. Such ratios and their confidence intervals are simple to derive for large datasets, but for small-N situations, the statistical issues are nontrivial.

In Section 3.4.4, we considered the proportions (ratios) of Bernoulli trials for large n where asymptotic normality gives a well-defined confidence interval (3.27). But the case for small-n samples is not at all clear; Brown *et al.* (2001 and its following discussion) review the literature. The standard maximum likelihood Wald estimator and its confidence interval is

$$E_{MLE}[p] = \hat{p} \pm z_{\alpha/2}\sqrt{\frac{\hat{p}(1-\hat{p})}{n}} \tag{4.8}$$

where $\hat{p} = X/n$ and $z_{\alpha/2}$ is the $100(1-\alpha/2)$th percentile of the normal distribution. For example, in a hardness ratio problem we may have $X = 7$ soft counts out of $n = 12$ total counts giving the estimated *Soft/Total* ratio of 0.58 ± 0.28 for $z = 1.96$ corresponding to the 95% confidence interval.

This standard estimator is both biased and gives incorrect confidence intervals where p is near the limits of 0 or 1. The statistical performance is chaotic in the sense that it does not smoothly approach the asymptotic normal limit at n increases. For datasets with $n < 40$, Wilson's estimator from the 1920s performs better,

$$E_{Wilson}[p] = \tilde{p} \pm z_{0.025}\sqrt{\frac{\tilde{p}(1-\tilde{p})}{\tilde{n}}} \quad \text{where}$$
$$\tilde{n} = n+4 \quad \text{and} \quad \tilde{p} = (X+2)/(n+4) \tag{4.9}$$

for the 95% confidence interval. For the example with $X = 7$ and $n = 12$, this gives 0.56 ± 0.24. This estimator is essentially identical to the Bayesian credible interval for a noninformative Jeffreys prior. This problem of estimation of proportions is a good example where there is a consensus among statisticians that the maximum likelihood estimator is not the best estimator.

4.2 Poisson

4.2.1 Astronomical context

Events follow approximately a **Poisson distribution** if they are produced by Bernouilli trials (Section 2.7) where the probability p of occurrence is very small, the number n of trials is very large, but the product $\lambda = pn$ approaches a constant. This situation occurs quite naturally in a variety of astronomical situations:

1. A distant quasar may emit 10^{64} photons s^{-1} in the X-ray band but the photon arrival rate at an Earth-orbiting X-ray telescope may only be $\sim 10^{-3}$ photons s^{-1}. The signal resulting from a typical 10^4 s observation may thus contain only 10^1 of the $n \sim 10^{68}$ photons emitted by the quasar during the observation period, giving $p \sim 10^{-67}$ and $\lambda \sim 10^1$. There is little doubt that the detected photons sample the emitted photons in an unbiased fashion, even if relativistic beaming is present, as the quasar cannot know where our telescope lies.

2. The Milky Way Galaxy contains roughly $n \simeq 10^9$ solar-type (spectral types FGK) stars, but only $\lambda \simeq 10^1$ happen to lie within 5 pc of the Sun. The Poisson distribution could be used to infer properties of the full solar-type stellar population if the solar neighborhood sample is unbiased with respect to the entire Galaxy (which is not obviously true).
3. A 1 cm^2 detector is subject to bombardment of cosmic rays, producing an astronomically uninteresting background. As the detector geometrically subtends $p \simeq 10^{-18}$ of the Earth's surface area, and these cosmic rays should arrive in a random fashion in time and across the detector, the Poisson distribution can be used to characterize and statistically remove this instrumental background.

4.2.2 Mathematical properties

The Poisson distribution was first discussed by A. De Moivre in 1718 and S. Poisson in 1837 as a limit of the binomial distribution. Applications ensued during the twentieth century in a number of fields including: radioactive decay and particle counting, telephone traffic and other queuing behaviors, rare events such as accidents or fires, industrial reliability and quality control, medical epidemics and demographics. No classical theory of the Poisson distribution appears to have been motivated by astronomical problems, and few of its many mathematical properties have been used in astronomical research. Results summarized here appear in Haight (1967) and the stimulating volume by Kingman (1993) as well as the comprehensive reference on distributions by Johnson *et al.* (2005).

A discrete, nonnegative, integer-valued random variable X follows a Poisson distribution when the probability that X takes on the value $x = 0, 1, 2, \ldots$ is

$$P(X = x) \equiv P_\lambda(x) = \frac{e^{-\lambda}\lambda^x}{x!} \qquad (4.10)$$

where λ is the **intensity** of the Poisson distribution. Successive probabilities are related by

$$\frac{P(X = x + 1)}{P(X = x)} = \frac{\lambda}{x + 1}, \qquad (4.11)$$

indicating that the probability of individual values increases with x, peaks at an integer x between $x - 1$ and λ, and falls inversely as $1/x$ as $x \to \infty$.

Most tabulations related to the Poisson distribution give probabilities $p_\lambda(x)$ or $P(X > x)$ assuming λ is known. However, astronomers more commonly want to know confidence limits of λ for a measured value of p_λ. These are available in the statistical literature but astronomers are most familiar with the tabulation by Gehrels (1986). Another quantity of interest in astronomy is the survival probability such as the probability Q that a measured value will exceed the value x is

$$Q(\lambda) \equiv \sum_{i=x+1}^{\infty} p_\lambda(i). \qquad (4.12)$$

The Poisson distribution is unusual because its mean and variance are equal to each other. If λ is unknown, the sample mean

$$\bar{x} = \sum \frac{x_i}{n} \tag{4.13}$$

of a sample $(x_1, x_2, \ldots, x_n)$ is the best unbiased estimate of λ using moments, maximum likelihood and least-squares techniques. The $(1 - \alpha)$ lower and upper boundaries of the confidence interval for λ can be expressed in terms of quantiles of the χ^2 distribution (Krishnamoorthy 2006, Section 5.6).

The sum of two independent Poisson random variables X and Y with intensities λ and μ follows a Poisson distribution with intensity equal to $\lambda + \mu$. This property can be written

$$P(X + Y = n) = \frac{(\lambda + \mu)^n e^{-(\lambda + \mu)}}{n!}$$

$$\sum_j p_\lambda(j) p_\mu(x - j) \equiv p_\lambda * p_\mu(x) = p_{\lambda + \mu}(x) \tag{4.14}$$

where $*$ denotes convolution. The converse is also true: if the sum of independent random variables has a Poisson distribution, then each of the variables has a Poisson distribution. This is called Raikov's theorem.

However, the difference of two independent Poisson variables cannot follow a Poisson distribution because negative values may result. Instead, the difference has a Skellam (1946) distribution. This is relevant to astronomical problems involving integer signals in the presence of integer background. If one independently measures signal-plus-background counts and background counts, then their difference gives an estimate of the signal. The Skellam distribution would apply only if the direct counts — without scalings or other variable transformations — are used and the signal and background counts are measured in independent datasets.

For two independent Poisson random variables X and Y with intensities λ and μ, the Skellam distribution is

$$P(X - Y = k) = f_{Skellam}(k) = e^{-(\lambda + \mu)} \left(\frac{\lambda}{\mu} \right)^r I_{|k|}(2\sqrt{\lambda \mu}) \tag{4.15}$$

where $I(x)$ is the modified Bessel function of the first kind. The mean and variance of the Skellam distribution are

$$E[\lambda, \mu] = \lambda - \mu$$

$$Var[\lambda, \mu] = \lambda + \mu. \tag{4.16}$$

For the special case where $\lambda = \mu$ and k is large, the distribution is asymptotically normal

$$f_{Skellam}(k) \sim \frac{e^{-k^2/4\lambda}}{\sqrt{4\pi \lambda}}. \tag{4.17}$$

Another generalization of the Poisson model for count data that is widely used in other fields involves the two-parameter negative binomial distribution

$$y(p, x_i) = \frac{\Gamma(x_i + r)}{x_i! \Gamma(r)} (1 - p)^r p^{x_i}. \tag{4.18}$$

The negative binomial can arise in a number of ways: the sum of Poisson processes; a combined birth-and-death process; a discrete process where the variance exceeds the mean; the waiting time between successes of a binomial trial process; or the product of certain Markov chains. Estimation and regression involving the negative binomial distribution is described by Hilbe (2011).

Other situations involving Poisson distributions have been studied. Bayesian estimation is valuable when several Poisson distributions are involved, such as the subtraction of background from signal, or when prior constraints (e.g. zero-truncated distributions where no values may occur at zero) are known. Maximum likelihood estimation for Poisson variates with left- and/or right-truncation values are discussed by Johnson *et al.* (2005). The Conway–Maxwell–Poisson distribution is a generalization of the Poisson for integer and event processes exhibiting a variance either smaller (underdispersed) or larger (overdispersed) than the mean (Schmueli *et al.* 2005).

Among astronomers, there has been a debate over the best representation of Poisson ratios, a limiting case of the binomial proportion discussed in Section 4.1.1. In the context of a hardness ratio where S represents the soft counts, H represents the hard counts and $S + H$ gives the total counts, options include: (a) $0 < S/H < \infty$; (b) $0 < H/(S+H) < 1$; and (c) $-1 < (H-S)/(S+H) < 1$. Generally, option (a) is avoided due to its extreme range and option (c) is often chosen for its symmetry around zero. It is equivalent to a binomial experiment as $(H-S)/(S+H) = 1 - 2S/n$. Statistical estimates for the confidence interval of these hardness ratios have not been investigated. A Bayesian treatment for the case where the background must be subtracted from both S and H is developed by Park *et al.* (2006).

Hypothesis tests relating to the Poisson distribution have been developed but are not commonly used. Several statistics have been proposed for testing whether a dataset is Poisson distributed. Bayesian tests are available for specified loss functions. Two-sample tests, such as the likelihood ratio test, have been investigated.

Multivariate Poisson distributions can be considered marked Poisson univariate distributions where the ancillary variables are themselves Poisson distributions. Such situations often occur in astronomical photon-counting observations; for example, each photon may be characterized by an arrival time, a wavelength, a polarization, and a sky location which may all follow (often inhomogeneous) Poisson distributions. The statistical treatment of the joint distribution of k mutually independent Poisson random variables is qualitatively similar to the treatment of univariate Poisson distributions.

4.2.3 Poisson processes

The most common origin of the Poisson distribution in astronomy is a counting process, a subset of stochastic point processes which can be produced by dropping points randomly along a line (Haight 1967, Kingman 1993, Grandell 1997). The parameter λ, the intensity of the Poisson process, need not be constant along the line. In this case, we have a **nonhomogeneous (or nonstationary) Poisson process** where the counting process is "mixed" with another distribution function. Poisson processes are a subclass of more general **Markov processes** and martingales, and are related to other ergodic and renewal processes. A vast mathematical theory has been established for processes of these types. In the Poisson case, for example, the distribution of intervals between successive events follows an exponential distribution, and the distribution of intervals between two events separated by n other events can be expressed in terms of the gamma distribution.

Marked Poisson processes refer to the association of additional random variables (say, mass and luminosities for galaxies) to a Poisson distributed variable (say, position in space). Theory has established that the distributions of the marked variables (or their functionals such as gravitational potential and radiation fields) are also Poisson distributed.

Many extensions of the simple homogeneous Poisson process have been investigated. If the observation process has a threshold preventing measurement of the events at $x = 0$ or below some value $x < n$, then one considers the **truncated Poisson distribution**. Double truncation occurs when a maximum value is also present, perhaps due to saturation of the detector. The **censored Poisson process** where only the k-th event is recorded, and other Poisson distributions with gaps, were studied during the 1940–60s. This was motivated in part by detectors such as Geiger counters suffering dead time after each trigger. **Stuttering** Poisson processes refer to situations where two or more events can occur simultaneously; each event thus has a discrete magnitude. This can occur in astronomy where the time tagging is not instantaneous due to detector readout constraints. During the 1950s, Takács studied the situation where every event of a Poisson process generates a continuous signal with a characteristic distribution function (e.g. a rise to amplitude A with exponential decline). His results may be relevant to stochastic shot-noise models of astrophysical systems such as accretion onto compact objects.

A major class of nonhomogeneous Poisson processes are called **mixed Poisson processes** where the intensity λ is viewed as the outcome of a separate random variable Λ with a known (or assumed) "structure" distribution U based on some property associated with the random events. If U is only a function of the intensity, $U(\lambda)$, this is called a **Pólya process**. These mixed Poisson distributions have a wider dispersion than the original Poisson distribution. A broader class where both U and λ are functions of time are called **Cox processes**. These can also be viewed as thinned Poisson processes where certain events are removed from the dataset. Most treatments consider well-studied mixing distributions U such as the gamma (which includes the exponential as a limit), normal or (another) Poisson distribution.

These types of generalized Poisson processes permit a wide variety of situations to be modeled. Cox processes, in particular, are particularly flexible and have proved useful in many fields of applications. The mixing distribution can be independent of the original Poisson distribution, representing some additional random factor in the situation under study, or can have parametric dependencies representing positive or negative correlation between events.

Mixed Poisson distributions may be helpful for modeling high-energy accretion disk emission such as the shot-noise model for Cyg X-1 and quasars and quasi-periodic oscillation (QPO) behavior seen in many X-ray sources (van der Klis 1989; see Section 11.13). Contagion models are closely related to the avalanche model for solar flares (Wheatland 2000). Here, solar physicists compare the waiting time distribution between consecutive solar flares to those of nonstationary Poisson models with intensities $\lambda(t)$. The goal of such statistical analysis might be to elucidate the nature of the shot, avalanche or quasi-periodic process by constraining the parameter structure distribution $U(\lambda)$. The extensive mathematical work on the exponential structure distribution $U \propto x^{k-1}e^{-\lambda x}$ with k integer, referred to as **Erlang distributions**, may be particularly relevant for astronomical applications.

4.3 Normal and lognormal

The normal, or Gaussian, distribution $N(\mu, \sigma^2)$ is the most widely used in statistics and is also well-known in astronomy. A great number of statistical procedures require that the data be normally distributed and, through the Central Limit Theorem (Section 2.10), many statistics are asymptotically normally distributed. While the frequency distribution of some astronomical variables (such as mass or luminosity) follow power-law and related asymmetrical relationships, other variables often show centrally concentrated distributions similar to the normal or lognormal relationships. These include distributions of color indices, emission-line ratios, densities, spatial locations, and residual distributions around relationships. Examples of normal and lognormal distributions are plotted in Figure 2.2, and normal fits to astronomical datasets are shown in Figure 4.2 below.

The normal distribution has many advantageous mathematical properties. The population mean and median are both μ, and the sample mean

$$\bar{X} = \frac{1}{n} \sum_{i=1}^{n} X_i \qquad (4.19)$$

is simultaneously the least-squares estimator and maximum likelihood estimator (MLE) for the normal mean. The standard deviation of the distribution is σ, the square root of the variance. A random variable $X \sim N(\mu, \sigma^2)$ can be **standardized** by the transformation

$$Z = \frac{X - \mu}{\sigma} \qquad (4.20)$$

where $Z \sim N(0, 1)$.

Astronomers often evaluate the 1σ confidence interval of the sample mean using a normal population,

$$\bar{x} \pm S/\sqrt{n}. \qquad (4.21)$$

This approximation is based on the Law of Large Numbers (Section 2.10) as S is an estimator of σ based on the data. However, the t distribution gives an exact confidence interval for all n. The $(1 - \alpha/2)$-th confidence interval (e.g. $\alpha = 0.025$ for a 95% confidence interval) for the mean is

$$\bar{X} \pm t_{n-1,1-\alpha/2} \frac{S}{\sqrt{n}}. \qquad (4.22)$$

For a two-sample comparison of samples drawn from normal distributions, the difference of the means and its asymptotic confidence interval are

$$\bar{X}_1 - \bar{X}_2 \pm \sqrt{S_1^2/n_1 + S_2^2/n_2}. \qquad (4.23)$$

This formula is a good approximation if the samples are sufficiently large $n_1, n_2 \geq 30$. For small samples, there is no good two-sample statistic if the variances of the two samples differ, so one must test for the equality of the sample variances using the t distribution or, for multiple samples, the F distribution.

As many statistics are asymptotically normal, a critical step is to evaluate whether a particular statistic for a particular dataset is consistent with a normal distribution. A vast literature emerged during the twentieth century on normality tests. The nonparametric goodness-of-fit tests based on the empirical distribution function discussed in Section 5.3.1 are often used for this purpose: the Anderson–Darling, Cramér–von Mises, and Kolmogorov–Smirnov tests (or the variant designed for normality tests, the **Lilliefors test**). The **Jarque–Bera test** measures departures from normality based on the sample skewness and kurtosis, the third and fourth moments of its distribution. The **Shapiro–Wilks test** for normality uses a combination of rank and least-squares measures. The χ^2 **test** can be used, but is not advised if it involves arbitrary choices in binning.

Graphical techniques are also commonly applied to compare a dataset with a normal distribution. In a **probability-probability (P-P) plot**, sorted data values are plotted on the ordinate while the normal distribution is plotted on the abscissa. For large samples ($n \gg 100$), the **quantile-quantile (Q-Q) plot** introduced in Section 2.6 can conveniently summarize a P-P plot that would otherwise have many points.

While the sum of independent normal random variables is another standard normal, the sum of squares of independent standard normals $X_1, X_2, \ldots, X_n$ is

$$X^2 = \sum_{i=1}^{n} X_i^2 \tag{4.24}$$

and follows a χ^2 distribution with n degrees of freedom, χ_n^2. An important application is that the sample variance follows a χ^2 distribution,

$$\frac{\sum_{i=1}^{n} (X_i - \bar{X})^2}{\sigma^2} \sim \chi_{n-1}^2 \quad \text{where } \bar{X} = \frac{1}{n} \sum_{i=1}^{n} X_i. \tag{4.25}$$

This situation also appears in hypothesis testing of categorical data. When $r \times c$ contingency tables are constructed, the test statistic $T_{\chi^2, r \times c}$ defined in Equation (5.19) is approximately χ^2 distributed, as discussed in Section 5.5.

The lognormal distribution inherits many of the mathematical properties of the normal after a logarithmic transformation of the variable, $Y = \log X$. Note that the lognormal is not the logarithm of a normal distribution, but the normal distribution of a logged random variable. It was first considered by F. Galton in the late-nineteenth century, and astronomer J. Kapteyn contributed to its early development (Hald 1998). The obvious restriction that $X > 0$ to permit the logarithmic transformation is natural for astrophysical variables (e.g. radius, mass, temperature, elemental abundance), but astronomers must be aware that a normal distribution for the transformed variable is not guaranteed. Tests of normality should be performed on Y prior to application of procedures that depend on the lognormal transformation. Just as normal distributions naturally arise from equilibrium stochastic processes where uniform random quantities are repeatedly added or subtracted, lognormal distributions arise from processes with repeated multiplication or division of random quantities. For example, lognormal size distributions might emerge from steady-state coagulation and fragmentation processes.

A random vector $\mathbf{X} = (X_1, X_2, \ldots, X_p)$ is said to follow a **multivariate normal (MVN) distribution** if every linear combination $\sum_{j=1}^{p} a_j X_j$ has a univariate normal distribution. If the variables are not linearly dependent, then the MVN p.d.f. is given by

$$P(\mathbf{x}) = \frac{|\mathbf{V}|^{1/2}}{(2\pi)^{-p/2}} e^{-\frac{1}{2}(\mathbf{x}-\mu)^{\mathbf{T}}\mathbf{V}^{-1}(\mathbf{x}-\mu)} \qquad (4.26)$$

where μ is the vector of mean values and $\mathbf{V}$ is the variance–covariance matrix of $\mathbf{X}$. When $p = 2$, the bivariate normal probability density can be expressed as

$$P(\mathbf{x}_1, \mathbf{x}_2) = \frac{1}{(2\pi\sqrt{1-\rho^2})} e^{-\frac{1}{2(1-\rho^2)}\left[\left(\frac{x_1-\mu_1}{\sigma_1}\right)^2 - 2\rho\left(\frac{x_1-\mu_1}{\sigma_1}\right)\left(\frac{x_2-\mu_2}{\sigma_2}\right) + \left(\frac{x_2-\mu_2}{\sigma_2}\right)^2\right]} \qquad (4.27)$$

where ρ is the correlation between the two random variables, X_1 and X_2. Zero correlation implies independence for the bivariate normal, although this is not generally true for other distributions.

Although the maximum likelihood (equal to the least squares) estimators for the moments μ and Var are best known, alternatives have been developed. These include minimum variance unbiased estimators, James–Stein estimators, and Bayesian estimators for different amounts of prior knowledge and different loss functions. Many of these alternatives differ from the MLEs significantly only for small samples. For example, the usual sample correlation between X_{im} and X_{jm},

$$\hat{\rho}_{ij}(MLE) = \frac{S_{ij}}{\sqrt{S_{ii}S_{jj}}} \quad \text{where}$$

$$S_{ij} = \sum_m (X_{im} - \bar{X}_i)(X_{jm} - \bar{X}_j), \qquad (4.28)$$

is biased for $n \leq 30$. A better MVUE correlation estimator is

$$\hat{\rho}_{ij}(MVUE) = \hat{\rho}_{ij}(MLE)\left(1 + \frac{1 - \hat{\rho}_{ij}^2(MLE)}{2(n-4)}\right). \qquad (4.29)$$

For the variance, some loss functions give Bayesian estimators of $\tilde{Var} = (1/(n-1))\mathbf{S}$ or $(1/(n+k))\mathbf{S}$ rather than the MLE $\widehat{Var} = (1/n)\mathbf{S}$. MVUE or Bayesian methods are superior to MLEs when prior information about μ or Var is available.

4.4 Pareto (power-law)

The **power-law distribution** that is very widely applied to astronomical phenomena is almost universally known in statistics as the **Pareto probability distribution**. Pareto (or in adjectival form, Paretian) distributions are named after Vilfredo Pareto, a nineteenth-century engineer and scholar who was one of the principal founders of modern quantitative economics. Pareto found that the distribution of incomes in modern market economies is strongly skewed with the great majority of men poor and a small fraction wealthy following a power-law distribution

$$P(X > x) \propto \left(\frac{x}{x_{min}} \right)^{-\alpha}, \quad \alpha > 0 \tag{4.30}$$

where $0 < x_{min} \leq x$ is a minimum income. Economists found that α, sometimes called Pareto's index, lies in the range 1.5–2 where larger values indicated greater inequality of income. The Gini, Lorenz, Hoover and Thiel indices, in addition to Pareto's index, are widely used statistical measures of income inequality. Figure 4.1 below plots a typical Pareto distribution encountered in astronomy, a simulation of the stellar initial mass function.

The Pareto probability distribution function (p.d.f.) can be qualitatively expressed as

$$P(x) = \frac{\text{shape}}{\text{location}} \left(\frac{\text{location}}{x} \right)^{\text{shape}+1}. \tag{4.31}$$

Applying the normalization necessary to integrate to unity over all x,

$$P = \begin{cases} \frac{\alpha b^{\alpha}}{x^{\alpha+1}} & \text{for } x \geq b \\ 1 & \text{for } x < b \end{cases} \tag{4.32}$$

where $\alpha > 0$ and $b > 0$ is the minimum data value. The Paretian shape parameter $\alpha + 1$ is typically called the **power-law slope** by astronomers. Other functions related to the p.d.f. also have simple forms: the **hazard function** $h(x) = f(x)/(1 - F(x)) = \alpha/x$ decreases monotonically with x, and the **elasticity** $d \log F(x)/d \log x = \alpha$ is constant. Equation (4.30) is the Pareto distribution of the first kind, Pareto(I), a special case of the more general **Pareto(IV) distribution**

$$P_{IV}(X > x) = \left[1 + \left(\frac{x - \mu}{x_{min}} \right)^{1/\gamma} \right]^{-\alpha} \tag{4.33}$$

which is itself a case of the large Burr family of distributions.

Power-law population distributions were repeatedly found and studied in many fields throughout the twentieth century: word frequencies in natural language; the population of cities; the sizes of lunar impact craters; the sizes of particles of sand; the areas burnt by forest fires; the number of links to Internet Web sites; the intensity of earthquakes; the intensities of wars; the frequencies of family names; and so forth (Reed & Hughes 2002, Newman 2005). In other cases, the power law relates two observable variables: the lengths of rivers and their basin areas; the body weight and metabolic rates of mammals; the perceived and actual magnitude of a sensory stimulus; and other applications. In different contexts, the distribution has been called Pareto's law, Zipf's law, Gibrat's law, Hack's law, Kleiber's law, Horton's law, Steven's power law, Gutenberg–Richter law, and so forth.

In astronomy, power-law distributions appear in the sizes of lunar impact craters, intensities of solar flares, energies of cosmic-ray protons, energies of synchrotron radio lobe electrons, the masses of higher-mass stars, luminosities of lower-mass galaxies, brightness distributions of extragalactic X-ray and radio sources, brightness decays of gamma-ray burst afterglows, turbulent structure of interstellar clouds, sizes of interstellar dust particles, and so forth. Some of these distributions are named after the authors of seminal studies: de

Vaucouleurs galaxy profile, Salpeter stellar initial mass function (IMF), Schechter galaxy luminosity function, MRN dust size distribution, and Larson's relations of molecular cloud kinematics, and others. The power-law behavior is often limited to a range of values; for example, the spectrum of cosmic rays breaks around 10^{15} eV and again at higher energies, the Schechter function drops exponentially brighter than a characteristic galaxy absolute magnitude $M^* \sim -20.5$, and the Salpeter IMF becomes approximately lognormal below $\sim 0.5 \, M_\odot$.

Hundreds of astronomical studies are devoted to precise measurement of the slope parameters α, characterization of deviations from a simple power-law distribution, and astrophysical explanations of the origin of these distributions. Despite these long-standing research enterprises, the astronomical community is generally unfamiliar with the mathematical and statistical properties of the Pareto and related distributions, and often uses ill-advised methods.

The mathematics of the Pareto distribution was studied throughout the twentieth century; detailed reviews are given by Arnold (1983) and Johnson *et al.* (1994). Interest in power laws has revived in the past decade with the recognition that they appear in an astounding variety of human and natural sciences (Slanina 2004, Mitzenmacher 2004, Newman 2005). If the variable under study is inherently discrete (e.g. integer or categorical values), then the relevant statistical distribution is the **Riemann ζ (zeta) distribution** which we do not discuss here. We will use the terms "power-law" and "Pareto" distribution interchangeably.

4.4.1 Least-squares estimation

Stellar mass functions and other power-law distributions in astronomy are most often fit using least-squares linear regression of the grouped (binned) data after logarithmic transformation. In IMF studies, star counts are binned in logarithmic mass units, plotted as a histogram in a $\log N$–$\log M$ graph, and fitted with a linear function by minimizing the sum of squared deviations. The regression is sometimes weighted by the number of points in each bin, and the centroid (rather than center) of each bin is sometimes used. The least-squares slope and intercept are given by Section 7.3.1. In economics, this method has been used in the analysis of Pareto charts since the early twentieth century.

While the least-squares estimator of the Pareto slope using grouped data is a consistent statistic (Section 3.3), recent numerical simulations have established that its bias and variance are considerably worse than that achieved in more symmetrical distributions, particularly when the slope α is steep (Goldstein *et al.* 2004, Newman 2005, Bauke 2007). Systematic biases appear even for large samples, caused by an inaccurate response of the least-squares statistic to the highly variable points near the tail of the distribution. For example, Bauke conducts Monte Carlo simulations of distributions with $N = 10000$ points and $\alpha = 2.5$ and finds that least-squares estimators from logarithmically grouped data underestimate slope values by several percent and overestimate the slope confidence interval several-fold compared to unbinned MLEs (Section 4.4.2).

Some of these problems are resolved by maximum likelihood estimation of α for grouped data as derived by Aigner & Goldberger (1970). But this estimator is not equal, even for large and well-behaved samples, to the unbinned MLE. MLEs of asymmetrical distributions (e.g. Pareto, exponential) using grouped data have complex mathematical properties that have been studied only for limited problems. Arnold (1983) discourages use of the Aigner–Goldberger likelihood.

Due to these various problems, it is ill-advised to estimate power-law parameters from grouped data if ungrouped data values are available. Least-squares linear regression from logarithmic grouped data is generally viewed as a heuristic graphical procedure (the traditional Pareto chart) producing biased and inaccurate results, rather than a statistical technique with a sound mathematical foundation.

4.4.2 Maximum likelihood estimation

In contrast to the performance of least-squares fitting, maximum likelihood techniques for ungrouped data provide asymptotically (for large n) unbiased, normal and minimum variance estimators of the power-law distribution. Derived by A. Muniruzzaman in 1957, the MLE is obtained by taking the logarithm of the Pareto likelihood

$$L = \prod_{i=1}^{n} \frac{\alpha b^{\alpha}}{X_i^{\alpha+1}}, \tag{4.34}$$

and setting the differential with respect to the parameters to zero. To avoid an unbounded likelihood, the minimum data value $X_{(1)}$ is needed for the estimator of b. The resulting MLEs of the slope α and scale parameter b are

$$\hat{\alpha} = n \left[\sum_{i=1}^{n} \ln \frac{X_i}{\hat{b}} \right]^{-1}$$
$$= 1/\text{mean}[\ln(X/\hat{b})]$$
$$\hat{b} = X_{(1)} \tag{4.35}$$

where $X_{(1)}$ is the smallest of the X values. These estimators apply if neither parameter is known in advance; if b is determined independently of the dataset under consideration, then a different MLE for α applies.

These MLEs have a bias in small samples which can be easily corrected. The uniform minimum variance unbiased estimators (MVUE) for the two parameters of the power-law distribution are

$$\alpha^* = \left(1 - \frac{2}{n} \right) \hat{\alpha}$$
$$b^* = \left(1 - \frac{1}{(n-1)\hat{\alpha}} \right) \hat{b} \tag{4.36}$$

with asymptotic variances

$$Var(\alpha^*) = \frac{\alpha^2}{n-3}, \quad n > 3$$
$$Var(b^*) = \frac{b^2}{\alpha(n-1)(n\alpha-2)}, \quad n\alpha > 2. \tag{4.37}$$

The estimators $\hat{\alpha}$ and $\hat{b}$ are independent, as are the estimators α^* and b^*.

The central moments of the Pareto distribution − mean, variance and skewness − are

$$E[X] \;=\; \frac{\alpha b}{\alpha - 1} \quad \text{for } \alpha > 1$$

$$Var(X) \;=\; \frac{b^2}{(\alpha - 1)^2} \frac{\alpha}{\alpha - 2} \quad \text{for } \alpha > 2$$

$$Skew(X) \;=\; 2\frac{\alpha + 1}{\alpha - 3} \sqrt{\frac{\alpha - 2}{\alpha}} \quad \text{for } \alpha > 3. \tag{4.38}$$

The mean and variance are often not useful for understanding the underlying scale-free power-law population as they depend critically on the lowest x value in the sample under study. However, the skewness and the ratio of the mean to the standard deviation are free from b, depending only on the shape parameter α. Other statistics of the Pareto distribution used by researchers in other fields include: the geometric mean $be^{1/\alpha}$; the median $2^{1/\alpha}b$; the Gini index $1/(2\alpha - 1)$; the fraction of the population exceeding a specified value x_0 $(x_0/b)^{1-\alpha}$; and the fraction of the total integrated value in the population above the specified value $(x_0/b)^{2-\alpha}$.

In Section 4.7.1 we compare several estimation methods for a simulated sample with a power-law distribution using an **R** script.

4.4.3 Extensions of the power-law

By variable transformations and choice of parameters, the Pareto distribution is linked to other statistical distributions such as the exponential, χ^2, gamma, and log-logistic distributions. Many of these can be considered generalizations of the power-law distribution. In addition, Pareto distributions of the second and fourth kind are variants with more complex power-law scalings and offsets. The Pareto distribution of the third kind includes an exponential cutoff; a limiting case is the gamma distribution that is closely related to the Schechter function known in astronomy (Section 4.5).

Perhaps the most common generalization in astronomy is the **double Pareto**, or **broken power-law**, distribution with two (or more) power-laws with different slopes that are normalized at a change point but with a discontinuity in the first derivative. Well-known examples include a three-power-law model for a universal stellar initial mass function with breaks at 0.8 and 0.2 $M_\odot$ (Kroupa 2001), the extragalactic cosmic-ray energy spectrum with a break at $10^{18.6}$ eV and suppression above $10^{19.5}$ eV (Abraham *et al.* 2008), and the gamma-ray burst afterglow light curve with breaks at times $t < 500$ s and $t \sim 10^4$ s (Nousek *et al.* 2006). Parameters are usually estimated by least squares with some *ad hoc* estimation of the **break point** where the two power laws meet. A better method is an MLE estimator where the break point is a parameter of the model; equations are given by the statisticians Chen & Gupta (2000) and physicist Howell (2002).

A flexible class of generalized Pareto distributions have been recently studied by statistician W. Reed including the product of two power-law distributions, a power law with a lognormal distribution, two power laws with a lognormal distribution, and a power law with a Laplacian distribution. He finds that these hybrid distributions provide effective models for the distribution of incomes in a capitalistic economy, sizes of cities, sizes of sand particles, capacity of oil fields, temporal behavior of stock prices, page numbers of World Wide Web sites, and other problems. For some parameters, this resembles a broken power law with a smooth rather than discontinuous change in slope. For other parameters, it resembles a lognormal with heavy power-law tails on one or both sides.

For yet other parameters, Reed's distribution consists of power laws at high and low levels with a lognormal at intermediate levels. A one-sided version of this model is commonly used for the stellar initial mass function (Chabrier 2003). Basu & Jones (2004) apply Reed's distribution to this astronomical problem. This double Pareto-lognormal distribution can be viewed as a mixture of power laws and lognormals with p.d.f. (Reed & Jorgensen 2004)

$$f_{dPlN}(x) = \frac{\beta}{\alpha + \beta} f_1(x) + \frac{\alpha}{\alpha + \beta} f_2(x) \quad \text{where}$$

$$f_1 = \alpha x^{-\alpha-1} A(\alpha, \nu, \tau) \Phi \left(\frac{\log x - \nu - \alpha \tau^2}{\tau} \right)$$

$$f_2 = \beta x^{\beta-1} A(-\beta, \nu, \tau) \Phi^c \left(\frac{\log x - \nu + \beta \tau^2}{\tau} \right) \tag{4.39}$$

where $A(\theta, \nu, \tau) = \exp \theta \nu + \theta^2 \tau^2 / 2$. The mean and variance (for $\alpha > 1$) are

$$E[X] = \frac{\alpha \beta}{(\alpha - 1)(\beta - 1)} e^{\nu + \tau^2/2}$$

$$Var(X) = \frac{\alpha \beta e^{2\nu + \tau^2}}{(\alpha - 1)^2 (\beta + 1)^2} \left[\frac{(\alpha - 1)^2 (\beta + 1)^2}{(\alpha - 2)(\beta + 2)} e^{\tau^2} - \alpha \beta \right]. \tag{4.40}$$

Reed & Jorgensen describe methods for obtaining the four parameters $(\alpha, \beta, \nu, \tau^2)$ using the method of moments and maximum likelihood using the EM algorithm.

Astronomers have proposed some mathematical procedures for treating Pareto distributions. In a prescient study, astronomers Crawford *et al.* (1970) found that least-squares estimators overestimated both the slope and confidence interval and strongly argued for maximum likelihood estimation of power laws in the flux distributions ($\log N$–$\log S$ relation) of radio sources. Nonstandard binning procedures to reduce bias in the least squares estimator are investigated by astronomers Maíz Apellániz & Úbeda (2005). Marschberger & Kroupa (2009) treat a number of problems: a modified MLE for a truncated power-law distribution; a goodness-of-fit test for truncation of a power-law distribution; and a large-sample algorithm for estimation of power-law parameters through its relationship to the exponential distribution. Koen & Kondlo (2009) consider a dataset subject to homoscedastic measurement error, and derive a MLE for the convolution of the power-law and Gaussian distributions. Astronomers have also considered the effect of faint undetected sources from a power-law flux distribution for an imaging detector: Maloney *et al.* (2005) study the effect on the noise characteristics, while Takeuchi & Ishii (2004) consider the effect of overlapping faint sources (the source confusion problem).

4.4.4   Multivariate Pareto

While most astronomical studies consider Paretian variables individually, the scientific questions are often bivariate or multivariate in nature and modeling of the joint distribution would be useful. For example, main-sequence stars with masses $M \geq 0.5$ M$_\odot$ exhibit power-law distributions in mass and luminosity with different shape parameters linked by the well-known relationship $L \propto M^{3.5}$. Alternatively, the mass distributions of two star clusters should have the same α shape parameters but different location parameters, b_1 and b_2.

For this latter situation, Mardia (1962) derived a bivariate p.d.f. with

$$P(x_1, x_2) = (\alpha + 1)\alpha(b_1 b_2)^{\alpha+1} (b_2 x_1 + b_1 x_2 - b_1 b_2)^{-(\alpha+2)} \tag{4.41}$$

that follows a Pareto(I) marginal distribution $P(X > x) = \alpha b^\alpha / x^{\alpha+1}$. Here the parameters are restricted to be $x_1 \geq b_1 > 0, x_2 \geq b_2 > 0$, and $\alpha > 0$. The MLEs of the parameters are

$$\hat{b}_1 = \min(X_{1i}), \quad \hat{b}_2 = \min(X_{2i})$$

$$\hat{\alpha} = \frac{1}{S} - \frac{1}{2} + \sqrt{\frac{1}{S^2} + \frac{1}{4}} \quad \text{where}$$

$$S = \frac{1}{n} \sum_{i=1}^{n} \log\left(\frac{X_{1i}}{\hat{b}_1} + \frac{X_{2i}}{\hat{b}_2} - 1\right). \tag{4.42}$$

This bivariate p.d.f. can be generalized to the multivariate case. A number of other approaches to multivariate Paretian distributions have also been studied.

4.4.5 Origins of power-laws

An early study of the origin of the Pareto distribution was Fermi's (1949) model for the acceleration of cosmic rays with a power-law energy distribution. Here, charged particles stochastically bounce across a strong magnetic shock from a supernova explosion, gaining a fractional increase in energy upon each passage, $E(t + 1) = \lambda E(t)$ where $\lambda > 1$ and $E(t) > E_{min}$ for all times. The Fermi particle acceleration process has both a low-energy cutoff (no electron can have negative energy) and a high-energy cutoff (sufficiently energetic particles will have Larmor radii larger than the shock size). Processes without truncation will produce lognormal distributions.

The association of a zero-avoiding multiplicative stochastic motion as a generative process of the power-law distribution has been widely discussed in economics, where wealth rather than energy is exchanged in binary interactions, as well as in statistical physics and other fields (Levy & Solomon 1996, Slanina 2004, Mitzenmacher 2004, Newman 2005). Deviations from a pure power law can emerge with additional constraints on a multiplicative stochastic process, such as a limit of wealth traded at low levels, the introduction of a tax on accumulated wealth, or a limit in the number of potential trading partners. A subfield nicknamed **econophysics** has emerged to study these phenomena (Chatterjee *et al.* 2005).

Both the distribution of wealth in market economies and the stellar initial mass function (IMF) exhibit a power-law distribution at high values and a downturn at low values, which is sometimes modeled as a lognormal (Chabrier 2003). Power-law and lognormal distributions are very similar over much of their range, and distinguishing them can be difficult. Debates whether populations follow Pareto, lognormal or both distributions, and over the precise nature of the stochastic process underlying these distributions, have arisen repeatedly in biology, chemistry, ecology, information theory, linguistics and Internet studies over the past century (Mitzenmacher 2004).

Reed & Hughes (2002) describe a class of generative models for Pareto-type distributions that, like Fermi's cosmic-ray acceleration mechanism, are multiplicative stochastic processes that are **killed** (i.e. stopped or observed) by a secondary process before equilibrium is established. **Geometric Brownian motion**

$$dX = \mu X dt + X\epsilon, \tag{4.43}$$

where $\epsilon = N(0, \sigma^2)$ is a Gaussian noise, produces an exponential growth in X which becomes a power law if it is killed at random times that are themselves exponentially distributed. In an astrophysical application of this idea, the power-law stellar IMF may arise from multiplicative stochastic growth with an exponentially growing star-formation process (Basu & Jones 2004). There is independent evidence for both of these effects in star formation: the accretion rate of protostars scales with protostellar mass, and star formation can accelerate. The logistic stochastic differential equation

$$dX = \mu X dt - \lambda X^2 dt \tag{4.44}$$

also leads to Pareto-like populations. Known as **Lotka–Volterra processes**, these are widely studied for applications in biology, demography and economics (Solomon & Richmond 2002).

4.5 Gamma

The **gamma distribution** is a particularly important generalization of the Pareto distribution that combines a power law at low values of x with an exponential at high values of x. The power-law index α is called the shape parameter and the exponential cutoff parameter β is called the scale parameter. A third location offset parameter is sometimes added. The standard gamma distribution with zero offset has p.d.f.

$$f_{Gamma}(x|\alpha, \beta) = \frac{1}{\Gamma(\alpha)}\beta^{-\alpha}x^{\alpha-1}e^{-x/\beta} \tag{4.45}$$

where $x > 0$, $\alpha > 0$ and $\beta > 0$. Its c.d.f. is a constant multiple of the incomplete gamma function,

$$F_{Gamma}(\beta x|\alpha) = \frac{1}{\Gamma(\alpha)}\int_0^{x/\beta} e^{-t}t^{\alpha-1}dt. \tag{4.46}$$

This distribution has been extensively discussed in extragalactic astronomy with respect to the distribution of luminosities of normal galaxies where it is known as the **Schechter (1976) function**. For example, from photometric and redshift measurements of $\sim$150,000 galaxies in the Sloan Digital Sky Survey and assuming standard cosmological parameters, Blanton *et al.* (2005) find that a maximum likelihood Schechter function fit to

$$f_{Schechter}(L) = C\left(\frac{L}{L_*}\right)^a e^{-L/L_*} \tag{4.47}$$

for a constant C gives $a = -1.05 \pm 0.01$ using r-band magnitudes and $a = -0.89 \pm 0.03$ using g-band magnitudes. Thus, the Schechter function form of the galaxy luminosity distribution found by Blanton *et al.* is formally a gamma probability distribution using g magnitudes but not a gamma distribution using r magnitudes.

The gamma distribution has an important link to the Poisson distribution: its c.d.f. gives the probability of observing a or more events from a homogeneous Poisson process. The gamma distribution also results from the sum of i.i.d. exponential random variables; if x is a time-like variable, it gives the waiting time for a specified number of independent events to occur. The sum of gammas is also distributed as a gamma if the scale parameters b are identical. From these properties, the gamma distribution is important in applications such as queuing theory, failure models and quality control. The χ_ν^2 **distribution** is another important special case of the gamma distribution when $b = 2$ and $\alpha = \nu/2$.

The maximum likelihood estimators for the scale and shape parameters of the gamma p.d.f. are found by solving

$$\psi(\hat{\alpha}) - \ln(\hat{\alpha}) = \frac{1}{n}\sum_{i=1}^{n}\ln(X_i) - \ln(\bar{X})$$

$$\hat{\beta} = \frac{\bar{X}}{\alpha} \tag{4.48}$$

where X_i are i.i.d. $Gamma(\alpha, \beta)$ random variables, $\bar{X}$ is the sample mean, and ψ is the digamma function $\psi(\alpha) = \Gamma'(\alpha)/\Gamma(\alpha)$. Note that the MLE of the shape parameter α does not involve the scale parameter β. Asymptotic variances of $\hat{\alpha}$, $\hat{\beta}$, and their correlation depend on the derivatives of the digamma function, $\psi'(\alpha)$.

Gamma distribution parameter estimators are also easily computed from the second and third central moments of the sample (variance and skewness) according to

$$\hat{\alpha}_{mom} = \frac{4m_2^3}{m_3^2} \quad \text{and} \quad \hat{\beta}_{mom} = \frac{m_3}{2m_2} \tag{4.49}$$

where $m_k = \frac{1}{n}\sum_{i=1}^{n}(X_i - \bar{X})^k$. If the scale parameter b is known and set equal to unity, then the shape parameter is estimated by

$$\hat{\alpha}_{mom} = \frac{m_1^2}{m_2^2} = \frac{\bar{X}^2}{S^2}. \tag{4.50}$$

However, these moment estimators have greater bias and variance than the MLEs. Johnson *et al.* (2005) give other mathematical properties characterizing gamma distributions.

A stochastic process generating the gamma distribution has been used to model biological populations (Dennis & Patil 1984). Here, a population $n(t)$ has a multiplicative growth rate g and multiplicative noise term $h(n)$ that depends on the current population level,

$$\frac{dn}{dt} = n[g(u, t) + h(n)\epsilon(t)] \tag{4.51}$$

where $\epsilon \sim N(0, \sigma^2)$ is a normal noise and the function $g(u, t)$ can incorporate processes such as predation and feeding. Analogous models for multiplicative growth of stars, planets or galaxies might be constructed.

Multivariate gamma functions may be useful in astronomy when considering, for example, galaxy luminosity functions in several spectral bands. However, there is no unique generalization of the univariate gamma. Over a dozen bivariate distributions proposed from the 1930s through 1980s have marginal univariate gamma distributions (Hitchinson & Lai 1990; Kotz *et al.* 2000, Chapter 48).

4.6   Recommended reading

Evans, M., Hastings, N. & Peacock, B. (2000) *Statistical Distributions*, 3rd ed., John Wiley, New York

Krishnamoorthy, K. (2006) *Handbook of Statistical Distributions with Applications*, Chapman & Hall, London
Short introductory handbook on 20–30 important distributions with basic mathematical properties and inference procedures. Much of this useful material can also be found on-line at Wikipedia (http://en.wikipedia.org/wiki/List_of_probability_distributions).

Johnson, N. L., Kemp, A. W. & Kotz, S. (2005) *Univariate Discrete Distributions*, 3rd ed., Wiley-Interscience, New York

Johnson, N. L., Kotz, S. & Balakrishnan, N. (1994) *Continuous Univariate Distributions*, 2nd ed., Vols 1 and 2, Wiley-Interscience, New York

Johnson, N. L., Kotz, S. & Balakrishnan, N. (1997) *Discrete Multivariate Distributions*, Wiley-Interscience, New York

Kotz, S., Balakrishnan, N. & Johnson, N. L. (2000) *Continuous Multivariate Distributions*, 2nd ed., Vols 1 and 2, Wiley-Interscience, New York
Comprehensive and authoritative reviews of the mathematical properties of statistical distributions including distributions, approximations, moments and estimation.

4.7   R applications

The built-in *stats* library of **R** provides four functions for over 20 common probability distributions. Consider the Poisson distribution: the function *dpois* gives the probability density function (p.d.f.), *ppois* gives the cumulative distribution function (c.d.f.), *qpois* gives the quantile function (inverse c.d.f.), and *rpois* generates random numbers drawn from the Poisson distribution. The prefixes "d", "p", "q" and "r" can be applied to "binom" for the binomial, "chisq" for the χ^2, "gamma" for the gamma, "lnorm" for the lognormal, "multinom" for the multinomial, "norm" for the normal, "pois" for the Poisson, "unif" for the uniform distribution, and similarly for the beta, Cauchy, exponential, F, geometric, hypergeometric, logistic, negative binomial, sign rank, Student's t, Tukey, Weibull and Wilcoxon distributions. The **R** reference manual provides algorithmic details for calculating each distribution. The parameters of the distribution must be given when calling the function. Several of these functions were used to generate Figure 2.2.

4.7.1 Comparing Pareto distribution estimators

While **R** does not provide a built-in computation for the Pareto (power-law) distribution, it is available from **CRAN**'s *VGAM* package (Yee 2010). However, in order to illustrate **R** scripting, we develop our own Pareto distribution functions using the *function* command.[1] While we could proceed by constructing a Pareto random variable X_{Par} from an exponential random variable X_{exp} according to $X_{Par} = be^{X_{exp}}$, we choose here to construct X_{Par} from the density in Table 4.1, $f_{Par}(x) = \alpha b^{\alpha}/x^{\alpha+1}$ under the constraints $\alpha > 0$, $b > 0$, and $b < x < \infty$. The Pareto c.d.f. is $F(x) = 1 - (b/x)^{\alpha}$, and the quartile function is obtained by inverting the p.d.f. The simulation of Pareto random deviates is based on the fact that whenever a distribution's quantile function is applied to a uniform random variable, the result is a random variable with the desired distribution. Note how **R** allows these constraints to be applied within the algebraic computation without use of if-statements. The default values like $\alpha = 1.35$ and $b = 1$ below are normally overridden by the user, but are used when the function is called without parameters specified.

We plot a simulated dataset with 500 data points made with these **R** functions in Figure 4.1 assuming a power-law slope of $\alpha = 1.35$ similar to a Salpeter stellar initial mass function. Note that this is a simple Pareto distribution and does not turn over below ~ 0.3 M$_\odot$ like a real stellar IMF. The first panel shows a dot-chart showing the individual values. The simulation is unit-free but, for astronomical convenience, we can assume units of solar masses (M$_\odot$). The second panel plots the log of the p.d.f. and c.d.f. against the log of the stellar masses. The p.d.f. shows the expected linear relationship. Recall that **R**'s *log* function gives a natural logarithm, and we use the *log10* function to get decimal logarithms. The third panel shows the quantile function; 10% of the stars have masses below ~ 1.2 M$_\odot$, 50% have $M < 2$ M$_\odot$, and 90% have $M < 9$ M$_\odot$. In addition to the usual *plot* and *lines* functions, we use the *legend* function to identify multiple curves on a single panel.

```
# Define Pareto distributions for Salpeter IMF

dpareto <- function(x, alpha=1.35, b=1) (alpha>0)*(b>0)^alpha / x^(alpha+1)
ppareto <- function(x, alpha=1.35, b=1) (x>b)*(1-((b>0)/x)^(alpha>0))
qpareto <- function(u, alpha=1.35, b=1) (b>0)/(1-u)^(1/(alpha>0))   # 0<u<1
rpareto <- function(n, alpha=1.35, b=1) qpareto(runif(n),alpha,b)


# Produce Pareto-distributed simulated dataset with n=500; plot

par(mfrow=c(1,3))
z <- rpareto(500)
```

---

[1] We are grateful to Prof. David Hunter (Penn State) who developed the ideas and **R** scripts in this section for the Summer School in Statistics for Astronomers.

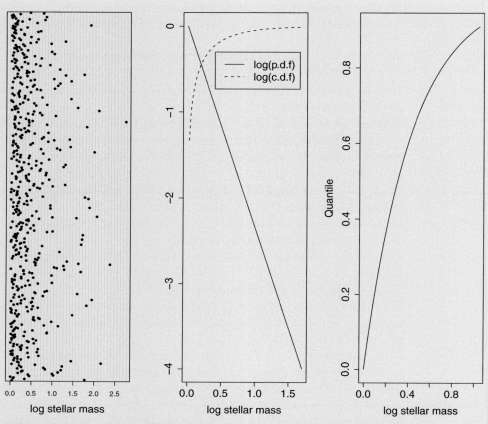

Fig. 4.1   A simulation of the initial mass function of 500 stars assuming a power-law (Pareto) distribution with $\alpha = 1.35$. Panels show: (left) individual values; (center) logged p.d.f. and c.d.f.; and (right) quantile function.

```
dotchart(log10(z), main='', xlab='log stellar mass', pch=20, cex.lab=1.5)

x <- seq(1, 50, len=1000)
plot(log10(x), log10(dpareto(x)), type="l", ylab='', xlab='log stellar mass',
   cex.lab=1.5, cex.axis=1.5)
lines(log10(x), log10(ppareto(x)), type="l", lty=2)
legend(0.4, -0.3, legend=c('log(p.d.f)', 'log(c.d.f)'), lty=c(1,2), cex=1.5)

u=seq(.0005, .9095, len=1000)
plot(log10(qpareto(u)), u, type="l", xlab='log stellar mass', ylab='Quantile',
   cex.lab=1.5, cex.axis=1.5)
par(mfrow=c(1,1))
```

We now proceed to compare six estimators of the power-law slope: the MLE defined by Equation (4.35); the MVUE defined by Equation (4.36); the method of moments; a least-squares fit to the e.d.f., $1 - F(x) = -\alpha \log x$; a least-squares fit to the histogram; and a least-squares fit to the histogram weighted by the counts in each bin. A histogram with 20 evenly spaced bins is arbitrarily chosen. The *function* function defines the different estimators; note the repeated use of the linear modeling function *lm* plotted using *abline*. *lm* stores the slope in its output at location "coef[2]", and *hist* stores the histogram values in location "counts"; this information is provided in the **R** help files for the *lm* and *hist* functions. The results are concatenated using the function *c* into a matrix called "six_alphas". The *return* function permits a complicated output from a function.

```
# Estimates of Pareto shape parameter (power-law slope)

npts <- 100
alpha.MLE <- function(x)   1 / mean(log(x))
alpha.MVUE <- function(x)  (1-(2/npts)) / mean(log(x))
alpha.mom <- function(x)  1 / (1-1/mean(x))
alpha.LS.EDF <- function(x) {
   lsd <- log(sort(x)) ; lseq <- log((npts:1) / npts)
   tmp <- lm(lseq~lsd)
#  plot(lsd, lseq) ; abline(tmp,col=2)
   -tmp$coef[2] }
alpha.LS.hist <- function(x) {
   hx <- hist(x, nclass=20, plot=F)
   counts <- hx$counts
   ldens <- log(hx$density[counts>0])
   lmidpts <- log(hx$mids[counts>0])
   tmp1 <- lm(ldens~lmidpts)
#  plot(lmidpts, ldens) ; abline(tmp1, col=2)
   alpha.LS.hist <-  -1 - as.numeric(tmp1$coef[2])
   return(alpha.LS.hist) }
alpha.LS.hist.wt <- function(x) {
   hx <- hist(x, nclass=20, plot=F)
   counts <- hx$counts
   ldens <- log(hx$density[counts>0])
   lmidpts <- log(hx$mids[counts>0])
   tmp2 <- lm(ldens~lmidpts, weights=counts[counts>0])
   alpha.LS.hist.wt <-  -1 - as.numeric(tmp2$coef[2])
   return(alpha.LS.hist.wt)  }
six_alphas <-  function(x) {
   out <- c(alpha.MLE(x), alpha.MVUE(x), alpha.mom(x),
         alpha.LS.EDF(x), alpha.LS.hist(x), alpha.LS.hist.wt(x))
   return(out)  }
```

Table 4.2 Performance of power-law slope estimators from simulated samples ($n = 100$, $\alpha = 1.35$)						
	MLE	**MVUE**	**Moments**	**LS [EDF]**	**LS [Hist]**	**LS [Hist wt]**
Bias	−0.331	−0.351	−0.152	−0.398	−1.158	−0.801
Variance	0.011	0.010	0.011	0.018	0.018	0.053
MISE	0.120	0.134	0.034	0.177	1.358	0.695

The simulations then proceed with the **R** script below using the generator of random deviates we created above, *rpareto*. The six estimators for the simulations are collected into a matrix "alpha_sim" using the row-binding function *rbind*. From the ensemble of simulations we compute the bias, variance and mean integrated square error (MISE) using formulae provided in Section 6.4.1. Note the efficient production of a summary matrix by consecutive use of **R**'s *apply* and *rbind* functions. The *apply* and related tools apply a simple operation (such as *mean*) on the rows or columns of a matrix, dataframe or table. The MISE serves as the summary statistic for the accuracy of an estimator. We have run this simulator for a variety of sample sizes (npts) and power-law slopes (α) in a fashion similar to previous studies (Crawford *et al.* 1970, Goldstein *et al.* 2004, Newman 2005, Bauke 2007).

The results of this simulation are given in Table 4.2. Surprisingly, the method of moments has the best performance due to its low bias. For all other estimators, the bias dominates the variance. The MLE and MVUE estimators also perform well. In all cases, the least-squares fits to the binned histogram have average MISE values ~10 times larger than the MLE and MVUE estimators; weighting reduced this to a factor of ~5. We thus confirm the studies cited that least-squares fits to binned distribution functions perform very poorly for small samples following a Pareto distribution.

```
# Comparison of alpha estimators on simulated datasets
# Construct nsim simulated datasets with npts data points and alpha=1.35

  nsim=500 ; alpha_sim = NULL
  for(i in 1:nsim) {
      xtmp = rpareto(npts)
      alpha_sim = rbind(alpha_sim,six_alphas(xtmp))  }
      colnames(alpha_sim)=c('MLE','MVUE','Moments','LS_EDF','LS_hist',
        'LS_hist_wt')

# Compute mean integrated square error

  bias_sim = apply(alpha_sim,2,mean) - 1.35
  var_sim = apply(alpha_sim,2,var) * (nsim-1)/nsim
  mise_sim = bias_sim^2 + var_sim
  rbind(bias_sim, var_sim, mise_sim)
```

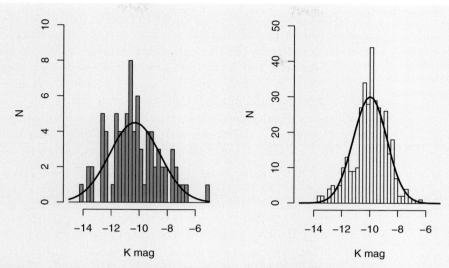

Fig. 4.2 Distribution of K-band absolute magnitudes of globular clusters in the Milky Way Galaxy and M 31 (Andromeda) Galaxy.

4.7.2 Fitting distributions to data

A maximum likelihood regression of a univariate dataset to some probability distribution functions is provided by the function *fitdistr* which resides in the *MASS* library from the monograph of Venables & Ripley (2002). The function treats the beta, Cauchy, χ^2, exponential, F, gamma, geometric, lognormal, logistic, negative binomial, normal, Poisson, Student's t and Weibull distributions. The normal, lognormal, exponential and Poisson parameters are calculated using analytic formulae, while the others are computed by optimization of the loglikelihood. **R**'s general-purpose *optim* program uses a simplex method, but other methods (variable metric Newton, conjugate gradient, simulated annealing, and golden section search) can be chosen. Probability distribution functions, particularly for multivariate datasets, can be fitted using **R**'s more general regression methods such as *lm* and *nlme* (Chapter 7).

We consider here the near-infrared K-band distribution of globular cluster luminosities in the Milky Way Galaxy and Messier 31, our neighboring Andromeda Galaxy. The question we address here is whether the distributions are consistent with normality. (Since the variable is an astronomical magnitude which is a logged flux, we are really testing for lognormality in flux.) We convert the M 31 photometry to absolute magnitudes using the distance modulus to M 31 ($DM = m - M = 24.44$) to place the values on the same scale as the Milky Way data (see Section 5.9.2). We first visualize the data in the form of histograms, and fit them using the *fitdistr* function (Figure 4.2). The fits show that the mean values are compatible ($\bar{K} = -10.32 \pm 0.20$ for the Milky Way and $\bar{K} = -9.98 \pm 0.06$ for M 31) but the M 31 distribution is substantially narrower than the Milky Way distribution ($s.d. = 1.79 \pm 0.14$ for the Milky Way and $s.d. = 1.19 \pm 0.04$ for M 31).

We proceed with a variety of **tests for normality**. The Milky Way sample is fully consistent with a normal ($P \sim 0.5$ for all tests) but the M 31 sample is inconsistent with a normal. Three tests in the **CRAN** package *nortest* (Gross 2010) discussed in Section 5.3.1 are based on the e.d.f.: the Lilliefors test, a specialized version of the Kolmogorov–Smirnov test ($P \sim 0.001$); the Cramér–von Mises test ($P \sim 0.0004$); and the Anderson–Darling test ($P \sim 0.0001$). Note that the Anderson–Darling test is much more sensitive than the Kolmogorov–Smirnov test. The *pearson.test*, implementing the binned χ^2 familiar to astronomers, gives $P \sim 0.006$; note again the loss of information and sensitivity due to binning. The default binning scheme is based on the rule-of-thumb $2n^{2/5} = 22$ bins for $n = 360$ globular clusters.

The M 31 distribution's deviation from normality might arise from large-scale asymmetries; for example, a loss of sensitivity at faint magnitudes (the right-hand side of the distribution) might produce a skewness in the observed sample that is not present in the true population. **CRAN**'s package *moments* provides tests for normality based on global moments of the distribution (Komsta & Novamestky 2007). The Agostino and Jarque–Bera normality tests based on skewness give $P \sim 0.05$ and $P \sim 0.004$, respectively. The Anscombe–Glynn and Bonnett–Seier tests for normality based on kurtosis show a less significant effect ($P \sim 0.2$ and 0.02).

However, examination of the histogram in Figure 4.2(b) suggests the effect may be due to a dip in the distribution around $K \sim -11$. We run the Shapiro–Wilks and Shapiro–Francia tests for normality that are based on small-scale correlations in the data; these give $P \sim 0.001$. Hartigan's **test for multimodality** from **CRAN**'s *diptest* package (Maechler 2010), discussed in Section 9.5, gives $P = 0.01$. We conclude that both large-scale skewness, possibly due to observational selection effects, and a small-scale dip around $K \sim -11$ contribute to the deviation of the M 31 distribution from normality (see also the discussion in Section 5.9.2).

```
# Read Milky Way Galaxy and M 31 globular cluster K magnitudes

GC_MWG <- read.table('http://astrostatistics.psu.edu/MSMA/datasets/
GlobClus_MWG.dat',header=T)
GC_M31 <- read.table('http://astrostatistics.psu.edu/MSMA/datasets/
GlobClus_M31.dat',header=T)
KGC_MWG <- GC_MWG[,2];KGC_M31 <- GC_M31[,2]-24.44
kseq <- seq(-15.0, -5.0, 0.25)

# Fit normal distributions

library(MASS)
par(mfrow=c(1,2))
hist(KGC_MWG, breaks=kseq, ylim=c(0,10), main='', xlab='K mag',
  ylab='N', col=gray(0.5))
normfit_MWG <- fitdistr(KGC_MWG,'normal') ; normfit_MWG
```

```
lines(kseq, dnorm(kseq, mean=normfit_MWG$estimate[[1]],
   sd=normfit_MWG$estimate[[2]]) * normfit_MWG$n*0.25,lwd=2)
hist(KGC_M31, breaks=kseq, ylim=c(0,50), main='', xlab='K mag', ylab='N',)
normfit_M31 <- fitdistr(KGC_M31,'normal') ; normfit_M31
lines(kseq,dnorm(kseq, mean=normfit_M31$estimate[[1]],
   sd=normfit_M31$estimate[[2]]) * normfit_M31$n*0.25, lwd=2)

# Test for normality

install.packages('nortest') ; library(nortest)
install.packages('moments') ; library(moments)
lillie.test(KGC_MWG) ; cvm.test(KGC_MWG) ; ad.test(KGC_MWG)

lillie.test(KGC_M31) ; cvm.test(KGC_M31) ; ad.test(KGC_M31)
pearson.test(KGC_M31)
skewness(KGC_M31) ; agostino.test(KGC_M31) ; jarque.test(KGC_M31)
kurtosis(KGC_M31) ; anscombe.test(KGC_M31) ; bonett.test(KGC_M31)
shapiro.test(KGC_M31) ; sf.test (KGC_M31)

# Test for multimodality

library(diptest)
dip(KGC_M31)
```

As discussed in Sections 4.4 and 7.4, statisticians do not recommend binning the data and using a χ^2 test. The choice of binning parameters is not guided by mathematical criteria, information within bins is lost, and parameters relying on the tails of the distribution are often inaccurate. It is also unclear what to use as weights in the least-squares summation of the χ^2 test: the expected value from the model distribution evaluated at bin centers, $\chi^2 = \sum (O_i - E_i)^2 / E_i$; or the measurement uncertainty of the data entering into each bin, $\chi^2 = \sum (O_i - E_i)^2 / \sigma_i$ where $\sigma_i \simeq \sqrt{n_i}$.

4.7.3 Scope of distributions in **R** and **CRAN**

The coverage of distributions is summarized in the updated **CRAN** "views" Web page http://cran.r-project.org/web/views/Distributions.html.

The **VGAM** package, for Vector Generalized Additive Models, has over 80 built-in distributions giving densities, quantiles and random deviates. Some of these are relevant to specialized areas of astronomy: bivariate gamma, log gamma, Dirichlet, doubly censored normal, generalized Pareto, Poisson-lognormal, Rayleigh, Rice, Skellam, von Mises, and zero-avoiding binomial and Poisson distributions. Another gives the distribution of distances from a fixed point to a Poisson spatial distribution in two and three dimensions. **VGAM** then provides a large suite of regression techniques including MLE, quantile regression, nonlinear regression, weighted regressions and treatment of categorical variables. A similar, though less extensive, package is *gamlss.dist* which provides generalized additive models for location, scale and shape.

Other **CRAN** packages provide distributions absent from base **R**. *SuppDists* gives distributions for Pearson's correlation coefficient, Kendall's τ, Spearman's ρ and Kruskal–Wallis tests. The variance gamma function, the weighted difference of two independent and shifted gamma variables, is treated in **CRAN**'s *VarianceGamma* package. The Skellam distribution for the difference of two Poisson variables is calculated in the *skellam* package. Multivariate normal and Student's t distributions are given in *mvtnorm* and other **CRAN** packages.

Truncated datasets in astronomy, described in Section 10.5, can be modeled with truncated distributions. A number of individual truncated probability distributions are provided by *VGAM*, and truncated multivariate normal distributions are treated in *tmvtnorm*. Generally useful **R** functions to truncate any p.d.f., c.d.f, quantile or random number distribution are provided by Nadarajah & Kotz (2006).

The package *Runuran*, or universal random deviate generator, provides efficient random numbers for a wide range of probability distributions based on the methods described by Hörmann *et al.* (2004). These include several dozen well-established probability distributions as well as user-supplied functions for discrete or continuous, univariate or multivariate distributions. This can be useful for a variety of Monte Carlo problems in astronomy.

New distributions, like any computation, can be created in **R** using the general *function* function as illustrated above for the Pareto distribution. However, the machinery associated with probability distribution functions – such as p.d.f., c.d.f., quantile, random numbers and moments – can be inherited if the new distribution is inserted into **CRAN**'s *distr* package. While the procedure is not simple, it may be worthwhile for important distributions that emerge in astronomy. An example might be the Navarro *et al.* (1997, NFW) "universal" density profile of dark matter galaxy halos found in N-body simulations of large-scale structure growth,

$$f_{NFW}(x) = \frac{a}{(x/b)(1 + x/b)^2}. \tag{4.52}$$

The NFW distribution has been discussed in over 2000 astrophysical papers on galaxy formation and structure.

5 Nonparametric statistics

5.1 The astronomical context

Our astronomical knowledge of planets, stars, the interstellar medium, galaxies or accretion phenomena is usually limited to a few observables that give limited information about the underlying conditions. Our astrophysical understanding usually involves primitive models of complex processes operating on complex distributions of atoms. In light of these difficulties intrinsic to astronomy, there is often little basis for assuming particular statistical distributions of and relationships between the observed variables. For example, astronomers frequently take the log of an observed variable to reduce broad ranges and to remove physical units, and then assume with little justification that their residuals around some distribution are normally distributed. Few astrophysical theories can predict whether the scatter in observable quantities is Gaussian in linear, logarithmic or other transformation of the variable.

Astronomers may commonly use a simple heuristic model in situations where there is little astrophysical foundation, or where the underlying phenomena are undoubtedly far more complex. Linear or loglinear (i.e. power-law) fits are often used to quantify relationships between observables in starburst galaxies, molecular clouds or gamma-ray bursts where the statistical model has little basis in astrophysical theory. In such cases, the mathematical assumptions of the statistical procedures are often not established, and the choice of a simplistic model may obfuscate interesting characteristics of the data.

Feigelson (2007) reviews one of these situations where hundreds of studies over seven decades modeled the radial starlight profiles of elliptical galaxies using simple parametric models, see king to understand the distribution of mass within the galaxies. A dataset of this type is analyzed in Section 7.10.5. Little consensus emerged on the correct model family; radial profiles were named for Hubble, de Vaucouleurs, King, Hernquist, Sérsic, and Navarro, Frenk & White. Astrophysical understanding gradually emerged that luminous elliptical galaxies are complicated triaxial structures dominated by dark matter formed by sequential galaxy mergers. No simple functional form for the radial profile is expected from this origin. Astrostatisticians Wang *et al.* (2005) show how modern nonparametric techniques may best uncover the underlying mass distribution of such galaxies without the simplifying, and physically unrealistic, assumptions of the analytical formulae.

Another common situation occurs when the information about an object is in the form of unordered categories rather than real-valued random variables. Quasi-stellar objects, Seyfert galaxies, radio galaxies, BL Lac objects, LINERS and ULIRGs are a nonordered classification of galaxies with some form of active nuclei. SA(r) and SBb(s) are classes of galaxy morphologies. Molecular cloud clumps in the Taurus, Perseus and Orion cloud complexes can be compared. Many astronomical findings are summarized with simpler binary variables of the form "A" and "not-A": young stars with and without dusty proto-planetary disks; intergalactic gas clouds with and without Mg II absorption lines; and so forth. Statistical methods for these problems often fall under the rubric of nonparametric methods because the distributions of objects within these categories is usually unknown.

Astronomers thus need nonparametric statistical tools that make no assumptions regarding the underlying probability distribution. Nonparametric methods treat deficiencies with models as well as problems with the data. An example of such a nonparametric procedure that is familiar to astronomers is the Kolmogorov–Smirnov (KS) two-sample test. Nonparametric approaches to data analysis are thus very attractive and, with a full suite of methods, can provide considerable capability for the analysis and interpretation of astronomical data.

This chapter also presents some results that do not involve the theory of nonparametric inference, but are nonetheless valuable for data analysis. These include the quantile function, and robust estimates of location and spread (Sections 2.6 and 5.3.3).

5.2   Concepts of nonparametric inference

Nonparametric statistics fall into two categories: (a) procedures that do not involve or depend on parametric assumptions, though the underlying population distribution may belong to a particular parametric family; and (b) methods that do not require that the data belong to a particular parametric family of distributions. Distribution procedures such as the KS test, rank statistics, the sign and Wilcoxon signed rank tests are in the first category. Contingency tables and the variety of density estimation methods discussed in Chapter 6 (histograms, kernel smoothing, nearest neighbor and nonparametric regressions) are in the second category. In the latter group of cases, the structure of the relationship between variables is treated nonparametrically, while there may be parametric assumptions about the distribution of model residuals. The term **semi-parametric** is sometimes used for procedures which combine parametric modeling with principles of nonparametrics.

The term nonparametric is not meant to imply that its modeling procedures completely lack parameters. Many nonparametric procedures are **robust**, or insensitive to slight deviations from model assumptions. A quantitative measure of robustness is the **breakdown point**, or the contaminated fraction of the dataset which destroys the effectiveness of a procedure.

Some nonparametric procedures are analogous to parametric procedures but operate on the ranks, or numbered position in the ordered sequence of data points, rather than the measured values of the data points. Rank tests are often the most powerful available for classificatory variables which are ordered but not with meaningful numerical values: Seyfert 1, 1.5, 1.9 and 2 galaxies; or Class 0, I, II and III evolutionary phases of pre-main-sequence stars. Even for continuous variables with normal distributions, the efficiency of rank methods can nearly match that of parametric methods. Note, however, that ranks cannot be reliably defined for multivariate datasets.

Bayesian nonparametrics can be viewed as an oxymoron as Bayesian inference requires a mathematical probability model for the data in terms of well-defined parameters. Nonetheless, Bayesian methodology is actively applied to problems addressed by nonparametric methods (Müller & Quintana 2004).

Nonparametric statistics has been a burgeoning approach in modern statistics with tendrils in many subfields. In this volume, nonparametric methods are a central theme in Chapter 6 on smoothing and density estimation, Section 3.6.2 on bootstrap and related resampling methods, Chapter 10 on methods for censored and truncated data, and Chapter 12 on spatial processes. This chapter should thus be viewed only as a first step in nonparametric statistics for astronomy.

While the boxplot is rarely used in astronomy, the scatterplot is a well-known and useful visualization tool. Usually restricted to two variables, the data are plotted as unconnected points on a diagram. Sometimes additional categorical information, such as a sample identifier or a stratified value of a third variable, modifies the color, size or shape of the plotting symbol. For multivariate problems, in which the covariance matrix summarizes linear relationships between the variables, a matrix of small scatterplots is often very effective in visualizing patterns and locating outliers. Examples of scatterplot matrices are shown in Chapter 8.

5.3 Univariate problems

5.3.1 Kolmogorov–Smirnov and other e.d.f. tests

The **empirical distribution function (e.d.f.)** is the simplest and most direct nonparametric estimator of the underlying **cumulative distribution function (c.d.f.)** for the underlying population. The univariate dataset $X_1, X_2, \ldots, X_n$ is assumed to be drawn as independently and identically distributed (i.i.d.) samples from a common distribution function F. The e.d.f. $\widehat{F}_n$ is defined to be

$$\widehat{F}_n(x) = \frac{1}{n} \sum_{i=1}^{n} I[X_i \leq x], \tag{5.1}$$

for all real numbers x. The e.d.f. thus ranges from 0.0 to 1.0 with step heights of $1/n$ located at the values X_i. This produces steps with heights that are multiples of $1/n$. For each x, $F_n(x)$ follows the binomial distribution which is asymptotically normal. The e.d.f.'s of two univariate astronomical datasets are shown in Figure 5.3.

The mean and variance of $\widehat{F}_n(x)$ are

$$E[\widehat{F}_n(x)] = F(x)$$
$$Var[\widehat{F}_n(x)] = \frac{F(x)[1 - F(x)]}{n}. \tag{5.2}$$

For a chosen significance level α (say, 0.05), the $100(1 - \alpha)\%$ asymptotic confidence interval for $\widehat{F}_n(x)$ is then

$$\widehat{F}_n(x) \pm z_{1-\alpha/2}\sqrt{\widehat{F}_n(x)[1 - \widehat{F}_n(x)]/n} \qquad (5.3)$$

where z are the quantiles of the Gaussian distribution.

Astronomers sometimes worry about the interdependency of the e.d.f. value at some x on data points at lower values of x, feeling it is safer to estimate the differential value of the distribution function using a histogram. However, histograms have a number of difficulties — arbitrary choice of bin widths and origin, loss of information within the bin — as explained in Chapter 6. On the other hand, well-established theory shows that the e.d.f. is an unbiased and consistent estimator of the population distribution function, and is the generalized maximum likelihood estimator of the population c.d.f. Thus, the e.d.f. $\widehat{F}_n$ uniquely, completely and accurately embodies all information in the measured X_i without any ancillary choices.

Several statistics have been developed to assist inference on the consistency of an observed e.d.f. with a model specified in advance. We want to test the null hypothesis that $F(x) = F_0(x)$ for all x, against the alternative that $F(x) \neq F_0(x)$ for some x, where F_0 is a distribution specified independently of the dataset under study. The test most well-known in astronomy is the one-sample **Kolmogorov–Smirnov (KS) test** that measures the maximum distance between the e.d.f and the model,

$$M_{KS} = \sqrt{n}\max_x |\widehat{F}_n(x) - F_0(x)|. \qquad (5.4)$$

A large value of the supremum M_{KS} allows rejection of the null hypothesis that $F = F_0$ at a chosen significance level. The distribution of M_{KS} is independent of the shape of F as long as it is a continuous function. For large n and a chosen significance level α (e.g. $\alpha = 0.05$), the cumulative distribution of the one-sample KS statistic is approximately

$$P_{KS}(M_{KS} > x) \simeq 2\sum_{r=1}^{\infty}(-1)^{r-1}e^{-2r^2x^2} \qquad (5.5)$$

with critical value

$$M_{KS}^{crit} > \left(-\frac{1}{2}\ln\left(\frac{\alpha}{2}\right)\right)^{1/2}. \qquad (5.6)$$

These results are based on advanced weak convergence theory (Billingsley 1999). For small samples, tables of M_{KS}^{crit} or bootstrap simulations must be used.

A common use of the KS one-sample test is to evaluate the goodness-of-fit of a model derived using some procedure. The statistic does indeed provide a measure of goodness-of-fit or model validation. However, it is crucial to recognize significant limitations of the KS test.

1. The KS test is not distribution-free — so the widely tabulated critical values of the KS statistic are not valid — if the model parameters were estimated from the same dataset being tested. The critical values are only correct if the model parameters (except for normalization which is removed in the construction of the dataset e.d.f. and model c.d.f.) are known in advance of the dataset under consideration (Lilliefors 1969). For example,

if one compares the distribution to a normal distribution, the mean and variance must have been specified in advance, perhaps from a similar dataset or from astrophysical theory.

2. The distribution of the KS statistic is also not distribution-free when the dataset has two or more dimensions. The reason is that a unique ordering of points needed to construct the e.d.f. cannot be defined in multivariate space (Simpson 1951). A two-dimensional KS-type statistic can be constructed and has been used fairly often in astronomy (Peacock 1983, Press *et al.* 2007). But the distribution of this statistic is not knowable in advance and is not distribution-free; probabilities would have to be calculated for each situation using bootstrap or similar resampling techniques.

3. The KS test is sensitive to global differences between two e.d.f.'s or one e.d.f. $\widehat{F}_n$ and the model c.d.f. F_0 producing different mean values. But the test is less efficient in uncovering small-scale differences near the tails of the distribution.

The Cramér–von Mises (CvM) statistic, T_{CvM}, measures the sum of the squared differences between $\widehat{F}_n$ and F_0,

$$
\begin{aligned}
T_{CvM,n} &= n \int_{-\infty}^{\infty} [\widehat{F}_n(x) - F_0(x)]^2 dF_o(x) \\
&= \frac{1}{12n} + \sum_{i=1}^{n} \left(\frac{2i-1}{2n} - F(X_{(i)}) \right)^2
\end{aligned}
\tag{5.7}
$$

It captures both global and local differences between the data and model, and thus often performs better than the KS test. Two-sample KS and CvM tests are also commonly used, where the e.d.f. of one sample is compared to the e.d.f. of another sample rather than the c.d.f of a model distribution where $X_{(i)}$ is the i-th entry when $X_1, X_2, \ldots, X_n$ are placed in increasing order.

But here again a limitation is seen: by construction, the cumulative e.d.f. and model converge at zero at low x values and at unity at high x values, so that differences between the distributions are squeezed near the ends of the distributions. Consistent sensitivity across the full range of x is provided by the **Anderson–Darling (AD) statistic** A_{AD}^2, a weighted variant of the CvM statistic:

$$
A_{AD,n}^2 = n \sum_{i=1}^{n} \frac{[i/n - F_0(X_i)]^2}{F_0(X_i)(1 - F_0(X_i))}.
\tag{5.8}
$$

Stephens (1974) has found that the Anderson–Darling test is more effective than other e.d.f. tests, and is particularly better than the Kolmogorov–Smirnov test, under many circumstances. Astronomers Hou *et al.* (2009), comparing the χ^2, KS and AD tests in a study of galaxy group dynamics, also validated that the AD test has the best performance, and we obtain a similar result in Section 4.7.2. While the Anderson–Darling test was historically used to test a dataset for normality, there is no reason why it cannot compare a dataset with any parametric model. Tables are available for AD critical values for different sample sizes n, or they can be obtained from bootstrap resampling. The limiting distribution for large n has a complicated expression. Like the KS and CvM tests, the AD test distribution is independent of the distribution of X_i provided the distribution of X_i is continuous.

5.3.2 Robust statistics of location

Concepts of **robustness** were known to astronomers Simon Newcomb and Arthur Stanley Eddington a century ago; for example, Eddington debated with R. A. Fisher on the relative merits of the MAD and standard deviation to estimate scatter (Huber & Ronchetti 2009). But a coherent mathematical theory for robust statistics did not emerge until the 1960s and 1970s.

There is a wide consensus that the **median** (*Med*), or central value, of a dataset constitutes a reliable measure of location for a univariate dataset. The **mean**, which historically has dominated location estimates, is not robust to outliers. The median of a univariate dataset $(X_1, X_2, \ldots, X_n)$ is the m-th data value where half the data is above and half the data is below. It is easiest to obtain when the data are sorted into increasing order, $x_{(1)} \leq x_{(2)}, \ldots, x_{(n)}$. When n is odd,

$$Med = \begin{cases} X_{(n-1)/2} & \text{when } n \text{ is odd} \\ \frac{X_{n/2} + X_{n/2+1}}{2} & \text{when } n \text{ is even.} \end{cases} \tag{5.9}$$

Confidence intervals around the median, and variants of the median and mean, are discussed in Section 5.3.3.

The median is a very robust measure of central location with breakdown at 50%; that is, its value is unaffected by outliers unless they constitute over half of the dataset. The mean, in contrast, has a breakdown at a single observation. The median is particularly efficient when the population has heavier tails than a normal distribution.

Despite their effectiveness and ease of calculation, the astronomer must recognize that both the mean and median values may be scientifically meaningless if the dataset is heterogeneous with physically different populations in the same dataset. In cases where the populations are partially or fully distinct, a mixture model is often used to derive different medians for distinct subsamples. This topic will be covered in Section 9.5 when we consider classification in multivariate analysis.

The mean and median can be viewed as limiting cases of a continuum of **trimmed means**,

$$\bar{x}_{trim}(m) = \frac{1}{n - 2m} \sum_{i=m+1}^{n-m} x_{(i)}. \tag{5.10}$$

A choice of $m = 0$ gives the untrimmed mean, and values $0 < m < 0.5n$ trim a range of values at both ends of the distribution. Here, the scientist chooses to eliminate a fraction of the data points symmetrically from the two ends of the ordered dataset, and uses the mean of the central points. This procedure shields the estimator from undue influence of outliers. A related estimate of central location, advocated by Tukey (1977) as a central tool of exploratory data analysis, is the **trimean (TM)**

$$\bar{x}_{TM} = \frac{x_{0.25n} + 2Med + x_{0.75n}}{4}, \tag{5.11}$$

a weighted mean of the 25%, 50% and 75% quartiles. Unfortunately, the distribution of trimmed means and the trimean are not established for arbitrary parent distributions, and thus their variances are not defined. However, the trimmed mean using the central three points is recommended for very small datasets, $n \leq 6$, where the large random deviations of the median due to the poor sampling of the parent distribution is stabilized by averaging with the neighboring points.

5.3.3   Robust statistics of spread

Just as the median and trimmed means are robust measures of location in a univariate dataset, we can also define measures of dispersion (also called spread, scale or variation) with different degrees of robustness. Several nonparametric measures of dispersion are more robust than the standard deviation (s.d.) about the mean (and its square, the variance) which is based on the normal distribution.

The **interquartile range (IQR)**, or distance between the 25% and 75% quantiles, is commonly used in many fields, though rarely in astronomy. The IQR has a breakdown point of 25% and can be scaled to the standard deviation of a normal distribution as $IQR \simeq 1.35 \, s.d.$

We recommend instead that astronomers use the **median absolute deviation (MAD)** as a preferred measure of dispersion about the median where

$$MAD = Med|X_i - Med|. \qquad (5.12)$$

That is, the MAD is the median of the absolute deviations of the data points about the median. The **normalized MAD (MADN)** scales the MAD to the standard deviation of a normal distribution,

$$MADN = Med|X_i - Med|/0.6745 = 1.4826 \times Med|X_i - Med|. \qquad (5.13)$$

Stated more formally, a random variable with a normal distribution $N(\mu, \sigma^2)$ has $MADN = \sigma$.

The MAD is an extremely stable estimator of dispersion and breakdown at 50%, but is not a very efficient (or precise) estimator. In particular, $MAD/\sqrt{N}$ is not a good estimate of the confidence interval of the median whereas $S.D./\sqrt{N}$ is a good estimate of the confidence interval of the mean. The MAD is asymptotically equal to half of the interquartile range for symmetric distributions.

In addition to nonparametric measures of location and dispersion discussed here, a significant field of statistical theory has been developed for a class of robust statistics known as **M-estimators**. M-estimates of location, for example, are weighted means where outlying points are downweighted according to carefully chosen functions such as **Huber's** ψ **function** or the bisquare weight function. Likelihoods can be constructed for these statistics, permitting formal evaluation of their confidence limits, influence function, breakdown and asymptotic normality (Maronna & Martin 2006). We discuss later some M-estimators in the context of regression (Section 7.3.4).

5.4   Hypothesis testing

If we think we know the central location of a distribution, we can apply a nonparametric hypothesis test to evaluate whether this is true. For example, we can check whether the median of a collection of Hubble constant measurements based on Cepheid variable distances in nearby galaxies is consistent with the $H_0 = 73$ km s^{-1} Mpc^{-1} obtained from modeling fluctuations in the Cosmic Microwave Background. Or we can check nonparametrically the validity of a parametric model by testing whether the residuals are centered around zero. This is a very powerful way that nonparametrics can assist in parametric modeling.

5.4.1 Sign test

A simple but important nonparametric test is the **sign test**. Consider the hypothesis that the median of a sequence of random variables $X_1, X_2, \ldots, X_n$ has a specified value d_0; that is, $H_0 : d = d_0$ against the alternative $H_1 : d \neq d_0$. The sign statistic is

$$
\begin{aligned}
S(d_0) &= \sum_{i=1}^{n} \mathrm{sgn}(X_i - d_0) \\
&= \#[X_i > d_0] - \#[X_i < d_0] \\
&= 2\#[X_i > d_0] - n.
\end{aligned}
\tag{5.14}
$$

The statistic simply compares the number of points above the assumed value d_0 with the number of points below the assumed value. If d_0 is the median of the population, then the expectation is $S(d_0) = 0$. The one-sided sign statistic $\#(X_i > d_0)$ follows the binomial distribution, $Binom(n, 1/2)$, and is free from the distribution of X_i. Binomial probabilities can be used to get critical values of the statistic. The sign test can also be used to formulate confidence intervals about the median.

5.4.2 Two-sample and k-sample tests

We describe here nonparametric methods available for comparing two or more univariate datasets. They test the null hypothesis that the two populations underlying the two samples are identical.

The **Mann–Whitney–Wilcoxon (MWW) test** (or the closely related **Wilcoxon rank sum test**) starts by merging a sample $X_1^1, \ldots, X_n^1$ with a sample of $X_1^2, \ldots, X_m^2$ values. The combined sample is sorted and ranked from 1 to $n + m$. The test statistic T_{MWW} is the sum of the ranks $R(X_i^1)$ of the first sample,

$$
T_{MWW} = \sum_{i=1}^{n} R(X_i^1).
\tag{5.15}
$$

This statistic is asymptotically normal with mean and variance

$$
\begin{aligned}
E[T_{MWW}] &= \frac{n(n + m + 1)}{2} \\
Var[T_{MWW}] &= \frac{nm(n + m + 1)}{12}.
\end{aligned}
\tag{5.16}
$$

This approximation based on the Central Limit Theorem applies for sufficiently large datasets (n and $m \geq 20$), and no ties with n and m of the same order. The MWW test is unbiased and consistent under the following assumptions: both samples are randomly selected from their respective continuous populations (i.e. X^1 and X^2 are i.i.d.), and they are mutually independent.

The MWW test can be viewed as a nonparametric substitute for the t **test** for the difference between sample means for normal populations. The asymptotic relative efficiency of the MWW test compared to the t test is 0.955 for normal distributions, indicating that it can be effectively used for both normal and nonnormal situations. An adjusted statistic with averaged ranks when many ties are present, and tables of T quantiles for small datasets, are provided by Conover (1999, Section 5.1).

Related to the MWW test is the **Hodges–Lehmann (HL) test** for shift which gives a confidence interval to a shift (or offset) between the populations underlying samples X^1 and X^2. The HL statistic is the median of the $n \times m$ differences between the values in the two samples, and thus gives a nonparametric estimate of the difference between two medians. Here the differences between the measured values in the two samples, not the difference between the ranks, are used.

The extension of the MWW test to $k > 2$ independent samples is called the **Kruskal–Wallis (KW) test**. The asymptotic distribution of this statistic under the null hypotheses that all k population distributions are identical is a χ^2 distribution with $k - 1$ degrees of freedom. Tables for small k and small n_i are available, but the χ^2 approximation is generally accurate or conservative even for small samples. Because ties are permitted in the KW test, it can be fruitfully used to test differences in populations in $r \times c$ contingency tables. The KW test can be viewed as the nonparametric substitute for **ANOVA**, or the **one-way F test** for comparing means of normal populations.

Another important class of two-sample tests is based on the e.d.f and the Kolmogorov–Smirnov statistic. Here, two e.d.f.'s are constructed on the same graph, and the maximum vertical distance between the two functions is recorded. Suppose we have two samples $X_1, \ldots, X_m$ and $Y_1, \ldots, Y_n$ from c.d.f.'s F and G and wish to test the null hypothesis that $F(x) = G(x)$ for all x against the alternative that $F(x) \neq G(x)$ for some x. For the e.d.f.'s $\widehat{F}_m$ and $\widehat{G}_n$ with m and n data points respectively, the **two-sample Kolmogorov–Smirnov (KS) statistic** is

$$M_{KS} = \max_x |\widehat{F}_m(x) - \widehat{G}_n(x)|. \tag{5.17}$$

For large m and n, asymptotic critical values of M_{KS} are available while tables are needed for small samples.

All of the two-sample tests outlined here — MWW, HL, KW and KS — are distribution-free. Astronomers are encouraged to use them in a very broad range of circumstances, even when little is known about the underlying behavior of the variable.

5.5 Contingency tables

Contingency tables present data when the variables are categorical rather than real numbers. The simplest is a 2×2 contingency table where a dataset with $n = a + b + c + d$ objects has

Table 5.1 2×2 contingency table.			
	Class I	Class II	Totals
Yes	a	b	$n_1 = a + b$
No	c	d	$n_2 = c + d$
Totals	c_1	c_2	$n = n_1 + n_2$

binary values "Class I" and "Class II" and "Yes" and "No" in two astronomical variables (Table 5.1).

The astronomer may seek to test hypotheses about the independence between attributes such as "Is the second property more likely to be present in Class II objects or are the properties independent?" The χ^2 **test** is commonly used for this problem with the statistic

$$T_{\chi^2} = \sqrt{\frac{n}{n_1 n_2}} \frac{ad - bc}{\sqrt{c_1 c_2}} + 0.5. \qquad (5.18)$$

An important generalization of the 2×2 contingency table is the $r \times c$ contingency table where the number of rows r (representing either different samples or categorical values of an astronomical property) and columns c (representing categorical values of another astronomical property) both exceed 2. Here the χ^2 test that all of the probabilities in the same column are equal can be tested using the statistic

$$T_{\chi^2, r \times c} = \sum_{i=1}^{r} \sum_{j=1}^{c} \frac{(O_{ij} - E_{ij})^2}{E_{ij}} \qquad (5.19)$$

where O_{ij} are the observed counts in the cells and $E_{ij} = n_i c_j / n$ are the expected counts under H_0. The null distribution of $T_{\chi^2, r \times c}$ for large n is the χ^2 distribution with $(r-1)(c-1)$ degrees of freedom.

This χ^2 test for $r \times c$ contingency tables can be used both when the samples are distinct with row subtotals known in advance, and when the rows and columns are generated from a single sample with subtotals unknown in advance. The meaning of the null hypothesis changes subtly; in the former case, one is testing for differences in probabilities within a column; and in the latter case, one is testing for independence between rows and columns. Users are cautioned against using the test in the small-n regime where any $E_{ij} < 0.5$ or many $E_{ij} < 1$. Categories might be combined to give larger count rates under these conditions.

A $2 \times c$ contingency table is useful to test whether multiple samples are drawn from populations with the same median. Here, the two rows give the number of objects above and below (or equal to) the median, and the $T_{\chi^2, r \times c}$ statistic is applied. Similarly, a $r \times 2$ contingency table can compare the observations of a binary variable (e.g. "Yes" and "No") for multiple samples.

Classical techniques for analyzing contingency tables have been extended in various ways. 2×2 contingency tables can also be generalized into three-way or k-way contingency tables if multiple variables are under study. More complex contingency tables — such as systems with multiple independent binomial response variables and multiple dependent variables with polytomous responses — can be studied with generalized linear models (GLM) outlined in Section 7.3.6 as described by Faraway (2006, Chapters 4 and 5). The statistical family may be binomial, multinomial or Poisson, and one can search for mutual, joint and conditional independence with hierarchical relationships between the variables. Statistical treatments for errors in contingency table classification are presented by Buonaccorsi (2010, Chapter 3).

5.6 Bivariate and multivariate tests

Astronomers often measure two or more properties for each object in a sample, and ask whether significant relationships exist between the properties in the underlying population. Chapter 8 presents classical parametric methods based on multinormal distributions such as Pearson's linear correlation coefficient r, least-squares linear regression and principal component analysis. However, it is often desirable to establish the existence of relationships without assuming linearity and normality, and without dependence on the choice of logarithmic transformations or physical units. Here, a few nonparametric methods are valuable.

For a dataset of paired i.i.d. measurements $(X_1, Y_1), \ldots, (X_n, Y_n)$, **Spearman's ρ_S rank test**, for independence between the two variables, is Pearson's r given in Equation (3.39) but applied to the ranks:

$$\rho_S = \frac{\sum_{i=1}^{n} R(X_i)R(Y_i) - n(n+1)^2/4}{\sqrt{\sum_{i=1}^{n} R(X_i)^2 - n(n+1)^2/4}\sqrt{\sum_{i=1}^{n} R(Y_i)^2 - n(n+1)^2/4}}, \qquad (5.20)$$

where $R(X_i)$ is the rank of X_i among the X values and $R(Y_i)$ is the rank of Y_i among the Y values. Spearman's ρ dates back to 1904. In the absence of ties, this simplifies to

$$\rho_S = 1 - \frac{6\sum_{i=1}^{n}[R(X_i) - R(Y_i)]^2}{n(n^2 - 1)}. \qquad (5.21)$$

For large samples ($n \geq 30$) and independence between X's and Y's, the distribution of ρ_S is Gaussian with mean and variance

$$E[\rho_S] = 0$$
$$Var[\rho_S] = \frac{1}{n-1} \qquad (5.22)$$

where z_p is the corresponding normal quantile. Probability tables for small n are available. The data values can be real numbers, ordinal integers, or nonnumeric categories providing a unique ranking can be performed. There is no requirement that X and Y have similar distributions.

Kendall's τ_K correlation measure produces nearly identical results to Spearman's ρ_S, but is sometimes preferable as its distribution approaches normality faster for small samples.

It is defined in several ways including

$$\tau_K = \frac{N_c - N_d}{N_c + N_d} \tag{5.23}$$

where N_c counts the concordant pairs where $(Y_i - Y_j)/(X_i - X_j) > 0$, and N_d counts discordant pairs where $(Y_i - Y_j)/(X_i - X_j) < 0$. Pairs in X are ignored, and pairs in Y contribute $1/2$ to N_c and $1/2$ to N_d. Under independence of X's and Y's, critical values of τ_K are tabulated for small n. For large n ($n \geq 60$), the statistical distribution is approximately normal with mean and variance

$$E[\tau_K] = 0$$
$$Var[\tau_k] = \frac{\sqrt{2(2n+5)}}{3\sqrt{n(n-1)}}. \tag{5.24}$$

5.7 Remarks

Astronomers commonly use a few well-known nonparametric techniques such as the Kolmogorov–Smirnov two-sample and goodness-of-fit tests and Kendall's τ measure of bivariate correlation. We believe that use of nonparametrics should be considerably expanded with a greater repertoire of methods and more frequent use; they are excellent components of the astronomers' analytic toolbox. Nonparametric approaches are philosophically attractive to astronomy because, except in situations where parametric astrophysical theory is well-defined and clearly applicable, parametric distributions of astrophysical quantities can rarely be convincingly established. Nonparametric methods can be used in an exploratory fashion, establishing with few assumptions whether or not an effect is present before parametric modeling is pursued. These procedures are also robust against non-Gaussian noise and outliers, whether they arise from data errors or relationships with hidden variables.

Many familiar parametric methods have nonparametric counterparts. The empirical distribution function is a valuable nonparametric estimator of univariate distributions. Computations involving the median, such as the median absolute deviation and the Hodges–Lehmann shift, can substitute for inferential procedures based on the mean. Different samples can be compared using Mann–Whitney–Wilcoxon and related tests rather than parametric tests, and contingency tables can be used when variables are categorical. When different statistical tests are available for the same purpose (e.g. KS, CvM and AD goodness-of-fit tests; Spearman's, Kendall's, and Cox–Stuart tests for correlation), scientists are encouraged to compare the results of all methods, as they have different efficiencies under various conditions.

Nonparametric statistics have broad but not universal applicability. Astronomers must become aware of erroneous practices, particularly concerning the KS statistic and tests. The KS extremum measure is often inefficient for two-sample discrimination; the AD statistic is more sensitive in many situations. The KS statistic is not identifiable for multivariate datasets, and the tabulated probabilities for the KS statistic are often inapplicable for modeling goodness-of-fit evaluation.

5.8 Recommended reading

Agresti, A. (2002) *An Introduction to Categorical Data Analysis* 2nd ed., John Wiley, New York

A comprehensive text on the statistical treatment of datasets with binary or categorical variables. It includes nonparametric and parametric inference for contingency tables, logistic and other linear models, and mixture models.

Conover, W. J. (1999) *Practical Nonparametric Statistics* 3rd ed., John Wiley, New York

A thorough and authoritative presentation of classical nonparametrics. The text covers binomial distribution tests, contingency tables, rank tests and Kolmogorov–Smirnov-type tests.

Higgins, J. J. (2004) *An Introduction to Modern Nonparametric Statistics*, Thomson/Brooks-Cole, Belmonk

A readable undergraduate-level text. It covers traditional topics (univariate and multivariate methods, one-sample and *k*-sample tests, tests for trends and correlation) as well as brief treatments of censoring, bootstrap methods, smoothing and robust estimation.

Huber, P. J. & Ronchetti, E. M. (2009) *Robust Statistics* 2nd ed., John Wiley, New York

An advanced and authoritative monograph covering maximum likelihood (M), linear combinations (L) and rank (R) estimation, minimax theory, breakdown points, regression, covariance and correlation, small samples, and Bayesian robustness.

5.9 **R** applications

5.9.1 Exploratory plots and summary statistics

In the **R** script below, we bring the dataset under study into **R** using the *read.table* function appropriate for ASCII tables. The "header=T" option, where "T" represents "TRUE", indicates that the first line of the dataset is reserved for variable names. All astronomical datasets used in this volume are available on the Web site http://astrostatistics.psu.edu/MSMA. We next manually assign individual columns to vectors; in future scripts, we will use the *attach* function to achieve this automatically.

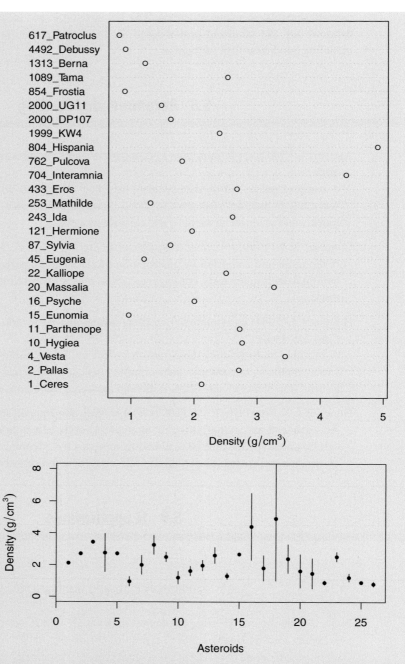

Simple plots for a small univariate sample of asteroid density measurements. *Top*: A dot chart. *Bottom*: Plot with heteroscedastic measurement errors.

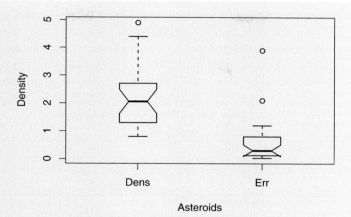

Fig. 5.2   Boxplots of asteroid densities (left) and their measurement errors (right).

See Appendix B for an introduction to the **R** programming language and commonly used functions. Basic **R** syntax features that are used here and throughout this volume include the symbol "$<-$" assigning contents to a new variable. In many cases, "$=$" can be used as a synonym; the symbol "$=$" always assigns values to predefined parameters in functions. To give a more compact presentation, the symbol ";" is sometimes used to separate distinct **R** statements; it replaces a carriage return.

Figure 5.1 shows two simple visualizations of the univariate dataset of 28 measurements of asteroid densities described in Appendix C.1. First, the Cleveland dot-chart displays small datasets directly with each point labeled. Second, the **R** function *plot* is used ubiquitously in data analysis for one- and two-dimensional graphical displays. Three asteroids clearly have very inaccurate density measurements; these might be deleted during later scientific analysis. **R** does not have a built-in plotting function for measurement errors; here we add error bars using the *segments* function that draws line segments between two specified points. Note that, throughout the **R** analysis, functions operate automatically on entire vectors and matrices unless their elements are individually specified or constrained.

Parameters like "cex", "cex.axis", and "cex.lab" specify scaling factors for the symbol and font sizes of various texts within the plots. These will differ with the device used. In this volume, we give values appropriate for the default Quartz device under the MacOS operating system. Our figures were then converted to encapsulated Postscript format using the function *dev.copy2eps*.

The **R** boxplot in Figure 5.2 shows a thick horizontal bar at the median, horizontal boundaries (or **hinges**) showing the interquartile range (IQR), and dashed line extensions (or **whiskers**) typically set at $1.5 * IQR$ above and below the 25% and 75% quartiles. Any data points lying outside the whiskers are plotted individually as **outliers**. The **notches** on the sides of the box are set at $\pm 1.58 * IQR/\sqrt{n}$ representing the standard deviation of the median if the distribution were normal. The box width here is scaled to the square root of the sample size n. These options are useful for comparing the distributions of several samples by plotting their boxplots together. A single sample can be divided into subsamples for boxplot comparison using **R**'s *split* or *cut* functions.

For the asteroid densities, the *summary* mean and standard deviation, 2.15 ± 1.04, is quite similar to the median and MAD, 2.00 ± 1.04. Note that the MAD function in *R* is the MADN function normalized to the standard deviation presented here. The boxplot of the densities shows a hint of asymmetry towards high values. The distribution of density errors again shows the three inaccurate points. The weighted mean of asteroid densities is slightly larger than the mean and median.

```
# Data plots and summaries for asteroid densities

asteroids <- read.table("http://astrostatistics.psu.edu/MSMA/datasets/
   asteroid.dens.dat", header=T)
astnames <- asteroids[,1] ; dens <- asteroids[,2] ; err <- asteroids[,3]
dotchart(dens, labels=astnames, cex=0.9, xlab=expression(Density~(g/cm^3)))
plot(dens, ylim=c(0,8), xlab="Asteroids", ylab=expression(Density~(g/cm^3)), pch=20)
num <- seq(1,length(dens))
segments(num, dens+err, num, dens-err)

# Boxplot and summary statistics for asteroid data

boxplot(asteroids[,2:3], varwidth=T, notch=T, xlab="Asteroids", ylab="Density",
 pars=list(boxwex=0.3,boxlwd=1.5,whisklwd=1.5,staplelwd=1.5,outlwd=1.5,
 font=2))
summary(dens)
mean(dens) ; sd(dens)
median(dens) ; mad(dens)
weighted.mean(dens,1/err^2)
```

Despite the fact that the nonparametric first and second moments of the measured densities are close to normal values, we can investigate further whether some more subtle deviation from normality is present. We are scientifically motivated by the possibility that subpopulations of asteroids are present with a range of densities due to different structure (solid vs. porous) and compositions (rock vs. ice).

```
# Tests for normality

shapiro.test(dens)
install.packages('nortest') ; library('nortest')
ad.test(dens) ; cvm.test(dens) ; lillie.test(dens) ; pearson.test(dens)
install.packages('outliers') ; library('outliers')
dixon.test(dens) ; chisq.out.test(dens)
grubbs.test(dens) ; grubbs.test(dens,type=20)
```

The nonparametric Shapiro–Wilks test for normality in base **R** gives a marginal significance, $P_{SW} = 0.05$, for a non-Gaussian shape in the asteroid density values. We then apply several other tests for normality from the **CRAN** package *nortest*: the Anderson–Darling, Cramér–von Mises, Lilliefors (a version of the Kolmogorov–Smirnov), and Pearson χ^2 tests. For the comparison of globular cluster samples, the P-values for nonnormality are not significant with $P = 0.14$–0.28 for the various tests. Using the **CRAN** package *outliers*, several tests give $P = 0.008$–0.15 that the highest density value, 4.9 g cm^{-3}, is an outlier. However, we know that that measurement has a large uncertainty. Altogether, we conclude that small deviations from normality may be present, but there is not enough evidence to convincingly indicate the presence of a skewed distribution or distinct subpopulations. The results also indicate that standard statistical procedures that assume normal populations are applicable to this sample.

5.9.2 Empirical distribution and quantile functions

Here we compare the univariate distributions of K-band magnitudes of globular clusters in the Milky Way Galaxy (MWG) and Andromeda Galaxy (M 31). The datasets are described in Appendix C.3, and were previously examined in Section 4.7.2.

```
# Read magnitudes for Milky Way and M 31 globular clusters

GC1 <- read.table("http://astrostatistics.psu.edu/MSMA//datasets/
   GlobClus_MWG.dat",header=T)
GC2 <- read.table("http://astrostatistics.psu.edu/MSMA/datasets/
   GlobClus_M31.dat",header=T)
K1 <- GC1[,2] ; K2 <- GC2[,2]
summary(K1) ; summary(K2)


# Three estimates of the distance modulus to M 31

DMmn <- mean(K2) - mean(K1) ;   DMmn
sigDMmn <- sqrt(var(K1)/length(K1) + var(K2)/length(K2)) ; sigDMmn
DMmed <- median(K2) - median(K1) ; DMmed
sigDMmed <- sqrt(mad(K1)^2/length(K1) + mad(K2)^2/length(K2)) ; sigDMmed
wilcox.test(K2, K1, conf.int=T)
```

The distribution summaries quickly show a large offset between the distributions: the MWG values are intrinsic (absolute) magnitudes while the M 31 values are observed (apparent) magnitudes. The difference between them is called the distance modulus of M 31. We calculate three estimates of the distance modulus using the mean, median and Hodges–Lehmann (HL) estimator for shift. In base **R**, the HL estimator is built into the Wilcoxon rank sum test function. Note how we assign the result of a calculation to a new variable (e.g. *DMmn* for the distance modulus based on the mean) and then

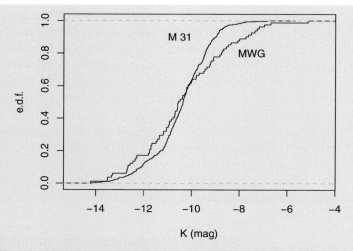

K (mag)

Fig. 5.3 Empirical distribution functions for the K-band magnitudes of globular clusters in the Milky Way Galaxy and the Andromeda Galaxy (M 31).

display the result on the **R** console by stating the variable. The resulting estimated distance modulus of Andromeda from the globular cluster K magnitudes are: 24.71 ± 0.21 using parametric means; 25.10 ± 0.20 using nonparametric medians; and $24.90^{+0.36}_{-0.37}$ (95% confidence interval) using the nonparametric HL estimator. These values are consistent with each other, but higher than the consensus 24.44 ± 0.1 based on a variety of astronomical considerations.

We can now offset the M 31 magnitudes by the estimated DM to place them on an absolute magnitude scale, and inquire into the similarity or differences in the shapes of K globular cluster luminosity functions of the Milky Way and M 31 samples. **R** has the function *ecdf* to construct a univariate empirical cumulative distribution function. A plot of confidence bands around the e.d.f. based on the Kolmogorov–Smirnov statistic is available in the **CRAN** package *sfsmisc* as function *ecdf.ksCI*.

The **R** function *quantile* produces sample quantiles with options of nine algorithms for interpolation when small samples are considered. There is little mathematical guidance regarding the best method. The function *qq* makes a quantile-quantile (Q-Q) plot for two samples. A common variant is the one-sample plot made with *qqnorm* which compares a dataset to the normal distribution. A straight line in a Q-Q plot indicates agreement between the distributions.

The resulting plots are shown in Figures 5.3 and 5.4. The first figure comparing the e.d.f.'s shows that the M 31 distribution is substantially narrower than the MWG distribution. (Inconveniently, the abscissa scale has bright clusters on the left; the axis direction can usually be reversed using the *plot.window* function but this option does not work with step function plots.) Of course, this finding may not reflect the underlying distribution; for example, the M 31 measurements may be truncated at faint magnitudes by observational limitations. For our purposes here, we will continue with the comparison of the datasets without scientific interpretation. The second figure shows Q-Q plots comparing the MWG

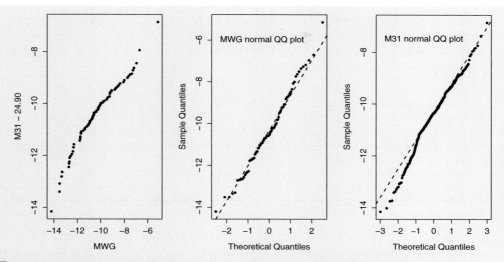

Fig. 5.4 Quantile-quantile plots for comparing univariate empirical distributions. *Left*: Comparison of Milky Way and M 31 globular cluster distributions. *Middle* and *Right*: Normal Q-Q plot comparing the Milky Way and M 31 globular cluster distributions to a Gaussian.

sample with the M 31 sample and both empirical distributions with a normal distribution. We see that the two galaxy samples are systematically different; the MWG cluster distribution is quite close to a Gaussian while the M 31 distribution is deficient in the lower tail.

```
# e.d.f., quantile and Q-Q plots for globular cluster magnitudes

plot(ecdf(K1), cex.points=0, verticals=T, xlab="K (mag)", ylab="e.d.f.", main="")
plot(ecdf(K2-24.90), cex.points=0, verticals=T, add=T)
text(-7.5, 0.8, lab="MWG")  ;  text(-10.5, 0.9, lab="M 31")

par(mfrow=c(1,3))
qqplot(K1, K2-24.90, pch=20, cex.axis=1.3, cex.lab=1.3,  xlab="MWG",
   ylab="M31 - 24.90", main="")
qqnorm(K1, pch=20, cex.axis=1.3, cex.lab=1.3, main="")
qqline(K1, lty=2, lwd=1.5)
text(-2.5, -6, pos=4, cex=1.3, 'MWG normal QQ plot')
qqnorm(K2-24.90, pch=20, cex.axis=1.3, cex.lab=1.3, main="")
qqline(K2-24.90, lty=2, lwd=1.5)
text(-3, -7.5, pos=4, cex=1.3, 'M31 normal QQ plot')
par(mfrow=c(1,1))

# Plot e.d.f. with confidence bands
```

```
install.packages('sfsmisc') ; library('sfsmisc')
ecdf.ksCI(K1,ci.col='black')

# Nonparametric tests for normality

install.packages('nortest') ; library(nortest)
cvm.test(K1) ; cvm.test(K2)
ad.test(K1) ; ad.test(K2)
```

5.9.3 Two-sample tests

The significance of the shape difference between the MWG and M 31 globular cluster magnitudes is evaluated nonparametrically using the MWW and KS two-sample tests. The two tests do not give quite the same significance level: $P = 0.2\%$ for MWW and $P = 6\%$ for KS that the null hypothesis of identical shapes is true. This is due to the poor sensitivity of the KS test in situations where the two distributions have the same means and thus their e.d.f.'s cross at the middle (Figure 5.3). The nonparametric two-sample **Mood test** for differences in scale (i.e. dispersion) most clearly discriminates the shape difference in the MWG and M 31 globular cluster distributions with $P < 0.01\%$. We also perform a two-sample CvM test from the **CRAN** package *cramer*. This test is not distribution-free and the $P = 1.6\%$ significance level for the difference between the MWG and M 31 distributions is based on bootstrap resamples.

If we ignore the non-Gaussianity, standard parametric two-sample tests can be applied. First, we examine the significance of the difference of the means using a pooled sample variance. Second, we apply the t test for the difference in means of two normally distributed samples. Both tests give $P > 10\%$ that the means are the same; this is expected as we purposefully offset the M 31 cluster values by 24.90 magnitudes to give the same mean as the MWG magnitudes. More relevant to the problem, we apply the F test for equal variances using **R**'s *var.test* function. Here, as with the nonparametric Mood test above, the spread of the two distributions is found to differ at the $P < 0.01\%$ significance level.

```
# Nonparametric two-sample tests for Milky Way and M31 globular clusters

ks.test(K1, K2-24.90) # with distance modulus offset removed
wilcox.test(K1, K2-24.90)
mood.test(K1, K2-24.90)
install.packages(cramer) ; library(cramer)
cramer.test(K1, K2-24.90)

# Parametric two-sample tests
```

```
pooled_var <- ((length(K1)-1)*var(K1) + (length(K2)-1)*var(K2)) /
    (length(K1)+length(K2)-2)
mean_diff <- (mean(K1) - mean(K2-24.90)) /
    sqrt(pooled_var * (1/length(K1)+1/length(K2)))
mean_diff ; pnorm(mean_diff)
t.test(K1, (K2-24.90), var.eq=T)
var.test(K1, K2-24.90)
```

5.9.4 Contingency tables

In Appendix C.2, we provide two contingency tables from the field of star formation and early stellar evolution. One is a 2×2 table concerning a possible relationship between the presence of jets and stellar multiplicity in protostars, and the other is a $r \times c$ table giving the observed population of five evolutionary classes of young stars in four star-forming regions. Our goal is to test the null hypotheses that the presence of jets has no relationship to multiplicity, and that the distribution of evolutionary classes is the same in all star-forming regions.

The **R** script begins with a direct input of the tables as a matrix. Note that the *matrix* function requires values by column not by row. **R**'s *ftable* function can also be used to create contingency tables. **R**'s function *chisq.test* is designed for contingency table inputs. It gives test probabilities both using the large-n approximation based on the χ^2 distribution (with Yates' continuity correction) and by random sampling from all contingency tables with the given marginals. The *fisher.test* function implements the Fisher exact test which is appropriate for small-n samples.

The tests give probabilities $P \simeq 20\%$ indicating that no significant effect of protostellar multiplicity on jet formation is present. The global tests applied to the entire 4×5 table on stellar populations in different star-forming regions show that strong differences are present ($P \ll 0.01\%$). Individual rows and columns can be tested; for example, the Fisher exact test shows that the Chamaeleon and Taurus samples differ marginally ($P = 5\%$) while the η Cha and Taurus samples differ strongly ($P < 0.01\%$). Note that the Fisher exact test sometimes fails due to the large memory required for the computation.

Figure 5.5 gives an informative display of the stellar population contingency table known as a **Cohen–Friendly association plot** created with **R**'s *assocplot* function. Each cell in the table is shown as a bar with height scaled to $(O_{ij} - E_{ij})/\sqrt{E_{ij}}$, where O represents the observed counts and E the expected counts assuming constant rates in all cells. The box widths are proportional to $\sqrt{E_{ij}}$ so that skinny boxes represent low expected counts and wide boxes represent high expected counts. The area of the box is then proportional to the difference in observed and expected frequencies. Excess observed counts appear above the dashed lines (black boxes), and deficient observed counts appear below the line (gray boxes). The numbers give the value of χ^2 when significant deviations from expected levels are present in a given row or column. The *mosaicplot* is also often used to visualize categorical data; many flexible options are provided by the **CRAN** *vcd* package (Meyer *et al.* 2006).

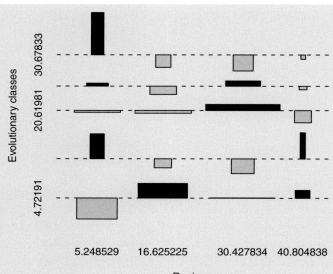

Fig. 5.5 Association plot for an $r \times c$ contingency table showing the population of young stars in five evolutionary classes and four star-forming regions.

From the association plot, we see that the strongest violation of the null hypothesis is that the η Cha region is deficient in Class 0-I-II stars and has excess Transition and Class III stars. This is readily understood because η Cha is known to be substantially older than the other star-forming regions. The Serpens region has an excess of the youngest Class 0 stars, reflecting its high current star-formation activity. There is strong scientific interest in the Transition class (fourth row of the plot), but no significant differences are seen between the four regions due to very low counts.

```
# Input star formation contingency tables

jets <- matrix(c(9,2,5,5), nrow=2)  ;  jets
evol <- matrix(c(21,1,2,0,16,14,42,1,61,90,179,2,17,4,5,5,22,95,126,10),nrow=4)
evol

# Test null hypotheses
chisq.test(jets)  ;  fisher.test(jets)
chisq.test(evol)
fisher.test(evol[2:3,]) # Chamaeleon vs. Taurus sample
fisher.test(evol[3:4,]) # Taurus vs. eta Cha sample

# Association plot
assocplot(evol,col=c("black","gray"),xlab='Regions',ylab='Evolutionary classes')
```

5.9.5   Scope of nonparametrics in **R** and **CRAN**

The base **R** package includes well-established nonparametric tests such as the Kolmogorov–Smirnov test (*ks.test*), the Shaprio–Wilks test for normality (*shapiro.test*), the Wilcoxon rank sum and signed rank tests (*wilcox.test*), the Ansari–Bradley, Bartlett, Fligner and Mood tests of variances, and Kendall's τ_K and Spearman's ρ_S correlation coefficients (*cor.test*).

A number of **CRAN** packages treat problems in nonparametric statistics. The package *Cramer* provides the Cramér–von Mises two-sample test for multivariate (as well as univariate) data with P-values estimated by permutation, bootstrap and eigenvalue methods. Distributions for Kendall's τ, Spearman's ρ and the Kruskal–Wallis tests are given by the package *SuppDists*. Packages for outlier detection include *mvoutlier* and *outliers*. Several packages including *moments*, *mvnormtest* and *nortest* provide tests for normality in a dataset (Anderson–Darling, Anscombe–Glynn, Bonett–Seier, Cramér–von Mises, D'Agostino, Kolmogorov–Smirnov, Pearson χ^2, Shapiro) relevant for the evaluation of the need for nonparametric procedures. The packages *ICSNP* and *SpatialNP* provide tools for multivariate nonparametrics. Packages treating ranks include: *crank* assisting with ranking incomplete data, *exactRankTests* performing Wilcoxon and permutation tests, *fuzzyRankTests* treating fuzzy confidence intervals, *Kendall* computing p-values for Kendall's τ_K, and *pspearman* gives p values for Spearman's ρ_S. The *vcd* package provides methods and graphics for categorical data.

Annotated links to a large body of **R** code for robust statistics are available from http://cran.r-project.org/web/views/Robust.html and http://www.r-project.org/robust. Two central packages are *robustbase* and *robust*. They include boxplots (*adjbox*), (non)linear regression (*lmrob*, *ltsreg*, and *nlrob*), estimates of multivariate location and scatter (*huberM*, *covMcd* and *covrob*), and bootstrap (*rb*) with associated summary plots and outputs.

6 Data smoothing: density estimation

6.1 The astronomical context

The goal of **density estimation** is to estimate the unknown probability density function of a random variable from a set of observations. In more familiar language, density estimation smooths collections of individual measurements into a continuous distribution, smoothing dots on a scatterplot by a curve or surface.

The problem arises in a wide variety of astronomical investigations. Galaxy or lensing distributions can be smoothed to trace the underlying dark matter distribution. Photons in an X-ray or gamma-ray image can be smoothed to visualize the X-ray or gamma-ray sky. Light curves from episodically observed variable stars or quasars can be smoothed to understand the nature of their variability. Star streams in the Galaxy's halo can be smoothed to trace the dynamics of cannibalized dwarf galaxies. Orbital parameters of Kuiper Belt Objects can be smoothed to understand resonances with planets.

Astronomical surveys measure properties of large samples of sources in a consistent fashion, and the objects are often plotted in low-dimensional projections to study characteristics of (sub)populations. Photometric color–magnitude and color–color plots are well-known examples, but parameters may be derived from spectra (e.g. emission-line ratios to measure gas ionization, velocity dispersions to study kinematics) or images (e.g. galaxy morphology measures). In these situations, it is often desirable to estimate the density for comparison with astrophysical theory, to visualize relationships between variables, or to find outliers of interest.

6.2 Concepts of density estimation

When the parametric form of the distribution is known, either from astrophysical theory or from a heuristic choice of some simple mathematical form, then the distribution function can be estimated by fitting the model parameters. This is the subject of regression discussed in Chapter 7. In this chapter, we make no assumption of the parametric form and are thus involved in **nonparametric density estimation**.

A simple estimator, very commonly used in astronomy, is the histogram, a plot of counts in grouped (or binned) data. But continuous density estimators with smooth transitions are also sought; this can often represent more realistic natural populations where discontinuities in properties are probably not present. The choice of an appropriate smoothing bandwidth, such as the width of a Gaussian in kernel density estimation, can be challenging. Sometimes the bandwidth is chosen subjectively or for astrophysical reasons independent of the dataset at hand. In other situations, the bandwidth can be estimated from the data based on a statistical criterion. The most desirable estimator approaches the underlying p.d.f. of the population as the sample increases with small bias and small variance. When an ideal estimator cannot be obtained, a balance between the bias and variance of a density estimator is often sought. For some methods, mathematical theory evaluates the convergence of the estimator to the true distribution under general conditions, allowing for bias correction and estimation of confidence bands along the smooth estimator.

This chapter starts with traditional and familiar density estimators, such as the histogram and convolution with simple kernels. We then outline some of the more sophisticated modern approaches to the problem.

6.3 Histograms

For a histogram of univariate i.i.d. random variables $X_1, X_2, \ldots, X_n$, a reference point and bin width $h(x)$ is first established by the scientist. Often the bins are chosen to have constant width, $h(x) = h_0$. The normalized **histogram estimator** is then

$$\hat{f}_{hist}(x) = \frac{N_m(x)}{nh(x)} \tag{6.1}$$

where $N_m(x)$ is the number of points in the m-th bin containing x. For equal bin widths h_0, the variable is divided into intervals $[mh_0, (m+1)h_0)$. Another choice is an adaptive bin width chosen to give a constant number of points in each bin; this is equivalent to an evenly spaced segmentation of the empirical quantile function (Section 2.6). Some datasets are intrinsically binned such as pixelated images, spectra derived from a multichannel spectrum analyzer, or time series obtained from a detector with a fixed readout interval. In these cases, the bin locations and widths are determined by the measurement instrument.

The critical issue for histograms and other techniques of nonparametric density estimation is the choice of bin width h. A general theorem (Freedman & Diaconis 1981) states that the optimal choice of bin width depends on the unknown first derivative of the density function being estimated and scales with the inverse cube root of the sample size. If the population resembles a normal distribution, simple rule-of-thumb estimates based on asymptotic normality and minimization of the mean square error are widely used (Scott 1979),

$$h_{Scott} = \frac{3.5 \, s.d.}{n^{1/3}} \quad \text{or} \quad \frac{2 \, IQR}{n^{1/3}} \tag{6.2}$$

where $s.d.$ is the sample standard deviation and IQR is the interquartile range or distance between the 25% and 75% quantiles. For a multivariate dataset with p dimensions,

$$h_{Scott} \simeq \frac{3.5 \, s.d}{n^{1/(2+p)}}. \tag{6.3}$$

As we will see in more detail below with kernel estimators, the mean square error has two components: a bias that shrinks as h is decreased, and a variance that shrinks as h is increased. The optimal bin widths here seek to balance the tension between these two components. Scott (1992) and Takezawa (2005) discuss the bias and variance in histogram estimators.

The histogram has several serious problems that have made it undesirable for statistical inference:

1. There is no mathematical guidance for choosing the reference point or origin for the histogram. The lowest point or a convenient rounded number just below this value is often chosen. But the apparent significance of a structure in the histogram may depend on this arbitrary choice. This problem can be mitigated with **average shifted histograms (ASH)**. The ASH procedure can be helpful in constructing smooth density estimators, as illustrated in Figure 6.5 below.

2. There is no mathematical guidance for choosing the method for grouping the data into bins: equal spacing but unequal number of data points (giving heteroscedastic errors); unequal spacing but equal number of data points (giving homoscedastic errors); or some other algorithm. The equal-points-per-bin option may appear attractive for later regression of other procedures, but performs poorly in reducing the mean square error compared to an optimal constant width option (Scott 1992, Section 3.2.8.4).

3. If the histogram is to be used for further quantitative analysis, the scientist must assign a fixed value for the bin's location. The common choice of bin midpoints leads to a bias compared to the preferred choice of group average of each bin. Alternatives such as the median or mean (centroid) location within the bin have not been well-studied.

4. As indicated above, the choice of optimal bin width requires knowledge of the true distribution function, which is precisely what is sought from the construction of the histogram. Various heuristic calculations are needed to obtain rules-of-thumb bin widths.

5. Bivariate and multivariate histograms can be constructed, but are difficult to visualize and interpret.

6. The discontinuities between bins in a histogram does not accurately reflect the continuous behavior of most physical quantities. The histogram thus cannot, by design, achieve the goal of density estimation approaching the true probability distribution function.

These drawbacks of the histogram estimator are so severe that statisticians often recommend they not be used for quantitative density estimation, although they are often helpful for exploratory examination and visualization. Silverman (1986) writes: "In terms of various mathematical descriptions of accuracy, the histogram can be quite substantially improved upon." Even for informal exploration, a variety of reference points, grouping algorithms and bin widths should be examined to assure that apparent effects are not dependent on these parameters. Scott (1992) writes: "The examination of both undersmoothed and oversmoothed histograms should be routine."

Histograms are useful for exploratory examination of univariate distributions; we use them often in this volume in this capacity (Figures 4.2, 6.1, 7.3, 8.1, and 11.1). But we perform quantitative analysis on the original data rather than the binned histograms.

Analysis of histograms is particularly dangerous for highly skewed distributions; for example, linear regression of binned data from a power-law (Pareto) distribution can be inaccurate even when large samples are available (Section 4.4). In a test of density estimators in a time series with multiple stochastic signals, the histogram was found to perform poorly compared to kernel estimators (Koyama & Shinomoto 2004).

Finally we note that direct analysis of histograms is appropriate when the original data arriving from the astronomical instrument has a binned structure. In such cases, it is sometimes desired to obtain a continuous estimator of the discontinuous histogram. Techniques for smoothing histograms based on maximum likelihood estimation or convolution with kernels are discussed by Takazawa (2005, Chpter 6).

6.4 Kernel density estimators

6.4.1 Basic properties

A broad and important class of smooth density estimators convolve the discrete data with a kernel function. Examples of one- and two-dimensional smoothing of astronomical datasets are shown in Figures 6.2 and 9.1, respectively. The scientist must make two decisions: the kernel's functional form and the kernel smoothing parameter or bandwidth. The mathematics of kernel density estimation has been extensively developed with results on consistency, asymptotic normality, rates of convergence, and so forth. Methods are thus available to assist the scientist in choosing the kernel function and bandwidth.

For i.i.d. random variables or vectors $X_1, X_2, \ldots, X_n$, a **kernel estimator** with constant bandwidth h has the form

$$\hat{f}_{kern}(x, h) = \frac{1}{nh} \sum_{i=1}^{n} K\left(\frac{x - X_i}{h}\right),   \tag{6.4}$$

where the kernel function K is normalized to unity, $\int K(x)dx = 1$.

To choose the kernel appropriate for a given problem, a criterion for the proximity of a kernel density estimator $\hat{f}_{kern}$ to the underlying p.d.f. $f(x)$ must be established. The most common criterion is to minimize the **mean integrated square error (MISE)**,

$$MISE(\hat{f}_{kern}) = E \int [\hat{f}_{kern}(x) - f(x)]^2 dx   \tag{6.5}$$

where E represents the expected or mean value. This quantity can be split into a bias term and a variance term,

$$
\begin{aligned}
MISE(\hat{f}_{kern}) &= \int E[\hat{f}_{kern}(x) - f(x)]^2 dx + \int Var\hat{f}_{kern}(x)dx \\
&= \int \text{Bias}^2[\hat{f}_{kern}(x)]dx + \int \text{Var}[\hat{f}_{kern}(x)]dx.
\end{aligned}   \tag{6.6}
$$

The bias term, measuring the averaged squared deviation of the estimator from the p.d.f., decreases rapidly as the bandwidth is narrowed, scaling as h^4. It does not generally improve as the sample size n increases. The variance, in contrast, decreases as the bandwidth is widened, scaling as h^{-1}. The optimal bandwidth is chosen at the value that minimizes the $MISE$, the sum of the bias and variance components.

Note that we choose here to minimize the squared errors (a least squares or L_2 statistic) but other choices can be made such as a robust minimization of the absolute errors (an L_1 statistic). In some engineering applications, the theory of kernel density estimation is viewed in terms of Fourier analysis. The convolution of a kernel with a dataset is readily performed in frequency space, and the bias–variance tradeoff is based on Fourier filters.

A very common choice of the functional form for the kernel K is the normal function

$$K(y) = \frac{1}{\sqrt{2\pi}} e^{\frac{-1}{2}y^2}, \tag{6.7}$$

a Gaussian density with mean $\mu = 0$ and variance $\sigma^2 = 1$. However, it is mathematically not quite optimal for minimizing the $MISE$. Other choices with good performance include the inverted parabola known as the **Epanechikov kernel**,

$$K(y) = \frac{3}{4}(1 - y^2) \tag{6.8}$$

over the range $-1 \le y \le 1$, and a similar inverted triangular function. The rectangular (or boxcar) kernel, $K(y) = 1/2$ over the range $|y| < 1$, has a somewhat weaker performance.

The choice of bandwidth h is more important than the functional form of the kernel for minimizing the $MISE$. As with the histogram, the scientist can subjectively choose to examine the data smoothed with different h, or use rule-of-thumb bandwidths designed for unimodal distributions. Silverman (1986) finds that

$$h_{r.o.t.} = 0.9An^{-1/5} \tag{6.9}$$

where A is the minimum of the standard deviation σ and the interquartile range $IQR/1.34$. This bandwidth will not perform well for p.d.f.'s with multiple peaks because the optimal bandwidth is narrower when the p.d.f. has large second derivatives. Note that optimal bandwidths for kernel density estimators become narrower only very slowly as the sample size increase ($h \propto n^{-1/5}$).

6.4.2 Choosing bandwidths by cross-validation

Except for very simple applications, there is insufficient information to estimate the optimal bandwidth that minimizes the $MISE$ by direct calculation. However, numerical evaluation is possible if the data are substituted for the p.d.f. The most common numerical approach to this problem is leave-one-out (or ordinary) **cross-validation** based on theory developed by Stone (1984) and others. Here we construct n datasets with $n - 1$ points by sequentially removing the i-th point and calculate the kernel density estimator for each new dataset, $\hat{f}_{-i,kern}$. The loglikelihood averaged over these datasets for each omitted X_i is then

$$CV(h) = \frac{1}{n} \sum_{i=1}^{n} \ln \hat{f}_{-i,kern}(X_i) \tag{6.10}$$

and the optimal bandwidth h_{CV} maximizes $CV(h)$. A normal kernel density estimator with bandwidth obtained from cross-validation is shown in Figure 6.2.

Another procedure based on cross-validation that is more resistant to outliers or heavy-tailed distributions is the least-squares cross-validation estimator. It is more complicated to write down, but can be evaluated in a computationally efficient manner using Fourier transforms. Other generalized cross-validation procedures include corrections for the degrees of freedom in the problem. In a likelihood formalism, these are related to the Akaike and Bayesian information criteria. These issues on cross-validation techniques are discussed by Silverman (1986) and Takezawa (2005).

Under some circumstances, kernel density estimators have asymptotically normal distributions allowing estimation of confidence bands. Providing the second derivatives of the underlying distribution are sufficiently small, kernel density estimators can be unbiased estimators with variance proportional to $f(x)/(nh)$ where $f(x)$ is the local value, h is the (uniform) bandwidth scaled to $n^{-1/5}$, and n is the sample size (Bickel & Rosenblatt 1973). Other approaches include bootstrap resampling and empirical likelihood (Chen 1996).

6.4.3 Multivariate kernel density estimation

Kernel density estimation techniques are readily extended to multivariate problems providing the dimensionality is low, although the computational burden for multivariate density estimation can become heavy. The kernel estimator can involve the convolution with a single multivariate density function, or as a product of univariate kernels. In the latter case, representing the data as $\mathbf{X}_i = (X_{i1}, X_{i2}, \ldots, X_{ip})$ and the bandwidth vector as $\mathbf{h} = (h_1, h_2, \ldots, h_p)$, the kernel estimator in Equation (6.4) can be written at the vector location $\mathbf{x} = (\mathbf{x}_1, \mathbf{x}_2, \ldots, \mathbf{x}_p)$

$$\hat{f}_{kern}(\mathbf{x}, \mathbf{h}) = \frac{1}{n \prod_{j=1}^{p} h_j} \sum_{i=1}^{n} \left[\prod_{j=1}^{p} K\left(\frac{\mathbf{x}_i - \mathbf{X}_{ij}}{\mathbf{h}_j} \right) \right]. \qquad (6.11)$$

For the simple case of a multivariate normal kernel in p dimensions, an approximate optimal bandwidth in each variable is

$$\mathbf{h}_{opt,j} = \frac{S_j}{n^{1/(p+4)}} \qquad (6.12)$$

where S_j is the sample standard deviation in the j-th variable. Note that the bandwidth becomes narrower more rapidly than in univariate cases as the sample size increases. Least-squares and likelihood cross-validation methods can be used for estimating optimal bandwidths in more general situations. However, they converge to optimal values only slowly as the sample size increases.

Finally, it is important to realize that multivariate sample sizes must be much larger than univariate samples to get comparable accuracy in calculating the $MISE$. For example, consider a histogram of a normal distribution in one dimension observed with a sample of $n = 100$. To match the $MISE$ of this histogram for a normal distribution in two dimensions, $n \sim 500$ is needed, and $n \sim 10,000$ is needed in four dimensions.

6.4.4   Smoothing with measurement errors

As with most statistical methodology, few density estimation methods treat the presence of heteroscedastic measurement errors with known variances. Solutions with demonstrated consistency, unbiasedness, asymptotic normality and convergence are difficult to obtain. This situation occurs quite often in astronomical data, as the acquisition of each flux measurement is often accompanied by acquisition of noise measurement for that particular observation. Current research efforts to treat this problem include Delaigle & Meister (2008), Staudenmayer *et al.* (2008), Delaigle *et al.* (2009) and Apanasovich *et al.* (2009). Schechtman & Spiegelman (2007) discuss the effect of measurement errors in estimating the quantile function. Figure 6.3 shows a univariate estimator of an astronomical dataset using the Delaigle–Meister (2008) deconvolution kernel density estimator.

6.5   Adaptive smoothing

While some astronomical datasets resemble clouds of points with simple, roughly unimodal structure, other datasets exhibit multiple peaks and/or structure on a range of scales and amplitudes. This often occurs in astronomical data involving fixed variables, such as spectra where well-separated emission lines are superposed on a smooth continuum, or images where well-separated galaxies are superposed on the dark sky. In such cases, smoothing with a constant-width bandwidth kernel does not optimally show the underlying structure. If a narrow bandwidth is used to reveal interesting structure in the bright regions, then the fainter (often noise-dominated) regions are broken up into insignificantly different clumps. If a broad bandwidth is used to appropriately smooth the faint regions, then details in the bright regions may be blurred. Astronomers thus have a strong interest in smoothing methods with bandwidths that vary, or adapt, to the local values of the density.

6.5.1   Adaptive kernel estimators

There is no single accepted approach to the problem of adapting local kernel bandwidths to variations in the data. An early suggestion with theoretical promise by I. Abramson is the local bandwidth

$$h(x) = h / \sqrt{\hat{f}(x)} \qquad (6.13)$$

where h is a globally optimal bandwidth and $\hat{f}$ is an approximation to the p.d.f.

A more flexible approach proposed by Silverman (1986, Section 5.3) is implemented in three stages. We consider only the univariate case here. First, a pilot estimate of the density $\tilde{f}$ is calculated using a standard kernel K with constant bandwidth h_{opt}. Second, a local bandwidth factor λ_i is obtained at each point x_i by

$$\lambda_i = [\tilde{f}(X_i)/g]^{-\alpha} \text{ where}$$
$$\ln g = n^{-1} \sum_{i=1}^{n} \ln \tilde{f}(X_i). \tag{6.14}$$

Here g is the geometric mean of the pilot estimator values and α is an undetermined sensitivity parameter with $0 \le \alpha \le 1$. Third, the adaptive kernel estimator at chosen values of the x variable is calculated from

$$\hat{f}_{adp,kern}(x) = \frac{1}{n} \sum_{i=1}^{n} \frac{1}{h_{opt}\lambda_i} K\left[\frac{x - X_i}{h_{opt}\lambda_i}\right]. \tag{6.15}$$

The final estimator is relatively insensitive to the choice of pilot estimator, but depends strongly on the α sensitivity factor. The adaptive estimator is increasingly sensitive to local variations in the data for larger values of α while the limit $\alpha = 0$ corresponds to the standard constant bandwidth estimator. The choice $\alpha = 1/2$ has advantageous mathematical properties, giving a smaller bias term in the $MISE$ than the best constant-bandwidth model, and is recommended for use in practice. The choice $\alpha = 1/p$ where p is the dimensionality of the dataset is equivalent to scaling by nearest-neighbor distances.

Research on these and other adaptive kernel estimators, particularly for multivariate data, is discussed in the review article by Scott & Sain (2005). Figure 6.4 shows an adaptive normal kernel density estimator with $\alpha = 0.5$ applied to an astronomical dataset.

6.5.2 Nearest-neighbor estimators

A simple algorithm for scaling a density estimator to the local density of data points uses the distance d (or in multivariate p-space, the Euclidean volume d^p) to the nearest neighbor of each point x. To reduce random noise, the distance d_k (or d_k^p) to the **k-th nearest neighbor (k-nn)** is often used. The density estimator is then

$$\hat{f}_{knn}(x) = \frac{k/2n}{d_k(x)} \text{ or } \frac{k/2n}{d_k^p(x)}. \tag{6.16}$$

This estimator is approximately 1 if the data are uniformly distributed, and scales with the inverse of the distance (volume) encompassing k nearby data points.

The k-nn estimator, straightforward to understand and efficient to compute, is widely used in data mining, pattern recognition and other computer science applications. However, from a statistical point of view, it has several disadvantages: it is not a p.d.f. because it does not integrate to unity; it is discontinuous; it is prone to local noise; and there is no choice of k which generally provides an optimal tradeoff between bias and variance. This last problem gives a tendency to oversmooth regions of low density. Locally, the k-nn estimator is equal to the kernel density estimator when the uniform kernel $K = 1/2$ for $|x| < d_k$ (in one dimension) is used.

In an astronomical context, J. Wang $et\ al.$ (2009) compare the performance of a $k = 10$ nearest-neighbor smoother with a uniform kernel smoother on the spatial distribution of young stars in a star-formation region. The two methods give similar results for the identification of star clusters.

6.6 Nonparametric regression

Many astronomical datasets consist of multivariate tables of properties (listed in columns) measured for a sample of objects (listed in rows). Some, but not all, scientific questions can be formulated in terms of regression where we seek to quantify the dependence of a single **response variable** y on one or more independent variables $\mathbf{x}$. In Chapter 8, we treat regression in the form $y = f(\mathbf{x})$ where each object is considered separately and f is a parametric function with specified mathematical relationships and parameters.

However, often we do not know (and choose not to guess) the true parametric functional relationship between y and $\mathbf{x}$. Also, it may be desirable to smooth the data before performing the regression to reduce the variance of individual points and, for megadatasets, to reduce the computational burden of the regression. In such circumstances, techniques of nonparametric regression come into play. When the astrophysical processes linking y and $\mathbf{x}$ are not understood, nonparametric regression is philosophically and practically more attractive than heuristic (typically linear or loglinear) parametric regression.

These methods are rarely applied in the astronomical literature, probably because they are not well known in the community. Methodological research in the field of **nonparametric regression** (or sometimes, **semi-parametric regression**) is active, and new developments are reviewed in several recently published monographs. Recent applications of nonparametric regression by astronomer–statistician collaborations include estimation of the fluctuation spectrum in the Cosmic Microwave Background by Miller *et al.* (2002) and estimation of the dark matter distribution in elliptical galaxies by X. Wang *et al.* (2005). These procedures can be very effective under many circumstances, and provide confidence intervals along the smoothed estimator.

6.6.1 Nadaraya–Watson estimator

For i.i.d. bivariate data (X_i, Y_i) with $i = 1, 2, \ldots, n$, the nonparametric regression function giving the expected value of y at specified values of x can be formulated in general terms as

$$r(x) = \frac{\int y f(x, y) dy}{\int f(x, y) dy} \tag{6.17}$$

where f is the unknown bivariate p.d.f. of the population. We estimate $f(x, y)$ by a bivariate product kernel density estimator as in Equation (6.11),

$$\hat{f}_{kern}(x, y) = \frac{1}{n h_x h_y} \sum_{i=1}^{n} K\left(\frac{x - X_i}{h_x}\right) K\left(\frac{y - Y_i}{h_y}\right), \tag{6.18}$$

where h_x and h_y are the bandwidths for each variable separately. The estimated regression function, named after E. Nadaraya and G. Watson (NW) who independently derived it in

1964, is then

$$\hat{r}_{NW}(x) = \frac{\sum_{i=1}^n Y_i K\left(\frac{x-X_i}{h_x}\right)}{\sum_{i=1}^n K\left(\frac{x-X_i}{h_x}\right)}. \tag{6.19}$$

The **NW estimator** can be viewed as a weighted local average of the response variable. Note that it does not depend on h_y which represents the structure in the dependent variable, and is typically evaluated for a chosen sequence of x values. Figure 6.4 shows an astronomical application of the NW smoother.

6.6.2 Local regression

The NW kernel estimator has another interpretation: it is the least-squares regression fit of y as a function of x when the y values are averaged over a local window $X_i - h_x < x < X_i + h_x$ around each X_i location weighted by the kernel $K[(x - X_i)/h_x]$. This concept has been generalized to fit higher order polynomials (usually lines or parabolas) in the local windows to more accurately model gradients in the density f. The solution can be readily generalized to low-dimensional multivariate problems. W. Cleveland's **LOWESS method** (a robust version of his LOESS method), which involves efficient algorithms for the repeated calculation of kernels for local polynomial regression, has been very widely disseminated and is used in many applications (Cleveland and Loader 1996). Here, the local bandwidth or **range** is determined by cross-validation on the local density of data points. We show a LOESS regression fit to an astronomical dataset in Figure 6.5. Figure 7.5 shows a similar adaptive local regression fit developed by Friedman (1984) with smoothing bandwidths estimated locally by cross-validation.

Spline smoothing is related to the LOESS solution to estimate the true functional relations from a sample of paired data points (X_i, Y_i). Here a chosen function, typically a cubic polynomial, is fitted to local portions of the data with a roughness penalty to constrain the smoothness of the curve. The curve is evaluated at a chosen sequence of **knots** rather than the data points themselves. Smoothing splines generally have small bias (the bias of a cubic smoothing spline under general conditions approaches 1/8 of the variance as $n \to \infty$), but do not necessarily minimize the $MISE$. An example of spline smoothing of an astronomical dataset is shown in Figure 6.4.

Local regression techniques with variable bandwidths cannot be fully characterized by a global measure of error such as the $MISE$. In particular, in regions of rapid curvature in the p.d.f., the bias locally can be large. We encountered this problem in our astronomical application of LOESS (compare the red and green curves in Figure 6.5). Locally calculated cross-validation may be needed to estimate errors of the smoother (Cummins *et al.* 2001).

Multivariate extensions of locally adaptive regressions have been developed (Takezawa 2005). These include cross-validation to find bandwidths in each dimension, thin plate smoothing splines, multivariate LOESS estimators, projection pursuit, and kriging. **Projection pursuit** is a sophisticated procedure whereby the data matrix is subject to a variety of transformations until an "interesting" projection is found that shows a high degree of structure in a reduced number of dimensions. **Kriging** is a least-squares interpolation method for irregularly spaced data points independently developed for applications in geology. The procedures give both a regression surface and a **variogram** showing the local uncertainty of the estimator. The theory and application of kriging is extensive with variants to treat different data spacings, intrinsic scatter and weightings (Wackernagel 1995). Kriging is discussed further, with an astronomical example, in Section 12.5.

6.7 Remarks

Nonparametric density estimation, or data smoothing, is a well-developed statistical enterprise with a variety of powerful methods. The histogram familiar in astronomy can be useful for data visualization but is not useful for statistical inference. The most well-developed smoothers are kernel density estimators with bandwidths chosen by rules-of-thumb or (preferably) by cross-validation from the dataset itself. Astronomers are often interested in adaptive smoothing, and a variety of techniques can be used for datasets with structure on different scales. These include Silverman's adaptive kernel and k-nearest-neighbor estimators.

Nonparametric regression is an approach to data smoothing that is rarely used in astronomy, yet can be of considerable use. The simple Nadaraya–Watson estimator is useful when a single smoothing scale suffices for the entire dataset. But in more complicated situations, local regressions can often be designed that can follow complicated multimodal distributions on various scales. With cross-validation and bootstrap resamples, confidence bands around the fits can be obtained.

Smoothing data with treatment of their heteroscedastic measurement errors is still in its infancy. However, the methods under development by Delaigle and others seem promising and are recommended for astronomical use.

The presentation here reviews only a portion of the large literature on density estimation. Some modern techniques such as structured local regression models, radial basis functions, local likelihood and mixture models are discussed by Hastie *et al.* (2009, Chapter 6).

6.8 Recommended reading

Bowman, A. W. & Azzalini, A. (1997) *Applied Smoothing Techniques for Data Analysis*, Clarendon Press, Oxford

A brief, practical presentation of kernel density estimation with extensive applications in **R** using the authors' *sm* package. It treats kernel methods up to three dimensions, adaptive bandwidths, nearest-neighbor methods, confidence bands, bandwidth selection, nonparametric regressions, comparing density estimates, testing parametric regression models and time series applications.

Scott, D. W. (1992) *Multivariate Density Estimation: Theory, Practice and Visualization*, John Wiley, New York

An authoritative monograph on nonparametric density estimation in one to several dimensions covering topics such as estimation criteria, histograms, kernel density estimators, dimensionality reduction, nonparametric regression and bump hunting.

Silverman, B. W. (1986) *Density Estimation for Statistics and Data Analysis*, Chapman & Hall, London

A short and readable account of nonparametric density estimation. It emphasizes kernel density estimation but also covers nearest-neighbor methods, cross-validation, adaptive smoothing and bump hunting.

Takezawa, K. (2005) *Introduction to Nonparametric Regression*, Wiley-Interscience, New York

A more mathematically advanced, integrated treatment of histograms, kernel density estimation, splines, local regressions and pattern recognition. Extensive annotated **R** code is provided.

6.9   R applications

The following illustrations of density estimation methods and displays in **R** use portions of the Sloan Digital Sky Survey dataset with 17 variables defined for ∼77,000 quasars. The dataset is presented in Appendix C.8.

6.9.1   Histogram, quantile function and measurement errors

In Figure 6.1 we show a histogram and quantile function of the SDSS quasar redshift distribution for the full sample of 77,429 quasars using **R**'s *hist* and *quantile* functions. There is no consensus binwidth for histogram displays; we show here the binwidth recommended by Scott (1979) rather than the default Sturges' binwidth used by **R** which is more coarse. Similarly, the quantile estimator is not uniquely defined for small discrete samples; **R**'s function gives nine options for the quantile estimator.

```
# Construct large and small samples of SDSS quasar redshifts and r-i colors

qso <- read.table("http://astrostatistics.psu.edu/MSMA/datasets/SDSS_QSO.dat",head=T)
dim(qso) ; names(qso) ; summary(qso) ; attach(qso)
z.all <- z ; z.200 <- z[1:200]
r.i.all <-  r_mag - i_mag ; r.i.200 <- r.i.all[1:200]
sig.r.i.all <- sqrt(sig_r_mag^2 + sig_i_mag^2)  ; sig.r.i.200 <- sig.r.i.all[1:200]
```

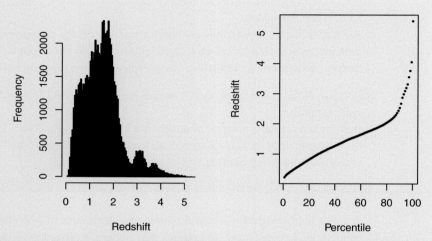

Fig. 6.1   Histogram and quantile function for the redshift distribution of 77,429 quasars from the Sloan Digital Sky Survey.

```
# Plot histogram and quantile function

par(mfrow=c(1,2))
hist(z.all, breaks='scott', main='', xlab='Redshift', col='black')
plot(quantile(z.all, seq(1,100,1)/100, na.rm=T), pch=20, cex=0.5,
   xlab='Percentile', ylab='Redshift')
par(mfrow=c(1,1))
```

6.9.2 Kernel smoothers

Base **R** has several functions that convolve kernels with a dataset allowing various options: *density*, *kernapply* and *convolve*. These are computed with the fast Fourier transform. The **CRAN** package *np* is a sophisticated environment for nonparametric kernel density estimation (Hayfield & Racine 2008). Data can be provided either using **R**'s dataframe format or a regression formula. *np* treats multivariate problems, mixed real and categorical variables, parametric and quantile regression, bootstrap confidence bands and bias correction, Gaussian and other kernel functions, fixed and adaptive bandwidths, iterative cross-validated backfitting bandwidth selection, and multivariate graphing. Estimation methods (Hall *et al.* 2004) include kernel convolution and *k*-nearest neighbors, and confidence bands can be obtained using cross-validation methods.

For the SDSS quasar $r - i$ vs. redshift relationship, the function *npregbw* allows the scientist to examine a variety of bandwidths. For example, the Nadaraya–Watson bandwidth for the 200-quasar sample calculated by likelihood cross-validation is 0.078 in redshift. The search for optimal bandwidths through cross-validation can be computationally intensive, especially for large multivariate datasets. The plotting function *npplot* flexibly displays various kernel estimators, training and evaluation datasets, three-dimensional perspectives,

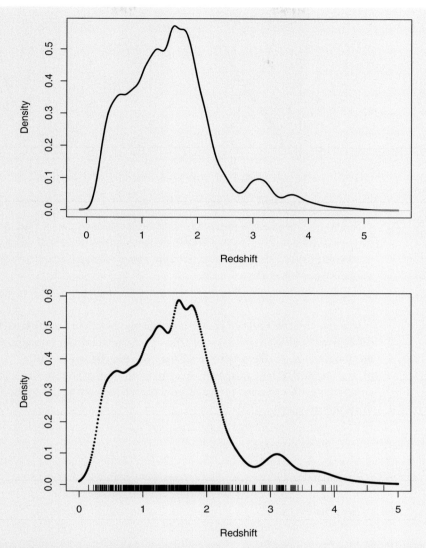

Fig. 6.2 Univariate kernel density estimators for the redshift distribuiton of SDSS quasars ($n = 77, 429$). *Top:* Constant kernel using Silverman's rule-of-thumb bandwidth. *Bottom:* Adaptive kernel bandwidths.

quantile functions, bootstrap confidence intervals and boxplots. The **CRAN** package *sm* (Bowman & Azzalini 2010) has a large suite of kernel smoothing functions and displays that are fully described in the volume by Bowman and Azzalini (1997). These include constant and variable bandwidths, nonparametric regression with confidence intervals, nonparametric checks on parametric regression models, comparing regression curves and surfaces, and other topics.

Figure 6.2 shows two similar versions of the kernel density estimator for the redshift distribution of the 77,429 SDSS quasars. The top panel is obtained using **R**'s *density*

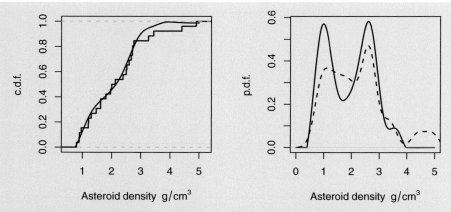

Fig. 6.3 Cumulative (left) and probability (right) distribution functions for the densities of 26 asteroids. The c.d.f. plot shows the empirical distribution function (step function) and the Delaigle–Meister (2008) estimator that incorporates measurement errors (solid curve). The p.d.f. plot shows a kernel density estimator of the asteroid densities (dashed) and the Delaigle–Meister estimator (solid).

function with Scott's rule-of-thumb constant bandwidth in Equation (6.2). The bottom panel gives Silverman's adaptive kernel density estimator from Equation (6.15) obtained with the function *akj* in the **CRAN** package *quantreg* (Koenker 2010). A random subsample of 500 redshift values is shown as a "rug" along the bottom axis. This operation is not computationally efficient for large samples. Here more detail in the peak of the redshift distribution is seen.

```
# Constant kernel density estimator

plot(density(z.all), bw=bw.nrd(z.all), main='', xlab='Redshift', lwd=2)

# Adaptive kernel smoother

install.packages("quantreg") ; library(quantreg)
akern.zqso <- akj(z.all, z=seq(0,5,0.01), alpha=0.5)
str(akern.zqso)
plot.window(xlim=c(0,5), ylim=c(0,0.6))
plot(seq(0,5, 0.01), akern.zqso$dens, pch=20, cex=0.5,  xlab="Redshift",
   ylab="Density")
rug(sample(z.all, 500))
```

The **CRAN** package *decon* (Wang & Wang 2010) implements the deconvolution kernel density estimator for a univariate dataset with heteroscedastic measurement errors developed by Delaigle & Meister (2008). We apply the method to the 26 measurements of asteroid densities described in Appendix C.1 with results shown in Figure 6.3. The asteroid density measurements have errors ranging over an order of magnitude. Scientific questions include whether all asteroids are consistent with a solid rock composition with densities around $3-4$ g cm^{-3}, or whether a separate population of denser (iron-rich) or less dense (porous) asteroids is present.

In the **R** script below, we first plot the discontinuous empirical distribution function (Section 5.3.1) using the *ecdf* function ignoring measurement errors. This is compared to the Delaigle–Meister (2008) kernel smoothed estimator using the heteroscedastic measurement errors in the right panel of Figure 6.3. Here we evaluate the estimator on a sequence of asteroid density values using the functions *DeconCdf* and *DeconPdf*. The estimator is compared to the empirical e.d.f. and density values smoothed without consideration of measurement errors. The choice of $h = 0.1$ g cm^{-3} bandwidth is arbitrary.

Two effects of possible astrophysical significance are seen. First, although two asteroids have nominal densities above 4 g cm^{-3}, these measurements are very uncertain and the Delaigle–Meister estimator does not find that any asteroids reliably have densities above ~3.5 g cm^{-3}. This accounts for the difference between the smooth and step functions in Figure 6.3 (left panel). Second, the Delaigle–Meister estimator shows a deep gap between asteroids around 1 and 3 g cm^{-3} due to large measurement errors around 1.5 g cm^{-3}, suggesting that distinct porous and solid rock populations may be present in the asteroid sample. However, the statistical significance of the gap is not readily estimated.

Altogether, the Delaigle–Meister estimator provides valuable results for interpreting this dataset with wildly different measurement errors. The *decon* package also includes the function *DeconNpr* that calculates a bivariate nonparametric regression estimator allowing heteroscedastic measurement errors in the response variable.

```
# Distribution with and without measurement errors

aster <- read.table('http://astrostatistics.psu.edu/MSMA/datasets/
   asteroid_dens.dat', head=T)
summary(aster)  ;  dim(aster)  ;  attach(aster)

install.packages('decon')  ;  library(decon)

par(mfrow=c(1,2))
plot(ecdf(Dens), main='', xlab=expression(Asteroid~density~~g/cm^3), ylab='c.d.f.',
   verticals=T, cex=0, lwd=1.5)
x <- seq(0, 6, 0.02)
cdf_sm <- DeconCdf(Dens,Err,x,bw=0.1)
lines(cdf_sm, lwd=1.5)

plot(DeconPdf(Dens,Err,x,bw=0.1), main='', xlab=expression(Asteroid~density~~
   g/cm^3), ylab='p.d.f.', lwd=1.5, add=T)
lines(density(Dens, adjust=1/2), lwd=1.5, lty=2)
par(mfrow=c(1,1))
```

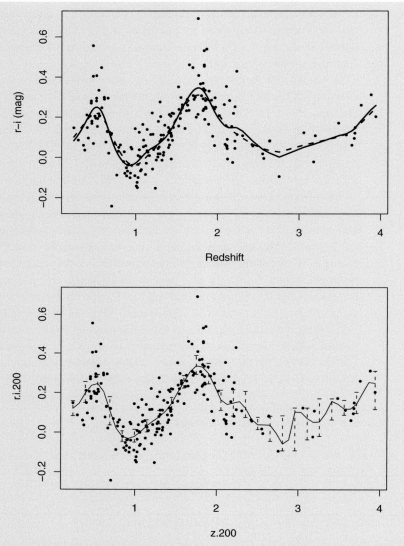

Fig. 6.4   Bivariate nonparametric regression estimators of quasar color index $r - i$ as a function of redshift distribution for a small sample of 200 SDSS quasars. *Top:* Spline fit (solid curve) and spline weighted by the variances of the $r - i$ measurement errors (dashed curve). *Bottom:* Nadaraya–Watson kernel-based regression with bandwidth $h = 0.08$ and 95% bootstrap confidence intervals.

6.9.3   Nonparametric regressions

R and **CRAN** have many spline fitting programs with different mathematical assumptions and capabilities. Figure 6.4 (top panel) illustrates a natural spline fit to the redshift–color relationship for a subset of 200 SDSS quasars. Spline fits with and without weighting by the measurement errors are shown; the weighted fit is less responsive to outlying uncertain $r - i$ values. The bottom panel of Figure 6.4 shows the Nadaraya–Watson regression estimator defined in Equation (6.19). This calculation was made using the function *npregbw* in the **CRAN** package *np*.

```
# Natural spline fit

install.packages("pspline") ; library(pspline)
fit=sm.spline(z.200, r.i.200)
plot(z.200, r.i.200, pch=20, cex=0.7, xlab="Redshift", ylab="r-i (mag)")
lines(fit$x, fit$y, lwd=2)

# Natural spline fit weighted by the variances of the measurement errors

fitw <- sm.spline(z.200, r.i.200, w=sig.r.i.200^2)
lines(fitw$x, fitw$y, lty=2, lwd=2)

# Bivariate Nadaraya-Watson regression estimator with bootstrap errors

install.packages("np") ; library(np)
bw.NW <- npregbw(z.200, r.i.200, regtype='lc', bwtype='fixed')
npplot(bws=bw.NW, ylim=c(-0.25,0.7), plot.errors.method="bootstrap",
plot.errors.bar='I', plot.errors.type='quantiles')
points(z.200, r.i.200, pch=20, cex=0.7)
```

Here we perform a *loess* local regression fit to a two-dimensional scatterplot of $r - i$ color index against redshift z from the full SDSS quasar sample. The plot we generate with **R** (Figure 6.5) can be compared to the plot published by Schneider *et al.* (2007). Producing the *loess* fit, the layered visualization, and the extraction of the *loess* solution require a multi-step **R** script which we outline here.

```
# Local regression
# 1. Read SDSS quasar sample, N=77,429. Clean bad photometry

qso <- read.table("http://astrostatistics.psu.edu/datasets/SDSS_QSO.dat",head=T)
q1 <- qso[qso[,10] < 0.3,] ; q1 <- q1[q1[,12] < 0.3,]
dim(q1) ; names(q1) ; summary(q1)
r_i <- q1[,9] - q1[,11] ; z <- q1[,4] ; r <- q1[,9]

# 2. Plot two-dimensional smoothed distribution

install.packages('ash') ; library(ash)
nbin <- c(500, 500) ; ab <- matrix(c(0.0,-0.5,5.5,2.), 2,2)
```

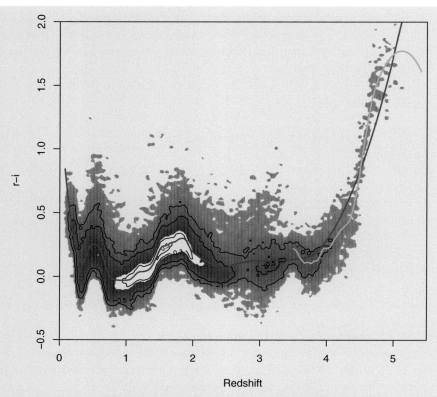

Fig. 6.5 Bivariate distribution of $r - i$ color index and redshift of SDSS quasars. Gray-scale and contours are calculated using averaged shifted histograms. Red and green curves are calculated using LOESS local regression. See text for details. For a color version of this figure please see the color plate section.

```
bins <- bin2(cbind(z,r_i), ab, nbin)
f <- ash2(bins, c(5,5)) ; attributes(f)
f$z <- log10(f$z)

image(f$x, f$y, f$z, col=gray(seq(0.5,0.2,by=-0.02)), zlim=c(-2.0,0.2),
    main='', xlab="Redshift", ylab="r-i")
contour(f$x, f$y, f$z, levels=c(-1.0,-0.5,0.3,0.5), add=T)
```

First, we read the SDSS dataset into a matrix and remove objects with poor photometry (errors > 0.3 mag) in r or i. Recalling that **R** variables cannot have a "$-$" in their names, we call the $r - i$ color index $r - i$. For convenience, we extract the $r - i$ and z vectors from the matrix.

Second, we plot the data. As with many very large samples, the standard *plot* function produces an uninformative black band where thousands of data point are superposed. We therefore convert the scatterplot into a gray-scale image using a two-dimensional averaged shifted histogram estimator in the **CRAN** package *ash* (Scott 2009). Similar tools with fewer options are available in the **R** functions *kde2d* and *bkde2D*. The use of the function *ash2* is described in its **R** help file, and some experimentation with smoothing parameters was needed. It illustrates use of the *c*, *matrix* and *cbind* functions which respectively concatenate values into a vector, create a matrix out of given values, and bind together vectors into a matrix by columns.

The product of our *ash2* run is a 500×500 image with a third dimension of real values giving the logarithm of the local surface density of quasars. This is displayed using the **R** functions *image* and *contour*; interactive adjustment of gray-scale and contour levels is again necessary. Note how the image is constructed from three elements of the **R** object produced by *ash2*. The functions *attributes* and *structure* are useful for showing the contents of **R** objects produced by various programs. The *image* function also illustrates the use of *seq* which creates a vector of evenly spaced sequence of values.

```
# 3. Construct loess local regression lines

z1 <- z[order(z)] ; r_i1 <- r_i[order(z)]
loct1 <- loess(r_i1~z1, span=0.1, data.frame(x=z1,y=r_i1))
summary(loct1)
lines(z1, predict(loct1), lwd=2, col=2)
z2 <- z1[z1>3.5] ; r_i2 <- r_i1[z1>3.5]
loct2 <- loess(r_i2~z2, span=0.2, data.frame(x=z2,y=r_i2))
lines(z2, predict(loct2), lwd=2, col=3)


# 4. Save evenly-spaced loess fit to a file

x1 <- seq(0.0, 2.5, by=0.02) ; x2 <- seq(2.52, 5.0, by=0.02)
loctdat1 <- predict(loct1, data.frame(x=x1))
loctdat2=predict(loct2, data.frame(x=x2))
write(rbind(x1,loctdat1), sep=' ', ncol=2, file='qso.txt')
write(rbind(x2,loctdat2), sep=' ', ncol=2, file='qso.txt', append=T)
```

Third, we apply the *loess* function after placing the data in order along the abscissa. The *loess* fit is performed by weighted least squares, and we adjust the *span* parameter that determines the smoothness of the fit by trial-and-error. Note the use of the tilde symbol in the *loess* function; this is **R** syntax for an algebraic formula. We examined the effect of applying an M-estimator (Tukey's biweight), but found that outliers do not affect the solution. The *loess* function produces an intermediate **R** object which is evaluated using the function *predict*, and these values can then be plotted using *lines* on the quasar surface density image.

The full loess fit, shown as the red curve in Figure 6.5, does a good job in following the ridge line of the quasar distribution in the high-surface-density region at low redshift ($z < 2.5$). But at higher redshifts, where only a small fraction of the 77,429 quasars are present and the $r - i$ vs. z relationship shows a steep gradient, the loess fit was poor. This is an example of large bias in the local regression solution in regions of rapid curvature (Section 6.6.2). A separate *loess* calculation in this region fitted the observed relationship reasonably well, as shown in the green dashed curve.

Fourth, we seek to save the two loess solutions (one at low redshift and the other at high redshift) to a file for later use. As it was originally calculated at each of the 77,429 abscissa values, we recalculate the loess fit using *predict* on a more convenient evenly spaced grid. We use *rbind* which binds to vectors by row to create a two-column matrix giving the new abscissa values and associated loess fit. These are then placed into an output file in ASCII format using *write*. Other **R** functions to output results to a file include *print*, *cat* and *format*. This tabulated distribution can then be used for photometric estimates of quasar redshifts, as a distribution for Monte Carlo simulations of quasar populations similar to the parametric bootstrap, or other purposes.

Finally, the reader is referred to Appendix B where some additional density estimation calculations and plots are produced (Figure B.2). The **R** function *smoothScatter* produces a gray-scale plot of a point process, *smooth.spline* gives a spline fit, and a polynomial fit is calculated using *lm* and plotted with *lines*.

6.9.4   Scope of smoothing in R and CRAN

R has incorporated the packages *KernSmooth* and *MASS* which provide kernel density estimates with a variety of bandwidth options based on the monographs by Wand & Jones (1995) and Scott (1992). For example, a two-dimensional kernel density estimator with cross-validation bandwidths is provided by the function *kde2d* from *MASS* and the function *bkde2D* from *KernSmooth*. Other functions in **R** include *smoothScatter* for color representations of kernel estimators of scatter plots, *ksmooth* giving a one-dimensional Nadaraya–Watson kernel estimator; and *smooth* for Tukey's running median estimator. Härdle–Steiger and Tukey running median smoothers for scatterplots are implemented in *runmed* and *smooth*, and Friedman's super-smoother for scatterplots is implemented in *supsmu*. **R** also has the package *splines* which calculates a variety of polynomial, interpolation and natural spline fits in one or several dimensions. The **R** function *smooth.spline* also gives a cubic spline smoother. Local polynomial regression fitting is given by *loess* and *locploy*.

Many capabilities are provided by **CRAN** packages. Kernel smoothing is treated in packages *ks*, *kzs*, *lokern*, *np*, *sm*. The *ks* package, for example, provides a variety of sophisticated bandwidth selection procedures, normal mixture models, kernel methods for two-sample tests and classification, and associated graphics for datasets of up to six dimensions. Adaptive smoothing for images is covered by *adimpro* and *aws*. Spline fits are given in packages *assist*, *cobs*, *DierckxSpline*, *earth* (implementing multivariate adaptive regression splines), *gss*, *lmeSplines*, *logspline*, *MBA*, *polspline*, *pspline*, and *sspline*. A sophisticated package implementing local regression and associated graphics is provided in *locfit* following the methods described by Loader (1999). The *simex* package is designed to treat measurement error models and can be used for density estimation. The *fields* package, developed for large spatial datasets in geostatistics and meteorology, provides smooth curve and surface fits from irregularly spaced multivariate data based on splines and kriging. The *tgp* package computes Bayesian nonlinear regression for Gaussian processes.

7   Regression

7.1   Astronomical context

Astronomers fit data both to simple phenomenological relationships and to complex nonlinear models based on astrophysical understanding of the observed phenomenon. The first type often involves linear relationships, and is common in other fields such as social and biological sciences. Examples might include characterizing the Fundamental Plane of elliptical galaxies or the power-law index of solar flare energies. Astrophysicists may have some semi-quantitative explanations for these relationships, but they typically do not arise from a well-established astrophysical process.

But the second type of statistical modeling is not seen outside of the physical sciences. Here, providing the model family truly represents the underlying phenomenon, the fitted parameters give insights into sizes, masses, compositions, temperatures, geometries and other physical properties of astronomical objects. Examples of astrophysical modeling include:

- Interpreting the spectrum of an accreting black hole such as a quasar. Is it a nonthermal power law, a sum of featureless blackbodies, and/or a thermal gas with atomic emission and absorption lines?
- Interpreting the radial velocity variations of a large sample of solar-like stars. This can lead to discovery of orbiting systems such as binary stars and exoplanets, giving insights into star and planet formation. Is a star orbited by two planets or four planets?
- Interpreting the spatial fluctuations in the Cosmic Microwave Background radiation. What are the best-fit combinations of baryonic, dark matter and dark energy components? Are Big Bang models with quintessence or cosmic strings excluded?

The goals of astronomical modeling also differ from many applications in social science or industry. In human affairs, regression is often used for forecasting or controlling the future behavior of a process, or for decision-making in complex environments. But in astronomy it is almost exclusively used to quantify apparent or causal relationships between the variables. We have little interest in predicting the redshift of the next galaxy to appear in the telescope, but a great deal of interest in characterizing and understanding how galaxies in general evolve with redshift.

The mathematical procedures used to link data with astrophysical models fall under the rubric of regression. We seek here to develop increased sophistication in regression methodology, discussing limitations of some traditional model-fitting methods and introducing procedures not commonly used in astronomical research.

7.2 Concepts of regression

Regression is a class of statistical analyses that involve estimating functional relationships between a dependent variable Y, also called the **response variable**, and one or more independent variables, $\mathbf{X}$,

$$E[Y|\mathbf{X}] = f(\mathbf{X}, \theta) + \epsilon \tag{7.1}$$

where ϵ is the random error. This equation can be read as "The expectation (mean value) of the response variable Y given the independent variable(s) $\mathbf{X}$ is a function of $\mathbf{X}$, where the function depends on a k-vector θ of model parameters, plus a random error (noise)." In general, $\mathbf{X}$ is an $n \times p$ matrix with n data points with p variables. For simplicity in this chapter, we will consider only the bivariate regression problem where $p = 1$; multivariate linear regression is briefly treated in Section 8.4.1. But it is important to recognize that nearly all of the methods discussed in the present chapter are designed for multivariate covariates. Note the intrinsic asymmetry in most regression analyses: we estimate $E[Y|\mathbf{X}]$, the conditional expectation of Y given $\mathbf{X}$, rather than the joint distribution of $\mathbf{X}$ and Y. The function f describing the dependence of Y on $\mathbf{X}$ can be a simple function such as a polynomial, or a complicated astrophysical model with many parameters. The goal of regression usually focuses on estimation of the parameters θ of the function and their confidence intervals. In most regression analysis, the independent variable is assumed to be fixed without error, although this assumption is relaxed in **measurement error models** (Section 7.5).

Regression is one of the older components of mathematical statistics starting with the nineteenth-century challenges of celestial mechanics where uncertain observations were combined to estimate parameters of deterministic Newtonian orbits. Least-squares regression for celestial mechanics originated with A. Legendre, S. P. Laplace, and F. Gauss in the early nineteenth century, further developed by F. Galton, K. Pearson, R. A. Fisher and many prominent astronomers during the late nineteenth and early twentieth centuries (Section 1.2.2). In recent decades, sophisticated extensions to regression have been developed for complex problems. Examples include **generalized linear modeling** (GLM, Nelder & Wedderburn 1972) for non-Gaussian multivariate datasets and **lasso regression** (Tibshirani 1996) for multivariate datasets applying **shrinkage** for selection of variables important to the response variable.

In the **linear regression model** used ubiquitously throughout statistics, Equation (7.1) has the form

$$Y = \beta_0 + \beta_1 X + \epsilon \tag{7.2}$$

where X is the independent variable, and Y is the dependent or response variable. The vector of β values contains the unknown intercept and slope of the linear model. The data have the form (x_i, y_i) where $i = 1, 2, \ldots, n$.

Such models fall under two classes. In **structural models**, the X's are random variables as described in Section 2.5, while in **functional models** the X's are deterministic. Both cases are commonly seen in astronomy. Modeling the dependence of oxygen abundance on nitrogen and other abundances in a sample of gaseous nebulae involves random variables. Modeling the shape of the oxygen emission line in a spectrum involves a deterministic X, the wavelength. Similar fixed variables occur in astronomical images, spectra and time series.

Another distinction of interest in astronomical regression is the origin of ϵ, the scatter about the regression line. When Y values do not fall exactly on the parametric curve $f(X, \theta)$, then ϵ represents the **intrinsic scatter**. These are intrinsic errors in the response variable, deviations of data points from the true line that are present even if all measurements were made with prefect precision and accuracy. In contrast, when the values Y lie precisely on the curve, the scatter represented by ϵ is attributed entirely to **measurement error**. In the social sciences, this distinction often cannot be made, for instance, by examining the noise level in feature-free portions of the dataset.

Regression models that treat various types of errors are discussed in Section 7.5.1. If the scatter varies with X, then the errors are **heteroscedastic** and the ϵ term is an n-vector with corresponding variances $\sigma_{\epsilon,i}^2$. The error variances, either homoscedastic or heteroscedastic, might be unknown parameters of the model, or might be part of the input dataset as with datasets of the form $(x_i, y_i, \sigma_{\epsilon,i})$ for $i = 1, 2, \ldots, n$. A typical astronomical dataset with heteroscedastic measurement errors in a regression context is shown in Figure 7.1 below.

In most of our discussion here, both X and Y are random variables that take on real values. But many problems in astronomy involve regression with different data structures. As mentioned above, X may be a fixed variable with locations predetermined by the experimental setup: location in an image, wavelength in a spectrum, or observation time in a light curve. If Y is an integer variable with small values based on a counting process, then the statistical problem is called **Poisson regression**. If Y is a binary variable with values 0 or 1 (or a multivalued categorical variable where the categories are not ordered), then the statistical problem is called **logistic regression**. In all cases, the goal of regression is to estimate the regression parameters, such as the intercept β_0 and slope β_1 in (7.2), by an estimation procedure using a homogeneous sample of objects with measured (x_i, y_i) values.

In most forms of regression, the scientist must specify a single variable (or, in some multivariate methods, a group of variables) to be the response variable. This choice is not dictated by any mathematics and, if the scientific problem does not demand the selection of a special dependent variable, then ordinary regression is not an appropriate tool. It is ill-advised to choose the dependent variable because it is newly measured in an observational study, as the statistical results then depend on the historical sequence of telescopic developments rather than on properties of the celestial population. Often the astronomer is seeking intrinsic relationships that treat the variables in a symmetrical fashion. We describe some symmetric bivariate regressions in Section 7.3.2, and symmetric multivariate methods such as principal components analysis in Chapter 8.4.2.

It is important to recognize that the statisticians' term "linear models" includes many models that the astronomer would consider nonlinear. The critical criterion is linearity in the model parameters, not in the model variables. Thus, the following models fall under the rubric of linear regression:

$$Y = \beta_0 + \beta_1 X + \beta_2 X^2 + \epsilon$$
$$Y = \beta_0 e^{-X} + \epsilon$$
$$Y = \beta_0 + \beta_1 \cos X + \beta_2 \sin X + \epsilon. \tag{7.3}$$

The first of these linear models is a second-degree polynomial, the second represents an exponential decay, and the third is a sinusoidal behavior with fixed phase. The sums, or **mixtures**, of linear models are themselves linear.

The following models are nonlinear:

$$Y = \left(\frac{X}{\beta_0}\right)^{-\beta_1} + \epsilon$$
$$Y = \frac{\beta_0}{1 + (X/\beta_1)^2} + \epsilon$$
$$Y = \beta_0 + \beta_1 \cos(X + \beta_2) + \beta_3 \sin(X + \beta_2) + \epsilon$$
$$Y = \begin{cases} \beta_0 + \beta_1 X & \text{for } X < x_0 \\ \beta_2 + \beta_3 X & \text{for } X > x_0. \end{cases} \tag{7.4}$$

In each case, one or more of the β parameters is not in a simple linear relationship to the response variable. The first of these models is the power-law (or Pareto) model, the second often arises in equilibrium distributions of cosmic stars or gases, the third is a sinusoidal behavior with unknown phase, and the fourth is a segmented linear regression model. Care must be taken when astronomical data are modeled with functions derived from astrophysical theory. Some astrophysical models (which may appear complicated) will be linear in the statistical sense, while other models (which may appear simple) will be nonlinear.

Once a model is defined for a chosen problem, then a method must be chosen for **parameter estimation**. These include the method of moments, least squares, maximum likelihood, and Bayesian inference, as outlined in Section 3.4. We concentrate here on least-squares methods because these are most familiar with analytic results, and often give results identical to maximum likelihood estimation. A reasonable and common procedure is to add parameters to an unsuccessful model, such as a segmented linear model if a simple linear regression does not fit, or adding more components to an astrophysical model. The principle of parsimony (Occam's Razor) is applied to avoid proliferation of model parameters, and several criteria are available to quantify comparison of models. This enterprise involves methods of **model validation** and **model selection** discussed in Section 7.7.

7.3 Least-squares linear regression

7.3.1 Ordinary least squares

We consider first the simple and common case introduced in Equation (7.2): bivariate
regression with an assumed linear relationship based on (x_i, y_i) data without ancillary
information or effects. Our goal is to estimate the intercept β_0 and slope β_1 without
bias and with low variance. The **ordinary least-squares (OLS) estimator** for the linear
regression model in Equation (7.2) minimizes the **residual sum of squares (*RSS*)** between
the observed response values y_i and the model predictions,

$$\min \; RSS = \min \; \sum_{i=1}^{n} (Y_i - \beta_0 - \beta_1 X_i)^2. \tag{7.5}$$

If the x_i values are plotted on the horizontal axis and the y_i values on the vertical axis, this
procedure chooses a line that minimizes the sum of squared vertical distances from the data
points to the line.

Differentiating *RSS* with respect to the two regression parameters and setting these
quantities to zero gives two linear equations with two unknowns that can be algebraically
solved. The resulting regression parameter estimators are

$$\hat{\beta}_{1,OLS} = \frac{\sum_{i=1}^{n}(X_i - \bar{X})(Y_i - \bar{Y})}{\sum_{i=1}^{n}(X_i - \bar{X})^2} = \frac{S_{XY}}{S_{XX}}$$

$$\hat{\beta}_{0,OLS} = \bar{Y} - \hat{\beta}_{1,OLS}\bar{X} \tag{7.6}$$

where $\bar{X}$ and $\bar{Y}$ are the sample means. The S_{XX} and similar designations are commonly
used to represent sums of squared residuals. The OLS estimator of the error variance σ^2 is

$$\hat{\sigma}^2 = S^2 = \frac{RSS}{n-2} \tag{7.7}$$

where the denominator incorporates a reduction by the two parameters of the fit. These OLS
estimators of the slope and intercept are also the maximum likelihood estimators (MLE)
(Section 7.3.6) when the errors are normally distributed, $\epsilon = N(0, \sigma^2)$.

Under the assumption that the errors are i.i.d. random variables, the slope and intercept
are unbiased and asymptotically (for large n) normally distributed according to

$$\hat{\beta}_{1,OLS} \sim N\left(\beta_1, \frac{\sigma^2}{S_{XX}}\right)$$

$$\hat{\beta}_{0,OLS} \sim N\left(\beta_0, \sigma^2\left(\frac{1}{n} + \frac{\bar{X}^2}{S_{XX}}\right)\right). \tag{7.8}$$

If the errors are also normally distributed with zero mean and known variance σ^2, then the
variance is χ^2 distributed according to

$$\hat{\sigma}^2 \sim \left(\frac{\sigma^2}{n-2}\right)\chi_{n-2}^2. \tag{7.9}$$

However, the true variance σ^2 is rarely known and only the sample variance S^2 is available. This changes the statistical properties of the OLS regression estimators; for example, the standardized slope now follows a t distribution rather than a normal distribution. That is,

$$T = \frac{\hat{\beta}_1 - \beta_1}{SE} \quad \text{where} \quad SE = \frac{S}{\sqrt{S_{XX}}} \tag{7.10}$$

is t distributed with $n - 2$ degrees of freedom. SE is known as the **standard error**. The $100(1-\alpha)\%$ confidence interval where (say) $\alpha = 0.05$ is $(\hat{\beta}_1 - t_{\alpha/2,n-2}SE, \hat{\beta}_1 + t_{\alpha/2,n-2}SE)$. The **confidence interval** for Y at a specified value of $X = x$ is

$$Y(x) = \hat{\beta}_0 + \hat{\beta}_1 x \pm t_{\alpha/2,n-2}S\sqrt{1 + \frac{1}{n}\frac{(x - \bar{X})^2}{S_{XX}}}. \tag{7.11}$$

For a range of x values, this forms a confidence band about the best-fit OLS line bounded by two hyperbolas.

The extension of these results to multivariate problems is mathematically straightforward, although the notation of linear algebra is now needed. For example, when $\mathbf{X}$ is an $n \times p$ matrix with a column for each of the p covariates, the OLS slope can be expressed as

$$\hat{\beta}_{1,OLS} = (\mathbf{X}^T\mathbf{X})^{-1}\mathbf{X}^T\mathbf{y}, \tag{7.12}$$

provided $\mathbf{X}^T\mathbf{X}$ is invertible.

7.3.2 Symmetric least-squares regression

The regression model presented in Section 7.2, and discussed through most of this chapter, requires that one variable (X), or a suite of variables $\mathbf{X}$, is a fixed independent variable without variability, while the other (Y) is a random response variable whose value depends on X with some error ϵ. This asymmetrical relationship between the variables does arise in some astronomical problems, such as the estimation of distance to a cosmic population from some distance indicator, say the period of a pulsating Cepheid variable star or the photometric colors of distant galaxies.

But often the scientific problem confronting astronomers does not specify one variable as uniquely dependent on the other variable(s); rather a joint symmetrical relationship between the variables is sought. One symmetrical least-squares procedure well-known in statistics is Pearson's **orthogonal regression** or **major axis** line, obtained by minimizing the summed squared residuals from orthogonal projections of each point onto the line. Here the slope is given by

$$\hat{\beta}_{1,orth} = \frac{1}{2}\left[\left(\hat{\beta}_{1,inv} - \frac{1}{\hat{\beta}_{1,OLS}}\right) + \text{Sign}(S_{xy})\sqrt{4 + \left(\hat{\beta}_{1,inv} - \frac{1}{\hat{\beta}_{1,OLS}}\right)^2}\right] \tag{7.13}$$

where $\hat{\beta}_{1,inv} = S_{YY}/S_{XY}$ is the inverse regression line with the X and Y variables switched. However, the orthogonal regression slope will change if one of the variables is multiplied by a scale factor, and its small-n performance has high variance. A second approach, the

reduced major axis, was proposed by astronomers and biologists in the 1940s and 1950s to alleviate this problem, but it measures only the ratio of the variances of the variables and not the underlying correlation between the variables.

During the 1980s, astronomers introduced another symmetrical regression, the line that bisects the standard and inverse OLS lines,

$$\hat{\beta}_{1,bis} = \frac{\hat{\beta}_{1,OLS}\hat{\beta}_{1,inv} - 1 + \sqrt{(1 + \hat{\beta}_{1,OLS}^2)(1 + \hat{\beta}_{1,inv}^2)}}{\hat{\beta}_{1,OLS} + \hat{\beta}_{1,inv}}. \tag{7.14}$$

However, astronomers erroneously assumed that all of these lines have the same asymptotic variances. The variances of these and other least-squares lines, using asymptotic formulae, are given by Isobe *et al.* (1990) and Feigelson & Babu (1992). They found that, particularly for small-n samples or large scatter, the OLS bisector performs better than other symmetrical regression lines.

7.3.3 Bootstrap error analysis

The standard regression model in (7.1) and the procedures of least-squares regression have some important limitations. The assumptions of homoscedastic and/or normal errors may not apply to the problem at hand, or the appropriate error model may be unknown. The assumption of normality can be relaxed for large samples but not for small-n datasets. This is often the case in astronomy. For example, the astronomer does not confidently know that the distribution of properties around the Fundamental Plane of elliptical galaxies (Djorgovski & Davis 1987) is Gaussian when measured in convenient but arbitrary physical units. Furthermore, the variance of the residuals will systematically underestimate the true variance of the model by a factor of $(1 - p/n)$ where p is the number of model parameters.

To treat these and other difficulties, resampling procedures such as the bootstrap, jack-knife and cross-validation are often used. Basic properties of the jackknife and bootstrap were presented in Section 3.6 and cross-validation is discussed below in the context of model validation (Section 7.7.2). Here we concentrate on bootstrap methods for linear regression.

Consider the simple linear regression model where the data $(X_1, Y_1), \ldots, (X_n, Y_n)$ satisfy Equation (7.2). Here the error variables ϵ_i need not be Gaussian, but are assumed to be independent with zero mean and variance σ_i^2 and may be either homoscedastic (σ_i^2 constant) or heteroscedastic (σ_i^2 different). Figure 7.1 below shows heteroscedastic errors in a linear regression problem for an astronomical dataset.

There are two conceptually separate models to consider, random and fixed design models. In the first case, the pairs $\{(X_1, Y_1), \ldots, (X_n, Y_n)\}$ are assumed to be random data points and the conditional mean and variance of e_i given X_i are assumed to be zero and σ_i^2. In the latter case, $X_1, \ldots, X_n$ are assumed to be fixed numbers (fixed design). In both the cases, the least squares estimators $\hat{\beta}_0$ and $\hat{\beta}_1$ of β_0 and β_1 are given by

$$\hat{\beta}_1 = S_{XY}/X_{XX} \quad \text{and} \quad \hat{\beta}_0 = \bar{Y}_n - \hat{\beta}_1 \bar{X}_n. \tag{7.15}$$

However the variances of these estimators are different for a random and fixed designs, though the difference is very small for large n. We shall concentrate on the fixed design case here.

The variance of the slope $\hat{\beta}_1$ is given by

$$\text{var}(\hat{\beta}_1) = \sum_{i=1}^{n} (X_i - \hat{X}_n)^2 \sigma_i^2 / S_{XX}^2, \tag{7.16}$$

and depends on the individual error deviations σ_i, which may or may not be known. Knowledge of $\text{var}(\hat{\beta}_1)$ provides the confidence intervals for β_1. Several resampling methods are available in the literature to estimate the sampling distribution and $\text{var}(\hat{\beta}_1)$. We consider two bootstrap procedures: a) the paired bootstrap, and b). the classical bootstrap.

For the **paired bootstrap**. A random sample $(\tilde{X}_1, \tilde{Y}_1), \ldots, (\tilde{X}_n, \tilde{Y}_n)$ is drawn from $(X_1, Y_1), \ldots, (X_n, Y_n)$ with replacement. Bootstrap estimators of slope and intercept are constructed as

$$\tilde{\beta}_1 = \frac{\sum_{i=1}^{n} (\tilde{X}_i - \bar{\tilde{X}})(\tilde{Y}_i - \bar{\tilde{Y}})}{\sum_{i=1}^{n} (\tilde{X}_i - \bar{\tilde{X}})^2} \quad \text{and} \quad \tilde{\beta}_0 = \bar{\tilde{Y}} - \tilde{\beta}_1 \bar{\tilde{X}}. \tag{7.17}$$

To estimate the sampling distribution and variance, the procedure is repeated N_{boot} times to obtain

$$\tilde{\beta}_{1,1}^*, \tilde{\beta}_{1,2}^*, \ldots, \tilde{\beta}_{1,N_{boot}}^* \quad \text{where} \quad N_{boot} \sim n(\log n)^2 \tag{7.18}$$

is the necessary number of bootstrap samples. The histogram of these $\tilde{\beta}_1^*$ values gives a good approximation to the sampling distribution of $\hat{\beta}_1$. The variance $\hat{\beta}_1$ is estimated by

$$Var_{\text{Boot}} = \frac{1}{N_{boot}} \sum_{j=1}^{N} (\tilde{\beta}_{1,j} - \hat{\beta}_1)^2. \tag{7.19}$$

The paired bootstrap is robust to heteroscedasticity and can thus be used when different points have different measurement errors, as in Figure 7.1.

Now the **classical bootstrap** is applied where resamples are obtained by random selection of residuals from the original model fit, rather than by random selection from the original data points.

The procedure starts with the model $Y_i = \hat{\beta}_0 + \hat{\beta}_1 X + \epsilon$ fitted to the original data. Let $\hat{e}_i$ denote the residual of the i-th element of $\hat{e}_i = Y_i - \hat{\beta}_0 - \hat{\beta}_1 X_i$ and define $\tilde{e}_i$ to be

$$\tilde{e}_i = \hat{e}_i - \frac{1}{n} \sum_{j=1}^{n} \hat{e}_j. \tag{7.20}$$

A bootstrap sample is obtained by randomly drawing $e_1^*, \ldots, e_n^*$ with replacement from $\tilde{e}_1, \ldots, \tilde{e}_n$. The classical bootstrap estimators β_1^* and β_0^* of the slope and the intercept are given by

$$\begin{aligned}
\beta_1^* - \hat{\beta}_1 &= \frac{\sum_{i=1}^{n} (X_i - \bar{X})(e_i^* - \bar{e}^*)}{\sum_{i=1}^{n} (X_i - \bar{X})^2} \\
\beta_0^* - \hat{\beta}_0 &= (\hat{\beta}_1 - \beta_1^*)\bar{X}_n + \bar{e}_n^*.
\end{aligned} \tag{7.21}$$

The variance of the original slope coefficient $\hat{\beta}$ is then estimated following Equation (7.19) with the β_1^* values replacing $\tilde{\beta}_1$ values. This variance estimator is efficient if the residuals are homoscedastic, but it is inconsistent (that is, does not even approach the actual variance) if the residuals are heteroscedastic.

7.3.4 Robust regression

Often a dataset has a small fraction of points with very large residuals from the regression curve. Typical reasons for outliers are long-tailed distributions of a homogeneous sample, contamination of extraneous populations into the sample, and erroneous measurements (in either the X or Y variable) during acquisition of the data. Many statistical techniques, including least-squares regression based on assumptions of normal distributions, can be very sensitive to even minor intrusion of outliers into the sample under study. If the identities of the erroneous or contaminating objects can be independently established, then they can be removed from consideration prior to statistical analysis. This removal would be based on scientific, rather than statistical, criteria.

Robust statistics provide strategies to reduce the influence of outliers when scientific knowledge of the identity of the discordant data points is not available. We focus here on robust regression; other topics in robust statistics are discussed in Chapter 5. Desirable techniques will give parameter estimators that are unbiased (i.e. converge to the true value as the sample size n increases) and consistent (diminish mean squared error as n increases) with a high breakdown point. The breakdown point of an estimator is the proportion of arbitrarily large data values that can be handled before giving an arbitrarily large estimator value. For example, the sample mean is sensitive to even a single outlier with a breakdown point of $1/n$, while the sample median is very robust with a breakdown point of 0.5. Several approaches to robust regression can be considered: iteratively moving discrepant points closer to the line; iteratively downweighting their contribution to the estimation procedure; or keeping all points but minimizing a function with less dependence on outlying points than the sum of squared residuals. Huber & Ronchetti (2009) give an authoritative presentation of robust regression; the slim volume by Andersen (2008) covers these methods also.

The most common method, called **M-estimation**, takes this last approach. For any function ψ, any solution $\hat{\beta}_M$ of

$$\sum_{i=1}^{n} \psi(y_i - \hat{\beta}_M x_i)x_i = 0 \tag{7.22}$$

is called an M-estimator. It may not be unique. MLE and least squares estimators are special cases of M-estimators.

Under certain restrictions, theory establishes that these regression M-estimators are asymptotically (i.e. for large n) consistent, unbiased and normally distributed. These restrictions include: the ρ function must be convex, differentiable, and nonmonotone; the ψ function should have no systematic effect on the noise, $E[\psi(\epsilon_i)] = 0$; the maximum leverage should not be too large; the errors should not be asymmetrical; and for multivariate problems, the number of points n should scale with the dimensionality as p^3.

The classical least-squares solution is recovered when $\psi(x) = x$; this is not robust to outliers. Three ψ functions which are robust with high breakdown points are in common use. The simplest is the trimmed estimator

$$\psi_{trim}(x) = \begin{cases} x & |x| < c \\ 0 & \text{otherwise} \end{cases} \tag{7.23}$$

which entirely removes (or **trims**) outliers with residuals above a chosen value. The **Huber estimator** keeps all points in the dataset but resets large residuals to a chosen value,

$$\psi_{Huber}(x) = \begin{cases} -c & x < -c \\ x & |x| < c \\ c & x > c. \end{cases} \tag{7.24}$$

For normal distributions, this estimator is optimal when $c = 1.345$. An application of this M-estimator to a linear regression involving an astronomical dataset is shown in Figure 7.1.

Tukey's bisquare (or biweight) function is one of several **redescending** M-estimators that trims very large outliers and downweights intermediate outliers in a smooth fashion,

$$\psi_{Tukey}(x) = \begin{cases} x(c^2 - x^2)^2 & |x| < c \\ 0 & \text{otherwise.} \end{cases} \tag{7.25}$$

For normal distributions, this estimator is optimal when $c = 4.685$. The optimal slope is typically computed using an iteratively reweighted least-squares algorithm with weights

$$w_i = \psi\left(\frac{Y_i - \beta X_i}{s}\right) \Big/ \left(\frac{Y_i - \beta X_i}{s}\right) \tag{7.26}$$

where s is a scale factor like the median absolute deviation (MAD, Chapter 5). Asymptotic variances for regression coefficients are similar to those of ordinary least-squares estimation weighted by sums involving $\psi(Y_i - \beta X_i)$. M-estimators have the disadvantages that the solution may not be unique.

A number of other robust regression estimators have been introduced. In the early nineteenth century, S. Laplace proposed the **least absolute deviation (LAD or L_1) estimator** which minimizes

$$\sum_{i=1}^{n} |Y_i - \beta_{LAD} X_i|. \tag{7.27}$$

More recently, P. Rousseeuw developed the **least trimmed squares (LTS)** that minimizes the quantity

$$\sum_{i=1}^{q} |y_i - \beta_{LTS} x_i|^2 \tag{7.28}$$

where the smallest $q = (n + p + 1)/2$ squared residuals are included in the sum. Here p is the number of parameters in the regression. The LTS estimator is thought to be robust even when nearly half of the data are outliers. Another approach is called **S-estimation** involving the search for an optimal scale factor for the residuals, and another is **MM-estimation** that combines the advantages of the M- and S-estimators at the cost of computational effort. These methods have high breakdown points; today, MM-estimators are popular robust regression procedures.

A final robust regression method is the **Thiel–Sen median slope** line with slope β_{TS} given by the median of the $n(n + 1)/2$ slopes of lines defined by all pairs of data points,

$$\beta_{ij} = \frac{Y_i - Y_j}{X_i - X_j}. \tag{7.29}$$

This method has its roots in work by R. Boscovich in 1750 and can be formulated in terms of Kendall's τ rank correlation coefficient (Sen 1968). The calculation can be computer-intensive for large-n samples, but $O(n \log n)$ algorithms have been developed (Brönnimann & Chazelle 1998). The Thiel–Sen estimator is conceptually simple, giving reliable parameter confidence intervals with a reasonably high breakdown point against outliers. It treats heteroscedastic data well and is effective when censoring is present (Section 10.4.2). Applications of the Thiel–Sen line to astronomical datasets are shown in Figures 7.1 and 10.3.

7.3.5 Quantile regression

Standard regression can be viewed as a method for obtaining the mean of Y values as a smooth function of X. However, this does not take cognizance of the distribution of Y values about this mean which may be approximately uniform, centrally concentrated (e.g. Gaussian), asymmetrical or multimodal. An alternative strategy is to estimate quantile curves (and surfaces, for multivariate problems) rather than mean curves. The obvious choice is to examine median curves, where the median serves as a measure of location that is more robust to outliers and non-Gaussianity than the mean (Section 5.3.2). But considerable insight into the relationship between X and Y accrues by examining the response of the (say) $\tau = 0.1, 0.25, 0.75, 0.9$ quantiles, as well as the median $\tau = 0.5$ quantile, of Y to variations in X.

The theory and computations involved in quantile regression are not straightforward because the construction of a quantile function from an empirical distribution function is mathematically nontrivial (Section 2.6). Quantile regression is equivalent to an optimization problem in linear programming using, for example, the simplex algorithm. The τ-th sample quantile estimator $\hat{q}_\tau$ of the data $(Y_1, Y_2, \ldots, Y_n)$ is obtained by finding the value of q in the range exhibited by Y that minimizes the quantity (Koenker 2005)

$$\hat{q}_\tau = (\tau - 1) \sum_{y_i < q} (Y_i - q) + \tau \sum_{y_i \geq q} (Y_i - q). \tag{7.30}$$

For a linear dependency on the independent variable X, the quantile regression slope estimator $\hat{\beta}_q$ associated with the τ-th quantile is found by minimizing a similar function where Y_i is replaced by $Y_i - \beta_q X_i$. This can be calculated for several values of τ to give a suite of regression curves corresponding to quantiles of the Y distribution.

Quantile regression gives considerably more detailed information about the $X-Y$ relationship than the single mean relationship provided by least-squares regression. Quantile regression captures effects such as outliers, asymmetrical and heteroscedastic errors that are missed by standard regression techniques. A suite of inferential methods is available for quantile regression. These include confidence intervals for sample quantiles, confidence intervals for the regression coefficients, hypothesis tests of location and scale, bootstrap methods, and more. An example of quantile regression for an astronomical dataset is shown in Figure 7.4.

7.3.6 Maximum likelihood estimation

Since the OLS solution is also the MLE under Gaussian assumptions for the linear regression model defined in Equation (7.2), there is no urgent need to derive the maximum likelihood solution. It is nonetheless instructive to see this simple case, as more difficult cases that benefit from an MLE approach are treated in a similar fashion.

Assuming the X_i are fixed rather than random variables, we can formulate (7.2) as a joint distribution of the Y_i values,

$$Y_i \sim N(\beta_0 + \beta_1 X_i, \sigma^2). \tag{7.31}$$

Assuming independence of errors, the likelihood is the joint probability distribution function

$$L(Y_1, Y_2, \ldots, Y_n | \beta_0, \beta_1, \sigma^2) = \prod_{i=1}^{n} L(Y_i | \beta_0, \beta_1, \sigma^2)$$

$$= \frac{1}{(2\pi\sigma^2)^{n/2}} \exp\left[\frac{-1}{2\sigma^2} \sum_{i=1}^{n} (Y_i - \beta_0 - \beta_1 X_i)^2\right]. \tag{7.32}$$

The loglikelihood, omitting constants, is then

$$-2 \ln L(\beta_0, \beta_1, \sigma^2 | Y_1, Y_2, \ldots, Y_n) = n\ln\sigma^2 + \frac{1}{\sigma^2} \sum_{i=1}^{n} (Y_i - \beta_0 - \beta_1 X_i)^2. \tag{7.33}$$

Differentiating this loglikelihood with respect to the three model parameters gives three linear equations which can be solved to give

$$\begin{aligned}
\hat{\beta}_{1,MLE} &= \frac{S_{XY}}{S_{XX}} \\
\hat{\beta}_{0,MLE} &= \bar{Y} - \hat{\beta}_1(MLE)\bar{X} \\
\hat{\sigma}^2_{MLE} &= \frac{RSS}{n}.
\end{aligned} \tag{7.34}$$

Note that the MLE of the variance is biased by a factor of $n/(n-2)$; the OLS variance in Equation (7.7) gives the correct value. This problem is similar to the bias found for estimation of the normal mean in Section 3.6.1.

7.4 Weighted least squares

Astronomers often obtain estimates of observational errors independently of the measurements themselves. For bivariate situations, the data are of the form $(X_i, \sigma_{X,i}, Y_i, \sigma_{Y,i})$. Here we set $\sigma_{X,i} = 0$; regression for problems with errors in both variables is discussed in Section 7.5. The errors are no longer i.i.d. — they are still independent, but are no longer identically distributed — and we make the standard assumption that they are normally distributed, $\epsilon_i \sim N(0, \sigma_{Y,i}^2)$. The regression model in Equation (7.2) is now

$$Y_i = \beta_0 + \beta_1 X_i + \epsilon_i. \tag{7.35}$$

The least-squares procedure, now weighted by the inverse of the heteroscedastic variances, minimizes the quantity

$$S_{r,wt} = \sum_{i=1}^{n} \frac{(Y_i - \beta_0 - \beta_1 X_i)^2}{\sigma_{Y,i}^2}. \tag{7.36}$$

The resulting weighted slope and intercept estimates are

$$\hat{\beta}_{1,wt} = \frac{\sum_{i=1}^{n}(X_i - \bar{X}_{wt})(Y_i - \bar{Y}_{wt})/\sigma_{Y,i}^2}{\sum_{i=1}^{n}(X_i - \bar{X}_{wt})^2/\sigma_{Y,i}^2}$$

$$\hat{\beta}_{0,wt} = \bar{Y}_{wt} - \hat{\beta}_{wt,1}\bar{X}_{wt} \quad \text{where}$$

$$\bar{X}_{wt} = \sum_{i=1}^{n} \frac{X_i}{\sigma_{Y,i}^2} \quad \text{and} \quad \bar{Y}_{wt} = \sum_{i=1}^{n} \frac{Y_i}{\sigma_{Y,i}^2}. \tag{7.37}$$

The residuals are also weighted according to

$$\hat{e}_{i,wt} = \frac{Y_i - \hat{\beta}_{0,wt} - \hat{\beta}_{1,wt}X_i}{\sigma_{Y,i}}. \tag{7.38}$$

An application of a weighted least-squares linear regression line to an astronomical dataset is shown in Figure 7.1.

In astronomy this weighted least-squares regression procedure, minimizing $S_{r,wt}$ given in Equation (7.36), is called **minimum χ^2 regression**. This is due to the resemblance of $S_{r,wt}$ to Pearson's χ^2 statistic,

$$X^2 = \sum_{i=1}^{k} \frac{(Y_i - M_i)^2}{M_i} \tag{7.39}$$

where Y_i are counts and M_i are model estimators, for $i = 1, 2, \ldots, k$ classes. Minimum χ^2 for regression and goodness-of-fit has long been the most common regression technique in astronomy. We now discuss some limitations of minimum χ^2 regression methodology.

The statistic (7.39) was proposed by K. Pearson in 1900 specifically for the multinomial experiment. Here a series of n trials is made of a variable that has $2 \le k < n$ distinct categorical outcomes. A common example is rolling a six-sided die many times with $k = 6$ random outcomes. The resulting observed distribution in the $i = 1, 2, \ldots, 6$ categories is summarized in a 1×6 contingency table. Pearson's X^2 statistic is designed to test the null hypothesis, H_0, that the distribution of outcomes is consistent with the predictions of a model M. In a more general framework that does not assume a linear model, X^2 is defined as

$$X^2 = \sum_{i=1}^{k} \left(\frac{[O_i - M_i(\theta_p)]^2}{M_i(\theta_p)} \right) \tag{7.40}$$

where n is the number of original observations, k is the number of categories (also called cells or bins), O_i are the observed number in each category such that $\sum_{i=1}^{k} O_i = n$, M_i are the expected number in each category from the model and θ_p are the p parameters of the model.

R. A. Fisher showed that this X^2 statistic is asymptotically (as $n \to \infty$) distributed as the χ^2 distribution with $k - p - 1$ degrees of freedom. Formally stated,

$$P[X^2 \geq x|H_0] \simeq P[\chi^2_{k-p-1} \geq x]. \tag{7.41}$$

This is the basis for **Pearson's χ^2 test**, and can be phrased as follows: "Given the null hypothesis H_0 that the data are drawn from a population following the model M, the probability that the X^2 statistic in Equation (7.40) for n observations exceeds a chosen value α (say, $\alpha = 0.05$) is asymptotically (for $n \to \infty$) equal to the probability that a χ^2 random variable with $k - p - 1$ degrees of freedom exceeds x." Approximate critical values $\chi^2_{k-p-1,1-\alpha}$ of X^2 for $0 < \alpha < 1$ are obtained from $P[\chi^2_{k-p-i} > \chi^2_{k-p-1,1-\alpha}] = \alpha$. The theory underlying Pearson's and Fisher's results are presented in Greenwood & Nikulin (1996, Chapter 1).

Use of Pearson's χ^2 test for regression and goodness-of-fit dates back to Slutsky (1913) and became widespread by the 1940s. X^2 is calculated repeatedly for a range of θ_p values, and perhaps for different model families M^*, to evaluate what range of models and parameters is consistent with the observed frequencies. J. Neyman proposed a variant known as the **restricted χ^2 test** allowing a model to be compared to a restricted set of alternative models rather than the entire space of possible alternative models.

The following four functions asymptotically (for large n) behave like χ^2 random variables (Berkson 1980),

$$
\begin{aligned}
\text{Pearson} \quad & X^2 = \sum (O_i - M_i)^2 / M_i, \\
\text{Neyman} \quad & X^2 = \sum (O_i - M_i)^2 / O_i \\
\text{Likelihood} \quad & X^2 = 2 \sum O_i \ln(O_i/M_i) \\
\text{Kullback} \quad & X^2 = 2 \sum M_i \ln(M_i/O_i).
\end{aligned} \tag{7.42}
$$

Asymptotically (for large n) the model parameter estimates obtained by minimizing these functions are all consistent and have the same χ^2 distribution.

It is important to realize that, in many cases, astronomers use yet another χ^2-like function based on heteroscedastic measurement errors $\sigma_{i,me}$,

$$X^2_{me} = \sum_{i=1}^{k} \frac{(O_i - M_i)^2}{\sigma^2_{i,me}}. \tag{7.43}$$

If the measurement errors are obtained from counts in the bins under consideration, then $\sigma_{O_i}^2 \simeq O_i$ and the function is asymptotically equal to Neyman's X^2 function. Under this situation, the astronomers' minimizing of (7.43) is equivalent to minimizing the other four functions. When the measurement errors are derived in a complicated fashion — for example, involving measurements from source-free regions of the dataset or calibration runs of the instrument — the astronomers' X_{me}^2 may not be asymptotically χ^2 distributed. In this case, the measurement errors in the denominator have an entirely different origin than the bin counts O_i or the model predictions M_i, and the distribution of X_{me}^2 is unknown without detailed study. A detailed treatment of these issues is provided by Greenwood & Nikulin (1996). They write: "χ^2 testing remains an art. No attempt at a complete table of set-ups and optimal tests can be made. Each practical situation has its own wrinkles." Even mild deviations from Pearson's multinomial experimental design can lead to difficulties. Chernoff & Lehmann (1954) consider the case where the exact data values are known prior to grouping them into k bins. Then the number of degrees of freedom of the X^2 statistic is not fixed, and incorrect significance probabilities may be obtained. This situation occurs frequently in astronomy where the binning is applied to real-valued data. Astronomers thus often use X_{me}^2 for regression when the number of bins is not determined by the experimental setup. The algorithm for bin boundaries and assignment of bin centroid may be arbitrary (see the discussion in the context of histograms in Section 6.3). In these cases, the convergence to the χ^2 distribution is not guaranteed.

The asymptotic theory of X^2 fails when bins have very few counts. For small-n samples, W. Cochran recommended in the 1950s that the asymptotic χ^2 result is acceptable at the $\alpha = 0.01$ significance level for smooth unimodal models if $k \geq 7$ bins are present and all bins have $O_i \geq 5$ counts (Greenwood & Nikulin 1996).

Finally, astronomers sometimes insert other quantities into the denominator of X^2 such as approximations to the variance. The impact of these substitutions in the denominator of X^2 on its distribution is largely unknown, but the resulting X^2 statistic is unlikely to have a limiting χ^2 distribution.

In summary, the use of weighted least squares, or minimum χ^2, as a regression method is fraught with difficulties. The statistic X_{me}^2 may be useful in some contexts, but it cannot be assumed to follow the χ^2 distribution.

7.5 Measurement error models

Throughout this volume we wrestle with the errors that astronomers often obtain to quantify the uncertainty of their measurements. They estimate measurement errors by carefully calibrating and simulating uncertainties within the detector, devoting telescope time to observing dark sky, and studying the statistical characteristics of featureless regions of their images and spectra. These measurement errors, often differing between data points (heteroscedastic), are established independently of, and are published along with, the actual measurements of scientific interest. The resulting datasets have the structure $(X_i, \sigma_{X,i}, Y_i, \sigma_{Y,i})$ rather than the commonly addressed (X_i, Y_i) structure where X is the measured surrogate variable for the unobserved true variable W and Y is the measured surrogate for the unobserved true variable V.

Modern statistics addresses regression problems involving measurement errors using a hierarchical structure that allows both known and unknown measurement errors to be incorporated into the model (Carroll *et al.* 2006, Buonaccorsi 2009). Regression models of this type are variously called **measurement error models**, **errors-in-variables models**, and **latent variable models**. A general additive measurement error model assuming linear relationships can be written as

$$V = \beta_0 + \beta_1 W + \epsilon$$
$$W = X + \mu$$
$$V = Y + \eta. \tag{7.44}$$

The underlying relationship is between unobserved **latent** variables V and W in the regression equation, while W and V are linked to the observed **manifest** variables X and Y with additional error terms in the measurement equations.

As in other regression models considered here, we assume that all scatter terms $-\ \epsilon,\ \mu$ and $\eta\ -$ are Gaussian with zero mean although with different variances. For homoscedastic measurement errors and no intrinsic scatter, $\epsilon = 0$, μ is distributed as $N(0, \sigma_\mu^2)$, and η is distributed as $N(0, \sigma_\eta^2)$. For the heteroscedastic case, the error terms are vectors of n elements each with their own known variances: $\mu_i \sim N(0, \sigma_{\mu,i}^2)$ and $\eta_i \sim N(0, \sigma_{\eta,i}^2)$. Heteroscedastic errors violate the "identical" distribution assumptions of i.i.d. random variables. The ϵ **formula noise** term represents **intrinsic scatter** about the line in the underlying population and is usually taken to be homoscedastic, $\epsilon_i \sim N(0, \sigma_\epsilon^2)$. The $\sigma_\epsilon, \sigma_\mu$ and σ_η terms can either be treated as unknown parameters of the model or as known inputs to the model.   While it may seem that the splitting of the regression model into two parts involving manifest and latent variables is artificial, it provides a very flexible formulation for a variety of cases that astronomers often encounter.

1. If the only source of scatter is in the response variable Y, then $\sigma_\mu = 0$ and σ_η can be incorporated into σ_ϵ. The model then becomes the traditional structural linear regression model of Equation (7.2) discussed in Section 7.3.
2. If the only source of variance is from measurement error such that $\sigma_\epsilon = 0$, then we have a functional regression model where the underlying population is assumed to precisely follow the linear (or other) regression function without any arror.
3. The case where $\sigma_\epsilon = \sigma_\mu = 0$ is the model commonly assumed by astronomers when they place $\sigma_{\eta,i}^2$ values in the denominator of the X^2 statistic and perform a weighted least-squares regression as in Equation (7.43).
4. If Y and X have homoscedastic measurement errors, then σ_μ and σ_η are constants and the regression is determined by the ratio $h = \sigma_\eta/\sigma_\mu$. When $h = 1$, the least-squares solution becomes Pearson's orthogonal regression line (Section 7.3.2).
5. If the scatter is dominated by heteroscedastic measurement errors in both Y and X, then $\sigma_\epsilon = 0$ and the μ_i and η_i are normally distributed errors. This is a common situation in astronomy because the independent variables are often random variables subject to similar measurement errors as the dependent variable.

6. Astronomers have widely studied the case where Y has two types of scatter, one intrinsic to the celestial objects (σ_ϵ) and the other arising from the astronomical measurement process (σ_η). Several models are outlined below (Akritas & Bershady 1996; Tremaine *et al.* 2002; Kelly 2007).

7. Other more complex relationships can be calculated within a measurement error model. The three error terms (ϵ, μ and η) may depend on each other or on the values of the measured variables X and Y. A common situation of this type is measurement errors that increase as the variable values decrease due to reduced signal-to-noise ratios in the detection process. It is possible to introduce these relationships as additional equations in the hierarchy shown in Equation (7.44), though this is rarely done.

The astronomer also encounters **systematic errors** when observations do not measure their underlying variables in an unbiased fashion, even if the stochastic measurement error is small. These are sometimes called **calibration biases**, as the situation arises when a poorly calibrated instrument is used. The relationship between the observed and true quantities can involve additive, multiplicative or other functional biases. They can be incorporated into a more elaborate measurement equation such as

$$W = \gamma_0 + \gamma_1 X + \mu \tag{7.45}$$

where γ_0 and γ_1 are additive and multiplicative errors, respectively, representing bias and scaling problems in the W measurements of the X variable. Regression estimators can be sought using the recalibrated variable $W^* = (W - \gamma_0)/\gamma_1$. Osborne (1991) reviews statistical approaches to calibration problems.

It can be difficult to distinguish random **measurement errors** and **systematic errors** in astronomical data analysis. Systematic errors need not arise from faulty instrumental calibration. For example, the "measurement" of the abundance of oxygen from a stellar spectrum emerges from a complex process involving instrumental corrections (flat fielding, wavelength calibration), estimation of a detrended continuum level to measure the oxygen line strength, and application of an astrophysical curve-of-growth analysis to elemental abundances. Each step may introduce systematic errors which are often fused into a single estimated error term. Another source of systematic error arises when an astrophysical model is used as the regression function, rather than a simple heuristic model like a linear function. Here the model uncertainties arise from imperfectly known physical quantities or processes.

Both modern and classical approaches to these and other cases are presented in two monographs on measurement error regression models by Carroll *et al.* (2006) and Buonaccorsi (2009). We summarize here a few of these results, particularly those familiar in physics and astronomy.

7.5.1 Least-squares estimators

The homoscedastic functional regression model (Case 4 above) where $\sigma_\epsilon = 0$ and the response and independent variables each have constant error can be treated as a weighted orthogonal regression line that minimizes the squared sum of distances

$$\sum_{i=1}^{n} \left[(Y_i - \beta_0 \beta_1 X_i)^2 + \delta (W_i - X_i)^2 \right] \text{ where } \delta = \sigma_{\eta}^2 / \sigma_{\mu}^2 \tag{7.46}$$

and X_i are the unknown true values corresponding to the measured W_i values. A well-known algorithm called **orthogonal distance regression** is available to perform this calculation for both linear and nonlinear functions (Boggs *et al.* 1987).

Physicist D. York (1966) derived a least-squares solution for a functional regression model when both variables are subject to heteroscedastic measurement errors, as in Case 5 above. This solution is sometimes used in the physical sciences but seems unknown in the statistics community. The slope of the regression line β_1 is obtained by numerical solution to the equation

$$\beta_1^3 \sum_{i=1}^{n} \frac{Q_i^2 (X_i - \bar{X})^2}{\sigma_{\mu,i}^2} - 2\beta_1^2 \sum_{i=1}^{n} \frac{Q_i^2 (X_i - \bar{X})(Y_i - \bar{y})}{\sigma_{\mu,i}^2} -$$

$$\beta_1 \left[\sum_{i=1}^{n} Q_i (x_i - \bar{X})^2 - \sum_{i=1}^{n} \frac{Q_i^2 (Y_i - \bar{Y})^2}{\sigma_{\mu,i}^2} \right] + \sum_{i=1}^{n} Q_i (X_i - \bar{X})(Y_i - \bar{Y}) = 0$$

$$\text{where } Q_i = \frac{\sigma_{\mu,i} \sigma_{\eta,i}}{(\beta_1^2 \sigma_{\eta,i} + \sigma_{\mu,i})}. \tag{7.47}$$

Geometrically, this solution is equivalent to minimizing the sum of square residuals from the data points to the regression line calculated along directions associated with the ratios $h_i = \sigma_{\eta,i} / \sigma_{\mu,i}$. York's approach to functional regression has recently been extended by Caimmi (2011).

A common treatment in astronomy of heteroscedastic errors in both variables (Case 5) involves alteration of the denominator of the X_{me}^2 statistic shown in Equation (7.43). Parameter estimation then proceeds by a weighted least-squares procedure. While sometimes a reasonable procedure, we discuss in Section 7.4 that this approach has various limitations. A popular algorithm appears in the code *FITEXY* of *Numerical Recipes* (Press *et al.* 1997) and has been revised by Tremaine *et al.* (2002). Here one minimizes

$$X_{me,FITEXY}^2 = \sum_{i=1}^{n} \frac{(Y_i - \beta_0 - \beta_1 X_i)^2}{\sigma_{\epsilon,i}^2 + \beta_1 \sigma_{\mu,i}^2}. \tag{7.48}$$

Akritas & Bershady (1996) give an alternative least-squares solution for the linear regression when heteroscedastic measurement errors are present in the variables and intrinsic scatter of the true variables about the line is present (Case 5). Known as the BCES estimator for **bivariate correlated errors and intrinsic scatter**, the slope is

$$\beta(OLS, BCES) = \frac{\sum_{i=1}^{n} (W_i - \bar{W})(V_i - \bar{V}) - \sum_{i=1}^{n} \rho_{\mu\eta,i}^2}{\sum_{i=1}^{n} (W_i - \bar{W})^2 - \sum_{i=1}^{n} \rho_{\mu\eta,i}^2} \tag{7.49}$$

where $\rho_{\mu\eta,i}$ measures the correlation between the measurement errors in the two variables (Case 6 above). The BCES estimator has been criticized in a number of respects (Tremaine *et al.* 2002); for example, the slope estimator neglects heteroscedasticity in the Y variable, and gives unreliable estimates when the spread of X does not greatly exceed the X measurement errors.

When the measured independent variable W has systematic or calibration errors as in Equation (7.45), the unbiased least-squares slope is

$$\hat{\beta}_1(OLS, sys) = \frac{\beta_1(OLS)\left[\gamma_1^2\sigma_\epsilon^2 + \sigma_\mu^2\right] - \rho_{\mu\eta}\sigma_\epsilon\sigma_\mu}{\gamma_1\sigma_\epsilon^2} \tag{7.50}$$

where *sys* refers to a systematic error model (Buonaccorsi 2009). Analogous treatments have been developed when the response variable Y is measured by a surrogate variable with calibration bias and scale error.

Kelly (2007) has shown that several of these least-squares estimators for measurement error regression models suffer higher bias and variance than estimators obtained from maximum likelihood estimation (Section 7.5.3).

7.5.2 SIMEX algorithm

The **simulation–extrapolation (SIMEX) algorithm** is a Monte Carlo procedure introduced in the 1990s to reduce least-squares regression biases arising from measurement errors. It has gained considerable interest in the statistics community as a powerful approach to regression with measurement errors; many variants and applications of SIMEX are described by Carroll *et al.* (2006, Chapter 5).

The idea of SIMEX is to simulate datasets based on the observations with a range of artificially increased measurement errors. The parameter of interest, such as the slope of a linear regression line $\hat{\beta}_1$, is estimated for each simulation. A plot is made relating the estimated parameter to the value of the total measurement error, the sum of the true and artificial measurement error. The resulting curve of the biased parameter is extrapolated to zero measurement error to estimate the parameter without the measurement error bias. The extrapolation is often performed using a quadratic fit. The procedure thus involves a sequence of **simulations** followed by a graphical **extrapolation**, whence the designation SIMEX.

For heteroscedastic measurement errors with known variance, the simulated datasets are generated with

$$\widetilde{W}_i = W_i + \sqrt{\zeta}\mu_i \tag{7.51}$$

for a range of $\zeta \geq 0$ where the W_i are the observed values in the measurement equation (7.44). The μ_1 values are heteroscedastic normal errors with variance σ_i^2. Analogous procedures are used for homoscedastic errors, multiplicative errors, non-i.i.d. errors, multivariate problems, classificatory variables, and other cases. The plot of the estimated parameter against ζ is then extrapolated to $\zeta = -1$ to estimate its unbiased value. Confidence intervals of SIMEX estimators are evaluated using jackknife resamples.

7.5.3 Likelihood-based estimators

Maximum likelihood procedures can be more reliable and flexible than least-squares procedures for regression, though more difficult to formulate and compute. In particular, assumptions of large samples with normal errors can be relaxed. MLEs have been developed for regression with measurement errors within the statistical community (Carroll *et al.* 2006, Chapter 8), although they are not often used. Astronomer B. Kelly (2007) created a likelihood measurement for model situations based on a hierarchical model similar to Equation (7.44), though for a more complex normal mixture model. Both the X and Y variables have known heteroscedastic measurement errors with a linear relationship. The distribution of the independent X variable is modeled as a sum of k Gaussian functions with different means and variances. This structure has several sources of scatter: the intrinsic scatter of the Y values about the linear relationship, the widths of the Gaussian components in X, and the measurement errors in both variables.

This likelihood can be maximized using the EM algorithm with confidence intervals estimated from the Fisher information matrix. Kelly compared the MLE procedure with minimum χ^2 procedures for measurement error problems commonly met such as the FITEXY and BCES estimators. In a variety of simulated datasets, the MLE exhibits less bias and variance than the OLS-based estimators. This likelihood can also be used in Bayes' theorem for Bayesian inference. MCMC computations give posterior distributions of all parameters, and allows marginalization over less interesting parameters (such as the Gaussian error components) to focus on the intrinsic regression parameters. The formulation can also be extended to include complicated selection effects or dependence between the errors and data values. The presence of data points may depend on critical values of the X or Y variables due to flux limits; objects below these limits will either be truncated from the sample entirely, or appear as left-censored data points (Chapter 10). Maximum likelihood treatments of regression with measurement error have also been considered by Wang (1998) and Higdon & Schafer (2001).

7.6 Nonlinear models

A common procedure for regression involving nonlinear models (7.2) is a least-squares estimation procedure that minimizes the residual sum of squares (RSS) with respect to the p-vector of parameters β,

$$RSS(\beta) = \sum_{i=1}^{n} (Y_i - f(X_i, \beta))^2 \qquad (7.52)$$

where $f(X, \beta)$ is the assumed model with nonlinear dependencies on the parameters like those in Equation (7.4). The minimization procedure typically requires some starting values for the β parameters and an optimization procedure such as the Gauss–Newton algorithm. With complicated functions and datasets, these procedures risk finding a local rather than a global minimum solution. The residual variance of the fit is then

$$s^2 = \frac{RSS(\hat{\beta})}{n - p}.$$

(7.53)

Figure 7.5 shows results from a nonlinear regression fit to astronomical data.

Maximum likelihood methods can be performed for nonlinear regression problems within the framework of **generalized linear modeling (GLM)** formulated by Nelder & Wedderburn (1972). GLM unifies and broadens linear and nonlinear regression into a single procedure, treating multivariate datasets with real, integer and/or categorical variables. The dependent variable Y can follow one of many common distributions. A **link function** is constructed that maps the linear function of the dependent variable(s) X to the mean of the Y variable. When the link function is the identity function, the method is equivalent to ordinary least squares with normal errors. A logarithmic link function, for example, gives Poisson regression (Section 7.6.1). The MLE estimates of the model's β coefficients are then obtained by iteratively reweighted least squares or Bayesian computations. The GLM formalism has been extended to treat correlated (non-i.i.d.) errors, hierarchical and mixture models. **Generalized additive models** is a related large body of regression methods that treat nonlinearities using smoothing functions.

Two simple nonlinear regression models, Poisson and logistic regression, play an important role in many fields, though they have rarely been used to date in astronomy.

7.6.1 Poisson regression

Poisson regression is a particular nonlinear model where the response variable Y is an enumeration of events following the Poisson distribution. This occurs often in astronomy. In X-ray and gamma-ray astronomy, spectra, images and light curves show the dependency of photon counts on wavelength, sky position and time. In other investigations, distributions of luminosities (known as luminosity functions), masses (mass functions), spectral lines or elemental abundances, distance or redshift, or many other astrophysical quantities may also follow Poisson distributions.

In regression involving a **Poisson process**, the counted events are independent of each other except for a specified dependency on covariates. This assumption of independence may be easily satisfied; astronomical spectra, for example, record a tiny random fraction of photons emitted by the celestial object. In other situations, however, the assumption of independence is not so obvious. A flux-limited sample from a large population uniformly distributed in space will oversample more luminous members of the population, giving a biased luminosity function. Here the luminosity function may depend on redshift, mass, color or some other properties of the population so that Poisson regression comes into play.

The advantage of Poisson regression is a correct treatment of the nonnegativity and integer nature of count data. Standard least-squares regression can treat count data when the counts are large and the Gaussian approximation is valid. The most common Poisson regression model has the integer response variable Y of the form

$$P_{\mu_i}(Y = y_i) = \frac{e^{-\mu_i}\mu_i^{y_i}}{y_i!}$$

(7.54)

where the mean values depend on the covariate X according to

$$E(Y|x) = e^{\beta X}. \tag{7.55}$$

This is also called a loglinear model because

$$\ln E[Y|X] = \ln \mu = \beta X. \tag{7.56}$$

The loglikelihood for the data $(X_1, Y_1), (X_2, Y_2), \ldots, (X_n, Y_n)$ is then

$$\ln L(\beta|X, Y) = \sum_{i=1}^{n} (Y_i \beta X_i - e^{\beta X_i} - \ln Y_i!) \tag{7.57}$$

and the maximum likelihood estimator is obtained by setting the derivative $\ln L/d\beta$ to zero. The resulting nonlinear equations require numerical solution. The uncertainty of this MLE is asymptotically normal,

$$\hat{\beta}_{P,MLE} \sim N\left(\beta, \left(\sum_{i=1}^{n} \mu_i X_i^2\right)^{-1}\right). \tag{7.58}$$

The volume by Cameron & Trivedi (1998) describes this Poisson model and its many extensions. For example, a bimodal count distribution would suggest a Poisson mixture model. Truncation or censoring of zero or other low values can be treated. Strategies for including non-Poissonian measurement errors, biased sample selection, model validation diagnostics, model comparison measures, hypothesis tests on regression coefficients, and other aspects of Poisson regression are available. Poissonian time series and multivariate datasets can be analyzed.

7.6.2 Logistic regression

A form of nonlinear regression widely used in the social and biomedical sciences is **logistic regression**. It treats the problem where the response variable is binary, taking only two values: "0" and "1"; "Yes" and "No"; or "Class I" and "Class II". Such problems arise in astronomy, although logistic regression has not been used. For example, one might regress the binary variable "AGN/Starburst" against galaxy spectral properties, or "Star/Galaxy" against image characteristics. While an integer counting variable may follow the Poisson distribution, a binary variable will follow the binomial distribution. Recall from Section 4.1 that a binomial process has n identical and independent trials with two possible outcomes with a constant probability of success, p. If we let Y represent the probability of success (for example, membership in Class I), then $E[Y] = p$ and $Y \sim Bin(n, p)$ with mean np and variance $np(1 - p)$.

Logistic regression is based on the model that the **logit** of the probability of success p is a linear function of the independent variable(s) X that can be expressed in two ways:

$$\text{logit}(p) = \ln\left(\frac{p}{1-p}\right) = \beta_0 + \beta_1 X$$

$$p = \frac{1}{1 + e^{-(\beta_0 + \beta_1 X)}}. \tag{7.59}$$

The logistic function has a sigmoidal shape that transforms an input value in the range $(-\infty, \infty)$ to an output value in the range $(0, 1)$. It is used in the perceptron algorithm in artificial neural networks; that is, a single-layer neural network is a logistic regression model (Chapter 9).

Regression coefficients for the logistic model can be obtained by maximum likelihood estimation, maximizing

$$L(\beta_1) = \frac{\exp \sum_{i=1}^{n} y_i x_i \beta_1}{1 + \exp x\beta}. \qquad (7.60)$$

A wide range of statistical inference follows: confidence intervals, goodness-of-fit, residual analysis, model selection, and extensions to multivariate $\mathbf{X}$ or polytomous (multi-category) Y variables. These are discussed by Sheather (2009, Chapter 8) and Kutner et $al.$ (2005, Chapter 14).

7.7 Model validation, selection and misspecification

Obtaining best-fit parameters from a regression procedure does not guarantee that the original model specification — perhaps a heuristic linear model or an elaborate astrophysical model — in fact fits the data. A "best fit" model does not guarantee an "adequate" fit and certainly not a "unique" fit. The choice of methodology for validating a model, and comparing one model with alternatives, is challenging. The major options are briefly reviewed by Kutner et $al.$ (2005, Chapters 9 and 14) and here in Section 3.7. More extensive discussions of model selection and goodness-of-fit tests in regression are presented in the monograph by Burnham & Anderson (2002) and the review articles collected by D'Agostino & Stephens (1986), Lahiri (2001), and Huber-Carol et $al.$ (2002).

Although a unique "cookbook" for model validation and selection is not available, two steps are recommended. First, global measures of model success should be calculated. These include residual-based statistics (such as R_a^2 and C_p discussed below), likelihood-based information criteria (such as AIC and BIC discussed in Section 3.7.3), and nonparametric goodness-of-fit statistics (such as the e.d.f. tests discussed in Section 5.3.1). Second, a variety of graphs based on the residuals between the data points and the fit should be examined to reveal details about the discrepancies between the data and model.

The exploration of alternative models can proceed in several ways. If the data show more features than available in the trial model, **nested models** with additional parameters can be considered. For example, one can add terms to a polynomial or add Gaussian components to a mixture model. For multivariate problems, model exploration often involves different choices of variables, sometimes emphasizing combining and omitting variables for dimensionality reduction (Chapter 8). If the scientific goal is to explain the astrophysical causes of an observed behavior, then alternative models arise from astrophysical theory. One might consider nested models such as adding more temperatures and ionization states to model an absorption line spectrum, or consider nonnested models such as multi-temperature blackbodies vs. nonthermal components to model a continuum spectrum. Global model validation procedures are then pursued for the alternative models, and model selection criteria are applied to compare the performance of the different model classes for explaining features in the data.

7.7.1 Residual analysis

A successful model should account for a large fraction of the original scatter in the data. A simple statistic measuring this effect is

$$R^2 = 1 - \frac{\sum_{i=1}^{n}(Y_i - \hat{Y}_i)^2}{\sum_{i=1}^{n}(Y_i - \bar{Y}_i)^2} \tag{7.61}$$

where $\bar{Y}_i = \sum Y_i/n$ is the mean of the response variable and $\hat{Y}_i$ are the model predictions at the X_i locations. R^2 is known as the **coefficient of determination** and involves the ratio of the **error sum of squares** to the **total sum of squares**. A successful model has R^2 approaching unity. For a suite of nested models with different number of parameters p, a plot R^2 vs. p can show the fractional reduction in variance achieved by the addition of new parameters.

However, model selection by maximizing R^2 is ill-advised as it leads to very elaborate models with large p. This violates the principle of parsimony. For the least-squares coefficient (7.61), the **adjusted R^2** that penalizes the number of parameters is

$$R_a^2 = 1 - \frac{n-1}{n-p}R^2 \tag{7.62}$$

which is related to Mallow's C_p statistic for multivariate problems. For best-fit models obtained using likelihood-based methods, the AIC and BIC information criteria discussed in Section 3.7.3 are commonly used. Note the lack of consensus on the strength of the penalty for increasing model parameters; for example, the BIC penalty scales with both the number of parameters and the sample size while the AIC penalty does not depend on sample size.

While these statistics give a global measure of the success in fitting a dataset, it is often helpful to examine the residuals graphically to gain insight into the nature of localized discrepancies and the possible influence of outliers. Graphical analysis of the residuals

$$\begin{aligned} \hat{e}_i &= Y_i - \hat{Y}_i \\ &= Y_i - \hat{\beta}_0 - \hat{\beta}_1 X_i \end{aligned} \tag{7.63}$$

around the regression curve is extensively discussed in standard texts (Kutner *et al.* 2005; Sheather 2009) and implemented in many software packages.

An important first step is to examine the distribution of errors compared to those expected from a Gaussian distribution using a **normal quantile-quantile (Q-Q) plot**. Strong deviations will indicate either inadequacy of the chosen model or non-Gaussianity in the error distribution. In either case, standard least-squares or MLE estimators can be invalid if the residuals fail this simple test.

If the assumption of normal errors is approximately valid, then each residual has an expected variance given by

$$Var(\hat{e}_i) = \sigma^2 (1 - h_{ii}) \qquad (7.64)$$

where h_{ii}, the diagonal elements of the **hat matrix** where the ij-th entry is

$$h_{ij} = \frac{1}{n} + \frac{(X_i - \bar{X})(X_j - \bar{X})}{S_{XX}}, \qquad (7.65)$$

are known as the **leverages**. The hat matrix measures the contribution of each point to the regression fit,

$$\hat{Y}_i = \sum_{j=1}^{n} h_{ij} Y_j. \qquad (7.66)$$

Consider, for example, a linear regression with most points in the range $0 < x < 1$ and a single point on the line around $x \simeq 3$. This point will have a high leverage (h_{ii} near unity) and its residual will have low variance. To remove this heteroscedasticity in the residual variances, it is advantageous to examine the standardized residuals

$$r_i = \frac{\hat{e}_i}{\sqrt{S_e^2 (1 - h_{ii})}} \quad \text{where} \quad S_e^2 = \sum_{i=1}^{n} \frac{\hat{e}_i^2}{(n-2)} \qquad (7.67)$$

is the sample variance of the residuals.

Cook's distance (Cook 1979) is based on the product of the standardized residual and the leverage of each data point with respect to the fitted curve,

$$D_{Cook,i} = \frac{r_i^2}{pS^2} \left(\frac{h_{ii}}{1 - h_{ii}} \right) \qquad (7.68)$$

where p is the number of parameters in the model and S^2 is the sample variance. Thus, a point with large residual near the center of the distribution may have a smaller Cook's distance than a point with a small residual near the tail of the distribution where the relative effect on the regression fit is greatest. Cook's distance is useful for discerning both large-scale discrepancies between the data and model and small-scale discrepancies due to localized features or outliers. Several residual plots from a linear regression on an astronomical dataset, including Cook's distance, are shown in Figure 7.2.

Systematic variations in residuals on small and intermediate scales can be found from measures for serial autocorrelation commonly used in econometric time series analysis (Section 11.3.2). These include the **nonparametric runs test** (Chapter 5), the **Durbin–Watson (1950) statistic**

$$d_{DW} = \frac{\sum_{i=2}^{n} (r_i - r_{i-1})^2}{\sum_{i=1}^{n} r_i^2} \qquad (7.69)$$

for short-term $AR(1)$ autocorrelation, and the **Breusch–Godfrey test** for $AR(k)$ autocorrelation involving k data points. Similarly, one can apply various statistics to detect trends in the residuals as a function of the independent X variables. A nonparametric correlation statistic, such as Spearman's ρ, can be applied to a plot of standardized residuals and X. The **Breusch–Pagan (1979) test** regresses the squared residuals against the X variable and uses a χ^2 test to demonstrate the presence of conditional heteroscedasticity.

7.7.2 Cross-validation and the bootstrap

Two closely related numerical methods have come into prominence to help validate best-fit regression models and choose among competing models. Both are simple in concept yet have strong foundations in mathematical theorems (Stone 1977, Singh 1981). As the bootstrap was discussed in Section 3.6.2, we present here the use of cross-validation for regression.

In **cross-validation**, a portion of the dataset is withheld from the original regression, and is later used to evaluate the model's effectiveness. A simple choice is to use 80% of the dataset in the original analysis with 20% reserved for evaluation. This strategy is related to machine learning techniques where some data serve as training sets and others as test sets (Chapter 9). In **k-fold cross-validation**, the data are divided into k equal subsets and the regression is performed k times with each subset removed. When $k = n$, the procedure is equivalent to jackknife resampling.

If $\hat{\mu}_k(X_i)$ is the predicted value for the i-th data point in the k-th subsample, then the k-fold cross-validation prediction error for that point is

$$\widehat{PE}_{CV} = \frac{1}{k}\sum_{j=1}^{k}\sum_{i=1}^{n_k}[Y_i - \hat{\mu}_{-k}(X_i)]^2 \tag{7.70}$$

which replaces the usual least-squares residual sum of errors (see Section 7.5),

$$PE_{OLS} = (1/n)\sum_{i=1}^{n}[Y_i - \hat{\mu}(X_i)]^2. \tag{7.71}$$

Here the subscript $-k$ denotes the full sample with the k-th subset withheld. The difference

$$\widehat{ME}_{CV} = \widehat{PE}_{CV} - PE_{OLS} \tag{7.72}$$

is then a cross-validation estimate of the model error. This is a measure of the model's systematic bias, a goodness-of-fit measure independent of the random scatter of the residuals. The number of learning sets to consider affects the balance between bias and variance; a recommended choice is $k = 10$.

There is considerable discussion about the relative effectiveness of cross-validation, jackknife and bootstrap for practical linear modeling (Wu 1986, Shao 1993, Kohavi 1995).

7.8 Remarks

Regression is a very important statistical activity in astronomy using both simple heuristic linear models and complicated models from astrophysical theory. But astronomers are not commonly taught to think deeply about the initial construction of the model. Does the scientific problem point to a single response variable dependent on the other variables, or is a symmetrical treatment appropriate? Is the response variable real, integer or binary? Are the independent variables fixed or random? Has a correlation analysis been made (for example, using Kendall's τ, Section 5.6) to show that a relationship between the variables is indeed present? Is the chosen functional form of the model known to apply to the population under study, or is model selection needed to find both the optimal model family and its optimal parameters? Is the model linear or nonlinear in the parameters? Is the formula error of the response variable normally distributed? Might outliers be present? Are there other sources of errors such as measurement errors in one or more variables with known variances? Answers to these questions lead to the different regression procedures discussed in this chapter.

Once the model is defined, then a method is chosen to estimate best-fit parameters. A number of approaches can be taken, but least squares is still effective for many situations. For complicated models involving nonlinearity, heteroscedasticity, non-Gaussianity or other features, maximum likelihood estimation is recommended. The choice of regression model is always crucial: the best-fit parameters for an ordinary least-squares regression will differ from a weighted least-squares or a Poisson or measurement error regression model.

Statisticians are nearly universally surprised, and sometimes disturbed, by astronomers' widespread use of minimum χ^2 fitting techniques for a wide range of regression problems (Section 7.4). Astronomers often place measurement error variances in the denominator of the X^2 statistic rather than the observed or model values (compare Equation 7.43 with 7.40). This alters the properties of X^2 such that it may no longer asymptotically follow a χ^2 distribution. Arbitrary choices in the binning procedure, often because the original data did not fall naturally into categories, may cause similar problems. The procedure also assumes that there is no intrinsic scatter; that is, $\epsilon = 0$ in Equation (7.2).

The consensus view on χ^2 fitting was expressed by a distinguished statistician: "Minimum chi-square estimates may be all right in certain forms of data analysis when very little is known about the data. In all other cases the likelihood is too useful a part of the data to be ignored." (J. K. Ghosh in discussion of Berkson, 1980). For datasets with measurement errors of the form $(X_i, Y_i, \sigma_{Y,i})$ or $(X_i, \sigma_{X,i}, Y_i, \sigma_{Y,i})$, hierarchical measurement error models are recommended and procedures like SIMEX used for parameter estimation (Section 7.5).

A final message is that, in some cases, the values of regression coefficients can be compared only within the context of a single model. The "slope" of a robust weighted M-estimator is not the same mathematical quantity as the "slope" of an unweighted least-squares bisector, despite the identical English word usage. The difference between "slope" values from different methods can considerably exceed the formal confidence intervals within a single method. Thus when the scientific goals depend on accurate and precise coefficients, as in evaluation of Hubble's constant (Freedman & Madore 2010) or other cosmological parameters, it is critical that the methodology be carefully specified.

7.9 Recommended reading

Carroll, R. J., Ruppert, D., Stefanski, L. A. & Crainiceanu, C. M. (2006) *Measurement Error in Nonlinear Models: A Modern Perspective*, 2nd ed. Chapman & Hall, London
An advanced monograph on the treatment of measurement errors in regression models. Topics include the concepts of measurement error, bias in linear regression, calibration, the SIMEX algorithm, score function methods, likelihood methods, Bayesian methods, hypothesis testing, longitudinal models, nonparametric estimation, survival data and response variable error.

Kutner, M. H., Nachtsheim, C. J., Neter, J. & Li, W. (2005) *Applied Linear Statistical Models*, 5th ed., McGraw-Hill, New York
A comprehensive, widely used, intermediate-level textbook with clear mathematical presentations, explanations, and examples. Least-squares methods are emphasized. Topics include linear regression with normal errors, inferences concerning the regression coefficients and predictions, regression diagnostics and residual analysis, simultaneous inference, multiple regression, model selection and validation, nonlinear regression and generalized linear models, experimental design and multi-factor studies.

Sheather, S. J. (2009) *A Modern Approach to Regression with R*, Springer, Berlin
A clear and practical presentation of elementary regression models that covers least-squares estimation, confidence intervals, ANOVA, residual diagnostic plots, weighted least-squares, multiple linear regression, variable selection, serially correlated errors, and mixture models. Useful **R** scripts are provided.

7.10 R applications

Only a few of the wide array of procedures and diagnostics for regression modeling in **R** and **CRAN** can be illustrated here. For illustrative purposes, we confine the examples here to a bivariate dataset with an approximately linear relationship of the form $Y = \alpha + \beta X$. However, the same codes can be used for multivariate problems with a matrix **X** of input values with p covariates, and for many linear models of the form $Y = f(X)$.

We select for study a comparison of the i-band and z-band magnitudes of quasars from the Sloan Digital Sky Survey. A subsample of 11,278 quasars with $18 < i < 22$ mag is chosen; the data are plotted in Figure 7.1. The regression results for this dataset are not

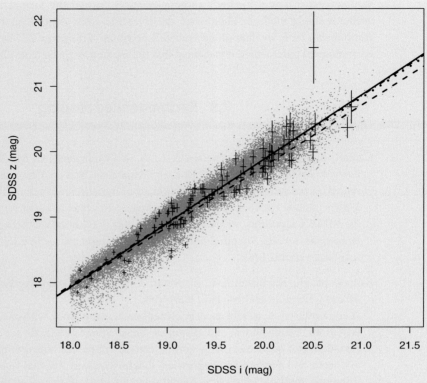

Fig. 7.1 Scatterplot of z- and i-band photometry for 28,476 quasars from the Sloan Digital Sky Survey (gray points). Heteroscedastic measurement errors in both variables are shown for illustrative objects. Four least-squares linear fits are shown: ordinary least squares (solid), weighted fit using z-band errors only (long-dashed), robust M-estimator weighted fit (short-dashed), and robust Thiel–Sen fit (dashed-dotted). This last line is nearly superposed on the first line.

of great scientific interest, as virtually any astronomical population will show a similar correlation between brightnesses in adjacent spectral bands. But the dataset exemplifies a number of nontrivial characteristics commonly seen in astronomical regression problems: deviations from linearity, non-Gaussian and asymmetrical residuals about the line, outliers that cannot lie on the line, and heteroscedastic measurement errors with known variances in both the X and Y variables.

Examples of the measurement errors are shown in Figure 7.1. The heteroscedasticity is enormous; the distance between points and a hypothetical linear fit can sometimes be entirely attributed to measurement errors, while in other cases it must be attributed to the discrepancies intrinsic to the data point. **R** plotting routines do not automatically plot error bars, so we construct them using the *lines* function. Error-bar plotting is supported by **CRAN**'s *ggplot2* package (Wickham 2009). A few points have near-zero error bars that will cause problems in weighted regressions, so we set a minimum error of 0.01 magnitudes. The reader may notice manipulation of the graphical devices using *dev.new* and *dev.set*, allowing us to add graphics to a particular window while other plots are produced on another window.

```
# Linear regression with heteroscedastic measurement errors
# Construct and plot dataset of SDSS quasars with 18<i<22

qso <- read.table('http://astrostatistics.psu.edu/MSMA/datasets/SDSS_QSO.dat',
   header=T)
summary(qso)
qso1 <- qso[((qso[,11]<22) & (qso[,11]>18)),c(11:14)]
dim(qso1) ; summary(qso1) ; attach(qso1)
sig_z_mag[sig_z_mag<0.01] <- 0.01

dev.new(2)
plot(i_mag, z_mag, pch=20, cex=0.1, col=grey(0.5), xlim=c(18,21.5),
   ylim=c(17.5,22), xlab="SDSS i (mag)", ylab="SDSS z (mag)")
for(i in 50:150) {
   lines(c(i_mag[i],i_mag[i]),c((z_mag[i]+sig_z_mag[i]),
      (z_mag[i]-sig_z_mag[i])))
   lines(c((i_mag[i]+sig_i_mag[i]),(i_mag[i]-sig_i_mag[i])),
      c(z_mag[i],z_mag[i]))   }
```

7.10.1 Linear modeling

R's *lm*, or linear modeling, function is a program with many capabilities for fitting and evaluating least-squares regressions as outlined in Section 7.3. The *lm* function is extremely flexible; its use is described in detail in the monographs by Fox (2002) and Sheather (2009). The statistical model is specified with an **R** *formula* with syntax "$y \sim model$". Examples include $y \sim x$ for a linear relationship, $y \sim x + 0$ for a line with zero intercept, $log10(y) \sim a + log10(x)$ for a loglinear relationship, and $y \sim x + x^2$ for a quadratic relationship. Sometimes an abbreviated form like $y \sim .$ is used where the dot refers to all columns in the attached dataframe. The independent variable can be real, integer, binary, categorical, or *R* factors. The solution to the system of linear equations is obtained by a QR decomposition of the data matrix.

The ordinary least-squares fit to the SDSS quasar dataset is plotted as the solid line in Figure 7.1 using the **R** function *abline*. Important results are given in the *summary* function for *lm*-class objects. The slope coefficient is 1.024 ± 0.007 where 0.007 is the standard error. The line accounts for 67% of the sample variance, the adjusted R^2 in Equation (7.62), and the residual standard error is 0.67 mag. The function *confint* gives confidence intervals on *lm* regression parameters; here we ask for the 3σ-equivalent (99.7%) bounds on the slope and intercept estimators. The hyperbolic bounds to the confidence band can be drawn using the function *linesHyperb.lm* in the **CRAN** package *sfsmisc* (Maechler *et al.* 2010).

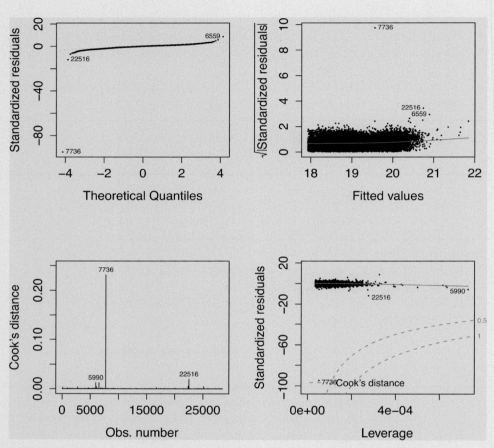

Fig. 7.2 Regression diagnostics for the ordinary least-squares fit of SDSS quasar photometry. *Top left:* Normal Q-Q plot of residuals. *Top right:* Square root of residuals plotted against the independent *i* magnitudes. *Bottom left:* Cook's distance values for the sample. *Bottom right:* Standardized residuals plotted against leverage.

The *plot* function for *lm*-class objects provides several diagnostic graphics; four of these are shown in Figure 7.2. In the first panel, the normal quantile-quantile (Q-Q) plot compares the distribution of standardized residuals with a Gaussian distribution. Aside from a few outliers, the distribution of residuals is approximately normal except at the low-X and high-X ends of the dataset. The second panel shows that the residual distribution of z-band magnitudes is correlated with the i-band magnitude values; this clearly indicates that the linear model is missing important structure in the $i-z$ relationship.

The bottom panels in Figure 7.2 involve leverage that gives more insight into the deviation of individual points from the regression fit. The points 7736 and 22,516 are found to be outliers affecting the fit. Cook's distance given in Equation (7.68) is a scale-free indicator of the influence of each point on the fitted relationship. The bottom panels isolate those points that simultaneously have high leverage and large residuals. Fox (2002) gives a detailed presentation of linear modeling and diagnostic plots in **R**. Bootstrap resampling of residuals is often valuable, as illustrated in Venables & Ripley (2002).

We next perform a weighted least-squares fit using the *lm* function, where each point is weighted by the inverse of the variance of the measurement error in the response *z*-band magnitude. Note that the errors in the independent *i*-band variable are ignored here. This line with slope 0.935 ± 0.004 is shown with the long-dashed line in Figure 7.1. The line now accounts for 86% of the sample variance, an improvement over the ordinary least-squares fit. The diagnostic plots (not shown) indicate some improvement in the fit: the Q-Q plot is more symmetrical, the systematic increase in residuals with *i* magnitude is reduced, and Cook's distance values have fewer outliers.

```
# Ordinary least squares fit

fit_ols <- lm(z_mag~i_mag)
summary(fit_ols)
confint(fit_ols, level=0.997)
dev.set(2) ;  abline(fit_ols$coef, lty=1 ,lwd=2)

dev.new(3) ; par(mfrow=c(2,2))
plot(fit_ols, which=c(2:5), caption='', sub.caption='' , pch=20, cex=0.3,
   cex.lab=1.3, cex.axis=1.3)
par(mfrow=c(1,1))


# Weighted least-squares fit

fit_wt <- lm(z_mag~i_mag, x=T, weights=1/(sig_z_mag*sig_z_mag))
summary(fit_wt)
dev.set(2) ; abline(fit_wt$coef,lty=2,lwd=2)
```

7.10.2 Generalized linear modeling

GLM procedures operate by maximum likelihood estimation rather than least squares, and give more flexibility in modeling (Section 7.6). A GLM fit assuming a normal distribution of residuals in *z*-band magnitude gives, as expected, the same slope as the ordinary least-squares regression above with slope 1.023 ± 0.007. The Akaike information criterion value for this case is 23,078. A better model is obtained assuming a gamma distribution for the residuals with AIC $= 22,465$. This is a nonlinear model, showing a curvature of ~ 0.3 mag amplitude around the ordinary least-squares line. As with the other models, a variety of graphical diagnostics is available to evaluate success and problems with a GLM fit.

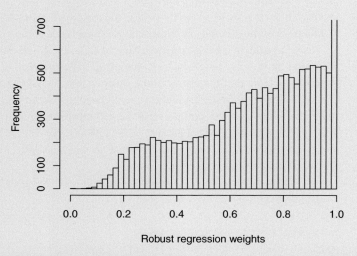

Fig. 7.3   Histogram of weights for 5608 downweighted quasars from a robust least-squares regression based on Huber's ψ function.

```
# Generalized linear modeling

fit_glm_gau <- glm(z_mag ~ i_mag) ; summary(fit_glm_gau)
fit_glm_gam <- glm(z_mag ~ i_mag,family=Gamma) ; summary(fit_glm_gam)
```

7.10.3 Robust regression

Figure 7.1 clearly shows that outliers are present that cannot be attributed to large measurement errors. We can run a robust regression to remove or reduce their influence on the fit (Section 7.3.4). The *MASS* library incorporated into base **R** has the *rlm* function for robust linear modeling using an *M*-estimator. Fitting is computed using an iteratively reweighted least-squares algorithm. Measurement error weighting is included, as some discrepant points have large errors while others do not. We use the default *M*-estimator based on Huber's ψ function with $k = 1.345$.

The result of this robust fit is a line with slope 0.948 ± 0.002 shown as the dotted line in Figure 7.1. The histogram in Figure 7.3 shows the distribution of weights; 2084 of the 11,278 points were downweighted by the procedure.

Our final regression fit is the Thiel–Sen line implemented in **CRAN** package *zyp* (Bronaugh & Werner 2009). The resulting fit has slope 1.018 ± 0.010 for the first 8000 quasars. The computation here is rather slow and inefficient in use of computer memory.

```
# Robust M-estimator

library(MASS)
fit_M <- rlm(z_mag~i_mag, method='M', weights=1/(sig_z_mag*sig_z_mag),
```

```
    wt.method='inv.var')
summary(fit_M)
length(which(fit_M$w<1.0))
dev.set(3) ; hist(fit_M$w, breaks=50, ylim=c(0,700), xlab='Robust regression
    weights',  main='')
aM <- fit_M$coef[[1]] ; bM <- fit_M$coef[[2]]
dev.set(2) ; lines(c(18,22), c(aM+bM*18, aM+bM*22), lty=3, lwd=3)


# Thiel-Sen regression line

install.packages('zyp') ; library(zyp)
fit_ts <- zyp.sen(z_mag~i_mag, qso1[1:8000,])
confint.zyp(fit_ts)
dev.set(2) ; abline(fit_ts$coef, lty=4, lwd=2)
```

7.10.4 Quantile regression

The linear regression line can be considered to be a fit to the mean of the Y values along
chosen values of the X variable. When the formula errors ϵ in Equation (7.2) are normal
and homoscedastic, then lines fitted to quantiles of the Y variable (say, the 10% and 90%
residuals from the fitted line) will be parallel to the mean line. But more can be learned from
regression along quantiles when the errors are non-Gaussian. We see such a situation in
Figure 7.1: at bright i magnitudes, the residuals show a distribution with a narrow peak and
a symmetric wide tail, while at faint i magnitudes, the residuals show a very asymmmetric
tail. This is thus a situation where quantile regression may be helpful.

Linear quantile regression is implemented by the *rq* function in the **CRAN** package
quantreg (Koenker 2010) with results shown in Figure 7.4. The *print.summary.rq* function shows that the slope increases from 0.92 to 1.23 at the 10% and 90% quantile
levels due to the asymmetry in the residuals. In the **R** script below, we display the resulting lines using a *for* loop, as this package does not provide the plot automatically.
The function *rqss* gives a nonlinear fit based on bivariate smoothing splines. Here we
see approximately linear behavior at bright values of i but strong nonlinearities at faint
values.

```
# Linear quantile regression

install.packages('quantreg') ; library(quantreg)
fit_rq <- rq(z_mag~i_mag, tau=c(0.10,0.50,0.90))
print.summary.rq(fit_rq)
par(mfrow=c(1,2))

plot(i_mag, z_mag, pch=20, cex=0.1, col=grey(0.5), xlim=c(18,22),
    ylim=c(17.5,22), xlab="SDSS i (mag)", ylab="SDSS z (mag)")
for(j in 0:2) {
```

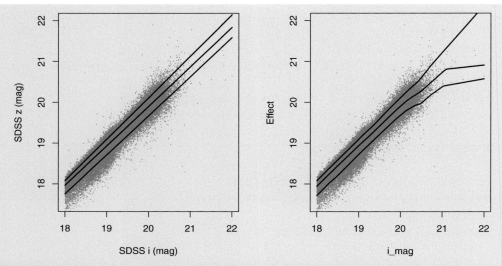

Fig. 7.4 Quantile regression for the z- and i-band photometry for 11,278 quasars from the Sloan Digital Sky Survey. *Left:* linear quantile regression with lines showing the fit at the 10%, 50% and 90% quantiles. *Right:* nonlinear quantile regression based on spline smoothing.

```
aquant=fit_rq$coef[[2*j+1]] ; bquant=fit_rq$coef[[2*j+2]]
  lines(c(18,22),c((aquant+bquant*18),(aquant+bquant*22)),lwd=2) }


# Nonlinear quantile regression

fit_rqss.1 <- rqss(z_mag~qss(i_mag), data=qso1, tau=0.10)
fit_rqss.5 <- rqss(z_mag~qss(i_mag), data=qso1, tau=0.50)
fit_rqss.9 <- rqss(z_mag~qss(i_mag), data=qso1, tau=0.90)
plot.rqss(fit_rqss.1, rug=F, ylim=c(17.5,22), titles='')
points(i_mag, z_mag, cex=0.1, pch=20, col=grey(0.5))
plot.rqss(fit_rqss.1, shade=F, rug=F, add=T, titles='', lwd=2)
plot.rqss(fit_rqss.5, shade=F, rug=F, add=T, titles='', lwd=2)
plot.rqss(fit_rqss.9, shade=F, rug=F, add=T, titles='', lwd=2)
par(mfrow=c(1,1))
```

7.10.5 Nonlinear regression of galaxy surface brightness profiles

To illustrate the capabilities of nonlinear regression in **R**, we examine the univariate radial brightness distributions of elliptical galaxies. These distributions are monotonic, bright in the center and fade with radial distance, but have proved difficult to parametrize and understand. Hubble (1930) proposed a simple isothermal ellipsoid projected onto the sky where the observed intensity profile would have the shape $I(r) \propto (1 + r/r_c)^{-2}$ where r_c is the core radius. As data improved, this model was found to be inadequate at both small and large radii, and de Vaucouleurs (1948) proposed a more accurate but less readily interpretable formula $\log_{10} I(r) \propto (r/r_e)^{1/4} - 1$. The parameter r_e is the galaxy's **effective radius** and, unfortunately, is not directly interpretable astrophysically like the r_c parameter.

Sérsic (1968) proposed a generalized de Vaucouleurs law that is still used today,

$$\log_{10} I(r) = \log_{10} I_e + b_n[(r/r_e)^{1/n} - 1] \tag{7.73}$$

where $b_n \simeq 0.868n - 0.142$ for $0.5 < n < 16.5$ and the observed surface brightness profile is measured in units of

$$\mu(r) = \mu_0 - 2.5 \log_{10} I(r) \text{ mag arcsec}^{-2}. \tag{7.74}$$

Two of the three model parameters, r_e and n, appear nonlinearly in the model, so standard least-squares methods cannot be reliably applied for parameter estimation.

Here we apply Sérsic's model to three elliptical galaxies in the Virgo Cluster for which the surface photometry and Sérsic model fits are presented by Kormendy *et al.* (2009) based on images obtained with NASA's *Hubble Space Telescope*. These surface brightness profiles are plotted in Figure 7.5; innermost and outermost regions are omitted due to deviations from the Sérsic model. We see that the galaxy NGC 4472 (=Messier 49) shows a gradual declining large extended envelope, NGC 4406 has a similar shape but is much smaller, while NGC 4551 has a much steeper brightness decline and smaller extent. The radial profiles are not congruent with each other so that a nonlinear model is needed for fitting.

R has an important function *nls* for computing nonlinear least-squares regressions. Its use is described in detail in the volume by Ritz & Streibig (2008). The call to *nls* starts with an **R** *formula* with syntax $y \sim f(x)$. Here the $f(x)$ combines Equations (7.73) and (7.74). The dataframe with the input data and user-provided starting values for the parameters are then provided. As discussed by Ritz & Streibig, we found that the minimization algorithm failed under poor choices of starting values.

We show three tools for extracting the results from Sérsic function fits to the galaxy profiles. The *summary* function gives the best-fit regression coefficients and their confidence intervals, the *deviance* function gives the minimized residual sum of squares from Equation (7.52), and the *logLik* function gives the maxmimum likelihood value that may be useful for model selection tools like the Bayesian information criterion. Contour maps showing the correlations between variable confidence intervals can be constructed using the *nlsContourRSS* function in the **CRAN** package *nlstools* (Baty & Delignette-Muller 2009).

```
# Fit Sersic function to NGC 4472 elliptical galaxy surface brightness profile

NGC4472 <-
   read.table("http://astrostatistics.psu.edu/MSMA/datasets/NGC4472_profile.dat",
   header=T)
attach(NGC4472)
```

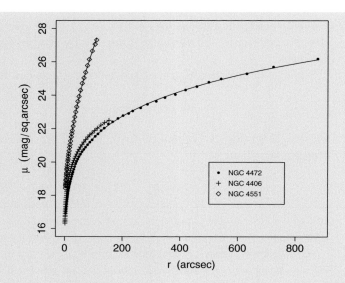

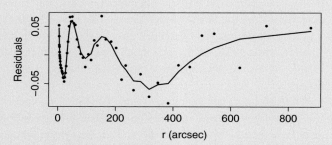

Fig. 7.5 *Top:* Radial surface brightness profiles for three elliptical galaxies in the Virgo Cluster (filled circles). Recall that fainter values correspond to larger μ values in magnitudes arcsec^{-2}. The lines show the best-fit nonlinear regression curves using **R**'s *nls* function. *Bottom:* The residuals between observed and fitted μ values for NGC 4472 with Friedman's super-smoother nonparametric density estimator.

```
NGC4472.fit <- nls(surf_mag ~ -2.5*log10(I.e * 10^(-(0.868*n-0.142)*
   ((radius/r.e)^{1/n}-1))) + 26, data=NGC4472, start=list(I.e=20.,
   r.e=120.,n=4.), model=T, trace=T)
summary(NGC4472.fit)
deviance(NGC4472.fit)
logLik(NGC4472.fit)


# Plot NGC 4472 data and best-fit model

plot(NGC4472.fit$model$radius, NGC4472.fit$model$surf_mag, pch=20,
   xlab="r (arcsec)", ylab=expression(mu ~~ (mag/sq.arcsec)), ylim=c(16,28),
   cex.lab=1.5, cex.axis=1.5)
lines(NGC4472.fit$model$radius, fitted(NGC4472.fit))
```

```
# Fit and plot radial profiles of NGC 4406 and NGC 4551

NGC4406 <-
   read.table("http://astrostatistics.psu.edu/MSMA/datasets/NGC4406_profile.dat",
   header=T)
attach(NGC4406)
NGC4406.fit <- nls(surf_mag ~ -2.5*log10(I.e * 10^(-(0.868*n-0.142)*
   ((radius/r.e)^{1/n}-1))) + 32, data=NGC4406, start=list(I.e=20.,
   r.e=120.,n=4.), model=T, trace=T)
summary(NGC4406.fit)
points(NGC4406.fit$model$radius, NGC4406.fit$model$surf_mag, pch=3)
lines(NGC4406.fit$model$radius, fitted(NGC4406_fit))

NGC4551 <-
   read.table("http://astrostatistics.psu.edu/MSMA/datasets/NGC4551_profile.dat",
   header=T)
attach(NGC4551)
NGC4551.fit <-  nls(surf_mag ~ -2.5*log10(I.e * 10^(-(0.868*n-0.142)*
   ((radius/r.e)^{1/n}-1))) + 26, data=NGC4551, start=list(I.e=20.,r.e=15.,n=4.),
   model=T, trace=T)
summary(NGC4551.fit)
points(NGC4551.fit$model$radius, NGC4551.fit$model$surf_mag, pch=5)
lines(NGC4551.fit$model$radius, fitted(NGC4551_fit))
legend(500, 20, c("NGC 4472","NGC 4406", "NGC 4551"), pch=c(20,3,5))
```

We can see from the top panel of Figure 7.5 that the Sérsic fit to all three galaxies seems excellent. Examination of the residual plot in the bottom panel, however, shows correlated structures, indicating that the model does not account for small-scale structures in the galaxy surface brightness profiles. The amplitude of the residuals, however, is around 0.05 mag arcsec^{-2} which is $< 1\%$ of the full range of magnitudes. To improve visualization of the residual patterns, we use **R**'s *supsmu* function implementing Friedman's **super-smoother** based on adaptive local regression methods discussed in Chapter 5. This autocorrelated behavior is present despite the normal quantile-quantile plot of the standardized residuals showing that the overall distribution is close to a normal distribution. This is validated with a nonparametric Shapiro–Wilks test for normality.

Very similar results were obtained for NGC 4472 by Kormendy *et al.* (2009) though with ~ 3 times larger parameter confidence intervals. The best-fit parameters from *nls* are $r_e = 267.6 \pm 7.1$ arcsec and $n = 5.95 \pm 0.10$ while Kormendy and colleagues find $r_e = 269 \pm 21$ arcsec and $n = 5.99 \pm 0.30$. The difference in confidence intervals is attributable to the different regression procedures. Both minimize *RSS* in a least-squares computation, but Kormendy *et al.* iteratively weight the squared residual of each point by the mean squared residuals of the final fit. This weighting is a nonstandard and heuristic procedure. Both methods found the same correlated residual pattern; Kormendy *et al.* remark on the surprisingly accurate fits of the Sérsic models given the lack of any astrophysical basis for the formulation.

We can also compare the analytical parameter confidence intervals based on least-squares theory to the parameter distributions found from bootstrap resampling. Using the *nlsBoot* function in the *nlstools* package, we find $r_e = 267.6 \pm 13.2$ and $n = 5.95 \pm 0.18$ where these are the 95% ranges. The distribution of bootstrapped values of n are approximately normally distributed without evident bias.

Finally, we make a model comparison between the three-parameter Sérsic model and the simpler two-parameter de Vaucouleurs model where the parameter n is set equal to 4. The de Vaucouleurs fit (not shown here) is less accurate on large scales with residuals several times higher than for the best Sérsic fit. The model selection can be evaluated with the Akaike information criterion using the **R** function *AIC*. It this case the difference in likelihood values is enormous and the Sérsic model is clearly much better than the de Vaucouleurs model.

The **CRAN** package *nls2* provides similar nonlinear least-squares regression to those in *nls* but with more extensive functionalities (Grothendieck 2010). Its capabilities are described in the brief volume by Huet *et al.* (2004). One option permits a grid of starting points so trapping in local minima can be reduced. The nonlinear models are defined in an ancillary subroutine, allowing models with arbitrary complexity (e.g. piecewise functions with thresholds). Hypothesis tests and parameter confidence intervals can be calculated in several ways. Heteroscedasticity in the variances is treated and other difficulties are treated with maximum likelihood estimation, including a multi-step procedure for high-dimensional problems. The package gives a variety of graphics to study model misspecification, and a suite of procedures deal with binary (following the binomial distribution), polytomous (multinomial), and count (Poisson) response variables.

```
# NGC 4472 analysis
# Residual plot

plot(NGC4472.fit$model$radius,residuals(NGC4472.fit), xlab="r (arcsec)",
   ylab="Residuals", pch=20, cex.lab=1.5, cex.axis=1.5)
lines(supsmu(NGC4472.fit$model$radius, residuals(NGC4472.fit), span=0.05),
   lwd=2)


# Test for normality of residuals

qqnorm(residuals(NGC4472.fit) / summary(NGC4472.fit)$sigma)
abline(a=0,b=1)
shapiro.test(residuals(NGC4472.fit) / summary(NGC4472.fit)$sigma)


# Bootstrap parameter estimates

install.packages('nlstools') ; library(nlstools)
```

```
NGC4472.boot <- nlsBoot(NGC4472.fit)
summary(NGC4472.boot)
curve(dnorm(x,m=5.95, sd=0.10)*58/5.95, xlim=c(5.6,6.4), ylim=c(0,50))
hist(NGC4472.boot$coefboot[,3], breaks=50, add=T)   # not shown
```

7.10.6 Scope of regression in **R** and **CRAN**

The coverage of regression in **CRAN** is too vast to summarize here, but we mention a few useful packages. Note that most methods are intrinsically multivariate, incorporating bivariate regression as a simple case. The *car* and *lmtest* packages provide suites of methods including tests for autocorrelation in residuals, tests for heteroscedasticity, leverage and influence diagnostic plots, and variable transformations to give normal residuals. The *nlme* package is a large collection of functions for nonlinear regression that are used by many other **CRAN** packages. The *chemometrics* package associated with the volume by Varmuza & Filzmoser (2009) includes methods of particular utility in the physical sciences. Linear structural equation modeling, allowing both observed and unobserved latent variables, is treated in the *sem* package. Advanced MS and MM robust regression estimators are given in the *FRB* (Fast and Robust Bootstrap) and *robustbase* packages. The Theil–Sen median regression estimator is implemented by the *mblm* package.

CRAN also extensively implements Bayesian methods for advanced modeling problems. The *arm* package gives multivariate regression with hierarchical models. A wide range of linear models is considered in the *bayesm* package. The *MCMCpack* performs Bayesian computations for a variety of models using C++ routines, and includes a general-purpose Metropolis sampling algorithm and tools for visualizing simulation results. *BMA* assists with variable selection using Bayesian model averaging.

Scientists interested in the applied mathematical methods by which optimal solutions are found in regression computations can find a listing of many relevant **CRAN** packages at http://cran.R-project.org/web/views/Optimization.html.

8 Multivariate analysis

8.1 The astronomical context

Whenever an astronomer is faced with a dataset that can be presented as a table — rows representing celestial objects and columns representing measured or inferred properties — then the many tools of multivariate statistics come into play. Multivariate datasets also arise in other situations. Astronomical images can be viewed as tables of three variables: right ascension, declination and brightness. Here the spatial variables are in a fixed lattice while the brightness is a random variable. An astronomical datacube has a fourth variable that may be wavelength (for spectro-imaging) or time (for multi-epoch imaging). High-energy (X-ray, gamma-ray, neutrino) detectors give tables where each row is a photon or event with columns representing properties such as arrival direction and energy. Calculations arising from astrophysical models also produce outputs that can be formulated as multivariate datasets, such as N-body simulations of star or galaxy interactions, or hydrodynamical simulations of gas densities and motion.

For multivariate datasets, we designate n for the number of objects in the dataset and p for the number of variables, the dimensionality of the problem. In traditional multivariate analysis, n is large compared to p; statistical methods for high-dimensional problems with $p > n$ are now under development. The variables can have a variety of forms: real numbers representing measurements in any physical unit; integer values representing counts of some variable; ordinal values representing a sequence; binary variables representing "Yes/No" categories; or nonsequential categorical indicators.

We address multivariate issues in several chapters of this volume. The present chapter on multivariate analysis considers datasets that are commonly displayed in a table of objects and properties. Here we assume that the sample is homogeneous; that is, the sample is drawn from a single underlying population. Chapter 9 treats problems where the sample is inhomogeneous with multiple populations. Here the goal is to partition the sample into homogeneous subsamples, using either clustering techniques based only upon the dataset at hand or classification techniques where prior information is available about the expected groupings.

Multivariate analysis seeks to understand the relationships between the measured properties of the sample. Mathematically, we seek to characterize and interpret the p-dimensional structure of the population. For example, the astronomical dataset might measure size, morphology, environment, optical luminosity, optical and infrared colors, and multi-waveband properties of a sample of spiral galaxies. We might find that some galaxies are interacting starbursts; that is, they exhibit a correlation between close interacting neighbors, excess blue emission from OB stars, excess infrared emission from heated dust, more X-ray emission from compact binary-star systems, and more radio continuum emission from supernova cosmic rays. Galaxies without close interactions will tend to have opposite properties of weak blue, infrared, X-ray and radio emission. Size and luminosity may be uncorrelated with this trend. Thus, study of the multivariate structure of spiral galaxy properties can give insights into star-formation processes in the Universe.

Multivariate methods are discussed in several chapters of this volume. In the present chapter, we discuss the relationships between the variables, assuming the dataset is drawn from a single population. When the dimensionality is low, typically $p = 2$ or 3, methods treated in other chapters come into play. Density estimation discussed in Chapter 6 smooths the cloud of individual data points into a continuous function for low-dimensional problems. The methods of spatial point processes discussed in Chapter 12 are designed for such situations. Many of the approaches to regression discussed in Chapter 7 were mathematically developed for multivariate problems, and come to immediate use for multivariate analysis. Multivariate clustering, discrimination and classification that study the possibility that the dataset is drawn from multiple populations are discussed in Chapter 9. A realistic multivariate data analysis project often involves interweaving of several methods such as multivariate methods, clustering and density estimation.

Astronomical images, light curves and spectra can often be viewed as multivariate analysis of datasets with combinations of fixed and random variables. In these cases, the location of an image pixel, the frequency in an astronomical spectrum, and the time of an observation are fixed rather than random variables. The measured brightness is a random variable but often does not satisfy the i.i.d. assumption; adjacent brightness values in an image, spectrum or time series are often correlated with each other due to characteristics of the measuring instruments or astrophysical process. For these reasons, multivariate problems in image and time series analysis are often treated separately from multivariate problems involving clouds of points in a p-space of random variables. For example, some multivariate time series methods are discussed in Chapter 11.

8.2 Concepts of multivariate analysis

The principal goals of multivariate analysis are to:

1. quantify the location and distribution of the dataset in p-space (multivariate mean locations, variances and covariances, and fits to probability distributions);
2. achieve structural simplification by reducing the number of variables without significant loss of scientific information by combining or omitting variables;
3. investigate the dependence between variables;
4. pursue statistical inference of many types including testing hypotheses, and calculating confidence regions for estimated parameters;

5. classify or cluster "similar" objects into homogeneous subsamples (this is covered in Chapter 9);
6. predict the location of future objects (this may not be important in astronomical research).

Multivariate random vectors can be written in terms of a matrix. Let X_{ij} be the measurement of the j-th variable ($j = 1, 2, \ldots, p$) for object i ($i = 1, 2, \ldots, n$), $\mathbf{X}_i$ be the vector of p measurements for the i-th object, and $\mathbf{X}$ be the matrix of the X_{ij} data. The sample mean values and variances for the j-th variables are

$$\bar{X}_j = n^{-1} \sum_{i=1}^{n} X_{ij}$$

$$S_{jj} = n^{-1} \sum_{i=1}^{n} (X_{ij} - \bar{X}_j)^2. \tag{8.1}$$

The sample variances are the diagonal terms of the sample **variance–covariance matrix**, also known as the **covariance matrix** or **dispersion matrix**, **S**, with elements

$$S_{jk} = \frac{1}{n} \sum_{i=1}^{n} (X_{ij} - \bar{X}_j)(X_{ik} - \bar{X}_k). \tag{8.2}$$

The covariance matrix **S** is very important in multivariate analysis, as its elements measure the linear association between the j-th and k-th variables. If the $\mathbf{X}_i$ are i.i.d. vectors, then, from an application of the Law of Large Numbers (Section 2.10), the sample covariance matrix will converge towards the population covariance matrix $\mathbf{\Sigma}$ as $n \to \infty$.

8.2.1 Multivariate distances

A fundamental choice needed for most (but not all) multivariate analyses is the metric defining the distance between objects. When the units in all p variables are the same, a natural metric is the **Euclidean distance**. A generalization is the **Minkowski metric** (or in mathematical parlance, the m-norm distance) where the distance d between two p-dimensional points $\mathbf{X}_i$ and $\mathbf{X}_j$ is

$$d_{ij} = d(\mathbf{X}_i, \mathbf{X}_j) = \left(\sum_{k=1}^{p} |X_{ik} - X_{jk}|^m \right)^{1/m}. \tag{8.3}$$

A value of $m = 1$ gives the **Manhattan distance** measured parallel to the parameter axes; $m = 2$ gives the Euclidean distance; and $m \to \infty$ gives the maximum or Chebyshev distance. When measurement errors are known and are responsible for much of the scatter, then weighted distances may be helpful. For datasets defined on a sphere, great-circle distances can be used. The matrix **D** with elements d_{ij} for all i and j is variously called the **distance matrix**, **dissimilarity matrix**, or **proximity matrix** of the sample.

But in astronomical datasets, the units of the different variables are often incompatible and the construction of a metric is not obvious. For example, a stellar astrometric catalog of the Galactic halo will have star streams associated with the past accretion of small galaxies which should appear as groupings in the six-dimensional phase space. But the observations are made in different units: right ascension and declination (in degrees), distance (in parsecs), proper motion rate (in milliarcseconds per year) and direction (in degrees), and radial velocity (in kilometers per second). A simple Euclidean metric can be highly distorted by the choice of units; for example, distance metric values will greatly differ if proper motion were measured in arcseconds per century rather than milliarcseconds per year.

A widely used solution to the problem of incompatible units and scales is to standardize the variables by offsetting each to zero mean and rescaling to unit standard deviation,

$$Z_{ij} = \frac{(X_{ij} - \bar{X}_j)}{\sqrt{S_{jj}}} \tag{8.4}$$

and then apply Euclidean distances. Multivariate analysis can then proceed with the standardized data matrix $\mathbf{Z}$. The results of the statistical calculation can be transformed back to the original variables if needed. Similarly, the covariance matrix can be standardized to give the matrix $\mathbf{R}$ where the elements are Pearson's bivariate linear correlation coefficients (Section 3.6.2)

$$R_{jk} = \frac{S_{jk}}{\sqrt{S_{jj}S_{kk}}} \tag{8.5}$$

where $-1 \leq R_{jk} \leq 1$. The **sample correlation matrix R** approximates the population correlation matrix ρ where the population standard deviations σ_{jk} replace the sample standard deviations S_{jk}. The determinants of the covariance and correlation matrices, $|\mathbf{S}|$ and $|\mathbf{R}|$, are measures of the total variance of the sample. If we define a variance matrix $\mathbf{V}$ to have diagonal elements σ_{ii}^2 and zero off-diagonal elements, then a useful relation is

$$\rho = (\mathbf{V}^{1/2})^{-1} \mathbf{\Sigma} (\mathbf{V}^{1/2})^{-1}. \tag{8.6}$$

Another approach to the metric distance between any two vectors $\mathbf{x}_1$ and $\mathbf{x}_2$ is to scale by the covariance matrix,

$$D_M(\mathbf{x}_1, \mathbf{x}_2) = \sqrt{(\mathbf{x}_1 - \mathbf{x}_2)'\mathbf{S}^{-1}(\mathbf{x}_1 - \mathbf{x}_2)}. \tag{8.7}$$

This is the **Mahalanobis distance** between two points and is useful in multivariate analysis and classification. It is closely related to Hotelling's T^2 and other statistics based on the multivariate normal distribution (Section 8.3).

Distances involving categorical variables, perhaps from a previous classification effort, are tricky to define. For binary or categorical variables, one can define $|X_{ik} - X_{jk}|$ to be zero if the values are the same, and some positive value if they differ. These values are called **similarity coefficients** and a number of choices can be considered. Furthermore, for some purposes in astronomy, a special metric could be constructed to incorporate the astrophysical process under study. For example, the "distance" between galaxies or stars in a gravitationally bound cluster might be scaled to mass/distance^2 to account for gravitational attraction, or the "distance" between an accreting body and its gaseous environments might be scaled to luminosity/distance^2 to account for irradiation effects.

For some astronomical variables with very wide ranges (say, luminosities of stars or galaxies), the variable is often first transformed by the logarithm. The logged variables can then be standardized for multivariate analysis. Astronomers typically make a logarithmic transformation as a matter of habit. It is critical to recognize that the choices of variable transformation − standardization, logarithmic, or both − and distance metric are essentially arbitrary and can strongly influence the results of many (though not all) multivariate analysis, clustering or classification procedures. Astronomers should test whether the scientific conclusions change when other reasonable choices of variable transformation and distance metrics are made.

8.2.2 Multivariate normal distribution

Many of the classical tests and procedures of multivariate analysis require that the variables, original or transformed, follow the p-dimensional **multivariate normal (MVN) distribution**,

$$f(\mathbf{x}) = \frac{1}{(2\pi)^{p/2}|\mathbf{\Sigma}|^{1/2}} e^{-(\mathbf{x}-\mu)'\mathbf{\Sigma}^{-1}(\mathbf{x}-\mu)/2} \tag{8.8}$$

where μ is the p-dimensional population mean and $\mathbf{\Sigma}$ is the population covariance matrix. This MVN distribution is often designated $\mathbf{N}_p(\mu, \mathbf{\Sigma})$. The multivariate factor $(\mathbf{x} - \mu)'\mathbf{\Sigma}^{-1}(\mathbf{x} - \mu)$ may appear complicated at first, but it is the natural multivariate generalization of the familiar scalar factor in the univariate Gaussian (Section 4.3) when the latter is expressed as

$$\frac{(x - \mu)^2}{\sigma^2} = (x - \mu)\sigma^{-2}(x - \mu). \tag{8.9}$$

Recall the notation of linear algebra: the superscript $\bullet'$ (also written $\bullet^T$) is the transpose of a vector or matrix, the superscript $\bullet^{-1}$ is the inverse of a matrix, and $| \bullet |$ denotes the determinant of a matrix.

If $\mathbf{X}$ is a multivariate normal random vector with mean μ and covariance matrix $\mathbf{\Sigma}$, then $(\mathbf{X} - \mu)'\mathbf{\Sigma}^{-1}(\mathbf{X} - \mu)$ is distributed as χ_p^2, a χ^2 distribution with p degrees of freedom. This allows calculation of probabilities that a new point is consistent with a chosen MVN distribution, and the calculation of probability ellipsoids in p-space giving the locus of points consistent with a chosen MVN. For example, the upper α percentile (say, $\alpha = 0.975$ for a two-sided 95% confidence interval) region is defined by all points $\mathbf{x}$ satisfying

$$(\mathbf{x} - \mu)'\mathbf{\Sigma}^{-1}(\mathbf{x} - \mu) \leq \chi_p^2(\alpha). \tag{8.10}$$

The elements of the covariance matrix $\mathbf{\Sigma}$, σ_{jk} where $j, k = 1, 2, \ldots, p$, determine the shape of the MVN distribution. If the diagonal elements σ_{jj} are the same and off-diagonal elements are zero, then a quantile of the MVN traces a p-dimensional hypersphere. If the diagonal elements are different and the off-diagonal elements are zero, then the MVN is a hyperellipsoid with axes parallel to the variables. Nonzero off-diagonal elements rotate the hyperellipsoid in p-space. Projections of a hyperellipsoid onto low dimensions can produce a cigar shape in some projections, a pancake or ellipse in others, and a sphere or circle in others.

As with the univariate normal, the MVN has advantageous mathematical properties: subsets with selected variables are MVN; conditional distributions of selected variables on others are MVN; sums and linear combinations of MVNs are MVN; the mean vector and covariance matrix are sufficient statistics; and so forth. The distribution of the sample covariance matrix follows a **Wishart distribution**, a generalization of the univariate χ^2 distribution.

The least-squares and maximum likelihood estimators for the $N_p(\mu, \Sigma)$ parameters are identical, given by

$$\hat{\mu} = \bar{X}$$
$$\hat{\Sigma} = \frac{1}{n} \sum_{j=1}^{n} (X_j - \bar{X})(X_j - \bar{X})'. \tag{8.11}$$

This MLE is obtained from the MVN likelihood

$$L(\mu, \Sigma) = \prod_{i=1}^{n} \left[\frac{1}{(2\pi)^{p/2}|\Sigma|^{1/2}} e^{-(X_i-\mu)'\Sigma^{-1}(X_i-\mu)/2} \right]. \tag{8.12}$$

8.3 Hypothesis tests

A variety of hypothesis tests based on the MVN distribution were developed in the 1930s that generalize classical univariate tests. The multivariate analog of **Student's t statistic** that a sample mean is consistent with a previously specified vector location μ_0 is **Hotelling T^2**,

$$T^2 = n(\bar{X} - \mu_0)'S^{-1}(\bar{X} - \mu_0). \tag{8.13}$$

For a MVN population, the distribution of T^2 can be expressed in terms of the F **distribution**,

$$F = n(n - p)T^2/(n - 1)p, \tag{8.14}$$

so that probabilities for the hypothesis test $H_0 : \mu = \mu_0$ can be obtained from tabulations of the F distribution. The F distribution (also known as the Fisher–Snedecor distribution) is defined in terms of the beta function studied by Euler; it applies to the ratio of two random variates that follow a χ^2 distribution. Hotelling's T^2 can be used with the population covariance Σ if it is known in advance or, more commonly, with the covariance matrix S estimated from the sample as in Equation (8.2).

The univariate likelihood ratio test of the null hypothesis $H_0 : \mu = \mu_0$ discussed in Section 3.7.2 is generalized with **Wilks' Λ** defined by

$$\Lambda = \left(\frac{|\hat{\mathbf{\Sigma}}|}{|\hat{\mathbf{\Sigma}}_0|} \right) = \frac{\sum_{i=1}^{n} (\mathbf{X}_i - \bar{\mathbf{X}})(\mathbf{X}_i - \bar{\mathbf{X}})'}{\sum_{i=1}^{n} (\mathbf{X}_i - \mu_0)(\mathbf{X}_i - \mu_0)'}. \tag{8.15}$$

Probabilities for values of Λ can be obtained through the distribution of Hotelling's T^2 statistic because

$$\Lambda^{2/n} = (1 + T^2/(n-1))^{-1}. \tag{8.16}$$

An important class of hypothesis tests evaluates whether two or more samples have physically distinct locations in p-space. If a p-dimensional dataset is divided into k groups and the groups can be considered to follow MVN distributions, then classical tests of the significance that the groups have different mean locations fall under the rubric of **multivariate analysis of variance (MANOVA)**. These can be viewed as multivariate, multisample generalizations of the t test. While these methods can be important in validating statistical clustering algorithms (Section 9.3), we consider them here due to their mathematical links to our discussions in the present chapter.

Consider first the case of two MVN populations $\mathbf{X}_1$ and $\mathbf{X}_2$ in p dimensions with respective means μ_1 and μ_2, and covariance matrices $\mathbf{\Sigma}_1$ and $\mathbf{\Sigma}_2$. We are testing the hypothesis that the population means are identical,

$$H_0 : \quad \mu_1 - \mu_2 = E[\mathbf{X}_1] - E[\mathbf{X}_2] = 0. \tag{8.17}$$

Let $\mathbf{X}_{1i}$, $i = 1, 2, \ldots, n$, be i.i.d. random vectors having mean μ and covariance matrix $\mathbf{\Sigma}$, and similarly for $\mathbf{X}_{2i}$. The sample mean and variance vectors for population 1 are then

$$\bar{\mathbf{X}}_1 = \frac{1}{n_1} \sum_{i=1}^{n_1} \mathbf{X}_{1i}$$

$$\mathbf{S}_1 = \frac{1}{n_1 - 1} \sum_{i=1}^{n_1} (\mathbf{X}_{1i} - \bar{\mathbf{X}}_1)(\mathbf{X}_{1i} - \bar{\mathbf{X}}_1)', \tag{8.18}$$

and similarly for population 2. A $100(1 - \alpha)\%$ confidence ellipsoid for the difference of the population means is the set of $\mu_1 - \mu_2$ satisfying

$$\left[(\bar{\mathbf{X}}_1 - \bar{\mathbf{X}}_2) - (\bar{\mu}_1 - \bar{\mu}_2) \right]' \left[\frac{\mathbf{S}_1}{n_1} + \frac{\mathbf{S}_2}{n_2} \right]^{-1} \le \chi_p^2(\alpha) \tag{8.19}$$

where $\chi_p^2(\alpha)$ is the upper 100α-th percentile of the χ^2 distribution with p degrees of freedom.

For the more general case of testing the equality of means of k MVN clusters, we use two matrices of the sums of squares and cross-products, $\mathbf{W}$ for **within** clusters and $\mathbf{B}$ for **between** clusters,

$$\mathbf{W} = \sum_{j=1}^{k} \sum_{i=1}^{n_j} (\mathbf{X}_{ji} - \bar{\mathbf{X}}_i)(\mathbf{X}_{ji} - \bar{\mathbf{X}}_i)'$$

$$\mathbf{B} = \sum_{j=1}^{k} n_j (\bar{\mathbf{X}}_j - \bar{\mathbf{X}})(\bar{\mathbf{X}}_j - \bar{\mathbf{X}})' \tag{8.20}$$

where $\bar{\mathbf{X}}$ is the global sample mean vector. The most common test of the equality of the cluster means is Wilks' lambda,

$$\Lambda^* = \frac{|\mathbf{W}|}{|\mathbf{B} + \mathbf{W}|} \tag{8.21}$$

which measures the ratio of total variances explained by the cluster structure. We saw Wilks' lambda in Equation (8.15) where it was used to test the compatibility of a sample with a specified mean vector and to compare the likelihoods of two models. The quantity $[n - 1 - (p + k)/2] \ln\Lambda^*$ follows a χ^2 distribution with $p(k - 1)$ degrees of freedom. It is a multivariate generalization of the F test of sample means, and is equivalent to the likelihood ratio test. Related tests of the equality of sample means are based on **Hotelling's trace**, $tr[\mathbf{BW}^{-1}]$ and **Pillai's trace**, $tr[\mathbf{B}(\mathbf{B} + \mathbf{W})^{-1}]$.

8.4 Relationships among the variables

8.4.1 Multiple linear regression

The bivariate linear regression model introduced in Section 7.2 had pairwise i.i.d. measurements (X_i, Y_i) and a model of the form

$$Y_i = \beta_1 + \beta_2 X_i + \epsilon_i \tag{8.22}$$

where $\epsilon_i \sim N(0, \sigma^2)$ is a normal error term. The average (or underlying) relationship between the predictor X and response Y variables is

$$Y = \beta_1 + \beta_2 X + \epsilon. \tag{8.23}$$

The extension of this linear regression model to multivariate data of the form $(X_{i1}, X_{i2}, \ldots, X_{ip}, Y_i)$ is straightforward (Johnstone & Titterington 2009):

$$Y = \mathbf{X}\beta + \epsilon \tag{8.24}$$

where β is a vector and $\mathbf{X}$ is a matrix with p columns. The first column of $\mathbf{X}$ has elements equal to 1 so that the first entry of β represents the intercept. $\mathbf{X}$ here is classically called the **design matrix**. The model can also be compactly represented as

$$Y \sim \mathbf{N}_p(\mathbf{X}\beta, \sigma^2\mathbf{I}) \tag{8.25}$$

where $\mathbf{N}_p$ is the MVN and $\mathbf{I}$ is the identity matrix of order p. A generalization of this **multiple regression** model is **multivariate regression** that models how a vector of independent variables $\mathbf{X}$ influences a vector of response variables $\mathbf{Y}$.

The multivariate extension of the bivariate least-squares estimator (which is also the MLE when normal distributions are involved) for the linear regression coefficients β is

$$\hat{\beta} = (\mathbf{X}'\mathbf{X})^{-1}\mathbf{X}'Y. \tag{8.26}$$

This requires that $(\mathbf{X}'\mathbf{X})$ be invertible; this is not true if the sample size n is smaller than the dimensionality p. The distribution of the best-fit coefficients is

$$\hat{\beta} \sim \mathbf{N}_p(\beta, (\mathbf{X}'\mathbf{X})^{-1}) \tag{8.27}$$

from which confidence intervals on the regression coefficients can be estimated if the model is correct.

A variety of problems can arise in multiple regression (Izenman 2008, Chapter 5). Linear regression calculations can be unstable when many variables are collinear; this situation is called **ill-conditioning** and is related to the the singularity of the $(\mathbf{X}^T\mathbf{X})$ matrix. When $p \gg n$ or ill-conditioning is present, ordinary least-squares calculations become impossible or very inaccurate. A variety of procedures can then be considered: generalized inverse, principal components, partial least squares and, most commonly, ridge regression. In **ridge regression**, the $(\mathbf{X}^T\mathbf{X})^{-1}$ matrix (Equation 8.26) is replaced by $(\mathbf{X}^T\mathbf{X} + \lambda\mathbf{I})^{-1}$ where λ is a scalar regularization parameter. As with bivariate regression (Section 8.6), residual analysis plays a critical role in evaluating model validity. Simple residual plots from regression analysis of an astronomical dataset are shown in Figure 7.2.

Another approach to ill-conditioned and collinear data is to remove redundant variables using **stepwise regression** methods. Variables are selected for removal due to their weak contribution to the reductions in the residual sum of squares. Various measures have been widely used since the 1970s based on the residual sum of squares with a penalty for the number of dimensions under consideration. These include Mallow's C_p, the Akaike information criterion (AIC), and the Bayesian information criterion (BIC). A plot of C_p values against number of variables $p^* \leq p$ is useful; subsets of variables giving the lowest C_{p^*} values are optimal, and the regression fit should be satisfactory if $C_p^{min} \simeq p^*$.

Recent developments of more sophisticated penalty functions are becoming popular. The **least absolute shrinkage and selection operation**, or **lasso**, is a least-squares regression model based on robust (L_1) constraints. In lasso regression, a combination of robust L_1 and least-squares L_2 criteria is applied to the likelihood maximization. The ridge and lasso regression estimators also emerge from Bayesian inference as the **maximum a posteriori** (MAP) estimators when the regression coefficients are assumed to have a particular prior distribution. **Least angle regression**, or LARS, is another advanced variable selection method.

Finally, we mention that a wealth of methodology has been developed for more general linear regression models than that shown in Equation (8.24).

8.4.2   Principal components analysis

Principal components analysis (PCA) models the covariance structure of a multivariate dataset with linear combinations of the variables under the assumption of multivariate normality. It finds the linear relationships that maximally explain the variance of the data. It addresses two of the goals of multivariate analysis outlined in Section 8.2: studying linear relationships between properties, and seeking structural simplification through reduction of dimensionality. For dimensionality reduction, the goal is not to obtain a meaningful understanding of the relationships between variables, but to address the practical problem of treating large numbers of variables with redundant or low information. PCA is the most commonly used multivariate technique in astronomy, particularly in studies of galaxies.

From an algebraic viewpoint, the first principal component, PC_1, is the linear combination

$$PC_1 = \sum_{i=1}^{n} a_{1i}\mathbf{X}_i \tag{8.28}$$

that has the largest variance of any linear combination, under the constraint $\mathbf{a}_1' \mathbf{a}_1 = 1$. Here $\mathbf{X}_i$ is the i-th p-vector and $\mathbf{a_i}$ is the p-vector of coefficients. The second component, PC_2, is similar with the additional constraint $\mathbf{a}_2'\mathbf{a}_1 = 0$ so that PC_1 and PC_2 are uncorrelated. This proceeds with additional PCs, each uncorrelated with all previous PCs. The method is also known as the **Karhunen–Loève transform**.

The PCA coefficients $\mathbf{a}_k$ for the k-th component are typically calculated using the method of Lagrangian multipliers; $\mathbf{a}_1$ is then the eigenvector associated with the largest eigenvalue of the sample covariance matrix S. The $\mathbf{a}_k$ vectors are often rescaled so the sum of squares of their elements equals the k-th eigenvalue. The k-th PC is associated with the k-th largest eigenvalue, and the variance of the k-th PC is the k-th eigenvalue. Stated another way, the k-th PC accounts for the fraction of the total variance equal to the k-th eigenvalue normalized by the trace of S. The PCA procedure can also be viewed as the iterative selection of axis rotations in p-space to maximize the variance along the new axes.

PCA can be remarkably effective for reducing a confusing high-dimensional multivariate problem to a much simpler low-dimensional problem. The procedure often automatically combines variables that redundantly measure the same property, and reduces the importance of variables that contribute only noise to the covariance structure. A seemingly intractable dataset in p-space can be reduced to a much simpler $k \ll p$ space of principal components. Sometimes, but not always, the a_{ki} PCA coefficients (or **loadings**) can be interpreted as scientifically meaningful relationships between properties.

Three important cautions should be heeded about PCA usage. First, the components derived from the original variables will differ from those derived from standardized (or otherwise transformed) variable. For standardized variables, the principal components of $\mathbf{Z}$ are associated with the eigenvectors of the correlation matrix ρ given in Equation (8.6), rather than the eigenvectors of the original covariance matrix $\mathbf{\Sigma}$. PCA is thus not a scale- or unit-invariant procedure.

The second difficulty is lack of mathematical guidance for evaluating the significance of a component, and thereby choosing an appropriate number of components to use for later study. One procedure is to apply a heuristic rule-of-thumb such as retaining components whose eigenvalues exceed 0.7 or 1.0. Alternatively, the analyst might state in advance an interest in components accounting for (say) 90% of the sample variance, and later find that (say) three components are needed. Another common procedure is to visually examine the variance associated with each component. If an elbow appears in this scree plot, then the components representing the strongest improvements in variance reduction are selected for use. Finally, the analyst might use scientific criteria to select the principal components to report. Often the first components have a clear scientific interpretation, while the later components are dominated by noise in the multivariate distribution and cannot be interpreted.

Given the paucity of clear statistical interpretations of PCA computations, visual examination of the results is essential. A normal quantile-quantile plot of each component can reveal outliers or strong structural deviations from normality that might be corrected through variable transformations. The **scree plot** shows the contributions to the sample variance by each component. PCA **biplots**, scatter plots of data points in the k-dimensional space of important components, will reveal multimodality and outliers (Gower & Hand 1996). Scree and biplots from a PCA of an astronomical dataset are shown in Figure 8.5 below. Explanatory variables are often superposed on biplots to assist with interpretation. **Coenoclines** are plots of explanatory variables along PCA axes. Principal component **rank-trace plots** show the loss of information in regressions with limited components.

While PCA is a mainstay of multivariate analysis, variants have been developed to treat a variety of issues (Rao *et al.* 2005, Izenman 2008). Principal components regression can treat cases with many nearly collinear variables. Functional PCA can treat multivariate datasets with dependencies on a fixed time-like variable. PCA can be effective for iterative denoising and segmentation image analysis. PCA procedures modified for nonlinearities and robustness against outliers are referenced in Sections 8.4.5 and 8.4.4, respectively.

8.4.3 Factor and canonical correlation analysis

We outline here two classical multivariate analysis procedures related to PCA that are widely used in various fields but not in astronomy.

Factor analysis is a regression model that seeks linear relationships among the observed variables that are indirectly caused by relationships among the unobservable or **latent variables**. Factor and PCA analyses both seek to reduce the dimensionality of a problem, but in different ways. PCA finds components that explain variances in the variables, while factor analysis finds factors that explain correlations.

Mathematically, the problem can be expressed as the matrix equation

$$\mathbf{x} = \mathbf{\Lambda f} + \mathbf{u}$$

where $\mathbf{x}$ is the vector of observed variables, $\mathbf{f}$ is a vector of k unobserved latent variables or **factors**, $\mathbf{\Lambda}$ is a matrix of regression coefficients or **factor loadings**, and $\mathbf{u}$ is a vector of Gaussian noise values. If we constrain the factors to be standardized and mutually uncorrelated, then the loadings give the correlations between the observed and latent variables. The loadings can be estimated from computations on the sample covariance matrix using an iterative least-squares or a maximum likelihood approach. The calculation is performed for a range of k, usually small. Various techniques are available for interpreting the factors, including rotation.

Canonical correlation analysis (CCA) is a generalization of PCA that treats variables in groups. Consider, for example, a multivariate analysis of results from NASA's Swift satellite that has characterized several hundred gamma-ray bursts. For each object, some variables are clearly associated with the stellar explosion (gamma-ray fluence, duration and hardness), other variables are associated with the fireball afterglow (X-ray flares and decay time-scales, radio emission), and yet other variables are associated with the host galaxy (redshift, luminosity, star-formation rate, cluster environment). CCA seeks a few correlations between linear combinations of one group of variables and linear combinations of another group of variables.

8.4.4 Outliers and robust methods

Outlier detection is an important step in the analysis of many multivariate datasets. These can arise from heterogeneous populations in a survey, extreme objects in a single population, or problems in the data acquisition or analysis process. Study of outliers can lead to radically new findings, as with the discovery of ultra-cool T dwarf stars as red outliers in the Sloan Digital Sky Survey photometric distributions (Leggett *et al.* 2000). Outlier detection and robust regression for multivariate problems, emphasizing efficient methods useful for large datasets, are reviewed by Hubert *et al.* (2005).

A classical, and often effective approach is to examine the distribution of Mahalanobis distances from each p-dimensional data point to the sample mean. This is the Euclidean distance from the mean standardized by the covariance matrix. Following Equation (8.7),

$$D_M(X_i) = \sqrt{(\mathbf{X}_i - \mu)'\mathbf{S}^{-1}(\mathbf{X}_i - \mu)} \qquad (8.29)$$

where μ is the sample mean. Assuming an MVN distribution for $\mathbf{X}_i$, an outlier threshold can be defined based on the χ^2 distribution, for example, at $\sqrt{\chi^2_{p,0.975}}$ for the 95% confidence interval for p degrees of freedom.

If the outliers are common or have very extreme values, then the ordinary mean and covariance matrix will be affected to reduce the significance of the outliers. An important robust estimator is the **minimum covariance determinant (MCD) estimator** developed by P. Rousseeuw. Here a large subsample of h of the n data points is obtained that minimizes the determinant of the covariance matrix. The subsample satisfying MCD can be found by iteratively omitting points with high Mahalanobis distances. A number of algorithms are available to seek the global minimum and to achieve computational efficiencies for large datasets.

Similarly, the bias in multiple regression lines produced by outliers can be reduced by iteratively removing or downweighting discrepant observations. Here the outlier criterion is based on residual distance from the line, $r_i = Y_i - \beta \mathbf{X}_i - \alpha$ (where α is the intercept) rather than the Mahalanobis distance. One such algorithm is Rousseeuw's **least trimmed squares estimator**.

Robust PCA procedures based on MCD and projection pursuit are discussed by Hubert *et al.* (2005). Budavári *et al.* (2009) present trimmed and robust PCA methods to treat outliers in a computationally efficient fashion for large astronomical surveys.

8.4.5 Nonlinear methods

An obvious limitation of multivariate methods discussed above is the assumption that structural relationships are linear. Simple nonlinear problems could be indirectly treated by variable transformations prior to analysis, but this is not a general solution to all nonlinear relationships.

A variety of advanced nonlinear approaches to multivariate analysis and dimensionality reduction are actively studied today (Gorban *et al.* 2007, Lee & Verleysen 2009). The procedures are often based on sophisticated mathematics involving manifold topology, geodesic distances, and spectral embedding. They include nonlinear principal components analysis, Kohonen's self-organizing map algorithm, the isomap algorithm, principal curves, principal manifolds, elastic maps, harmonic topographic maps, and diffusion maps. These procedures are often interrelated to each other and to various data mining classification methods. The greatest challenges are to discover and characterize structures that resemble highly nonlinear structures like p-dimensional snakes, spirals and torii.

We outline here three of the more well-established nonlinear multivariate techniques: principal curves, projection pursuit, and multivariate adaptive regression splines (MARS).

Principal curves are a generalization of PCA where the component lines are replaced by smoothly varying curves. A family of curve shapes is established with a functional form $f(\lambda)$ where the parameter λ regulates the degree of permitted curvature. Starting with the first linear PCA line, λ and curve parameters are iteratively estimated from kernel smoothed data to minimize the summed distances of the points projected onto the curve. Principal surfaces follow the same procedure by fitting functions in two or more dimensions. Projections of the data points onto the λ_1, λ_2 plane will often discriminate grouping of data points, as well as fit the sample-averaged structure. In this application, the result is similar to classification with the Self-Organizing Map.

Projection pursuit was developed by J. Friedman and J. Tukey in the 1970s primarily as an interactive tool to visually explore high-dimensional datasets. Here the analyst constructs one or more **projection indexes** which highlight "interesting" departures from normal structures in the multivariate dataset when it is projected onto (typically) two dimensions. The kurtosis is an example of a projection index. Projection pursuit then uses an optimization algorithm to find the projection of the p-space data that maximizes the index. As a tool to uncover structure in a multivariate dataset, it is related to **independent component analysis** which, in turn, is a generalization of factor analysis.

MARS is an adaptive regression technique also developed by Friedman in the 1970s. Rather than fitting a single line as in classical multiple regression, MARS fits piecewise linear or polynomial functions over localized regions around **knot** locations. After an initial fit by linear least squares in each dimension, terms with the smallest contributions to the residual squared errors are deleted. A generalized cross-validation criterion gives the final fit. Computationally efficient strategies have been developed to quickly arrive at a reasonable model. MARS has been adapted to classification problems where it has similarities to CART procedures.

Both projection pursuit and MARS can be implemented in a three-layer neural network learning algorithm. Neural net backpropagation is often easier than the forward-learning algorithms. Thus, nonlinear multivariate analysis is often conducted in the context of modern data mining techniques.

8.5 Multivariate visualization

Examination of visual displays of multivariate data is essential for gaining insight into large datasets, judging the reliability of statistical findings and communicating the results in an intelligible fashion to the wider community. Visualization is an informal, exploratory tool that complements the mathematical and algorithmic procedures of multivariate analysis, clustering and classification. Both types of analysis are needed: visualization allows the scientist's mind's eye capability to find patterns and identify problems, while statistical analysis gives quantitative evaluations for specified tests and procedures. Three examples of multivariate visualization of an astronomical dataset are shown in Figure 8.7.

Scientific visualization is a major field of study closely tied to computer science techniques, perceptual and cognitive psychology, and the capabilities of graphical displays. Real-time interaction with the display is essential including zooming, projecting, filtering, brushing and linking. While astronomers are most familiar with histograms (Figure 8.1), bivariate scatter plots and contour plots (Figure 8.3), other displays are often useful: box-and-whisker plots (Appendix B.6), parallel coordinate plots, 2.5-dimensional surfaces, dendrograms, network graphs, and pixelated visualizations such as Self-Organizing Maps.

A common starting point for visualizing a multivariate dataset is a matrix of univariate and bivariate plots. Favored univariate displays include the **boxplot** and the **histogram**. A matrix of pairwise scatterplots can be enhanced with local regression fits and univariate histograms. **Conditioning plots**, or coplots, are arrays of bivariate scatterplots displayed in strata of a third variable. Interactive multipanel graphical systems allow the user to brush interesting subsamples with a distinctive color or symbol. Changes made on one plot are linked onto other plots. Projected surface, contour and volumetric plots or images are commonly available to visualize three variables simultaneously.

Several visualization methods are well-established to study more complex relationships that may not be revealed in bivariate relationships (Theus 2008). **Mosaic plots** show the number of objects conditioned on different variables, while **trellis displays** use arrays of bivariate scatterplots that are organized by dependency ("conditioning") on additional variables. Trellis graphics are called **lattice** plots in the **R** software system (Sarkar 2008). The **grand tour** is a continuous one-parameter family of low-dimensional projections of a p-dimensional cloud of data points (Cook & Swayne 2007). Data tours are essentially a movie of the dataset intended for interactive exploration; user intervention can highlight interesting subsamples, zoom in scale, and investigate interesting projections in more detail. The **parallel coordinates plot** is an effective two-dimensional representation where the p

axes, usually plotted radially from a single origin, are instead plotted horizontally on p individual parallel axes (Inselberg 2009). A parallel coordinates plot for an astronomical dataset is shown in Figure 8.7.

Trellis plots, grand tours and parallel coordinate plots are effective for moderate-size problems; say, $3 < p < 15$ and $n < 10^4$. Large-n displays require smoothing dense regions where data points overlap or using **α-blending** translucent displays. Parallel coordinate plots, in particular, should be very advantageous in astronomy both for exploration and presentation of multivariate data. Trellis displays are also recommended for clear presentation of low-dimensional data.

For high-p problems, direct visualization is difficult. Smoothing, as with kernel density estimation, also is ineffective unless the sample size is also very large. Some techniques have been suggested including average shifted histograms (Section 6.9.3) and projection pursuit density estimation.

Multidimensional scaling (MDS) is a nonparametric method for displaying higher-dimensional data in a lower-dimensional space in a fashion that is not dependent on multivariate normality (MVN) and is somewhat insensitive to outliers. Here one constructs the matrix of pairwise distances between objects in the original p-space, and places these distances in an ordinal sequence from smallest to largest. MDS uses these ranks rather than the actual distances. The goal of MDS is to find a q-dimensional configuration of the data, $q < p$, in which the rank order of pairwise distances closely matches that seen in the original p-dimensional space. Proximity of the two ranked distance distributions is measured with a **stress function**. The smallest value of q is sought that reproduces the observed rank sequence to a desired precision (usually within a few percent).

Multidimensional statistical and graphical visualization is an active area of research in the computer science community. Areas of interest include visualization of vector and tensor fields, volume rendering, geographic information systems, and very large datasets (Hansen & Johnson 2004, Telea 2007, Chen *et al.* 2008).

8.6 Remarks

Astronomers encounter situations involving multivariate data very often, yet are often unschooled in both traditional and newer methods. Tables with columns of properties and rows of celestial objects are extremely common. Other datasets, such of collections of spectra or time series, can be reformulated as a multivariate dataset. Mathematically, a vast range of statistical methods can be applied to data with arbitrary dimensionality, although many require very large n and become computationally burdensome when the dimensionality is high.

As with most applications of statistical inference, a crucial first step is to define the scientific question under study. If the goal is to establish the dependence of a single specified variable on other variables, then multivariate regression should be applied. If the goal is to seek low-dimensional structures without specifying a response variable, then principal components analysis and its extensions should be applied. If the dataset is heterogeneous such that different classes (or outliers) are present, then multivariate clustering or classification (Chapter 9) might be interwoven with multivariate analysis. If the underlying distributions and relationships are not known to be linear, then nonparametric or nonlinear methods might be preferred over those assuming normality and linearity.

Finally, scientists in other fields experienced with multivariate datasets consistently encourage use of advanced visualization techniques to complement statistical analysis. Initial examination of the data, perhaps through a grand-tour movie or smoothed and rendered projections, can be essential to focus analysis on chosen aspects of the datasets. Graphical residual analysis after statistical analysis can reveal the success or limitations of the chosen procedures.

8.7 Recommended reading

Chen, C., Härdle, W. & Unwin, A., eds. (2008) *Handbook of Data Visualization*, Springer, Berlin
A collection of review articles on modern data visualization techniques. Topics include static, interactive and linked graphs; histograms; scatterplots; graph-theoretic approaches; data glyphs; dendrograms; mosaic plots; trellis plots; parallel coordinate plots; projection pursuit; and multivariate tours.

Izenman, A. J. (2008) *Modern Multivariate Statistical Techniques: Regression, Classification, and Manifold Learning*, Springer, Berlin
A comprehensive, advanced text that integrates classical with modern multivariate methods. Coverage includes: nonparametric density estimation; multiple and multivariate regression; principal component analysis; projection pursuit; multidimensional scaling; independent component analysis; principal curves; nonlinear manifold learning; and correspondence analysis.

Johnson, R. A. & Wichern, D. W. (2007) *Applied Multivariate Statistical Analysis*, 6th ed. Prentice Hall, Englewood Cliffs
A widely used and respected undergraduate textbook. Topics include matrix algebra, the multivariate normal distribution, inferences and comparisons of multivariate means, linear regression models, principal components analysis, factor analysis, structural equation models, canonical correlation analysis, linear discrimination and classification, hierarchical clustering, and multidimensional scaling.

Varmuza, K. & Filzmoser, P. (2009) *Introduction to Multivariate Statistical Analysis in Chemometrics*, CRC Press, Boca Raton
A well-written and useful book for physical scientists with extensive applications in **R** and **CRAN**. Coverage includes multivariate data, principal components analysis, ordinary and partial least-squares regression, robust regression, variable selection, classification (linear discriminant analysis, trees, neural networks, Support Vector Machines), clustering (hierarchical clustering, partitioning, model-based clustering).

8.8   R applications

To illustrate a range of multivariate analysis methods, we examine a small dataset presenting global properties of Galactic globular clusters (Webbink 1985) described in Appendix C.7. This dataset has 147 data points with 20 variables. In Section 5.9.2 we examined their univariate magnitude distribution; here we examine a variety of properties concerning their location in the Galaxy, photometry, metallicity, and dynamical properties such as core radius, tidal radius, crossing time-scales, escape velocities and eccentricity.

8.8.1   Univariate tests of normality

We first read the ASCII table using *read.table* with options for a header line with variable names and filling in incomplete rows with *NA* values. A useful early step in multivariate studies is to examine the distribution of each variable, both its range and its similarity to a normal distribution. Strong differences in either property will affect results of many multivariate analyses. **R**'s *summary* function gives the minimum, 25% quartile, median, mean, 75% quartile, and maximum for each variable. We construct a display of univariate histograms, *manyhist*, using **R**'s *function* function (Figure 8.1). The number of panels is evaluated using *n2mfrow* from the dimensionality of the dataset. Note that the Friedman–Diaconis algorithm for choosing the number of bins in each histogram does not lead to a consistent view of the variables.

```
# Overview of globular cluster properties

GC = read.table("http://astrostatistics.psu.edu/MSMA/datasets/GlobClus_prop.dat",
   header=T, fill=T)
summary(GC)
manyhist <- function(x) {
      par(mfrow=n2mfrow(dim(x)[2]))
      for (i in 1:dim(x)[2]) { name=names(x)[i]
            hist(x[,i], main='', breaks='FD', ylab='', xlab=name) }}
      par(mfrow=c(1,1))
manyhist(GC[,2:13])
```

Figure 8.2 shows boxplots for two of the variables. Cluster core radii range over 0–25 parsecs with a heavily skewed distribution; most are smaller than 3 pc with a tail to larger radii. The central surface brightnesses distribution is more symmetrical. More detailed information is provided by normal quantile-quantile plots produced by the function *qqnorm*. A variable with an approximately normal distribution will appear as a straight line in these plots; the *qqline* function draws a line through the first and third quartiles. We see that the globular cluster core radii, as expected from the boxplot, are very non-Gaussian. The central surface brightnesses are closer to normal but exhibit a heavy tail at low values that was not evident from the boxplot.

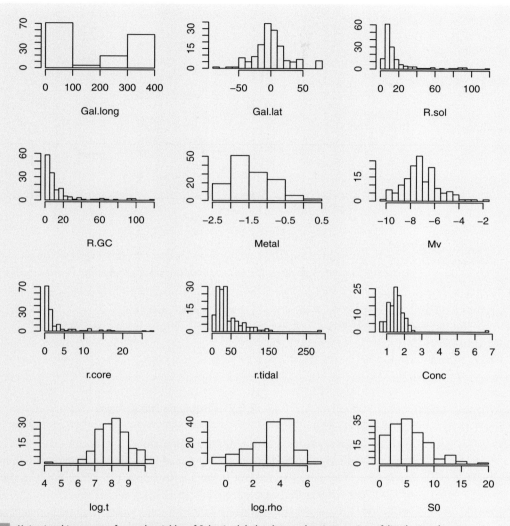

Fig. 8.1 Univariate histograms of several variables of Galactic globular clusters showing a variety of distribution shapes.

```
# Univariate boxplots and normal quantile-quantile plots for two variables

par(mfrow=c(1,4))
boxplot(GC[,8], main='', ylab=expression(Core~r~~ (pc)), pars=list(xlab='',
   cex.lab=1.3, cex.axis=1.3, pch=20))
qqnorm(GC[,8], main='', xlab='', ylab='', cex.lab=1.3, cex.axis=1.3, pch=20)
boxplot(GC[,20], main='', ylab=expression(CSB ~~(mag/arcmin^2)),
   cex.lab=1.3, cex.axis=1.3, pars=list(xlab='', pch=20))
```

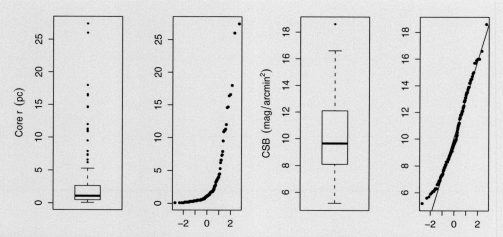

Fig. 8.2 Univariate distributions of two Galactic globular cluster variables. *Left panels:* Boxplot and normal quantile-quantile plot for the core radius. *Right panels:* Boxplot and normal quantile-quantile plot for the central surface brightness.

```
qqnorm(GC[,20], main='', xlab='', ylab='', cex.lab=1.3,
   cex.axis=1.3, pch=20)
qqline(GC[,20])
par(mfrow=c(1,1))
```

8.8.2 Preparing the dataset

We now prepare the dataset for multivariate analysis in several ways. Not all steps are needed for all later operations, but we proceed in a uniform fashion for simplicity. First, to compute variances, we remove objects with empty ("NA") cells using the **R** function *na.omit*; this reduces the sample from 147 to 113 globular clusters.

Second, the inconsistent ranges of the different variables hinder many analyses. For example, the first component of a PCA applied to the original dataset is dominated by Galactic longitude because it has a wider range (0–360°) than other variables. We therefore standardize the variables to unit variance using the *R* function *scale*.

Third, we find it scientifically advantageous to divide the variables into two groups: dynamical variables that describe the internal properties of each globular cluster, and locational properties describing where it lies in the Galaxy. Note the syntax "Dataset[, -c(1:4)]" that removes columns from a multivariate dataset.

Fourth, we examine the dataset for outliers which can strongly affect some analyses. From the univariate distributions of the standardized variables, we find two discrepant points: #12 has a very large tidal radius due to its location in the outer Galactic halo, and #63 has an unusually high absorption due to its location near the Galactic Center. These have 6σ deviations from the mean in one variable. Although these outliers can be scientifically important, here we artificially remove these points using the **R** function *which.max* that identifies the row of the maximum value for a chosen variable. To treat outliers in an automated and algorithmic fashion, robust methods should be used. Robust estimation of covariance matrices, ANOVA, multivariate regression, and PCA is provided by the **CRAN** package *robust* using the MCD and other methods. The chemometrics package *pcaPP* provides a robust PCA calculation.

```
#  Prepare the data
# 1. Remove objects with NA entries, remove labels

dim(GC) ; GC1 <- na.omit(GC[,-1]) ; dim(GC1)

# 2. Standardize variables

GC2 <- scale(GC1)

# 3. Separate locational and dynamical variables

GCloc <- GC2[,+c(1:4)]
GCdyn <- GC2[,-c(1:4)]

# 4. Remove two bad outliers

GCdyn1 <- GCdyn[-c(which.max(GCdyn[,4]), which.max(GCdyn[,11])),]
GCloc1 <- GCloc[-c(which.max(GCdyn[,4]), which.max(GCdyn[,11])),]
```

8.8.3 Bivariate relationships

A common and useful early procedure in multivariate analysis, providing the dimensionality is not too large, is to examine pairwise scatterplots of the dataset. This is provided by **R**'s *pairs* function (Figures 8.3 and 8.4), and the significance of correlations can be assessed using Kendall's τ with the *cor* function. The significance of a single bivariate relationship is calculated using *cor.test*. The **R** functions *var* and *cov* both give the covariance matrix of a multivariate dataset.

Concentrating on the dynamical variables, we first visually examine bivariate relationships from a *pairs* plot. We show both a standard pairs plot, and one with a variety of colors and useful annotations: a least-squares regression line using the *abline* function, a local regression curve using *lowess* (Section 6.6.2), one-dimensional *rug* plots along the axes, and a two-dimensional contour map based on a normal kernel density estimator. The names of 500 colors in **R** can be found at http://cloford.com/resources/colours/500col.htm.

A variety of bivariate relationships, both correlations and anticorrelations, is seen in the globular cluster dynamical variables. Distributions often do not have the MVN shapes and

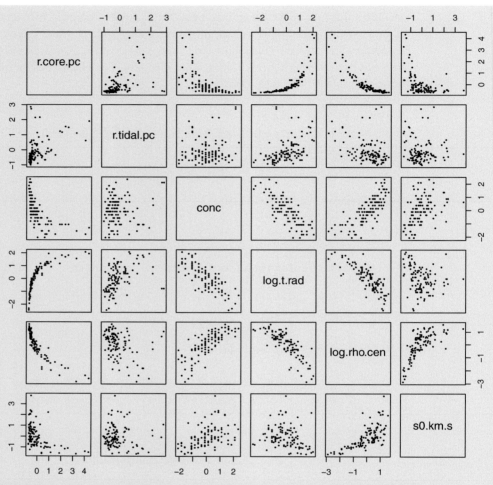

Fig. 8.3   Pair plot of six Galactic globular cluster dynamical variables showing a variety of bivariate relationships.

relationships are sometimes nonlinear. Linear procedures based on MVN distributions will thus not fully model the structure in the dataset, but traditional methods often provide useful insight. Much of the bivariate structure is statistically significant: over half of the bivariate relationships are significant at a level $|\tau| > 0.20$, equivalent to $P < 0.003$ for $n = 111$ or $> 3\sigma$ deviation assuming a normal distribution. Some correlations show little scatter, while others are broad. Some show threshold effects.

```
# Bivariate relationships

cor(GCdyn1, method='kendall')
var(GCdyn1)
pairs(GCdyn1[,3:8],pch=20,cex=0.4)
```

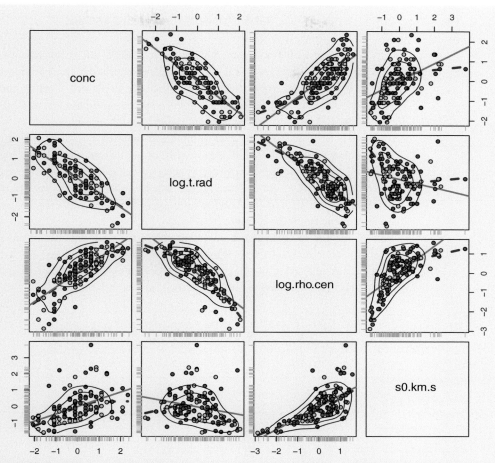

Fig. 8.4 An elaborated pairs plot of Galactic globular cluster dynamical variables with colored points, linear regression (red lines), local regression (dashed blue curves), contours of a kernel density estimator, and rug plots along the axes. For a color version of this figure please see the color plate section.

```
# Elaborated color pairs plot

library(MASS)
pairs(GCdyn1[,5:8],main='',labels=names(GCdyn1[5:8]), panel=function(x,y) {
+     abline(lsfit(x,y)$coef,lwd=2,col='deeppink2')
+     lines(lowess(x,y),lwd=3,col='blue3',lty=2)
+   points(x,y,pch=21,bg = c("red", "green3", "blue"))
+   rug(jitter(x,factor=3),side=1,col='lightcoral',ticksize=-.05)
+   rug(jitter(y,factor=3),side=2,col='cornflowerblue',ticksize=-.05)
+   contour(kde2d(x,y)$x, kde2d(x,y)$y, kde2d(x,y)$z, drawlabels=F,add=T,
    col='darkblue',nlevel=4)
+   })
```

8.8.4 Principal components analysis

The PCA solution is obtained using **R**'s *princomp* function, and various results are presented using *summary*, *loadings*, *plot* and *biplot*. **R**'s plot function for PCA produces a scree plot, shown in Figure 8.5(a), which presents the variance explained by each component. The biplot shows the location of each data point in a scatterplot of the first and second principal components, and displays arrows relating to the strength of each variable's loading in each component. There are many variants of biplots as detailed by Gower & Hand (1996).

Here we see some relationships expected from our astrophysical knowledge of globular clusters. S_0 and V_{esc} both measure stellar velocity dispersion, and the central star density is inversely correlated with the core radius. But the associations of the [Fe/H] metallicity index with absorption $E(B-V)$ and of luminosity M_V with ellipticity are less obvious and may be valuable scientific results.

Finally, we add the principal component values for each data point into a multivariate dataset for later analysis using **R**'s *data.frame* function. For example, if the dataset is heterogeneous, clustering and classification procedures (Chapter 9) are often valuable when performed on the components rather than the original variables.

```
# PCA for dynamical variables.

PCdyn <- princomp(GCdyn1)
plot(PCdyn, main='')
summary(PCdyn)
loadings(PCdyn)
biplot(PCdyn, col='black', cex=c(0.6,1))


# Add principal component values into the dataframe

PCdat <- data.frame(names=row.names(GCdyn1), GCdyn1, PCdyn$scores[,1:4])
```

Table 8.1 shows the loadings of the first three components which together account for 73% of the sample variance in the standardized variables. The first component is dominated by the expected close relationship between core radius, concentration, central star density, and escape velocity. Less obvious is the inclusion of total luminosity (M_V) and ellipticity in this component. The second component is dominated by measures relating to cluster metallicity: the metal line strengths ([Fe/H]), color indices $E(B-V)$ and $B-V$, and the height of the horizontal branch in the Hertzsprung–Russell diagram. The third component links a combination of dynamical, luminosity and metallicity variables in ways that are difficult to interpret. This component may be dominated by noise rather than astrophysically meaningful relationships.

Some common characteristics of PCA are seen here. First, PCA has effectively reduced the dimensionality of the globular cluster dynamical dataset from $p = 15$ to $p \sim 2$–3. This is achieved both by combining variables that measure almost the same physical property, and by combining independent variables that are linearly correlated. Second, the component

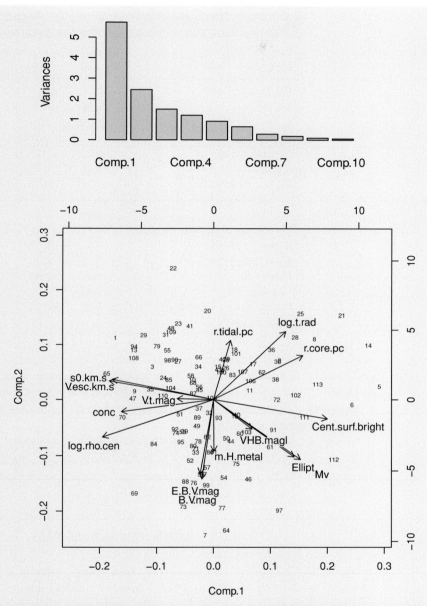

Fig. 8.5 PCA results for 15 dynamical variables of Galactic globular clusters: (a) scree plot showing the variance reduction by the first 10 components; and (b) biplot of the first two components showing individual cluster locations and variable relationships.

Table 8.1 Principal components analysis of Galactic globular clusters.

Property	Comp.1	Comp.2	Comp.3
Variance	44%	19%	11%
[Fe/H]	...	−0.27	−0.35
M_V (mag)	0.30	−0.32	0.33
r_{core} (pc)	0.31	0.24	−0.18
r_{tidal} (pc)	...	0.31	...
Concentration	−0.32	...	0.20
$\log t_{rad}$ (s)	0.25	0.37	−0.35
$\log \rho_{cent}$ (stars pc^{-3})	−0.38	−0.20	...
S_0 (km s^{-1})	−0.35	0.11	−0.38
V_{esc} (km s^{-1})	−0.36	0.11	−0.33
Vertical Horizontal Branch (mag)	0.13	−0.16	−0.31
$E(B-V)$ (mag)	...	−0.39	−0.22
$B-V$ (mag)	...	−0.42	−0.30
Ellipticity	0.28	−0.30	...
V_t (mag)	−0.12	...	0.11
Central surface brightness (mag arcmin^{-2})	0.39	−0.10	−0.10

loadings are sometimes readily interpretable scientifically, but in other cases are difficult to understand. Third, unless a distinct elbow appears in the scree plot, it is often unclear how many principal components should be used for further analysis and scientific interpretation. Components contributing only small fractions of the total variance in the sample are likely fitting noise as well as real structure.

We applied canonical correlations, **R**'s *cancor* function, to see whether linear correlations in the Galactic location variables might explain correlations in the dynamical variables. However, the results did not have any obvious astronomical interpretation.

8.8.5 Multiple regression and MARS

While PCA is designed to find the intrinsic relationships between globular cluster properties, regression tells of the dependencies between a chosen response variable and other dynamical properties. Here we investigate the central surface brightness (CSB) as our variable of interest. **R**'s linear modeling function *lm* gives unweighted least-squares regression coefficients using QR matrix decomposition. The output file, here called *CSB_fit*1, contains a variety of useful results and diagnostics. The resulting linear regression formula is

$$CSB = -0.94 \log \rho_{cen} + 0.33 V_{esc} - 0.33 S_0 - 0.18 VHB + 0.33 E(B-V)$$
$$+ 0.31(B-V) + 0.24 Ellipt + 0.09 M_V - 0.09 [Fe/H]$$
$$+ 0.07 Conc + 0.07 \log t_{rad} \tag{8.30}$$

with negligible dependence on r_{core} and V_t. Recall that the units of CSB are magnitudes arcmin^{-2} so that negative coefficients represent positive correlations between the covariate and the central surface brightness. If we restrict the regression to the six variables with coefficients more important than 0.1, the result is

$$CSB = -0.96\log\rho_{cen} + 0.62V_{esc} - 0.61S_0 + 0.59E(B-V) - 0.23VHB + 0.30Ellipt. \quad (8.31)$$

```
# Multiple regression to predict globular cluster central surface brightnesses

attach(GCdyn)
CSB_fit1 <- lm(Cent.surf.bright~.-Cent.surf.bright,data=GCdyn) ; CSB_fit1
CSB_fitt <- lm(Cent.surf.bright~.,data=GCdyn[,c(7:11,13)]) ; CSB_fit2
str(CSB_fit2)
summary(CSB_fit2)
sd(CSB_fit2$residuals)
par(mfrow=c(2,1))
plot(CSB_fit2$fitted.values, Cent.surf.bright, pch=20)
qqnorm(CSB_fit2$residuals, pch=20, main='')
qqline(CSB_fit2$residuals)
par(mfrow=c(1,1))


# MARS nonlinear regression

install.packages('earth') ; library(earth)
CSB_fit3 <- earth(Cent.surf.bright~.-Cent.surf.bright, data=GCdyn) ; CSB_fit3
sd(CSB_fit3$residuals) qqnorm(CSB_fit2$residuals) ; qqline(CSB_fit2$residuals)
```

This regression fit is excellent; the predicted values are very close to the observed values, as shown in Figure 8.6. The standard deviation of the residuals is only 5% of the original scatter in the central surface brightness values, and the distribution of residuals from **R**'s *qqnorm* function show only small deviations from normality. Qualitatively, the dependencies found in the regression are readily seen on the pairwise bivariate scatter plots (Figure 8.3). Some are easily understood astrophysically − the central surface brightness of a cluster scales with the central star density and with the cluster gravitational potential measured by two nearly redundant variables, V_{esc} and S_0. The inverse dependency on absorption, measured by $E(B-V)$ and $B-V$ may reflect the dynamical expansion of the cluster from tidal stresses near the Galactic Center. Perhaps for similar reasons, highly elliptical clusters have lower central surface brightnesses than spherical clusters.

In light of curved and thresholded relationships in the pairs plots, we examine the performance of the MARS nonlinear regression method. It is implemented in the **CRAN** package *earth* with function *earth* (Milborrow 2010). The regression relationships are considerably different from those in the linear regression. Seven terms are obtained involving only three covariates: $\log\rho_{cen}$, $E(B-V)$, and $\log t_{rad}$. But the performance of the fit is very similar with standard deviation of the residuals 6% of the original scatter. The quantile-quantile

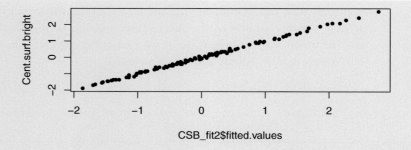

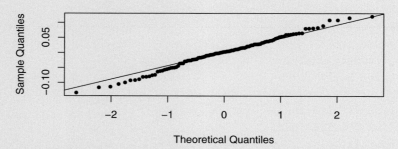

Fig. 8.6 Diagnostic plots of multiple linear regression for predicting globular cluster central surface brightness involving six covariates. *Top*: Comparison of observed and predicted values. *Bottom*: Normal quantile-quantile plot of residuals.

plot of residuals now shows no deviation from normality. We thus find that a nonlinear prediction of globular cluster central surface brightness need not depend on dynamical escape velocity, vertical height of the horizontal branch, or ellipticity. As we do not want our astrophysical interpretation to depend on whether the analysis used linear or nonlinear models, further investigation is needed before a clear understanding of the dataset is obtained.

8.8.6 Multivariate visualization

R has a wide range of visualization capabilities for three or more dimensions of multivariate datasets. While base **R** has simple display functions, the most effective tools provide user interactions that are available when powerful graphical interfaces are linked to **R**. We mention a few other multivariate graphics capabilities here.

The *lattice* package incorporated into base **R** includes *cloud*, a static display function for three-dimensional data points. The **CRAN** package *rgl* is a three-dimensional visualization device for **R** based on the well-established *OpenGL* application programming interface. The *rgl* rendering environment is widely used in other **CRAN** packages. It displays a cloud of three-dimensional data points or a surface, permitting interactive changes in aspects and scale, but does not have a brushing capability. Outputs from *cloud* and *rgl* are shown in Figure 8.7. **CRAN**'s *scatterplot3d* also gives a variety of three-dimensional plots.

The parallel coordinate plot (PCP, Inselberg 2009) is implemented with the **R** functions *parallel* and *parcoord*. **R**'s *lattice* package includes the *splom* package that gives an informative combination of conditional scatterplots and PCPs. The variables are represented as a series of evenly spaced parallel lines, each a real line representing the range of a single variable. Each p-dimensional data point is represented as a line segment with a node at each of the lines. The result is that the multivariate behavior of a dataset can be compactly represented on a single large graph. The tool is particularly effective for identifying clusters and outliers, and is less effective for showing multivariate relationships because the sequence of the variables is not determined by any statistical process. The PCP for the globular cluster dynamical variables is shown in Figure 8.7. Other PCP tools are provided in **CRAN**'s *iplots* and *rggobi* packages.

Finally, the **CRAN** package *abind* provides the useful service of binding together two matrices of arbitrary dimensions, generalizing **R**'s *cbind* and *rbind* functions.

```
# Some multivariate display techniques:
# interactive 3-dim scatter plot; 4-dim bubble plot; parallel coordinates plot

library(rgl)
open3d() ; plot3d(GCdyn1[,5:7])
snapshot3d(file='GlobClus3D.png')

library(lattice)
cloud(GCdyn1[,5]~GCdyn1[,6]*GCdyn[,7], screen=list(z=60,x=45,y=20),
    xlab='log.t.rad', ylab='log.rho.cen', zlab='conc', col=1, pch=1, cex=GCdyn1[,8]+1)

parallel(~GCdyn1, col=c('darkred','darkgreen','orange','blue','black'))
```

8.8.7 Interactive graphical displays

Several packages in **CRAN** are independent GUIs linked to **R** that provide extremely useful visualizations for exploring a multivariate dataset. Five of these are outlined here. We do not provide associated **R** scripts as they are best used in an interactive mode.

The *rgl* package (Adler & Murdoch 2010) is a three-dimensional rendering device driver for **R** based on geometric shapes (e.g. points, lines, polygons, spheres, terrain) with visualization techniques (e.g. lighting, shading, coloring, wiring, surface interpolation, textual annotation). The user can dynamically change the orientation of the projection and, when a desired visualization is achieved, export the result in PostScript format or a gif movie.

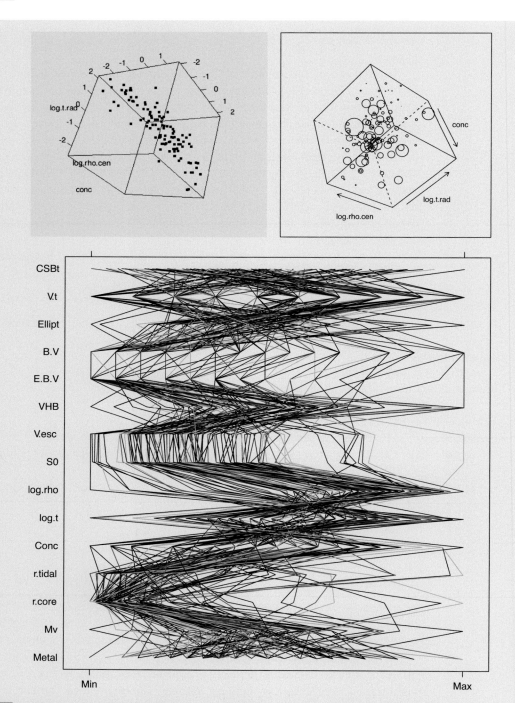

Fig. 8.7 Various multivariate displays of globular cluster dynamical variables: (a) snapshot from the *rgl* three-dimensional real-time rendering device; (b) "cloud" scatterplot of three variables with symbol size scaled to a fourth variable; and (c) 1+1 dimensional parallel coordinate plot of all variables. For a color version of this figure please see the color plate section.

The **CRAN** package *latticist* (Andrews 2010) for interactive exploration of multivariate datasets is based on *Lattice*, a comprehensive and sophisticated graphical package incorporated into **R**. Lattice is described in detail by Sarkar (2008). Three underlying GUIs for *latticist* can be chosen: *gWidgets*, *Gtk2*, or *GTK+*. In addition to standard **R** plots like the one-dimensional histogram and two-dimensional scatterplot, *latticist* provides three-dimensional contour plots, perspective plots, and multidimensional parallel coordinate plots. The user selects variables, or formulae involving variables, for dynamic display, zooming, subsetting (e.g. by formulaic filtering, interactive brushing or labeling), and conditioning. When data points are selected by brushing in one plot, they are similarly highlighted in other plots.

The **CRAN** package *RGGobi* brings the independent GGobi multivariate visualization package into **R** (Lang *et al.* 2010). GGobi, described in detail by Cook & Swayne (2007), is a long-standing public-domain multivariate data visualization tool written in C. In addition to a variety of static plots such as parallel coordinate plots, GGobi provides brushing and dynamic tours or movies of three-dimensional projections of higher-dimensional datasets. The **grand tour** provides a smooth sequence of random projections of the points in hyperspace, while the **projection pursuit tour** distorts the axes to enhance class separations. The user can display principal components, Fisher's linear discriminants, CART or support vector machine classifications in addition to the original variables. At any moment, the user can stop the movie, brush interesting points, and save the projection to a file.

The package *clustTool* provides a GUI for exploring the effects of different multivariate clustering algorithms (Templ 2010). Its function *clust* is a wrapper for a wide variety of clustering algorithms provided in **R** and other **CRAN** packages.

The *rattle* ("**R** Analytic Tool To Learn Easily") package is a particularly broadly capable GUI devoted to the exploration, analysis and data mining of multivariate datasets (Williams 2009). It is somewhat more complicated than most **CRAN** packages to install as it depends on the Gnome (RGtk2) external library as well as dozens of other **CRAN** packages. Interfacing RGtk2 to **R** can be tricky on MacOS systems. We describe *rattle*'s functionality is some detail here, as it illustrates common steps in a multivariate clustering or classification analysis.

A *rattle* analysis of a dataset proceeds by progressively clicking nine tabs, each revealing a selection of buttons, and pull-down menus. After choosing a procedure, the user clicks an "Execute" button, thereby producing graphical and tabular outputs from **R** and **CRAN** functions. A log file gives the **R** scripts used so the user can pursue the analysis outside of *rattle*. In the first tab, data are imported from ASCII, **R** dataframes or other formats. The data can be partitioned into training, testing and validation subsamples. The user can assign roles to particular variables: input variables for modeling, output from modeling (often a categorical classification variable), and ignoring variables.

The second tab allows the user to explore the dataset in a variety of ways, including: the **R** *summary* function; boxplots, histograms, and empirical distribution function plots; a matrix of bivariate correlation coefficients using the *cor* function; principal components analysis; interactive exploration using *prcomp*; and interactive dynamic displays using the independent GUI **CRAN** packages *Latticist* and *RGGobi*. The third tab in the *rattle* GUI gives simple statistical tests which we have discussed in Chapters 3 and 5. These include the

Kolmogorov–Smirnov and Wilcoxon rank nonparametric two-sample tests, the parametric two-sample t and F tests, and correlation coefficients. The fourth tab allows the user to standardize and rescale variables, impute missing values, group the data, and clean up data problems.

The fifth through seventh tabs implement a variety of data mining tools. Unsupervised clustering algorithms include k-means partitioning, *clara* and *hclust* hierarchical clustering. Supervised classification methods include an association rule analysis for large datasets of categorical variables, recursive classification and regression trees (CART), adaBoost and related models, random forests, support vector machines, generalized linear modeling, neural networks, and Cox regression for censored data. The final tab provides a wide variety of data mining model evaluation tools including an error matrix, risk chart, cost curve, lift chart, ROC curve or sensitivity chart, and precision chart. These can use the training, test or validation datasets.

8.8.8 Scope of multivariate analysis R and CRAN

As with most statistical packages, base **R** has extensive coverage of multivariate analysis, particularly in its *MASS* library. The *cancor* and *corresp* functions give canonical correlations between two datasets, *cmdscale* gives multidimensional scaling, *cor* and *cor.test* give correlation matrices, *cov* and *cov.wt* give unweighted and weighted covariance matrices, *dist* and *mahalanobis* compute the distance and Mahalanobis matrices, *factanal* provides a maximum-likelihood factor analysis, *gam* fits generalized additive models, *glm* fits generalized linear models, *lm* fits linear models, *lqs* and *rlm* provide robust regression estimators, *manova* computes multivariate analysis of variance tests, *mvrnorm* simulates from a MVN distribution, *nls* gives a weighted nonlinear least-squares regression, *parcoord* provides a parallel coordinate plot, *ppr* fits projection pursuit regression, *princomp* and *prcomp* give principal components analyses, *proj* returns projections of a matrix onto a linear model, and *sammon* provides a nonmetric multidimensional scaling. Many of these functions are accompanied by specialized summary, plot and print functions to assist in understanding the results.

Over a hundred **CRAN** packages provide specialized treatments of multivariate data; an updated annotated list is available at http://cran.r-project.org/web/views/Multivariate.html. Attractive biplots for PCA are given by *bpca*. The *dr* package assists with reduction of dimensionality for large-dimensionality problems. The *glmnet*, *lars* and *lasso2* packages compute lasso and LARS linear regression models. The *ICNP* and *MNM* packages implement methods of nonparametric multivariate anlaysis. The *kpca* package gives nonlinear principal components analysis with kernel smoothing. The *mvnormtest* package gives a nonparametric test for multivariate normality. The *mvoutlier* package provides a suite of robust methods including outlier identification, robust estimates of location and scatter, and robust Mahalanobis distances. The *pcaPP* package computes principal components with project pursuit. The *princurve* package gives principal curves through a cloud of multivariate data points.

CRAN's *chemometrics* package contains a useful collection of functions for physical scientists from the text of Varmuza and Flizmoser (2009). Several functions divide the sample into learning and test subsets to compute cross-validation estimators for multiple regression, lasso regression, partial least squares, and ridge regression. PCA is computed using an efficient NIPALS algorithm with cross-validation for selecting significant components. Helpful graphical outputs are provided. A wide range of multivariate techniques for ecology and environmetrics are described at http://cran.r-project.org/web/views/Environmetrics.html.

9 Clustering, classification and data mining

Multivariate analysis discussed in Chapter 8 seeks to characterize structural relationships among the p variables that may be present in addition to random scatter. The primary structural relations may link the subpopulations without characterizing the structure of any one population.

In such cases, the scientist should first attempt to discriminate the subpopulations. This is the subject of multivariate clustering and classification. Clustering refers to situations where the subpopulations must be estimated from the dataset alone whereas classification refers to situations where training datasets of known populations are available independently of the dataset under study. When the datasets are very large with well-characterized training sets, classification is a major component of data mining. The efforts to find concentrations in a multivariate distribution of points are closely allied with clustering analysis of spatial distribution when $p = 2$ or 3; for such low-p problems, the reader is encouraged to examine Chapter 12 along with the present discussion.

9.1 The astronomical context

Since the advent of astrophotography and spectroscopy over a century ago, astronomers have faced the challenge of characterizing and understanding vast numbers of asteroids, stars, galaxies and other cosmic populations. A crucial step towards astrophysical understanding was the classification of objects into distinct, and often ordered, categories which contain objects sharing similar properties. Over a century ago, A. J. Cannon examined hundreds of thousands of low-resolution photographic stellar spectra, classifying them in the OBAFGKM sequence of decreasing surface temperature. E. Hubble classified galaxy images into a tuning fork diagram of elliptical, spiral and barred spiral morphologies (e.g. E3, Sa, SBb) while G. de Vaucouleurs refined the classification with additional parameters like rings and lenses (e.g. Sbc(r)). Dozens of classes of variable stars are in use, often named after a bright prototype (e.g. RR Lyr, W Vir, R CrB, PG1159 stars). Various groups of astronomers discovered types of active galactic nuclei during the latter half of the twentieth century, leading to a plethora of overlapping classes: Fanaroff–Riley Class I and II radio galaxies; quasars and quasi-stellar objects; Seyfert 1, 1.5, 1.9 and 2 galaxies; BL Lac objects, blazars and flat-spectrum radio sources; and so forth. Supernova events are classified by their optical light curves and spectra into Type Ia, Ib, Ic, IIP, IIL, IIb and pec (peculiar). Protostars are classified by their infrared spectral energy distributions into Class 0, I, II, transition, and III systems.

Astronomers often divide such populations using simple quantitative criteria based on one or two observed variables: "Short gamma-ray bursts have durations < 2 seconds", or "Abell galaxy clusters with richness 2 have 80–129 galaxies within two magnitudes of the third brightest member", or "Class II young stellar objects have infrared colors $0.4 < [5.8] − [8.0] < 1.1$ and $0.0 < [3.6] − [4.5] < 0.8$". A class may later be divided into subclasses with secondary splits based on variables which may, or may not, have been used to create the original classification.

In all of these cases, the classifications were constructed by astronomers based on their subjective evaluation of the data. Sometimes quantitative criteria define the classes, but these criteria were established subjectively. We do not know of any important classification which was established using objective statistical procedures applied to quantitative measures of source properties. Whitmore (1984) performed a principal components analysis of spiral galaxy properties to find the properties responsible for Hubble's classification sequence, but this study did not examine the boundaries between classes. When quantitative classification procedures are used, they are typically based on rather simple techniques like cuts in a color–color diagram, without quantitative evaluation of the classification scheme. Mukherjee *et al.* (1998) implemented one of the first applications of quantitative clustering techniques, showing that a six-dimensional analysis of NASA's BATSE catalog of gamma-ray bursts revealed three classes, not just two classes well-known from earlier bivariate study.

Consider a more sophisticated problem arising from contemporary astronomical surveys. An optical survey telescope obtains spectra of a million extragalactic objects that, after correcting for redshift, can be divided into a thousand variables representing brightness at preselected wavelengths. We wish to classify these into three groups: normal galaxies, starburst galaxies, and active galactic nuclei. Normal galaxy spectra comprise a family of univariate distributions of correlated continuum shapes with many stellar absorption lines. Starburst galaxies have the normal galaxy spectra superposed with narrow emission lines from heated interstellar gas. Active galaxies have normal or starburst galaxy spectra superposed with broad emission lines from the nuclear region around the massive black hole. Reddening can distort all of these spectra in well-defined ways, as well as suppress broad emission lines. Other variables can be added to assist in the classification: X-ray luminosity, infrared luminosity, radio spectrum and polarization, and indicators of dynamical interaction with nearby galaxies.

Astronomers have developed a suite of effective, but mostly heuristic, procedures for conducting such classifications. Some objects at the extremes are easily classified without difficulty, such as normal elliptical galaxies and quasars. But other objects may appear superficially similar yet be placed in intrinsically different classes. For example, curved functions in scatterplots of the reddening-corrected line ratio log [O III]/Hβ plotted against log [N II]/Hα discriminate otherwise similar starburst and Type 2 active galaxies (Veilleux & Osterbrock 1987). In many cases, properties of both classes are mixed in the spectrum and classification can be quite difficult. In hundreds of astronomical studies of this problem, almost none have tried multivariate classification procedures.

The need for automated classification tools has become increasingly important with the growth of megadatasets in astronomy, often from wide-field surveys of the optical sky. Here, either from digitizing older photographic plates such as the Palomar Sky Survey or from CCD surveys such as the Sloan Digital Sky Survey, billions of faint objects are located in petabytes of images. The most elementary need is to discriminate galaxies, which are typically resolved blurry objects, from stars, which are unresolved. But the unresolved population is itself heterogeneous with main-sequence, giant, white dwarf and brown dwarf stars mixed with quasars where the active galactic nucleus outshines the host galaxy. Spectroscopic surveys have reached millions of objects, soon to increase to hundreds of millions. Recent progress in the use of classification and data mining techniques in astronomy is reviewed by Ball & Brunner (2010) and by Way *et al.* (2011).

While the human mind–eye combination can uncover previously unknown groupings and patterns in complicated datasets, statistical and algorithmic procedures can be effective in defining the number and boundaries of classes, and in applying established classifications to new objects. Automated quantitative techniques are essential when the dimensionality (p variables) and/or size (n objects) of the datasets are high. These advanced data analysis and visualization techniques can be applied in conjunction with mind–eye examination and scientific judgment to discover clusters and make classifications in a consistent fashion. Monte Carlo procedures can provide information on the reliability of a classification or the assignment of an individual object to a chosen class.

9.2 Concepts of clustering and classification

9.2.1 Definitions and scopes

The datasets we consider here consist of n objects with p properties. We further assume that the dataset includes members from k subpopulations with distinct locations or patterns in p-space. Many multivariate methods assume that all n objects lie in one or another cluster, but other methods permit some objects to remain unclassified.

When the number, location, size and morphologies of the groupings are unknown, **unsupervised clustering techniques** search for groupings in a multivariate dataset based on the proximity of objects using some distance measure in p-space. The results are often uncertain and highly dependent on the distance measure chosen. When a training set of well-studied objects is available to characterize the groupings in the dataset, discrimination rules are established on the training set and they are applied to a new test set. These two steps are together called **supervised classification**. Classification procedures of the form "Class II young stellar objects have mid-infrared colors $0.4 < [5.8] − [8.0] < 1.1$ and $0.0 < [3.6] − [4.5] < 0.8$" are called **classification trees** or **decision trees**.

For both clustering and classification, procedures can be model-based or distance-based. Model-based methods assume the cluster morphologies follow a predefined model where its parameters are established from the dataset. The most common parametric approach is the **normal mixture model** where the dataset is assumed to consist of $k \geq 2$ multivariate Gaussians. Likelihood-based procedures, either maximum likelihood estimation or Bayesian inference, are then used to obtain the number of groups in the mixture, their mean and variance values in each variable (McLachlan & Peel 2000). Large datasets are often needed to establish the model parameters with reasonable reliability.

Other clustering and classification procedures make no assumptions regarding the parametric form of the distributions and, as a consequence, the results can be very sensitive to the mathematical procedure chosen for the calculation. A considerable body of recent procedures, both numerical and visualization, has emerged to assist the scientist in establishing and evaluating the merits and reliability of nonparametric clustering and classification. As astronomers have no scientific basis to assume that the properties of astronomical populations follow normal distributions, distance-based approaches are emphasized in this chapter Robust methods insensitive to non-Gaussianity and outliers can be effective.

While some methods were developed by statisticians during the mid-twentieth century, other methods oriented towards classifying objects in very large datasets were more recently developed by computer scientists. Many of the latter procedures are loosely collected under the rubric **data mining**. The distinction between computationally intensive data mining and statistical classification techniques is blurry, and developments of a given method often proceed from both perspectives. A related theme is broadly called **machine learning** where computational strategies allow the system to improve its performance as more data are treated. These include both unsupervised and supervised classification using techniques like decision trees, neural networks, Support Vector Machines, and Bayesian networks. **Classifier fusion** or **metalearning** is the effort to combine the results from different classifiers into an integrated decision process.

9.2.2 Metrics, group centers and misclassifications

Certain fundamental choices must be made to conduct any clustering or classification algorithm. First, as discussed in Section 8.2.1, a **metric** defining the distance between objects is usually defined. When the units in all p variables are the same, a natural metric is the Euclidean distance based on standardized variables. Another common choice is the unit-free Mahalanobis metric.

A second important choice implicit in many clustering and classification algorithms is the definition of the **center** of a group of objects. An obvious choice is the centroid, the vector of mean values of each variable. But it is well-known that medians are more robust than means when outliers and non-Gaussian distributions are present (Chapter 5). Consequently, clustering or partitioning around **medoids** for each cluster is often preferred to clustering around means, However, here the choice of the central representative point may not be unique. Other definitions of group center are often used, and sometimes no center is defined at all.

The clustering algorithm most familiar to astronomers is nicknamed the **friends-of-friends algorithm** popularized in the 1970s by the search for galaxy groups in redshift surveys. Here an object within a specified distance from *any* object in a group (rather than the group center) is considered to be a new group member. This procedure has many designations in clustering methodology including **single-linkage agglomerative hierarchical clustering** and the **pruned minimal spanning tree**. Astronomers should realize that the choice of defining group membership by a friends-of-friends criterion is considered ill-advised in other fields, as it leads to elongated rather than compact clusters and is sensitive to noise and outliers.

A third aspect of any classification or discrimination technique is validation of the procedure's success in achieving correct classifications and avoiding misclassifications, both for the training set and the dataset of new objects. The rate of misclassification can be viewed as the conditional probability of classifying an object into group 2 when it is truly a member of group 1, $P(2|1)$. The total probability of misclassifying an observation into group 2 is $P(2|1)p_1$ where p_1 is the probability (or fraction) of the dataset in group 1. This can be combined with possible asymmetries in the **cost** of misclassifying an object into group 2 rather than group 1, $c(2|1)$. For example, if a multi-epoch optical survey seeks to find Type Ia supernovae for spectroscopic follow-up, if the sample is large it may be more costly to waste telescope time on erroneous spectroscopic measurements, while if the sample is small it may be more costly to lose important scarce targets. Any classification procedure into two groups will thus have a total **expected cost of misclassification (ECM)** of

$$ECM = c(2|1)P(2|1)p_1 + c(1|2)P(1|2)p_2. \tag{9.1}$$

The sum can be extended to k groups. If prior knowledge is weak, then it is common to assume that the prior probabilities and misclassification costs are equal, so that

$$ECM_0 = P(2|1) + P(1|2). \tag{9.2}$$

Classification procedures can be compared, or even combined, to obtain the smallest ECM value.

The **apparent error rate (APER)**, or proportion of misclassified objects in the training set, is a measure of classification quality that does not require knowledge of prior probabilities and costs. Here one can split the training set into two subsamples, one used to construct the classification rules and the other to validate it using the APER. Methods for conducting such tests include the jackknife and cross-validation. The APER can thus be used to compare a wide range of classification procedures.

9.3 Clustering

9.3.1 Agglomerative hierarchical clustering

Hierarchical methods are commonly used to find groupings in multivariate data without any parametric assumptions (except for choices of metric, variable transformation, and group center discussed above) or prior knowledge of clustering. Agglomerative clustering of n points $(x_1, x_2, \ldots, x_n)$ is based on the $n \times n$ matrix $\mathbf{D} = ((d(\mathbf{x_i}, \mathbf{x_j})))$ of pairwise distances between the points given in Equation (8.3). The procedure starts with n clusters each with one member. The clusters with the smallest value in the distance matrix are merged, and their rows and columns are removed and replaced with a new row and column based on the center of the cluster. This merging procedure is repeated $n - 1$ times until the entire dataset of n points is contained in a single cluster. The result is plotted as a **tree** or **dendrogram**, with n leaves at one end, and grouped into twigs and branches of increasing size until the entire sample is grouped into a single trunk. The ordinate of the dendrogram gives the distance at which each group merger occurs. Hierarchical clustering algorithms usually require $O(n^2 \log n)$ or $O(n^2)$ operations.

The cluster hierarchy resulting from this procedure depends strongly on the definition of the distance between a cluster and a new point. Let us designate this distance as $d_{C,k} = d_{1,2,\ldots,j,k}$ where k represents the new point and the cluster C has j members at some stage in the clustering calculation. Four choices are commonly considered. In **single linkage** clustering,

$$d_{C,k} = \min(d_{1,k}, d_{2,k}, \ldots, d_{j,k}); \tag{9.3}$$

that is, the distance to the closest member of the cluster is used. This is the astronomers' friends-of-friends algorithm, sometimes called the **nearest-neighbor algorithm**, and is the oldest of the hierarchical clustering methods (Florek *et al.* 1951, Sneath 1957). An inversion of the single linkage clustering is mathematically identical to the hierarchy of clusters obtained by constructing the **minimal spanning tree (MST)** of the points in p-space (using the chosen metric) and successively pruning the tree of its longest branches. A variety of computational algorithms is available for constructing the MST; the earliest was developed by O. Boruvka in the 1920s.

Single linkage clustering can result in groupings that are asymmetrical and elongated with low surface density. **Complete linkage** (or McQuitty) clustering gives groupings that are more symmetrical and compact using the group-point distance measure

$$d_{C,k} = \max(d_{1,k}, d_{2,k}, \ldots, d_{j,k}). \tag{9.4}$$

Here the distance to the furthest, rather than closest, member of a group is used. A reasonable and commonly used intermediate choice is the mean distance provided by **average linkage** clustering,

$$d_{C,k} = \frac{1}{j} \sum_{i=1}^{j} d_{i,k}. \tag{9.5}$$

Another common choice is **Ward's minimum variance** method where the clusters to be merged are chosen to minimize the increase in the sum of intra-cluster summed squared distances. Here the group-to-point distance is a weighted Euclidean distance and the cluster hierarchy gives roughly spherical or ellipsoidal cluster shapes. Median distances are not appropriate here because they can lead to inversions (crossings) in the resulting dendrogram.

There are no formal criteria available for evaluating the reliability of a hierarchical cluster dendrogram, either in its entirety or around a single branch. Formal validation of a decision to merge two clusters in a hierarchy can be made with a hypothesis test based on the sum of intra-cluster variances before and after the merger. This is useful if the situation warrants the assumption of asymptotic normality. The cophenetic correlation coefficient gives a measure of the similarity between an estimated hierarchy and the dataset's distance matrix, but its utility is not clear. Computationally sophisticated refinements to hierarchical clustering involve sequential improvements to the cluster pattern; these procedures include competitive learning, leader–follower clustering, adaptive resonance theory, and graph-theoretic treatments of the minimal spanning tree (Duda *et al.* 2001, Section 10.10).

The performances of traditional hierarchical clustering methods have been evaluated for a variety of simulated situations. Single linkage clustering is simple to understand and compute, but has the tendency to build unphysical elongated chains of clusters joined by a single point, especially when unclustered noise is present. Figure 12.4 of Izenman (2008) illustrates how a single linkage dendrogram can differ considerably from the average linkage, complete linkage and divisive dendrograms, which can be quite similar to each other. Kaufman & Rosseeuw (1990, Section 5.2) report that "Virtually all authors agreed that single linkage was least successful in their [simulation] studies." Everitt *et al.* (2001, Section 4.2) report that "Single linkage, which has satisfactory mathematical properties and is also easy to program and apply to large data sets, tends to be less satisfactory than other methods because of 'chaining.'" Ward's method is successful with clusters of similar populations, but tends to misclassify objects when the clusters are elongated or have very different diameters. Average linkage is generally found to be an effective technique in simulations, although its results depend on the cluster size. Average linkage also has better consistency properties than single or complete linkage as the sample size increases towards infinity (Hastie *et al.* 2009, Section 14.3).

Finally, we note that divisive hierarchical clustering is analogous to agglomerative hierarchical clustering but operating in the opposite direction, starting with the full dataset as a single cluster and dividing it progressively until each data point is a distinct cluster. Although not commonly used, divisive procedures may be computationally effective for megadatasets where the large-scale structure is under study. The calculation can then be terminated after a few steps, avoiding construction of the full dendrogram.

9.3.2 *k*-means and related nonhierarchical partitioning

A different approach to dividing a multivariate dataset into groups is ***k*-means partitioning** and related methods. Developed by J. MacQueen in the 1960s, this popular method is conceptually very simple: starting with k **seed** locations representing the cluster centroids, one iteratively (re)assigns each object to the nearest cluster to reduce the sum of within-cluster squared distances, the W matrix used in MANOVA testing (Section 8.3). The centroids of the clusters receiving and relinquishing the object are recalculated after every operation. In k-means partitioning, the centroid is the mean value of each variable for the current members of the cluster. This process continues until no reassignments are made.

k-means calculations can be computationally rapid if the dimensionality and number of clusters are not too large; k-means is thus an important data mining tool for large datasets. The operations are not complicated; for example, a centroid of the j-th cluster can be updated by the addition of a new data point $\mathbf{x}_i$ by

$$\bar{\mathbf{x}}_j{}' = \frac{n_j \bar{\mathbf{x}}_j + \mathbf{x}_i}{n_j + 1}, \tag{9.6}$$

where n_j is the current number of objects in the j-th cluster. Unlike hierarchical methods, the pairwise distance matrix does not have to be calculated and stored in memory. Computational algorithms include the Lloyd algorithm, the EM algorithm used in parametric mixture models (Section 3.4.5), and a filtering algorithm using kd-trees. Convergence of the cluster centroids is usually rapid, but there is no guarantee that a globally optimal solution is achieved.

Three choices must be made. First, the number of clusters k is predetermined by the scientist and does not change through the computation. Hastie *et al.* (2009, Section 14.3) outline strategies to choose a reasonable value for k; for example, one can calculate the partition for a range of $k = 1, 2, \ldots, k_{max}$ and choose a k giving the smallest sum of within-cluster variances. Second, the initial seed locations can influence the result. If the problem does not have prior estimates of the cluster centroids, it is common to compare results from a number of trials with random seed locations. In some cases, the final cluster assignments depend sensitively on the seed configuration, and for large datasets, it may be impossible to try enough initial conditions to insure that the global optimum within-cluster variance has been found.

Third, as with the hierarchical methods outlined above, a mathematical definition of **centroid** is needed. The original choice of mean values in each variable can be sensitive to outliers, and median values are often preferred. This is called **k-medoid partitioning**. Here the summed distances, not squared distances, are minimized. Another variant is k-center clustering which minimizes the maximum cluster size. Unlike k-means methods, each k-medoid centroid must coincide with a data point. k-medoid procedures are less computationally efficient than k-means procedures but often exhibit excellent performance.

k-means and k-medoid procedures have gained prominence in addressing important machine learning problems. In computer vision, they are used for image segmentation, edge detection and other aspects of pattern recognition. The **Linde–Buzo–Gray (LBG) algorithm**, a generalization of a standard algorithm for k-means clustering, is an important tool in lossy **vector quantization** image compression and speech recognition techniques. The LBG algorithm starts with the entire dataset as a single cluster, then progressively splits it into more clusters while updating the cluster centroids using the k-means procedure. Tree-structured vector quantization is a related algorithm that examines the hierarchy of multiscale structures resulting from repeated application of the LBG algorithm; it is important for data compression of images and other large datasets.

9.4 Clusters with substructure or noise

The above methods, both clustering and partitioning, assume that all of the objects in the sample belong in a cluster. However, in astronomy, it is not uncommon that one or more clusters are superposed on a featureless background of objects, often due to foreground and background populations that pervade wide-field surveys. Equivalently, the clustering structure may be multiscale with small denser regions superposed on large low-density regions. In model-based clustering, the background or low-density regions could be mimicked by a constant or polynomial component in addition to Gaussian clusters. In hierarchical clustering, one can probe the dendrogram branches with compact structures. But mathematically, these methods are not well-adapted to composite structures with both clustered and unclustered components.

Hastie *et al.* (2009, Section 9.3) describe a tree-based **patient rule induction method (PRIM)** that locates localized density enhancements in a multivariate dataset. This **bump hunting** algorithm starts with the full dataset, and progressively shrinks a box (hyper-rectangles) with sides parallel to the axes. The box face is contracted along the face that most increases the mean density, or other chosen criterion. After locating the peak location with a preset minimum population, the box may be expanded if the density increases. The box is removed from the dataset, and the process is repeated to find other bumps in the distribution.

Another clustering algorithm that locates peaks in a background distribution is the **Density-Based Spatial Clustering of Applications with Noise (DBSCAN)** procedure developed by computer scientists Ester *et al.* (1996). DBSCAN starts with the requirement that μ data points lie within the vector distance ϵ (the **reach**) of a data point. The user must supply the μ and ϵ values; examination of a kernel smoothed estimator of low-dimensional projections of the dataset may help in choosing values. A cluster around an object with $\geq \mu$ objects in its ϵ vicinity expands to include all objects that are density-reachable from any other object in the cluster. When the algorithm has converged, data points that are not in clusters are labeled as noise.

The DBSCAN algorithm has strong capabilities and is widely used in many fields. Unlike k-means, the user does not need to know the number k of clusters in advance. Unlike normal mixture models, the shape of a cluster is not restricted to MVN ellipsoids. DBSCAN is less vulnerable than single link hierarchical clustering to accidental chaining of clusters by intervening points, and is computationally efficient with $O(n \log n)$ operations. However, like most other methods, it depends on the appropriate choice of a metric and its results will thus change with variable units and transformations. It also is not well-adapted to complicated multiscale structures where μ varies across the p-space. Generalizations of DBSCAN for data with measurement errors are being developed (Volk *et al.* 2009).

Additional unsupervised algorithms are emerging to cluster datasets with background populations or complex structure. The **Ordering Points To Identify the Clustering Structure (OPTICS)** algorithm extends DBSCAN, ordering the objects by **core distance**, the smallest ϵ with μ points, and **reachability distance**, the pairwise maximum of the core distance or Euclidean distance to another object (Ankerst *et al.* 1999). This permits graphics that show density-based clusters over a range of ϵ values. The **Balanced Iterative Reducing and Clustering using Hierarchies (BIRCH)** algorithm is a thresholded tree-based clustering algorithm primarily designed to treat very large datasets in a single pass; its parameters can also be set to consider some objects as unclustered noise points (Zhang *et al.* 1997). **Density-based clustering (DENCLUE)** uses kernel-like influence functions associated with each object to find gradients and local maxima in the smoothed dataset (Hinneburg & Keim 1998). The **CHAMELEON** clustering algorithm first divides the dataset into small subclusters using a k-nearest-neighbor procedure, and then recombines them using the product of two dynamically evolving criteria, relative connectivity and relative closeness (Karypis *et al.* 1999). This method is effective in avoiding fragmentation of denser clusters while achieving sensitivity to discover less dense clusters.

9.5 Mixture models

An astronomer sometimes asks whether a dataset is consistent with white noise, has a single peak, or has two or more peaks. The distinction between white noise and peaked distributions can be called **bump hunting** while the distinction between one or more peaks can be called a test for **multimodality**. This issue is addressed by a variety of methods drawn from different branches of statistics: parametric inference, nonparametric inference, time series and image analysis. We briefly outline some approaches here.

A common method for distinguishing a constant value from a peaked distribution, or a multimodal distribution from a unimodal distribution, is to apply parametric **mixture models** and model selection techniques to quantify how many components are needed to fit the data. In normal (Gaussian) mixture models, the data are assumed to be collected into k ellipsoidal clusters with multivariate normal (MVN) distributions. These methods date back to K. Pearson in the 1890s and are presented in detail by McLachlan & Peel (2000). The model has $2kp + k + 1$ parameters: k means and variances in p dimensions; k subpopulations in each cluster; and k itself. Each cluster has a mean vector μ_k and covariance matrix Σ_k. Note that model-based clustering can assign an object to more than one cluster in a probabilistic fashion. This is a **soft** classification procedure unlike k-means and agglomerative methods which give **hard** classifications assigning each object to only one cluster.

The parameters of the normal mixture model are usually obtained by maximum likelihood estimation using the EM algorithm (Section 3.4.5). First, an estimate of k and cluster locations is made to start the calculation. These initial values are often obtained from procedures such as k-means clustering or from random locations. The EM algorithm here is quite similar to the k-means algorithm. In the expectation step, the likelihood of each object lying in each cluster is calculated using the MVN distribution; for some objects, these weights can be approximated as unity for the closest cluster and zero for more distant clusters. In the maximization step, the cluster parameters are optimized with contributions from all objects. One must take care in maximizing the likelihood of normal mixtures; in particular, it is necessary to pre-assign a lower bound to the variances.

Two implementations of the normal mixture model clustering method using the EM algorithm are in common usage: EMMIX (EM-based MIXture model) described by McLachlan & Peel (2000); and MCLUST with Bayesian regularization developed by Fraley & Raftery (2002, 2007). In EMMIX, confidence intervals on the number of clusters and cluster parameters are estimated from parametric and nonparametric bootstrap. In MCLUST, the likelihood is maximized for a range of k clusters, and the k-value giving the maximum Bayesian information criterion (BIC, Section 3.7.3) is selected as the best model. Auto-Class is another method with Bayesian optimization developed by astronomers Cheeseman

& Stutz (1995). Here the maximum a posteriori (MAP), or mode of the posterior, solution is found using a uniform prior.

Alternative procedures based on the assumption of MVN clusters are available. A mixture discriminant analysis analogous to linear discriminant analysis (Section 9.6.2) can be calculated using the EM algorithm, including variants such as rank constraints for robustness or penalties for roughness (Hastie *et al.* 2009, Chapter 12). Everitt *et al.* (2001, Chapter 5) discuss various clustering methods based on the within-cluster and between-cluster variances, **W** and **B**, discussed in Section 9.6.2. Rules for global selection of the number of groups, and local criteria for subdividing a group, have also been proposed, though often without formal mathematical validation. For MVN clusters, MANOVA tests such as Wilk's Λ and Pillai's trace can be used to evaluate whether two clusters obtained from a clustering algorithm are consistent with a single population. There is no accepted test that gives significant levels for the existence of a weak cluster.

Nonparametric approaches have also been proposed to test for multimodality. The most well-known is Hartigan & Hartigan's (1985) **dip test** that uses a Kolmogorov–Smirnov-type maximum distances between the data and some **taut string** envelope around a binned distribution (Davies & Kovac 2004). Silverman (1981) proposes a kernel density estimator together with hierarchical bootstrap resampling to estimate the number of modes in a distribution. Other tests are based on measures of local data concentration (Müller & Sawitzki 1991; Polonik 1999). No consensus has emerged on the effectiveness of these methods under general conditions.

9.6 Supervised classification

9.6.1 Multivariate normal clusters

One of the best-studied models for supervised classification is the multivariate normal distribution where **X** has the multivariate normal (MVN) probability density function $\mathbf{X} \sim N_p(\mu, \Sigma)$. The goals of a parametric clustering procedure are to estimate the number of clusters k, the cluster means and variance parameters with confidence intervals, and prediction probabilities for assigning new data points to clusters. Recall that the joint confidence region for the mean vector μ cannot be derived from the collection of univariate confidence intervals without considering the covariance structure.

An early result by A. Wald in the 1940s is a classification rule for assigning a new object to two MVN populations in a fashion that minimizes the *ECM* in Equation (9.1). Here the class means and variances – $\bar{\mathbf{X}}_1, \bar{\mathbf{X}}_2, \mathbf{S}_1$ and $\mathbf{S}_2$ – must be known in advance. A new object with vector $\mathbf{x}_0$ should be assigned into group 1 with mean μ_1 and covariance Σ_1 if

$$\mathbf{x}_0' \left(\frac{1}{\mathbf{S}_1} - \frac{1}{\mathbf{S}_2} \right) \mathbf{x}_0 + \left(\frac{\bar{\mathbf{X}}_1'}{\mathbf{S}_1} - \frac{\bar{\mathbf{X}}_2'}{\mathbf{S}_2} \right) \mathbf{x}_0 - \frac{1}{2} \ln \frac{|\mathbf{S}_1|}{|\mathbf{S}_2|}$$
$$-\frac{1}{2}(\bar{\mathbf{X}}_1' \mathbf{S}_1^{-1} \bar{\mathbf{X}}_1 - \bar{\mathbf{X}}_2' \mathbf{S}_2^{-1} \bar{\mathbf{X}}_2) \geq -2 \ln \left(\frac{c(1|2)}{c(2|1)} \frac{p_2}{p_1} \right) \tag{9.7}$$

where $\bar{\mathbf{X}}$ and $\mathbf{S}_i$ are sample means and covariances, and c and p are the cost functions and prior probability presented in Equation (9.1). If the inequality does not hold, the object is placed into group 2. A simpler formula arises if one knows that the covariances of the two groups are the same, $\mathbf{S}_1 = \mathbf{S}_2$. Hotelling's T^2 statistic (Section 8.3) is used to measure the distance between a collection of new objects and a predefined cluster.

9.6.2   Linear discriminant analysis and its generalizations

R. A. Fisher proposed a method in the 1930s for multivariate discrimination that is still widely used today. **Linear discriminant analysis (LDA)** for two classes can be viewed geometrically as the projection of the p-dimensional data point cloud onto a one-dimensional line in p-space that maximally separates the classes. Algebraically, linear combinations of the data points are constructed such that the separation of the two classes is maximized, where separation is measured by the ratio of the between-cluster variance $\mathbf{B}$ to the within-cluster variance $\mathbf{W}$ for the population. The maximum separation occurs for

$$Sep = \frac{\mathbf{B}}{\mathbf{W}} = \frac{\mathbf{a}^2(\bar{\mathbf{X}}_2 - \bar{\mathbf{X}}_1)^2}{\mathbf{a}'(\mathbf{S}_1 + \mathbf{S}_2)\mathbf{a}} \quad \text{where}$$
$$\mathbf{a} = \frac{\bar{\mathbf{X}}_2 - \bar{\mathbf{X}}_1}{\mathbf{S}_1 + \mathbf{S}_2}. \tag{9.8}$$

The vector $\mathbf{a}$ is perpendicular to the discriminant hyperplane. LDA requires that the covariance matrices of the two classes are the same; this assumption is relaxed in **quadratic discrimination analysis (QDA)**.

LDA is similar to principal components analysis (Section 8.4.2) but with a different purpose: PCA finds linear combinations of the variables that sequentially explain the variance for the sample treated as a whole, while LDA finds combinations that efficiently separate classes within the sample. The LDA vector $\mathbf{a}$ is usually obtained from a training set with known class memberships, and then applied to new data. The computation of $\mathbf{a}$ requires $O(p^2 n)$ operations.

LDA can also be formulated with likelihoods for maximum likelihood estimation and Bayesian inference. The assumption that each of the class probability density functions are products of marginal densities (i.e. that the variables are independent) corresponds to a **naive Bayes** classifier, while interactions between the variables can be treated with a more general approach. When the training set is small and the dimensionality is high, then regularized maximum likelihood or Bayesian analysis is needed. Srivastava *et al.* (2007), for example, demonstrate good performance for a Bayesian quadratic discriminant analysis with a prior based on the covariance of the data. LDA can also be derived by least-squares multiple regression of a categorical response variable Y representing the two classes.

Many generalizations of Fisher's LDA have been developed; these methods are important in contemporary statistical discrimination and data mining. Multiple discriminant analysis treats the case of more than two classes. The procedure outlined above is generalized to find a subspace of the original p-space that maximizes the separation, and the vector $\mathbf{a}$ becomes a matrix $\mathbf{A}$ of dimension $p \times (k-1)$ for k classes. QDA and polynomial discriminant analysis treat nonlinear problems; for example, QDA expands a bivariate

(X_1, X_2) to consider $(X_1, X_2, X_1 X_2, X_1^2, X_2^2)$. In canonical discriminant analysis, one uses LDA to define subspaces of fewer than p dimensions, giving a reduction in dimensionality emphasizing class structure. To treat outliers, the sample mean and covariance values in Equation (9.8) can be replaced by robust estimators such as the minimum covariance determinant estimators.

A criterion analogous to the minimum ECM is needed to implement LDA and related methods. The **perceptron algorithm** is a linear classification procedure with binary responses determined by a weight vector that minimizes the sum of the distances (or squared distances) from misclassified samples to the discriminant boundary (Minsky & Papert 1969). The weights are updated sequentially when classification errors of the training set are made. Other gradient descent procedures for finding linear discriminant functions include least-mean-square, Ho–Kashyap, and the simplex algorithms.

An important class of generalized LDAs are the **Support Vector Machines** (SVMs) presented by Vapnik (2000). Here the dataset is mapped by nonlinear functions onto a higher dimensional space with $p^* > p$ such that classes in the training set can be cleanly separated using linear discrimination (i.e. separated by flat hyperplanes in p^*-space). The **support vectors** straddle the optimal hyperplane, giving the distance or **margin** to the edges of the clusters on each side. The method is extended to nonlinear hypersurfaces with the use of kernel smoothers, and computation is performed using quadratic programming. These methods can be viewed as an optimization problem to find the hyperplane margin separating the two classes, and are efficient even in high-dimensionality spaces. From this perspective, LDA uses a quadratic loss function while the SVM loss is linear, estimating the mode of posterior class probabilities. The optimization calculation of SVMs differs from those of classification trees or ANNs in that the SVM iterations cannot be trapped at a local maximum that is not the global maximum. A variety of computational strategies have been developed. The computations are often efficient and well-adapted to large datasets. SVMs have thus become important procedures for challenging multivariate classification problems.

Duda *et al.* (2001, Chapter 5), Hastie *et al.* (2009, Chapter 12) and Izenman (2008, Chapters 7, 8 and 11) discuss these and other generalizations of LDA that seek to separate multivariate classes under nontrivial conditions. Marsland (2009) is a good introductory text to these and other aspects of machine learning.

9.6.3 Classification trees

Formal procedures for dividing a multivariate training sample into classes by progressive splitting based on the original variables are well-established in statistics and data mining under the rubric of **classification trees** or **decision trees**. L. Breiman was the leading figure in the development of these methods over several decades; he called the methods **Classification and Regression Trees (CART)**. A variety of procedures are available to create classification trees, grow and prune the tree, combine weak criteria (**bagging** and **boosting**), and quantify the importance and reliability of each split (often with bootstrapping). The construction and evaluation of classification trees are discussed by Duda *et al.* (2001, Chapter 8), Izenman (2008, Chapter 9), and Breiman *et al.* (1998).

Classification trees are broadly applicable as they are nonparametric and nonmetric. The latter characteristic is unusual as it readily treats variables with incompatible units or type (e.g. continuous, ordinal, categorical) and removes the need to bring variables onto similar scales. No distance matrix is used. Simple classification trees have a hierarchy of splits of the type "Variable 7 is less than or greater than 0.2", producing hyperplanes parallel to the axes in p-space. **Oblique decision trees** produce hyperplanes with arbitrary orientations.

CARTs are constructed to minimize the inhomogeneity or **impurity** along branches. At node m of a tree, let P_j be the fraction of objects in the training set with class j. Several measures of impurity are used:

$$i(m) = \begin{cases} 1 - \max_j P_j & \text{misclassification impurity} \\ P_j P_k & \text{variance impurity} \\ -\sum_j P_j \log_2 P_j & \text{entropy impurity} \\ \frac{1}{2}\left[1 - \sum_j P_j^2\right] & \text{Gini impurity.} \end{cases} \qquad (9.9)$$

At each node m, we seek to create a split that maximizes the drop in impurity along the left and right, m_L and m_R, descendent nodes. This drop is

$$\Delta i(m) = i(m) - P_L i(m_L) - (1 - P_L) i(m_R). \qquad (9.10)$$

We **grow** the classification tree with a progression of optimizations of the one-dimensional $\Delta i(m)$ quantities. The calculation is local, and the choice of impurity measure is usually not critically important; the entropy and Gini measures are often used. Unless the tree is grown to saturation with $m = n$ terminal nodes, the training set objects in many terminal nodes will probably not all have the same class. In such situations, each terminal node is assigned to the most common class of its members. This is called the **plurality rule**; a similar procedure is used in Bayes rule classifiers. Important implementation algorithms for CART include the ID3 algorithm (Iterative Dichotomiser 3) and its extension, the C4.5 algorithm (Quinlan 1993). Computational complexity typically scales as $O(pn_t[\ln n_t]^2)$ where p is the dimensionality of the problem and n_t is the size of the training set.

Several strategies are available for stopping tree growth, or **pruning** a tree that extends to terminal nodes with a prespecified number of members. The goal is to avoid proliferation of branches with little predictive power. One strategy is to prune the tree to minimize misclassifications of the training set with some penalty for tree complexity. Breiman suggested a penalty of αT where T is the number of terminal nodes in the subtree and $\alpha \geq 0$ is a chosen complexity factor.

The selection of the final decision tree then depends on the choice of the α complexity factor. If α is low, then the tree has many splits with fewer objects in the terminal nodes, while a higher α gives a simpler tree with more objects in the terminal nodes. A recommended procedure is to use the training set again to estimate the optimal α. If the training set is sufficiently large, then portions can be withheld from the tree construction calculation and used as validation sets to minimize misclassifications. For smaller training sets, cross-validation outlined in Section 6.4.2 can be used; the training set is repeatedly subdivided and misclassification rates are found for each subset using the different subtrees. Alternative strategies for pruning the classification tree include dropping variables with low importance, constraining the tree size, or testing the significance of each split.

The reliability of a classification tree, or a branch within a tree, can be estimated using a set of trees constructed from bootstrap replications of the original dataset. Predictions for a given class can be averaged from many bootstrapped trees; this is called **bootstrap aggregation** or **bagging**. Often complex classifiers from decision trees or automated neural networks are unstable due to correlations between the variables responsible for the branchings. Bagging often considerably reduces the mean square errors of classification of the training sample, giving more reliable classification rules for the test sample.

Boosting is another important innovation for improving CART and other classification procedures. Here, a large number of weakly predictive classifiers is combined to give a strong classification procedure. These methods are sometimes called **ensemble classifiers** or **committee classifiers**. The original algorithm developed by Freund & Schapire (1997), AdaBoost, iteratively applies the classifier to the training set, reweighting data points to emphasize those with incorrectly predicted classes. The misclassification fraction decreases as the algorithm proceeds. Boosting a simple classification tree can sometimes give better predictions than a more complex tree. Boosted classifiers are often very effective in data mining applications, and a large methodology based on boosting has emerged (Hastie *et al.* 2009, Chapter 10).

Random forests are another new strategy for enhancing classification trees by conducting **votes** among many trees generated from a single training set. First, many trial datasets are constructed by bootstrap resamples. Second, for each trial dataset, trees are constructed, choosing the best of a random selection of $m \ll p$ variables to split at each node. An optimal value of m is found by balancing the error rate and the classification strength of the forest. The method is presented by Breiman (2001) and is applied to an astronomical dataset by Breiman *et al.* (2003).

The random forests procedure has proved to be highly effective. It gives a high-accuracy classification, measured using the objects omitted from the initial tree construction. It does not require decisions about variable deletion, and can be efficiently applied to very large datasets with many variables. Random forest results are displayed and explored using three-dimensional scatterplots and α-blended parallel coordinates plots. Developments of random forests allow application to unsupervised clustering, outlier detection and measuring variable interactions.

9.6.4 Nearest-neighbor classifiers

k-nearest-neighbor (k-nn) and **automated neural networks (ANN)** are two broad approaches to difficult classification problems where interrelations between the variables and classes are very complicated. The most common case involves nonlinearities such that hyperplanes in p-space are poorly adapted to separate the classes; that is, their best results do not approach the minimum possible *ECM* given by Equation (9.1). Linear methods are also less effective when the classes are not simply connected so that members of a single class can appear in distinct regions of the p-space.

k-nn classifiers are related to nearest-neighbor and local regression density estimation techniques discussed in Section 6.5.2. Cluster membership of a new object is determined by the memberships of the k nearest-neighboring points in the training set; these methods are thus known as k-nn or **memory-based** classifiers. When the k neighbors have inconsistent memberships, the new value can take on the majority class, or another classification rule can be adopted. The value of k here plays the role of the kernel bandwidth in density estimation; a 1-NN classifier will respond to rapid gradients in cluster memberships but is likely to be noisy, while 100-NN classifiers produce a smooth distribution of cluster membership that may miss small-scale features.

While the choice of k can be made subjectively to balance bias and variance, it is best to choose k with quantitative techniques. If the training set is large, it can be divided into test and validation subsamples so that the choice of k can be based on optimizing a misclassification rate similar to the ECM. Bootstrap resampling or cross-validation of the training set can also assist in evaluating the stability of cluster numbers and shapes. k-nn procedures are important in machine learning applications with complicated clustering patterns (such as optical character recognition and remote sensing image analysis) due to the computational simplicity and responsiveness to small-scale clustering structures.

As with kernel density estimators discussed in Section 6.4 and illustrated in Figure 9.1 below, constant-bandwidth smoothers cannot follow both small-scale substructures in densely populated regions and large-scale sparsely populated regions of p-space. Such problems are particularly severe in high-p spaces. **Discriminant Adaptive Nearest Neighbor (DANN)** classification (Hastie & Tibshirani 1996) is a modified k-nn classifier to treat this problem, analogous to the density-based unsupervised clustering methods such as DBSCAN (Section 9.4). Here the standardization of the variables is performed locally rather than globally, so that the range of the k-nn averaging can expand or shrink in different variables as the cluster concentrations and shapes change. The adaptive nearest-neighbor procedure has three parameters: the number of neighbors used initially to estimate the local metric; the number of neighbors in the later k-nn computation; and a metric softening parameter. This localized distortion of the distance metric allows the simplistic k-nn procedure to adapt to elongated, curved and highly inhomogeneous clusters in p-space.

9.6.5 Automated neural networks

While k-nn methods can treat difficult classification problems, the results are often strongly dependent on the choice of k. A single value of k may not be optimal for all regions of the p-space. **Automated neural networks (ANNs)** often give more accurate classifications in these cases. They provide an environment where linear discriminants are obtained in a space that has been nonlinearly mapped from the original variables, where the nonlinear weightings are simultaneously learned with the linear discrimination function from the training set. The resulting mapping from training set to classifications is then applied to the test set. Effective ANNs often have two layers of weightings that map the measured variables to the known classes of the training set. Essentially, they are algorithms to find heuristic nonlinear rules for distinguishing classes in multivariate datasets.

A typical ANN is constructed as a $3 \times p$-dimensional structure: the $n \times p$ input data reside in the first layer, weights between the variables and the discrimination function lie in a second hidden layer, and the discrimination function for classification is developed in the third layer. This is an extension of the perceptron formulated by F. Rosenblatt in the 1950s, a model with a binary response when the weighted sum of inputs lie above or below some threshold. Training starts with setting random weights in the hidden layer and forward calculating classifications for the training set data. These initial weightings will give very poor class predictions. A **backpropagation** step using the known classifications then corrects the weights using a gradient descent procedure applied to each input vector sequentially in order to reduce the least-mean-square error in classifications. The process is iterated until the least-mean-square differences between the predicted and measured classes in the training set are achieved. Finally, the ANN is applied to the test set to obtain new classifications. Neural elements can be connected in complex topologies or ensembles, and coupled to genetic algorithms or other learning environments to evolve for dynamic problems.

A wide range of techniques has been developed to improve learning rates convergence and training accuracy of the network. These are reviewed and discussed by Duda *et al.* (2001, Chapter 6) and Izenman (2008, Chapter 10). Algorithmically, many choices must be made: network architecture (e.g. number of hidden layers and nodes), nonlinear filters (called activation functions), optimality criteria (e.g. least-squares classification errors) and stopping rules. Convergence is not guaranteed. ANNs were originally developed to mimic a simplistic view of human learning and are now actively used in practical problems such as speech recognition, handwriting recognition and predictive analytics. Particularly complex problems may require two or more hidden network layers. As with PCA and other multivariate procedures, it is advised to standardize the variables prior to analysis to avoid scale incompatibilities. Indeed, an ANN can be run on principal components that have reduced the dimensionality of the calculation.

ANNs are often run within a Bayesian framework. The posterior distribution obtained by solving Bayes's theorem gives the weights of the networks. Computations can follow maximum a posteriori (MAP) or variants of Markov chain Monte Carlo (MCMC) methods. In simple cases, the ANN result can be considered a least-squares fit to the Bayes discriminant function. A critical aspect of ANNs is that, except in simple problems that are more readily treated by other methods, the hidden layer of weights is usually uninterpretable. In astronomy, for example, a well-trained ANN applied to optical images could give effective discrimination between stars, and elliptical and spiral galaxies, but the scientist is unlikely gain insight into the observed properties leading to the classification by examining the ANN weightings.

9.6.6 Classifier validation, improvement and fusion

Duda *et al.* (2001, Chapter 9) discuss the difficult problem of choosing and evaluating machine learning classifier algorithms: LDA, SVMs, k-nn, CART, ANNs, and so forth. No one method performs best for all problems, and no probabilistic measure of goodness-of-fit is available except in the simplest cases such as MVN mixtures. Indeed a formal **no free**

lunch theorem has been proved showing that no single machine learning algorithm can be demonstrated to be better than another in the absence of prior knowledge about the problem (Wolpert & Macready 1997).

Nonetheless, a number of strategies for evaluating and improving classifiers have been developed. When the training set is sufficiently large, a portion of the training set can be withheld from the construction of the classifier and later used to estimate the classifier accuracy by cross-validation. For nontrivial problems, some aspects of the classifier may be unstable while other aspects are well-established. Bootstrap resamples of the training dataset can be constructed to examine the variance in the classifier. Boosting and bagging, discussed above in the context of classification trees (Section 9.6.3), can be applied to other classification methods. If likelihoods are computed, then Bayesian model selection criteria like the BIC can be used.

The results of different classifiers can be combined to give an improved result. For example, Chaudnuri *et al.* (2009) develop hybrid classifiers that combine the advantages of parametric LDA-type classifiers and nonparametric k-nn-type classifiers. A number of classifiers can themselves be subject to learning algorithms; this approach to classifier fusion is sometimes called **metalearning** (Brazdil *et al.* 2009).

One desirable characteristic of a classification is simplicity such that the sum of the algorithmic complexity and the description of the training set is minimized. This **minimum description length (MDL)** principle gives classifiers that do not become more elaborate as the training set size increases. The MDL principle can be grounded both in information theory and in Bayesian theory where the best classifier maximizes the posterior probability. It is a more sophisticated restatement of the bias–variance tradeoff we discussed concerning data smoothing (Section 6.4.1).

9.7 Remarks

Clustering, partitioning and classification are critically important operations in the scientific interpretation of astronomical observations. These procedures are important components of modern data mining and machine learning. They provide a taxonomy of subpopulations which can then be individually studied, both statistically using multivariate analysis and astronomically for astrophysical understanding. Except in a few situations such as the spatial clustering of galaxies, astronomers historically used subjective rather than algorithmic procedures to define classes. But today, with increasing awareness of statistical procedures and the growth of megadatasets, clustering and classification procedures are entering into wider use in astronomy (Ball & Brunner 2010, Way *et al.* 2011).

Many of the methods outlined here have a different character from methods with strong roots in the principles of statistical inference. For example, once a parametric family is specified in regression (Chapter 7) then the likelihood can be constructed, maximum likelihood estimation and Bayesian analysis give mathematically sound solutions, and the reliability of the solution can be evaluated with confidence intervals and credible regions. But, except for a limited class of parametric procedures such as normal mixture models, clustering and classification methods are not based on likelihoods. Clustering and classification are thus based mostly on computational algorithms rather than mathematical statistics, and their outcomes can be optimized only within the context of the chosen procedure and specific dataset under study.

The popularity in astronomy of single linkage hierarchical clustering (the friends-of-friends algorithm), to the near-exclusion of other methods, raises some concern. The procedure is rarely used in other fields due to its tendency to produce spuriously elongated structures and to chain together distinct clusters. Altogether, we recommend that astronomers examine a variety of hierarchical clustering results but give more weight to the average linkage results. Unless there is a specific astrophysical expectation of elongated structures, use of the familiar friends-of-friends algorithm is discouraged.

Choosing a classification method (including bagging and boosting options) can be confusing even with a clear goal, such as minimizing a training set misclassification measure like Equation (9.1). Hastie *et al.* (2009, Table 10.1) provide some guidance. CART and other decision trees are best for treating variables with incompatible units and are readily calculated for large-n problems, but often do not give the most accurate predictions. ANNs and SVM are generally more effective when linear combinations of variables dominate the classification. CART and k-nn classifiers are more robust to outliers. Sutton (2005) provides insight on when and why bagging and boosting are effective. Bagging is advantageous with adaptive techniques like classification trees, MARS and ANNs when the estimator has small bias but suffers from high variance. Boosting similarly is typically applied to weak and unstable learners. Neither bagging nor boosting are effective for regression techniques like LDA.

Practitioners in other fields have found that SVMs, AdaBoost and random forests can give very reliable classifications even for difficult problems. A large community of methodologists are actively developing improved methods. These new, sometimes hybrid, methods can give lower misclassification rates for a variety of complex, real-life problems.

Given this complicated situation, it is advised to try several classifiers for nontrivial astronomical problems, as the performances of different methods cannot be reliable predicted for different datasets. Adherence to a single method, such as neural networks or Bayesian classifiers, for all problems often will not give optimal results.

Despite these difficulties, the objectives of clustering and classification are essential for many astronomical endeavors and, in practice, their procedures are often effective. Modern methodology provides a number of validation procedures: cross-validation, bootstrap, bagging and random forests, and other procedures can show which clusters are fragile and which are robust to reasonable variations in the data. Graphical user interfaces like the *rattle* (Williams 2009) and *clustTool* (Templ 2010) packages in **CRAN** allow the user to quickly compare the results of alternative clustering and classification methods. Our classification of main-sequence stars, white dwarfs and quasars in a four-dimensional color space from the Sloan Digital Sky Survey using LDA, k-nn and SVM procedures (Section 9.6.2) is an example of how established modern techniques can give good results even when the population distributions are highly non-Gaussian and overlapping.

9.8 Recommended reading

Duda, R. O., Hart, P. E. & Stork, D. G. (2001) *Pattern Classification*, John Wiley, New York
An excellent and comprehensive volume at the graduate level by industrial statisticians with both classical and modern machine learning techniques. Topics include Bayesian decision theory, maximum likelihood and Bayesian parameter estimation, hidden Markov models, nonparametric and unsupervised clustering, linear discriminant analysis, neural networks, backpropagation, support vector machines, bagging and boosting, stochastic methods, tree classifiers, and classifier comparisons.

Everitt, B. S., Landau, S. & Leese, M. (2001) *Cluster Analysis*, 4th ed., Arnold, London
A brief, less mathematical, yet thorough review of clustering methodologies. Coverage includes distance measures, hierarchical clustering, model-based and other optimization techniques, mixture models, density-based clustering, overlapping clusters, fuzzy clustering, clustering with constraints, and strategies for selecting clustering methods.

Hastie, T., Tibshirani, R. & Friedman, J. (2009) *The Elements of Statistical Learning: Data Mining, Inference, and Prediction*, 2nd ed., Springer, Berlin
An advanced monograph by three distinguished statisticians with many insights. General statistical topics include linear regression, spline fitting, kernel density estimation, wavelet smoothing, local regressions, maximum likelihood estimation, Bayesian inference, bagging and model averaging, bootstrap and MCMC procedures, general additive models, bump hunting, and MARS. Cluster and classification topics include linear discriminant analysis, separating hyperplanes, regression and classification trees, boosting methods, projection pursuit and neural networks, naive Bayes classifiers, Support Vector Machine classifiers, *k*-nearest-neighbor classifiers, unsupervised association and clustering, and independent component analysis.

Izenman, A. J. (2008) *Modern Multivariate Statistical Techniques: Regression, Classification, and Manifold Learning*, Springer, Berlin
A comprehensive, advanced text that integrates classical with modern multivariate methods. Coverage includes: linear discriminant analysis; classification trees; neural networks; Support Vector Machines; hierarchical clustering; partitioning; multidimensional scaling; bagging and boosting. **R** scripts are available.

Johnson, R. A. & Wichern, D. W. (2007) *Applied Multivariate Statistical Analysis*, 6th ed., Prentice-Hall, Englewood cliffs
An established undergraduate textbook giving clear presentations of standard methods including similarity measures, hierarchical and nonhierarchical clustering, multidimensional scaling, correspondence analysis, biplots, procrustes analysis, multivariate normal mixtures, Fisher's discriminant function, and classification trees.

Marsland, S. (2009) *Machine Learning: An Algorithmic Perspective*, CRC Press, Boca Raton

A well-written volume on modern data mining methods with clear explanations and applications to classification, regression, time series and data compression. Applications in Python are provided. Coverage includes perceptron learning, backpropagation, splines and radial basis functions, Support Vector Machines, CART and decision trees, bagging and boosting, naive Bayes classifier, normal mixture models, nearest-neighbor methods, k-means neural networks, Self-Organizing Maps, linear discriminant analysis, optimization algorithms, genetic algorithms, reinforcement learning, Markov chain Monte Carlo methods, Bayesian networks, and hidden Markov methods.

9.9 R applications

R scripts illustrating a wide variety of clustering and classification techniques are given by Cook & Swayne (2007), Everitt *et al.* (2001) and Izenman (2008).

9.9.1 Unsupervised clustering of COMBO-17 galaxies

We illustrate unsupervised clustering algorithms using a two-dimensional color–magnitude diagram constructed from the COMBO-17 (Classifying Objects by Medium-Band Observations in 17 Filters) photometric survey of normal galaxies (Wolf *et al.* 2003). The **R** script below starts with the *which* function to filter the dataset, keeping only low-redshift galaxies with $z < 0.3$ and removing a few points with bad data values. Most of the original 65 variables are ignored, and we keep only the galaxy absolute magnitude in the blue band, M_B, and the ultraviolet-to-blue color index, $M_{280} - M_B$. The resulting color–magnitude diagram (Figure 9.1, left panel) shows the well-known concentrations of luminous red galaxies around $(M_B, M_{280} - M_B) \simeq (-16, -0.2)$ and fainter blue galaxies around $(-13, -0.9)$.

```
# Color-magnitude diagram for low-redshift COMBO-17 galaxies

COMBO_loz=read.table('http://astrostatistics.psu.edu/MSMA/datasets/COMBO17_lowz.dat',
   header=T, fill=T)
dim(COMBO) ; names(COMBO)
par(mfrow=c(1,2))
plot(COMBO_loz, pch=20, cex=0.5, xlim=c(-22,-7), ylim=c(-2,2.5),
   xlab=expression(M[B]~~(mag)), ylab=expression(M[280] - M[B]~~(mag)),
   main='')

# Two-dimensional kernel-density estimator

library(MASS)
COMBO_loz_sm <- kde2d(COMBO_loz[,1], COMBO_loz[,2], h=c(1.6,0.4),
   lims = c(-22,-7,-2,2.5), n=500)
```

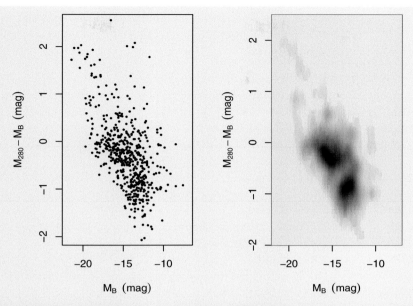

Fig. 9.1 A color–magnitude diagram of nearby, normal galaxies from the COMBO-17 survey. The left panel shows individual galaxies, and the right panel shows a kernel smooth estimator highlighting the **red sequence** (lower right) and **blue sequence** (upper left) concentrations. The grayscale is linear in surface density of points in the plot.

```
image(COMBO_loz_sm, col=grey(13:0/15), xlab=expression(M[B]~~(mag)),
   ylab=expression(M[280] - M[B]~~(mag)), xlim=c(-22,-7), ylim=c(-2,2.5),
   xaxp=c(-20,-10,2))
par(mfrow=c(1,1))
```

The clusters are more clearly seen after smoothing the point process with a two-dimensional kernel density estimator using *kde2d* in **R**'s *MASS* library, as described in Section 6.9.3. The blue galaxies are spirals and irregular galaxies that have experienced recent active star formation, while the red galaxies are mostly ellipticals that have only older stars formed early in the Universe's history. Note that many galaxies have properties distributed around the major concentrations. For example, a few extremely luminous and red galaxies are seen around $(-20, 1.5)$; these are nearby examples of the **luminous red galaxies** that are very important in cosmological studies (Eisenstein *et al.* 2005).

While the kernel density estimator provides a valuable visualization of the clustering pattern, it does not assign individual galaxies to specific clusters. We illustrate unsupervised clustering of the dataset using three methods in the following **R** script. For the nonparametric procedures where we assume Euclidean distances between points in the 2-space, we first standardize the variables by removing the means and dividing by the standard deviations. In **R**, standardization can be performed using the *scale* function.

R and **CRAN** have a variety of agglomerative hierarchical clustering algorithms. We start with the most commonly used procedure, the function *hclust* in **R**. The procedure runs on the matrix of pairwise distances between points constructed using the function *dist*. As the structures in the smoothed distribution seem roughly spherical, we choose the **complete linkage** definition of group locations. The product of *hclust* is a dendrogram which can be displayed using *plclust*.

```
# Standardize variables

Mag_std <- scale(COMBO_loz[,1])
Color_std <- scale(COMBO_loz[,2])
COMBO_std <- cbind(Mag_std,Color_std)


# Hierarchical clustering

COMBO_dist <- dist(COMBO_std)
COMBO_hc <- hclust(COMBO_dist, method='complete')
COMBO_coph <- cophenetic(COMBO_hc)
cor(COMBO_dist, COMBO_coph)


# Cutting the tree at k=5 clusters

plclust(COMBO_hc, label=F)
COMBO_hc5a <- rect.hclust(COMBO_hc, k=5, border='black')
str(COMBO_hc5a)
COMBO_hc5b <- cutree(COMBO_hc, k=5)
str(COMBO_hc5b)
plot(COMBO_loz, pch=(COMBO_hc5b+19), cex=0.7, xlab=expression(M[B]~~(mag)),
   ylab=expression(M[280] - M[B]~~(mag)), main='')
```

There is no formal procedure to select branches of the dendrogram as physically valid clusters. The cophenetic correlation coefficient, a measure of the similarity of the hierarchical structure and the data, is 0.52 using functions *cophenetic* and *cor*, but this cannot readily be converted to a probability. We investigated the tree by trial-and-error, and found that "cutting the tree" at $k = 5$ clusters separates the red and blue galaxies. Two procedures are shown here: *rect.hclust* which shows rectangles in the dendrogram (Figure 9.2, top panel), and *cutree* which gives an output with individual galaxy memberships of the five clusters. These are shown as different symbols in the color–magnitude diagram of Figure 9.2 (bottom panel); the open triangles show the red galaxies and the open circles show the blue galaxies. These clusters include many outlying galaxies, and examination of smaller clusters in the hierarchy does not cleanly discriminate the cluster extents seen in the smoothed distribution of Figure 9.1 (left panel).

Our second clustering method attempts to alleviate this problem with the hierarchical clustering results by using the density as a starting point for the clustering algorithm. We use the DBSCAN (density-based cluster analysis) in **CRAN** package *fpc* (**fixed point clusters**, Hennig 2010) which implements the procedure of Ester *et al.* (1996). DBSCAN is widely used, particularly for problems where compact clusters of interest are embedded

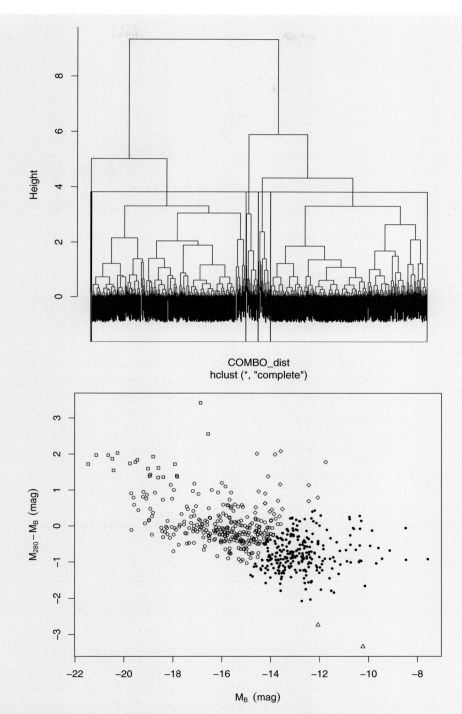

Agglomerative hierarchical clustering of the COMBO-17 color–magnitude diagram with complete linkage. *Left*: Dendrogram cut at five clusters. *Right*: Location of the five clusters.

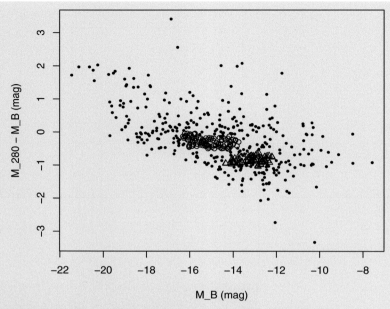

Fig. 9.3   Density-based clustering of the COMBO-17 color–magnitude diagram using the DBSCAN algorithm.

in a multiscale structure. The *dbscan* function requires user input of two parameters: the minimum number of points in a cluster, and the maximum radius (or **reach**) of a cluster. By trial-and-error, we found that a minimum of five points within 0.3 standardized magnitude units provided a useful result as shown in Figure 9.3. Here only the fraction of galaxies lying within the regions satisfying this local density criterion are classified; red and blue galaxy groups are clearly discriminated, and intermediate galaxies are not classified.

```
# Density-based clustering algorithm

install.packages('fpc') ; library(fpc)
COMBO_dbs <-  dbscan(COMBO_std, eps=0.1, MinPts=5, method='raw')
print.dbscan(COMBO_dbs) ; COMBO_dbs$cluster
plot(COMBO_loz[COMBO_dbs$cluster==0,], pch=20, cex=0.7, xlab='M_B (mag)',
   ylab='M_280 - M_B (mag)')
points(COMBO_loz[COMBO_dbs$cluster==2,], pch=2, cex=1.0)
points(COMBO_loz[COMBO_dbs$cluster==1 | COMBO_dbs$cluster==3,], pch=1, cex=1.0)
```

9.9.2 Mixture models

Both statisticians and astronomers often find that a dataset contains more than one class of objects and is thus a mixture of distributions. In parametric modeling, the most common assumption is that each class can be represented by a normal (Gaussian) distribution or, for multivariate problems, a multivariate normal distribution. Normal mixture modeling is thus

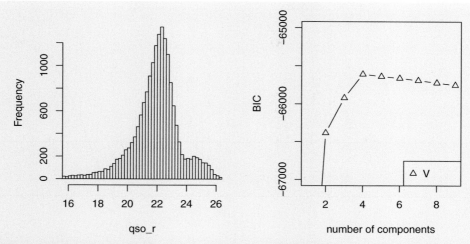

Fig. 9.4 Setting up the normal mixture model of the *r*-band magnitude distribution of SDSS quasars. *Left:* Histogram of the magnitude distribution. *Right:* Bayesian information criterion values from maximum likelihood estimation showing that four Gaussian components best fit the data.

an important and well-studied procedure (McLachlan & Peel 2000). Estimation is usually based on maximum likelihood methods computed using the EM algorithm.

In the **R** script below, we consider the univariate distribution of *r*-band magnitudes from a sample of 17,650 quasars from the Sloan Digital Sky Survey. The histogram shows a smooth but complicated distribution (Figure 9.4, left panel). Recall that there is little mathematical guidance regarding the number of bins to use in a histogram (Section 6.3). A well-known algorithm for fitting normal mixtures is known as **model-based clustering** implemented in **CRAN** package *mclust* (Fraley & Raftery 2002, 2007). The function *Mclust* will calculate maximum likelihood fits assuming $1, 2, \ldots, 9$ normal components are present. Figure 9.4 (right panel) shows the Bayesian information criterion for different numbers of components, and shows that the model with four components has the highest BIC value. Of course, models with five or more components fit better, but the gain in likelihood does not compensate for the increase in parameters of the model.

To further elucidate the results of the maximum likelihood normal mixture solution, we look at the structure of the output from *Mclust* using the *str* function. The output is an **R** *list* with many entries. Assisted by the help files, we can extract the mean, standard deviations and proportions (intensities) of the four Gaussian components. The remainder of the script creates the plot shown in Figure 9.5. First we plot the unbinned empirical cumulative distribution function (Section 5.3.1) of the quasar magnitudes using **R**'s *ecdf* function. Next, we create an artificial sequence of magnitude values using the *seq* function at which the Gaussian components can be evaluated for plotting. The first *for* loop sums the cumulative distribution of the four components using *pnorm* (Chapter 4). This is plotted as a thick dashed line superposed on the empirical distribution; they agree very closely with small deviations at the bright magnitudes. The second *for* loop plots the differential contribution of each component using the *dnorm* function.

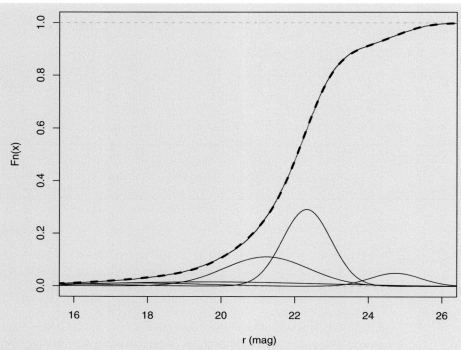

Fig. 9.5   Results of the quasar normal mixture model. The thin upper curve is the empirical distribution function of the data. The dashed curve, nearly superposed on the data, is the four-component normal mixture model. The lower curves show the contributions of each component.

```
# r-band distribution of Sloan quasars

SDSS_qso <- read.table("http://astrostatistics.psu.edu/MSMA/datasets/SDSS_17K.dat",
   header=T)
dim(SDSS_qso) ; summary(SDSS_qso)
qso_r <- SDSS_qso[,5] ; n_qso <- length(qso_r)
par(mfrow=c(1,2))
hist(qso_r, breaks=100, main='', xlim=c(16,26))

#  Normal mixture model

install.packages('mclust')  ; library(mclust)
fit.mm <- Mclust(qso_r,modelNames="V")
plot(fit.mm, ylim=c(-67000,-65000))

str(fit.mm)
fit.mm$parameters$mean
sqrt(fit.mm$parameters$variance$sigmasq)
fit.mm$parameters$pro
```

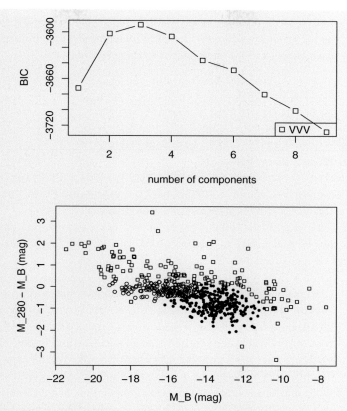

Fig. 9.6   Model-based clustering of the COMBO-17 color–magnitude diagram assuming a multivariate normal mixture. The top panel shows the Bayesian information criterion values for $k = 1$–9 clusters. The bottom panel shows the cluster membership for $k = 3$ clusters.

```
par(mfrow=c(1,1))
plot(ecdf(qso_r), xlim=c(16,26), xlab=' r (mag)', main='')

r.seq <- seq(10,30,0.02)
sum.comp <- rep(0,length(r.seq))
for (i in 1:fit.mm$G) { sum.comp <- sum.comp +  pnorm(r.seq, mean=
    fit.mm$parameters$mean[[i]], sd=sqrt(fit.mm$parameters$variance$sigmasq[[i]])) *
    fit.mm$parameters$pro[[i]] }
lines(r.seq, sum.comp, lwd=3, lty=2)
for (i in 1:fit.mm$G) {lines(r.seq, dnorm(r.seq, mean=fit2$parameters$mean[[i]],
   sd=sqrt(fit2$parameters$variance$sigmasq[[i]])) * fit2$parameters$pro[[i]])  }
```

        In the function *mclustBIC*, the "VVV" model name specifies multivariate ellipsoidal Gaussians with arbitrary orientations.

        We now run *mclust* for the COMBO-17 data set with the result shown in Figure 9.6. The likelihood for the COMBO-17 color–magnitude diagram is maximized for three clusters, two of which distinguish the red and blue galaxy sequences. Detailed results are provided by the *summary.mclustBIC* function including the probabilities of cluster membership for each galaxy and the uncertainties in these probabilities.

```
# Model-based clustering

library(mclust)
COMBO_mclus <- mclustBIC(COMBO_loz,modelNames='VVV')
plot(COMBO_mclus, col='black')
COMBO_sum_mclus <- summary.mclustBIC(COMBO_mclus,COMBO_loz,3)
COMBO_sum_mclus$parameters ; COMBO_sum_mclus$classification
COMBO_sum_mclus$z ; COMBO_sum_mclus$uncertainty
plot(COMBO_loz, pch=(19+COMBO_sum_mclus$classification), cex=1.0, xlab='M_B (mag)',
    ylab='M_280 - M_B (mag)', main='COMBO-17 MVN model clustering (k=3)',
    cex.lab=1.3, cex.axis=1.3)
```

Points lying between two clusters can be investigated using a visualization tool known as **shadow** and **silhouette** plots coupled to centroid-based partitioning cluster analysis (Leisch 2010). Each data point has a shadow value equal to twice the distance to the closest cluster centroid divided by the sum of distances to closest and second-closest centroids. Points with shadow values near unity lie equidistant from the two clusters. Silhouette values measure the difference between the average dissimilarity of a point to all points in its own cluster to the smallest average dissimilarity to the points of a different cluster. Small values again indicate points with ill-defined cluster memberships. These plots can be constructed using **CRAN**'s *flexclust* package (Leisch 2006).

9.9.3   Supervised classification of SDSS point sources

The Sloan Digital Sky Survey (SDSS) has produced some of the most impressive photometric catalogs in modern astronomy with hundreds of millions of objects measured in five optical bands. A selection of 17,000 SDSS point sources, along with training sets for three spectroscopically confirmed classes (main-sequence plus red-giant stars, quasars and white dwarfs), is derived in Appendix C.9. These are four-dimensional datasets with variables representing the ratios of brightness in the five SDSS photometric bands ($u-g, g-r, r-i$ and $i-z$). The resulting color–color scatterplots (Figures C.1 and C.2) show distributions that cannot be well-modeled by multinormal distributions, and distributions that are distinct in some color variables but overlapping in others. The analysis here starts with SDSS_train and SDSS_test obtained using the **R** script in Section C.9.

Unsupervised clustering fails to recover the known distributions in the SDSS photometric distribution. We show, for example, the result of a k-means partitioning in Figure 9.7. Using the input data and the user's desired number of partitions, **R**'s *kmeans* function gives the p-dimensional cluster centers, the within-cluster sum of squared distances, cluster

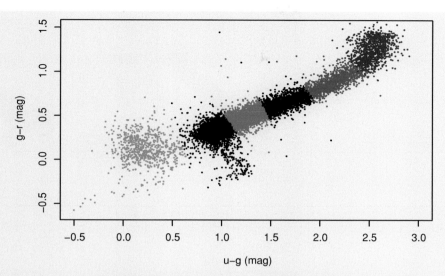

Fig. 9.7 The $(u - g) - (g - r)$ diagram for the test dataset of 17,000 SDSS point sources with classifications from k-means partitioning. This illustrates the failure of unsupervised classification methods for this problem.

populations and the n-dimensional vector of cluster memberships. The k-means partitioning of the SDSS data divides the main sequence into segments, even though there are no gaps in the distribution. Even a supervised k-means partitioning with three initial clusters roughly centered on the three training classes does not lead to a correct result. Similar problems arise when unsupervised hierarchical clustering (**R** function *hclust*) or model-based clustering (function *Mclust* in package *mclust*) are applied to the SDSS distributions.

```
# R script for constructing SDSS test and training datasets is given
# in Appendix C.


# Unsupervised k-means partitioning

SDSS.kmean <- kmeans(SDSS_test,6)
print(SDSS.kmean$centers)
plot(SDSS_test[,1], SDSS_test[,2], pch=20, cex=0.3, col=gray(SDSS.kmean$cluster/7),
   xlab='u-g (mag)', ylab='g-r (mag)', xlim=c(-0.5,3), ylim=c(-0.6,1.5))
```

9.9.4 LDA, k-nn and ANN classification

The classification of SDSS objects is much improved when the training set is used. We show here the result of linear discriminant analysis (LDA) using the function *lda* in base-**R**'s *MASS* package. The LDA classification from the training set is applied to the test set using

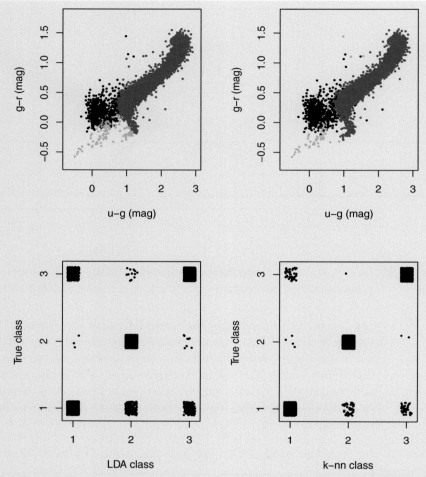

Fig. 9.8 Two supervised classifications of the SDSS test dataset: linear discriminant analysis (left) and k-nearest-neighbor classification with $k = 4$ (right). The top panels show the classification in the $(u - g) - (g - r)$ color–color diagram for the test dataset, and the bottom panels show the misclassifications of the training dataset. We follow the plotting scheme of Figure C.2 in the top panels where black symbols represent quasars, red symbols represent main-sequence and red-giant stars, and green symbols represent white dwarfs. For a color version of this figure please see the color plate section.

R's *predict* function. Figure 9.8 (top left panel) shows the result for the test sample; a similar plot can be inspected for the training sample.

Linear discriminant analysis gives a reasonable classification of stars (Class 2), quasars (Class 1) and white dwarfs (Class 3), with no difficulty following the elongated and curved distributions in 4-space. However, some classification errors are evident: a few main-sequence stars are mislabeled as quasars (black dots), and the white-dwarf class (green dots) is truncated by the quasar distribution. The closely related quadratic discriminate analysis using function *qda* has additional problems, classifying some main-sequence and red-giant stars as white dwarfs.

The **CRAN** package *class* implements a *k*-nearest-neighbors classifier where a grid of classifications is constructed from the training set (Venables & Ripley 2002). The application to the SDSS test set is shown in Figure 9.8 (top right panel) and shows good performance. Here we use $k = 4$ neighbors, but the result is not sensitive to the range of *k* values.

We consider two ways to examine the reliability of these classifiers by applying them to the training set. First, a cross-validation experiment can be made (e.g. using the function *knn.cv*) where leave-one-out resamples of the test dataset give posterior probabilities for the classification of each object. Second, the class obtained by the classifier can be plotted against the true class known for the training set objects. We show this in the bottom panels of Figure 9.8; the LDA clearly makes more misclassifications than the *k*-nn algorithm. For *k*-nn, misclassification of stars is rare (0.1%) but confusion between quasars and white dwarfs occurs in about 2% of cases. This is understandable given the overlap in their distributions in color–color plots (Figure C.2). Note the use of **R**'s *jitter* function to facilitate visualization of categorical data for scatterplots.

```
# Linear discriminant analysis

library(MASS)
SDSS_lda <- lda(SDSS_train[,1:4], as.factor(SDSS_train[,5]))
SDSS_train_lda <- predict(SDSS_lda, SDSS_train[,1:4])
SDSS_test_lda <- predict(SDSS_lda, SDSS_test[,1:4])

par(mfrow=c(2,1))
plot(SDSS_test[,1],SDSS_test[,2], xlim=c(-0.7,3), ylim=c(-0.7,1.8), pch=20,
   col=SDSS_test_lda$class, cex=0.5, main='', xlab='u-g (mag)', ylab='g-r (mag)')

# k-nn classification

install.packages('class') ; library(class)
SDSS_knn4 <- knn(SDSS_train[,1:4], SDSS_test,
   as.factor(SDSS_train[,5]), k=4, prob=T)
plot(SDSS_test[,1], SDSS_test[,2], xlim=c(-0.7,3), ylim=c(-0.7,1.8), pch=20,
   col=SDSS_knn4, cex=0.5, main='', xlab='u-g (mag)', ylab='g-r (mag)')

# Validation of k-nn classification

SDSS_train_lda <- lda(SDSS_train[,1:4], as.factor(SDSS_train[,5]))
SDSS_train_knn4 <- knn(SDSS_train[,1:4], SDSS_train[,1:4], SDSS_train[,5],k=4)
```

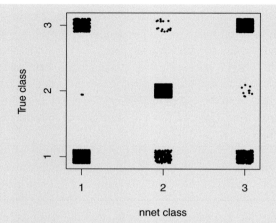

Fig. 9.9 Misclassifications from a single-hidden-layer automated neural network classifier applied to the SDSS point source training set.

```
plot(jitter(as.numeric(SDSS_train_lda$class), factor=0.5), jitter(as.numeric
    (SDSS_train[,5]), factor=0.5), pch=20, cex=0.5, xlab='LDA class',
    ylab='True class', xaxp=c(1,3,2),yaxp=c(1,3,2))
plot(jitter(as.numeric(SDSS_train_knn4), factor=0.5), jitter(as.numeric
    (SDSS_train[,5]), factor=0.5), pch=20, cex=0.5, xlab='k-nn class',
    ylab='True class', xaxp=c(1,3,2),yaxp=c(1,3,2))
par(mfrow=c(1,1))


# Single layer neutral network

options(size=100, maxit=1000)
SDSS_nnet <- multinom(as.factor(SDSS_train[,5]) ~ SDSS_train[,1] + SDSS_train[,2] +
    SDSS_train[,3] + SDSS_train[,4], data=SDSS_train)
SDSS_train_nnet <- predict(SDSS_nnet,SDSS_train[,1:4])
plot(jitter(as.numeric(SDSS_train_nnet), factor=0.5), jitter(as.numeric
    (SDSS_train[,5]), factor=0.5), pch=20, cex=0.5, xlab='nnet class',
    ylab='True class', xaxp=c(1,3,2),yaxp=c(1,3,2))
```

R provides only a simple automated neural network function, *nnet* in library *nnet*, that fits a single-hidden-layer network. More elaborate ANNs are available in **CRAN** packages. We compute the network with 100 hidden-layer weightings using the function *multinom* using the SDSS training set. Figure 9.9 shows that the predictions are rather poor. The stars (class $= 2$) are nicely identified but a large fraction of the quasars (class $= 1$) and white dwarfs (class $= 3$) are misclassified. A multi-layer ANN with backpropagation would likely perform better.

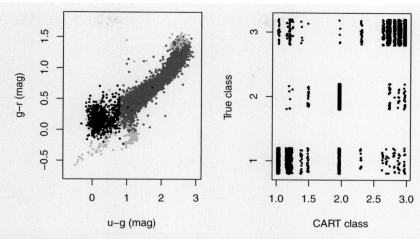

Fig. 9.10 CART classification of the SDSS point source dataset. *Left:* $(u - g) - (g - r)$ color–color diagram of the test dataset with colors as in Figure 9.8. *Right:* Validation of classifier on the training dataset. For a color version of this figure please see the color plate section.

9.9.5 CART and SVM classification

In the following **R** script, we apply CART using *rpart* (acronym for recursive partitioning and regression trees) in **R**'s *rpart* library. The procedure for running these and similar classifiers is straightforward. The "model" is produced by *rpart* or *svm* with a formula like "Known_classes $\sim$." to the training set. Examining the model using *summary* and *str* shows that the classifier output can be quite complicated; e.g. CART will give details on the decision tree nodes while SVM will give details on the support vectors. But the model predictions can be automatically applied to the training and test datasets using **R**'s *predict* function without understanding these details.

We plot the predicted classes against the known classes for the training set in Figure 9.10. CART does not perform as well as the k-nn shown in Figure 9.8. Figure 9.11 shows a portion of the regression tree with the splitting criteria, and provides guidance for pruning the tree to a parsimonious level based on cross-validation simulations. The scientist has control over various details in the CART computation to test the effects on the classification outcomes.

Finally we run a Support Vector Machine classifier on the SDSS point source dataset. Here we use function *svm* implemented in **CRAN**'s *e1071* package (Dimitriadou *et al.* 2010). This classifier is more computationally intensive than others we have examined, but does a good job (Figure 9.12).

We show the SVM classifications of the test SDSS sample in projections of the full four-dimensional color–color diagram of the SDSS dataset (Figure 9.13). This figure can be compared to the similar Figure C.2 showing the quasar (black), star (red) and white dwarf (green) training dataset. It is also helpful to examine the parallel coordinates plot with semi-transparency (alpha-blending) using the **R** function *parallel* or the interactive

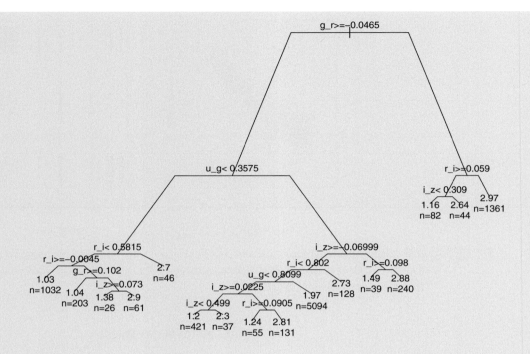

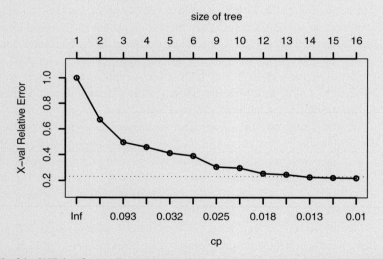

Fig. 9.11 Details of the CART classification. *Top:* Dendrogram for the SDSS training set. The criteria for each split and the number of quasars at terminal nodes (leaves) are given. *Bottom:* Plot of classification error for different prunings of the tree based on cross-validation. The dashed line is drawn one standard error above the minimum of the curve.

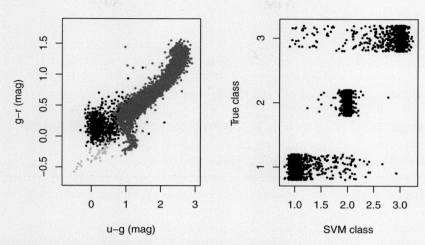

Fig. 9.12   Support Vector Machine classification of the SDSS point source dataset. *Left:* $(u-g)-(g-r)$ color–color diagram of the test dataset with colors as in Figure 9.8. *Right:* Validation of classifier on the training dataset. For a color version of this figure please see the color plate section.

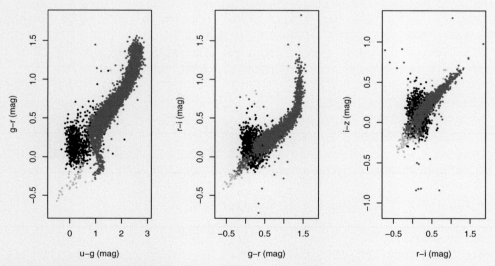

Fig. 9.13   SVM classification of SDSS test sources shown in the $(u-g)-(g-r)$, $(g-r)-(r-i)$, and $(r-i)-(i-z)$ color–color diagrams. For a color version of this figure please see the color plate section.

plot *ipcp* in **CRAN**'s *iplots* package. The **R** script ends with the function *write.table* to write the test dataset to an ASCII output file with a new column giving the SVM classifications. Note that **R**'s *write* function produces tables that are difficult to read; we use the *format* function and other options in *write* to improve the appearance of the ASCII output.

```
# Classification And Regression Tree model, prediction and validation

library('rpart')
SDSS_rpart_mod <- rpart(SDSS_train[,5] ~., data=SDSS_train[,1:4])
SDSS_rpart_test_pred <- predict(SDSS_rpart_mod, SDSS_test)
SDSS_rpart_train_pred <- predict(SDSS_rpart_mod, SDSS_train)
summary(SDSS_rpart_mod) ; str(SDSS_rpart_mod)

plot(SDSS_test[,1], SDSS_test[,2], xlim=c(-0.7,3), ylim=c(-0.7,1.8), pch=20,
   col=round(SDSS_rpart_test_pred), cex=0.5,
   main='', xlab='u-g (mag)', ylab='g-r (mag)')
plot(jitter(SDSS_rpart_train_pred, factor=5), jitter(SDSS_train[,5]), pch=20,
   cex=0.5, xlab='CART class', ylab='True class',yaxp=c(1,3,2))

plot(SDSS_rpart_mod, branch=0.5, margin=0.05)
text(SDSS_rpart_mod, digits=3, use.n=T, cex=0.8)
plotcp(SDSS_rpart_mod, lwd=2, cex.axis=1.3, cex.lab=1.3)


# Support Vector Machine model, prediction and validation

install.packages('e1071') ; library(e1071)
SDSS_svm_mod <- svm(SDSS_train[,5] ~.,data=SDSS_train[,1:4],cost=100, gamma=1)
summary(SDSS_svm_mod) ; str(SDSS_svm_mod)
SDSS_svm_test_pred <- predict(SDSS_svm_mod, SDSS_test)
SDSS_svm_train_pred <- predict(SDSS_svm_mod, SDSS_train)

plot(SDSS_test[,1], SDSS_test[,2], xlim=c(-0.7,3), ylim=c(-0.7,1.8), pch=20,
   col=round(SDSS_svm_test_pred), cex=0.5, main='',
   xlab='u-g (mag)', ylab='g-r (mag)')
plot(SDSS_svm_train_pred, jitter(SDSS_train[,5]), pch=20, cex=0.5,
   xlab='SVM class', ylab='True class', yaxp=c(1,3,2))


# Final SVM classification of the test set

par(mfrow=c(1,3))
plot(SDSS_test[,1], SDSS_test[,2], xlim=c(-0.7,3), col=round(SDSS_svm_test_pred),
   ylim=c(-0.7,1.8), pch=20, cex=0.5, main='', xlab='u-g (mag)',ylab='g-r (mag)')
plot(SDSS_test[,2], SDSS_test[,3], xlim=c(-0.7,1.8), col=round(SDSS_svm_test_pred),
   ylim=c(-0.7,1.8), pch=20, cex=0.5, main='', xlab='g-r (mag)',ylab='r-i (mag)')
plot(SDSS_test[,3], SDSS_test[,4], xlim=c(-0.7,1.8), col=round(SDSS_svm_test_pred),
   ylim=c(-1.1,1.3), pch=20, cex=0.5, main='', xlab='r-i (mag)',ylab='i-z (mag)')
par(mfrow=c(1,1))
```

```
SDSS_test_svm_out <- cbind(SDSS[,6], SDSS[,7], SDSS_test, SDSS_svm_test_pred)
names(SDSS_test_svm_out)[c(1,2,7)] <- c('R.A.', 'Dec', 'SVM Class')
write.table(format(SDSS_test_svm_out), file='SDSS_test_svm.out',sep='\\t',quote=F)
```

9.9.6 Scope of **R** and **CRAN**

The principal clustering methods available in base **R** are based on the classic book by Kaufman & Rousseeuw (2005). In addition to *hclust*, *kmeans* and *dendrogram* discussed in Section 9.9.1, they include *agnes* (agglomerative nesting, similar to *hclust*), *pam* (partitioning around medoids), *clara* (clustering large applications, also a k-medoid partitioning procedure), *diana* (divisive analysis clustering), *fanny* (fuzzy analysis clustering) and *mona* (monothetic analysis clustering for binary variables). The **R** function *clusplot* provides visualizations of clustered and partitioned data. The *class* package in **R** includes *SOM* for a Kohonen Self-Organizing Map and *knn* for k-nearest-neighbor classification with leave-one-out cross-validation (*knn.cv*). For supervised classification, **R** includes, in addition to the CART function *rpart* discussed in Section 9.9.5, the function *nnet* for single-hidden-layer neural networks. Most of these functions are accompanied by specialized graphical and textual outputs.

The methodology available in **CRAN** for both clustering and classification is vast. Annotated, up-to-date package lists on clusters and machine learning are available at http://cran.r-project.org/web/views. We mention only a fraction of the packages here.

Enhancements to hierarchical clustering include *flashClust* and *nncluster* for high computational efficiency, *pvclust* and *LLAhclust* for estimating probabilities for clusters based on multiscale bootstrap resampling and likelihood analysis, and *dynamicTreeCut* for adaptive pruning of dendrograms. A variety of modern and flexible clustering algorithms is provided in packages *flexmix fluxclust* and *apcluster*. Several packages provide Bayesian estimation of parametric mixtures including *bayesmix*, *mclust* and *mixAK*. This last package illustrates its use on a univariate galaxy redshift dataset. The *party* package automatically iterates between regression and partitioning of a multivariate dataset. Variable selection with k-nearest-neighbor classifiers is treated in package *knnTree*. Minimal spanning tree tools are provided in package *spdep*. The *prim* package implements the patient rule induction method for bump hunting in high-dimensional data. The *clusterGeneration* package provides simulations of cluster patterns.

Some packages provide broad suites of classification and machine learning techniques. The *chemometrics* package described by Varmuza & Filmoser (2009) includes functions for cluster validity (Equation 9.8), k-nearest-neighbor classification, automated neural networks, and support vector machines. The *RWeka* package brings the well-established independent Weka package (Witten *et al.* 2011) into **R** giving a wide variety of classification procedures including CART with bagging and boosting. The *mda* package gives functions for mixture and flexible discriminant analysis, adaptive spline fitting, and multivariate adaptive regression splines described by Hastie *et al.* (2009).

Other data mining and machine learning classification packages are available. The **CRAN** packages *randomForest* and *clv* evaluate cluster stability and the validity of a classification tree, *gbm* provides tools for boosting classifiers, *BayesTree* combines weak learners, and *ROCR* calculates receiver operating characteristic (ROC) curves to evaluate and compare classifiers. The *kernlab* package provides detailed capabilities for kernel-base classifiers, particularly Support Vector Machines.

A number of **R** and **CRAN** functions treat bump hunting and multimodality tests. The **R** function *Mclust* in the *mclust* library computes maximum likelihood estimates of normal mixture models. Extensions are implemented in **CRAN**'s *flexmix* package. Other unsupervised clustering procedures are discussed in Section 9.3. Hartigan's univariate dip test is implemented in the **CRAN** package *diptest* and the related Davies & Kovacs (2004) taut string procedure is implemented in **CRAN** package *pmspec*.

10 Nondetections: censored and truncated data

10.1 The astronomical context

Observational astronomers always struggle, and often fail, to characterize celestial populations in an unbiased fashion. Many surveys are flux-limited (or, as expressed in traditional optical astronomy, magnitude-limited) so that only the brighter objects are detected. As flux is a convolution of the object's intrinsic luminosity and the (often uninteresting) distance to the observer according to $Flux = L/4\pi d^2$, this produces a sample with a complicated bias in luminosity: high-luminosity objects at large distances are over-represented and low-luminosity objects are under-represented in a flux-limited survey. This and related issues with nondetections have confronted astronomers for nearly 200 years.

A blind astronomical survey of a portion of the sky is thus truncated at the sensitivity limit, where **truncation** indicates that the undetected objects, even the number of undetected objects, are entirely missing from the dataset. In a supervised astronomical survey where a particular property (e.g. far-infrared luminosity, calcium line absorption, CO molecular line strength) of a previously defined sample of objects is sought, some objects in the sample may be too faint to detect. The dataset then contains the full sample of interest, but some objects have upper limits and others have detections. Statisticians refer to **upper limits** as **left-censored** data points.

Multivariate problems with censoring and truncation also arise in astronomy. Consider, for example, a study of how the luminosity function of active galactic nuclei (AGN) depends on covariates such as redshift (as a measure of cosmic time), clustering environment, host galaxy bulge luminosity and starburst activity. If the AGN sample emerges from a flux-limited survey at a single spectral band, then the population will be truncated. If the AGN sample was pre-selected at another band but is only partially detected in a new band, then the sample will be censored. Similar problems can arise with populations of Kuiper Belt Objects, Galactic halo stars, dwarf galaxies, and so forth. However, it is important to realize that most methods developed in statistics for other fields permit sample bias only in the single response variable (AGN luminosity in this case), whereas some astronomical problems have sample bias in several variables.

Truncation and censoring prevent the application of simple and familiar statistical techniques in the analysis. For example, neither the mean nor the median of the detections is an unbiased estimator of the central location of the underlying population. While no statistical procedure can magically recover all of the information that was never measured at the telescope, there is frequently important information implicit in the failure to detect some objects which can be partially recovered under reasonable mathematical assumptions.

Statistical methods dealing with censored data were developed over a long period due to the prevalence of censoring in social, biological and industrial sciences. The seventeenth-century astronomer Edmund Halley (1693), in his later years while President of the Royal Society, initiated the statistical study of life tables for demography; his work laid the foundations for actuarial science for the insurance industry which is very dependent on censored datasets. Similar (and sometimes identical) statistical methods were being developed in other fields. A wide suite of statistical methods for recovering from censoring emerged during the 1960–80s under the rubric of **survival analysis** where the longevity of a population was a common theme in many applications.

To illustrate how survival problems arise in diverse fields, consider the following regression problems:

1. An epidemiologist wishes to determine how the human longevity or **survival time** (dependent variable) depends on the number of cigarettes smoked per day (independent variable). The experiment lasts five years, during which some individuals die and others do not. The survival time of the living individuals is only known to be greater than their age when the experiment ends; these lower limits to longevity are called **right-censored** data points. The technique of **Cox regression** is typically used to quantify the dependence of mortality on cigarette smoking.

2. An industrial manufacturing company wishes to know the average time between breakdowns of a new engine as a function of axle rotation speed to determine the optimal operating range. Test engines are set running until 20% of them fail. The average lifetime dependence on speed is then calculated with 80% of the data points right-censored. The technique of **accelerated life testing** is typically used, often based on an assumed Weibull (a parametric **U**-shaped) distribution to model the prevalence of early and late failures.

3. An economist seeks the relationship between education and income from a census survey. Household income is recorded down to the poverty level below which the value is not recorded. In some cases, even the number of impoverished households is unknown. In such econometric applications, censored and truncated variables are called **limited dependent variables** and methods like **Tobit regression** are used.

4. An astronomer observes a sample of high-redshift galaxies with infrared and submillimeter telescopes to determine the relation between dust and molecular gas. A fraction of the observed sample is not detected at infrared and/or submillimeter wavelengths. The astronomer then seeks the relationship between infrared luminosities and molecular line luminosities. They use correlation coefficients and linear regression methods designed for doubly left-censored datasets (Isobe *et al.* 1986).

Astronomers have dealt with their upper limits (left-censored data points) in a number of fashions. Some procedures are clearly biased or incorrect. One considers only detections in the analysis, omitting censored data points. While possibly acceptable for some purposes (e.g. correlation), this will clearly bias the results in others (e.g. estimating luminosity or other distribution functions). Another procedure is to assign arbitrary low data values to censored data points; this has unfortunately been a common procedure in environmental sciences (EPA 2002, Helsel 2005). A somewhat better procedure used in some astronomical studies involves grouping the data into bins and considering the ratio of detected objects to observed objects in each bin. These sequences of binned **detection fractions** can be mathematically similar to actuarial life tables (London 1997, Chapter 3); however, astronomers are generally unaware of the many statistical methods designed for actuarial life tables.

Basic survival analysis methods were brought to the attention of astronomers in the 1980s (Feigelson & Nelson 1985, Schmitt 1985, Isobe *et al.* 1986, LaValley *et al.* 1992). A few established methods of survival analysis had been independently derived by astronomers (Avni *et al.* 1980, Avni & Tananbaum 1986), and a few new methods for censored data were derived to address astronomical problems (Akritas *et al.* 1995, Akritas & Siebert 1996).

The best procedure is to consider all of the data points, including individual detected values and the individual upper limits of the nondetections, to model the properties of the parent population under certain mathematical constraints. This is the approach presented here in Section 10.3. However, astronomers must study the assumptions underlying each method to determine which are applicable to a given astronomical situation.

The statistical treatment of truncation has been more limited than the treatment of censoring because it is more difficult to uncover properties of the underlying population when nothing, not even the number of objects, is known beyond the truncation limit. A variety of estimation methods were developed during the 1940–70s for truncated data by assuming the population follows a normal or other simple statistical distribution (Cohen 1991). These are outlined in Section 10.3.1. The important nonparametric maximum likelihood estimator for a truncated dataset was derived by astrophysicist Lynden-Bell (1971) and rediscovered by statistician Woodroofe (1985). However, astronomers commonly use statistics such as Schmidt's (1968) $1/V_{max}$ distribution and other treatments for **Malmquist bias** whose mathematical foundations are not well-established. We review methods for treating truncation in Section 10.5.

Biomedical and astronomical surveys sometimes suffer simultaneous truncation and censoring. For example, in evaluating the spread and mortality associated with the AIDS epidemic, public health officials can readily measure the time between diagnosis and death. Many individuals diagnosed with AIDS are still alive, giving right-censored data points. But many individuals carry the AIDS virus without yet showing clinical symptoms, giving a left-truncation to the observed population. (In biomedical applications, this source of left-censoring is called **delayed entry**.) Similarly, in evaluating the role of accretion onto supermassive black holes across cosmic time, the astronomer seeks the X-ray luminosity function of quasars as a function of redshift. Many quasars are too faint for detection by current X-ray telescopes, leading to left-censored data points. But the sample examined for X-ray emission is often derived from a magnitude-limited optical survey, leading to an additional left-truncation in the observed population.

10.2 Concepts of survival analysis

As discussed above, **censoring** refers to cases where the object is known to exist, the observation has been made, but the object is undetected in the desired property so that its value is fainter than some known limit. **Truncation** refers to situations where the object, even the number of objects, is unknown when the property value is fainter than some known limit. These objects cannot be displayed at all on graphs. Astronomical nondetections produce **left-censored** and **left-truncated** data points; this nomenclature is important, as the most common situation treated in statistics texts assumes right-censoring and left-truncation. Methods for right-censoring can readily by used for left-censored data by multiplying the measured values by -1.

Type 1 left-censoring refers to situations where objects above a fixed, known limit have measured values, and objects below that limit are counted but not measured. (Type 2 left-censoring, where one examines the sample with increasing sensitivity until a fixed fraction is detected, will not generally occur in astronomy.) This situation occurs when one considers the flux variable from a flux-limited astronomical survey. **Random left-censoring** occurs when some extraneous factor unrelated to the censoring process causes objects to be undetected. An example would be the detection of some faint stars when sky conditions at the telescope are excellent, and the nondetection of other faint stars when sky conditions are poor. Note that the censoring distribution need not be uniform; it can have dependencies on ancillary variables, but the distribution of the measured variable must be independent of these ancillary variables for random censorings.

The common situation in astronomy mentioned in Section 10.1 does not readily fit any of these categories. If an astronomer surveys the sky for galaxies with luminosities L_i that lie at random distances d_i, then a fixed sensitivity limit at flux $Flux_{lim}$ will give a censoring pattern that depends on whether $Flux_i = L_i/4\pi d_i^2$ is above or below $Flux_{lim}$ where d is the distance to the object. The censoring of L is then partially dependent on the value of L, but is partially randomized by the distances. Other astronomical properties − an absorption line equivalent width, velocity dispersion, angular separation, spectral index, and so forth − that become undetectable when signals become weak also suffer censoring.

This particular example of **dependent censoring** has not been well-studied, although a few simulations in Isobe *et al.* (1986) show that the deleterious effect of this dependency may sometimes be small. Some survival analysis methods will still be effective with dependent censoring, but others may give inaccurate results.

The statistical methodology of survival analysis has a preferred nomenclature slightly different from that used elsewhere in statistics. Consider a random variable X_i measured for $i = 1, \ldots, n$ objects where each measurement has a limiting sensitivity C_i. The censoring status is recorded as a binary variable δ: $\delta_i = 1$ refers to a detection where X_i is observed, and $\delta_i = 0$ to a nondetection where only C_i is recorded. (Note that there is no consensus on this designation, and some books and computer codes will use $\delta = 1$ for nondetections and $\delta = 0$ for detections.) The observed data points are then the paired variables (T_i, δ_i) where $T_i = \min(X_i, C_i)$. The X_i are assumed to be independent and identically distributed (i.i.d.) with probability density distribution f and cumulative distribution F.

The **survival function** $S(x)$ (or in industrial contexts, the **reliability function**) is defined as

$$S(x) = P(X > x)$$
$$= 1 - F(x). \tag{10.1}$$

$S(x)$ is the probability that an object has a value above some specified level. The **hazard rate** (or in industrial contexts, the **conditional failure rate**) $h(x)$ is given by

$$h(x) = \frac{f(x)}{S(x)} = -\frac{d \ln S(x)}{dx} \tag{10.2}$$

$$= \lim_{\Delta x \to 0} \frac{P(x < X \le x + \Delta x | X > x)}{\Delta x}. \tag{10.3}$$

As $f(x) = S(x)h(x)$, the value of the p.d.f. at a chosen value can be viewed as the product of the survival and hazard functions at that value.

The information in the p.d.f., survival function and hazard rate is the same, but each addresses a different question. The hazard rate answers questions important in human affairs like "What is the chance that a person of a given age will die?" or "What is the chance that a manufactured object will break after a given usage?" While the astronomer typically does not ask the equivalent question "What is the chance that a star of a given luminosity will appear in the telescope?", the mathematics of censoring and truncation is often formulated in terms of survival and hazard functions. Once survival analysis methods are applied to estimate survival functions, or to compare survival distributions of different samples, or to establish the dependencies of survival functions on covariates, then the astronomer can easily convert from survival to e.d.f or p.d.f. functions using Equation (10.1). Then astronomical questions such as "Based on flux-limited surveys, can we infer that the brown-dwarf luminosity function is a continuation of the stellar luminosity function?" can be addressed.

While these functions contain the same information, the mathematical modeling of some problems such as censoring and truncation is often more easily developed with survival and hazard functions. A distribution with constant hazard rate $h(x) = h_0$ has an exponential survival function, $S(x) = e^{-h_0 x}$, and a mean value $E[x] = 1/h_0$. Covariates representing dependencies of the distribution on other properties are often treated with hazard rates of the form $h(x) = h_0 e^{-\beta x}$; this is called the **proportional hazard** model and estimation of the β dependencies is called **Cox regression**. This is one of the most widely applied methods in all of modern statistics.

To illustrate these quantities, consider a population of astronomical objects with a power-law (or Pareto) distribution. While a common astronomical formulation might be $N(x) = N_0(x/x_0)^{-\alpha}$, the correctly normalized probability distribution function is defined as (Section 4.4)

$$f(x) = \frac{\theta \lambda^\theta}{x^{\theta+1}} \tag{10.4}$$

with $\theta > 0$, $\lambda > 0$ and $x \ge \lambda$. Then the survival function and hazard rate are

$$S(x) = \frac{\lambda^\theta}{x^\theta}$$

$$h(x) = \frac{\theta}{x}. \tag{10.5}$$

Note that the hazard rate decreases linearly for all power laws.

Most survival and hazard functions are not so simple. If the population distribution follows a gamma distribution (or the astronomers' Schechter function commonly written as $N(x) = N_0(x/x_0)^{-\alpha}e^{-x/x^*}$, Section 4.5), then

$$f(x) = \frac{\lambda^\beta x^{\beta-1} e^{-\lambda x}}{\Gamma(\beta)}$$
$$S(x) = 1 - I(\lambda x, \beta) \tag{10.6}$$

where $\Gamma(\beta)$ is the gamma function, I is the incomplete gamma function, and the hazard rate is $f(x)/S(x)$. For $\beta > 1$ characteristic of astronomical distributions, the hazard rate starts at 0 and rises monotonically to λ in a nonlinear fashion for a gamma distribution. If the population distribution is normal then

$$f(x) = \frac{1}{\sqrt{2\pi}\sigma}\exp\left[-\frac{1}{2}\left(\frac{x-\mu}{\sigma}\right)^2\right]$$
$$S(x) = 1 - \frac{1}{2}\left[1 + erf\left(\frac{x-\mu}{\sqrt{2}\sigma}\right)\right] \tag{10.7}$$

where *erf* is the error function. Here the hazard function starts at 0, rises to a maximum at a value above μ, and falls slowly in an asymmetrical manner. An exponential probability distribution gives a constant hazard rate. Distribution functions which produce U-shaped hazard rates, such as the Weibull and Gompeertz distributions, are often applied to human mortality or industrial reliability because they model both early (infant mortality) and late (old age) deaths. This is unlikely to occur for astronomical distributions.

10.3   Univariate datasets with censoring

10.3.1   Parametric estimation

If the distribution of a variable is known in advance, and Type 1 censoring or truncation is present (that is, the measurements are limited at a single value for the entire sample), then the parameters of the distribution can be estimated in a straightforward manner using maximum likelihood methods. During the 1940s and 1950s, likelihoods for censored or truncated datasets were written for singly censored and truncated datasets following the normal, lognormal, Poisson, Pareto (power-law), gamma (which includes the astronomers' Schechter function), and other statistical distributions. These findings are presented in detail by Cohen (1991) and Lee & Wang (2003).

This situation arises in various situations in astronomy. Estimation of the Schechter function parameters for the galaxy luminosity function from a magnitude-limited survey will be singly truncated if the galaxies are all at the same distance, as in a rich galaxy cluster. Power-law mass distribution functions of asteroids, planets or stars are often inferred from samples that are truncated in mass from dynamical or photometric studies. In some exploratory studies, astronomers may plot singly truncated fluxes against covariates where the distributions are assumed to be multivariate normals.

Likelihoods for parametric survival distributions can be constructed by combining detected, censored and truncated data points using combinations of the p.d.f. $f(x)$, the survival function $S(x)$, and the truncation levels x_{trunc}. For left-censoring and left-truncation, these three components of the likelihood are (Klein & Moeschberger 2005)

$$L \propto \prod_{det} f(X_i) \prod_{cens} [1 - S(X_i)] \prod_{trunc} \frac{f(X_i)}{S(x_{trunc})} \tag{10.8}$$

where the three products are over-detected, censored and truncated data points, respectively. For example, consider the paired variables (T_i, δ_i) where $T_i = \min(X_i, C_i)$ is subject to random left-censoring; that is, the distribution of censoring times C_i is independent of the p.d.f. and other parameters of interest relating to X. Then the likelihood at value T based on paired (T_i, δ_i) data points is

$$L = \prod_{i=1}^{n} P[T_i, \delta_i] = \prod_{i=1}^{n} [f(T_i)^{\delta_i}][1 - S(T_i)]^{1-\delta_i}. \tag{10.9}$$

Once the likelihood function is established, the variety of tools described in Chapter 3 become available: parameter estimation with confidence estimation via maximum likelihood estimation or Bayesian inference; hypothesis tests on the model parameters; model selection with the likelihood ratio test or Bayesian information criterion; Bayesian posterior distributions with marginalization of uninteresting variables; and so forth. The volumes by Cohen (1991) and Lawless (2002) provide detailed explanations and examples for constructing likelihoods in the presence of censoring and truncation.

The likelihoods for singly truncated and censored distributions can be written in a straightforward fashion, and sometimes have solutions in closed form. For example, a Poisson-distributed sample with n objects and intensity λ truncated at $x = a$ (that is, the integer values $x = 0, 1, \ldots, (a - 1)$ are missing from the dataset) has a loglikelihood function

$$\ln L = -n\lambda + n\bar{X}\ln\lambda - n\ln \sum_{x=a}^{\infty} \frac{e^{-\lambda}\lambda^x}{x!} - \sum_{i=1}^{n} \ln X_i. \tag{10.10}$$

This can be maximized numerically, but for the special case of $a = 1$ where only zeros are missing, the intensity can be obtained by solving

$$\frac{\hat{\lambda}}{1 - e^{-\lambda}} = \bar{X}. \tag{10.11}$$

A galaxy luminosity function following a Schechter function, which is a subset of gamma distributions, will have the distribution

$$f(L, \alpha, L^\star) = \frac{L^{\star - \alpha}}{\Gamma(\alpha)} L^{\alpha-1} \exp -(L/L^\star), \tag{10.12}$$

where $L > 0$ and $L^* > 0$. If the dataset is singly truncated at L_{max}, the loglikelihood is

$$\ln L = -n\ln\Gamma(\alpha) - n\alpha\ln L^\star - \frac{n\bar{L}}{L^\star} + (\alpha - 1) \sum_{i=1}^{n} \ln L - n\ln F(L_{max}) \tag{10.13}$$

where $F(L_{max})$ is the cumulative distribution function for the gamma distribution up to the truncation level. This can be maximized numerically.

10.3.2 Kaplan–Meier nonparametric estimator

Nonparametric estimation of censored distribution functions are straightforward general-izations of the e.d.f. discussed in Section 5.3.1. For left-censored situations, the cumulative hazard function is

$$\hat{H}_{KM}(x) = \sum_{x_i \geq x} \frac{d_i}{N_i} \tag{10.14}$$

where x_i are ordered values of the observations, N_i is the number of objects (detected or undetected) $\geq x_i$ and d_i is the number of objects at value x_i. If no ties are present, $d_i = 1$ for all i. The ratio d_i/N_i is the conditional probability that an object with value above x will occur at x. The associated survival function estimator is

$$\hat{S}_{KM}(x) = \prod_{x_i \geq x} \left(1 - \frac{d_i}{N_i}\right). \tag{10.15}$$

This is the **Kaplan–Meier (KM)** or **product-limit estimator** derived by Kaplan and Meier (1958). The estimator does not change values at censored values. When left-censoring is present, the size of the jumps grows at lower values as the number of "at risk" objects N_i rapidly becomes small. The Kaplan–Meier estimator of an astronomical dataset is shown in Figure 10.1 below.

An intuitive procedure to obtain the Kaplan–Meier estimator of the survival function is the **redistribute-to-the-right** algorithm (or **redistribute-to-the-left** for left-censoring) presented by Efron (1967). Here the weight of each upper limit is redistributed equally among the detections at lower values:

$$\hat{S}_{KM}(x_i) = \hat{S}_{KM}(x_{i-1}) \frac{N_i - d_i}{N_i}. \tag{10.16}$$

The KM estimator can thus be viewed as the inverse of the e.d.f. of the detected values weighted by the redistributed upper limits.

For sufficiently large samples, the KM estimator is asymptotically normal with variance

$$\widehat{Var}(\hat{S}_{KM}) = \hat{S}_{KM}^2 \sum_{x_i \geq x} \frac{d_i}{N_i(N_i - d_i)}. \tag{10.17}$$

This is known as **Greenwood's formula** as it is the same variance used for censored life tables in actuarial science where the data are grouped into equal width bins. This variance is often used to construct confidence bands around the estimator, although other variance estimates have been proposed.

If the censoring pattern is random, the KM estimator has desirable mathematical proper-ties: under broad conditions, it is the unique, consistent, asymptotically normal, generalized maximum likelihood estimator. **Random censoring** does not require that the censoring distribution be uniformly random; a functional relationship to the underlying variable is permitted providing it is independent of the functional relationships determining the sur-vival distribution. This is discussed in terms of likelihoods by Lawless (2002, Section 2.2). In the estimation of luminosity functions from astronomical flux-limited surveys discussed in Section 10.2, the distances are independently random but the dependence on luminosity violates the independence of the censoring distribution from the luminosity. Thus, the KM estimator is not a true maximum likelihood estimator for most astronomical luminosity function situations. A full treatment of the problem in astronomical censoring problem has not yet been made.

10.3.3 Two-sample tests

We now compare two univariate datasets asking: "What is the chance that two censored datasets do not arise from the same underlying distribution?" Using the formulation outlined in Section 3.5, this is a hypothesis test evaluating the null hypothesis $H_0 : S_1 = S_2$ against the alternative $H_1 : S_1 \neq S_2$. In Section 5.4.1, we presented the nonparametric sign test and Mann–Whitney–Wilcoxon test for such comparisons when no censoring is present. The methods outlined here are often generalizations of these classical nonparametric two-sample tests with weights, or **scores**, assigned to the censored data.

It is important to note that the theory of these two-sample tests, unlike the theory of the KM estimator, does not make assumptions of random censoring, and they are thus more generally applicable than the KM estimator. This arises because we are asking only to evaluate a hypothesis of similarity or difference, while the KM seeks to estimate the shape and location of the underlying distribution.

The **Gehan test**, developed in 1965 as a generalized Wilcoxon test for survival data, is perhaps the simplest of the censored two-sample tests. For a left-censored Sample 1 with n objects, $x_1^1, x_2^1, \ldots, x_n^1$, and Sample 2 with m objects, $x_1^2, x_2^2, \ldots, x_m^2$, compute the pairwise quantity

$$U_{ij} = \begin{cases} +1 \text{ if } x_i^1 < x_j^2 \text{ (where } x_i^1 \text{ may be censored)} \\ -1 \text{ if } x_i^1 > x_j^2 \text{ (where } x_j^2 \text{ may be censored)} \\ 0 \text{ if } x_i^1 = x_j^2 \text{ or if the relationship is ill-determined.} \end{cases} \quad (10.18)$$

Gehan's test statistic is

$$W_{Gehan} = \sum_{i=1}^{n} \sum_{j=1}^{m} U_{ij}. \quad (10.19)$$

W_{Gehan} is asymptotically normal with zero mean. A common estimate of its variance based on permutations is

$$\widehat{Var}(W_{Gehan}) = \frac{mn \sum_{i=1}^{n+m} U_i^2}{(n+m)(n+m-1)} \quad (10.20)$$

where U_i is $\sum_{j=1}^{m} U_{ij}$. The probability of no difference between the two samples can then be estimated by comparing $Z = W_{Gehan}/\sqrt{W_{Gehan}}$ to a normal distribution. In statistical notation, this is stated as $P(Z > z_\alpha | H_0) = \alpha$ where α is a chosen significance level like 0.05.

From examination of this statistic, we note that the effect of censoring on the two-sample comparison is not always the same. Increased censoring reduces W as more U_{ij} values become zero because the relationships between more points in the two samples become indeterminate. But the U_i values, and hence $Var(W)$, may or may not also decrease depending on the location of the censored points. Thus, the presence of censoring does not necessarily reduce the scientist's ability to discriminate differences between two samples; it depends on the location of the censored values in each situation.

A general formulation of the k-sample problem is described by Klein & Moeschberger (2005, Chapter 7). Let y_{ij} be the number of objects at risk (not yet detected) and d_{ij} be the number of ties in the j-th sample at value x_i in the pooled sample, where $i = 1, 2, \ldots, N$ is the total number of objects in the pooled sample and $j = 1, 2, \ldots, k$ denote the individual samples. Let $y_i = \sum_{j=1}^{k} y_{ij}$ and $d_i = \sum_{j=1}^{k} d_{ij}$ be the pooled number at risk and ties at x_i. Then, a general statistic for testing the null hypothesis that the k hazard rates are the same can be formulated as

$$Z_j = \sum_{i=1}^{N} W(y_i) \left[d_{ij} - \frac{y_{ij} d_i}{y_i} \right], \quad \text{for } j = 1, 2, \ldots, k \qquad (10.21)$$

with approximate variance

$$\widehat{Var}(Z_j) = \sum_{i=1}^{N} W^2(y_i) \frac{y_{ij}}{y_i} \left(1 - \frac{y_{ij}}{y_i} \right) \left(\frac{y_i - d_i}{y_i - 1} \right) d_i, \quad \text{for } j = 1, 2, \ldots, k. \qquad (10.22)$$

Z_k is a measure of the difference in hazard rates between the k-th sample and the pooled sample weighted by a function $W(y_i)$.

Various two-sample and k-sample tests for censored data can then be constructed with different assumed weights $W(y_i)$. The **logrank test** assumes $W(y_i) = y_i$, the number of objects not yet detected. **Gehan's test** assumes $W(y_i) = y_i^2$. The **Peto–Peto test**, which like Gehan's test is a generalized Wilcoxon test, uses the weights $W(y_i) = y_i \hat{S}_{KM}$ where $\hat{S}_{KM}$ is the Kaplan–Meier estimator. **Fleming–Harrington tests** use various functions of the Kaplan–Meier estimator as weights.

Although the logrank and Gehan's test are the most commonly used two-sample tests for survival problems, it is difficult to know which test may be more effective in a given situation. Studies have shown that the logrank test is more powerful than the generalized Wilcoxon tests when the two hazard functions are parallel (that is, the proportional hazard model applies), while the generalized Wilcoxon tests are more effective than the logrank when sample differences are greater at the beginning (uncensored) side of the distribution. When the samples have similar means but different shapes, Fleming–Harrington tests are recommended. Gehan's test performs poorly under heavy censoring or when the censoring patterns of the two samples greatly differ. Generalizations of the Cramér–von Mises and t-test, and median test have also been developed. In the absence of prior information about the underlying population distributions, we recommend that several tests be applied to give a sense of the importance of different weightings of the censored values.

Extensions of two-sample tests are widely used for more complicated survival problems. The general formulation of Equation (10.21) readily extends standard two-sample tests to multiple samples. For example, in astronomy, one might compare censored X-ray luminosities of radio-quiet and radio-loud quasars in several redshift strata to adjust for the possible confounding effect of cosmic evolution. This test is related to the χ^2 test applied to a sequence of contingency tables (Section 5.5).

We finally note that some information can be obtained even for samples that are 100% censored. While the population distribution function cannot be estimated, some quantities can be obtained based on binomial probabilities. For example, one can estimate the fraction of objects that might exceed a sensitivity limit at a chosen significance level based on an observed sample with no detections (Helsel 2005, Chapter 8). A completely censored sample can be compared to a partially censored sample in two-sample tests like the Gehan test.

10.4 Multivariate datasets with censoring

10.4.1 Correlation coefficients

When considering bivariate and multivariate problems, it is important to discriminate problems where censoring is confined to a single response variable, y, or can also be present in the x covariates. In the bivariate case, data can be observed points, half-lines in one dimension for censoring in one variable, or quadrants of the plane for censoring in both variables. Various bivariate estimators have been suggested, but suffer variously from lack of consistency, nonuniqueness, nonmonotonicity, negativity, or inconsistency with univariate projections (Gentleman & Vandal 2002). We thus do not encourage use of two-dimensional Kaplan–Meier distributions like that proposed by astronomer Schmitt (1985).

The first step in establishing the nature of a relationship between two variables is often a hypothesis test for correlation based on a measure like Pearson's linear correlation coefficient or rank measures like Spearman's ρ and Kendall's τ (Section 5.6). When censoring is present, the best-developed measures are generalizations of Kendall's τ that have the additional advantage of permitting censoring in either or both variables. The ASURV package implemented the method proposed by Brown *et al.* (1973), but Helsel (2005) advocates a different procedure.

Kendall's τ rank correlation coefficient with correction for ties can be viewed in terms of the collection of lines joining pairs of (x, y) data points. Helsel's coefficient is then

$$\tau_H = \frac{n_c - n_d}{\sqrt{\left(\frac{n(n-1)}{2} - n_{t,x}\right)\left(\frac{n(n-1)}{2} - n_{t,y}\right)}} \tag{10.23}$$

where n_c is the number of **concordant** pairs with a positive slope in the (x, y) diagram, n_d is the number of **disconcordant** pairs with negative slopes, and $n_{t,x}$ and $n_{t,y}$ are the number of ties or indeterminate relationships in x and y respectively. For example, if $x_i < x_j$ and $y_i < y_j$ for pair (i, j), then it is included in n_c if both points are detected or if the lower i-th point is censored in x and/or in y. However, if the upper j-th point is censored so that the relationship between the ranks is not certain, it does not contribute to either n_c or n_d, but the pair contributes to $n_{t,x}$ and/or $n_{t,y}$. As the censoring fraction increases, fewer points contribute to the numerator of τ_H, but the denominator measuring the number of effective pairs in the sample also decreases. Thus, increased censoring decreases the effective sample size, but may or may not decrease the value of τ_H. The significance of a measured τ_H value can be estimated from the normal distribution as indicated in Section 5.6 for Kendall's τ in the absence of ties, but the effective sample size must be adjusted for the increased numbers of ties and indeterminate pairs.

For trivariate problems, Akritas & Siebert (1996) derive a partial correlation coefficient based on a generalized Kendall's τ for each of the three pairs of variables that removes the effect of the third variable. Using the Kendall's τ statistic for censored data derived by Brown *et al.* (1973), they show that the partial correlation coefficient relating variables 1 and 2 in the presence of variable 3 is

$$\hat{\tau}_{AS,12,3} = \frac{\hat{\tau}_{12} - \hat{\tau}_{13}\hat{\tau}_{23}}{[(1 - \hat{\tau}_{13}^2)(1 - \hat{\tau}_{23}^2)]^{1/2}} \tag{10.24}$$

which is asymptotically normal with variance

$$\widehat{Var}(\hat{\tau}_{AS,12,3}) = \frac{576 \sum [\star]}{n(n-1)^3(n-2)^2(n-3)^2(1 - \hat{\tau}_{13}^2)(1 - \hat{\tau}_{23}^2)} \tag{10.25}$$

where the quantity $\star$ is a complicated sum of products of the pairwise indicators n_c and n_d for the censored dataset.

10.4.2 Regression models

Quite a variety of approaches has been taken to linear regression when censoring or truncation is present. We outline six methods here.

1. For a censored dependent variable y and a vector of covariates $\mathbf{x}$, the **accelerated failure-time model** is

$$\log y = \alpha + \beta \mathbf{x} + \epsilon \tag{10.26}$$

where ϵ has a normal distribution with mean zero and variance σ^2. This **exponential regression** model, a standard linear regression model after a logarithmic transformation of the dependent variable, might apply if a censored variable with a wide range of values (such as the luminosity of stars) is correlated with uncensored covariates with narrow ranges of values (such as color index). The explanatory variables change the original survival function, $S(y)$ for $\beta = 0$, by a scale factor dependent on the covariates, $S(ye^{-\beta x})$. If the covariate is binary (that is, having values 0 or 1), the regression problem reduces to a two-sample problem where the ratio of the median values of the two subsamples is

$$\frac{Med(x=1, \beta)}{Med(x=0, \beta)} = e^{\beta}. \tag{10.27}$$

In survival terminology, the median survival time is **accelerated** by the factor e^{β} in group 1 compared to group 0.

2. The **proportional hazards** model is another multiplicative regression method for censored data. Introduced by Cox (1972) and often called **Cox regression**, this model

is extremely popular in biomedical applications. Here the hazard rate has an exponential dependence on the covariates,

$$h(y|\mathbf{x}) = h_0(y)e^{\beta \mathbf{x}}. \tag{10.28}$$

The descriptor "proportional hazards" refers to the characteristic of this model that, for two objects with different covariate values, $\mathbf{x}_1$ and $\mathbf{x}_2$, the hazard rate functions are proportional to each other according to

$$\frac{h(y|\mathbf{x}_1)}{h(y|\mathbf{x}_2)} = \frac{e^{\beta \mathbf{x}_1}}{e^{\beta \mathbf{x}_2}}. \tag{10.29}$$

Another way to view the effect of the regression parameter β is that the change in the hazard function from x_1 to x_2 is $e^{(x_1-x_2)\beta}$. If the proportional hazard assumption applies to a dataset (and this assumption can be readily checked), then Cox regression can be an effective tool.

Cox regression has been extensively developed; an accessible presentation is given by Hosmer *et al.* (2008). The likelihood of this model can be readily computed and maximized, allowing MLE estimation and inference. The validity of the assumption of proportional hazards for each covariate can be assessed with tests on residuals which measure the distance between the data and the MLE. One can evaluate goodness-of-fit with residual analysis, determine which β values are significantly different from zero, estimate regression confidence intervals from the Fisher information matrix, choose variables using the Bayesian information criterion, seek regression coefficients that depend on x, and so forth. For problems with high-dimensional covariates, methods are well-established for selecting covariate subsets that significantly affect the dependent variable. One can consider β coefficients that are not constants but rather functions of x. The proportional hazards model can also be applied to left-truncated data.

3. The familiar additive linear regression model (Section 7.2)

$$y = \alpha + \beta \mathbf{x} + \epsilon \tag{10.30}$$

may be appropriate if the censored variable and covariates have similar ranges. This model, developed as an **iterative least squares** regression method by engineers Schmee & Hahn (1979), is surprisingly rarely used in survival studies. A code implementing the MLE solution for this model using the EM algorithm under the assumption of random censoring in the dependent variable was presented by Wolynetz (1979). In astronomy, Deeming (1968) gives a least-squares moments approach to the problem.

4. Buckley & James (1979) proposed a variant to this MLE linear regression with censoring in the dependent variable that permits non-Gaussian residuals around the line. Here, a trial regression line is found, and the nonparametric KM estimator of the y variable censored residuals around the line is computed and used as the distribution of the scatter term ϵ. The y_i locations of censored data points are moved downwards (for left-censoring) to new locations based on incomplete integrals of the KM estimator. The likelihood is maximized to obtain new trial regression estimates, and the procedure is iterated using the EM algorithm until convergence (Section 3.4.5). The **Buckley–James line estimator** is generally consistent and unbiased.

5. The linear regression problem under Type 1 (constant value) censoring is very well studied in econometrics under the rubric of the **Tobit**, or **Tobin's probit**, model (Greene

2003, Chapter 22). MLE methods are commonly used, and treatments for heteroscedasticity, dependence of the censoring level on covariates, and nonnormality of residuals are available. Approaches to this problem have been addressed in the astronomical literature under the rubric of regression in the presence of **Malmquist bias of the second kind**; this problem was important for estimation in studies of the extragalactic cosmic distance scale (Teerikorpi 1997).

6. Regression based on nonparametric methods is recommended when doubly censored data or bad outliers are present (Helsel 2005). Recall from Section 7.3.4 that the Thiel–Sen regression line slope is the median of the slopes from every pair of data points. Pairwise slopes involving censored data points lie in a range of possible values, and a procedure similar to the KM estimator can estimate the distribution function of slopes with these interval-censored values. The median of this distribution becomes a slope estimator for the censored data. Akritas *et al.* (1995) recommend that Thiel–Sen-type slopes are subtracted from the y values until the generalized Kendall's τ correlation coefficient becomes zero. A simulation study indicates that the **Akritas–Thiel–Sen line** can have lower bias and greater precision than the Buckley–James line. These methods can treat doubly censored problems (where both x and y are subject to censoring) and both large- and small-n problems. Figure 10.3 shows an Akritas–Thiel–Sen line fitted to an astronomical dataset.

Finally, we discuss a question that has arises repeatedly in the context of regression in censored data from two flux-limited astronomical surveys. Consider an observation where the fluxes at two wavelengths, F_1 and F_2, are measured at the telescope for a previously defined sample of objects at different distances, both subject to censoring. Some researchers had advocated that correlations and regression be performed in the plane of observed $F_1 - F_2$ fluxes, rather than the plane of intrinsic luminosities $L_1 - L_2$ where $L = 4\pi d^2 F$ for each object at its distance d. The reason given for this choice is that, when no relationship is present between L_1 and L_2, the detections in the $L_1 - L_2$ plane get stretched into a spurious correlation due to multiplication by different distances.

This analysis is better performed in the intrinsic luminosity plane for two reasons. First, the problem of spurious correlation disappears when censored data points are included and are treated using survival analysis techniques. Simulations show that, if no intrinsic correlation is present between L_1 and L_2, then survival analysis tools (e.g. a generalized Kendall's τ or Buckley–James regression) will show that the data are consistent with no trend, even if the detections alone show a trend (Isobe *et al.* 1986). Second, any nonlinearity in the relationship between L_1 and L_2 will get blurred when transformed into the $F_1 - F_2$ plane. Assuming a power-law relationship, $L_1 \propto L_2^\beta$, the flux–flux relationship is then $F_1 \propto F_2^\beta d^{2(\beta-1)}$ where the distance term acts like a noise component suppressing the significance of the correlation unless $\beta = 1$.

10.5 Truncation

When the sampling of a celestial population is subject to the failure to detect many members due to sensitivity limits, the sample is truncated where the truncation limit(s) play a role analogous to the censoring limits discussed above. The main difference is that the scientist has no knowledge of how many undetected sources exist. This ignorance is a profound problem for statistical inference, and many of the tools available for survival studies involving censoring cannot be used directly in the presence of truncation.

Nonetheless, under specific assumptions, statistical methods related to those of survival analysis can be applied to truncated samples. The astronomical situation of truncation from flux-limited surveys gives left-truncated data. Inference requires that the distribution of the truncated distribution accurately represents the underlying distribution above the truncation limit(s).

10.5.1 Parametric estimation

If the underlying population has a known distribution, then parametric analysis of truncated samples can proceed based on the likelihood (Lawless 2003, Section 2.4)

$$L = \prod_{i=1}^{n} \left[\frac{f(x_i)}{S(u_i)} \right] \qquad (10.31)$$

where $f(x)$ is the population p.d.f., S is the survival function, x_i are the data points and u_i are the corresponding truncation limits. Moments and MLEs for several truncated distributions are provided by Cohen (1991) for the singly truncated case where $u = u_0$. The most commonly treated parametric case, particularly in econometrics, is the singly truncated normal (Gaussian) distribution (Greene 2003, Chapter 22).

Parametric approaches to bivariate regression involving truncated data have also been discussed. Simple least-squares methods applied to truncated data will be biased. Methods from econometrics, such as Heckman's two-step estimation, have been extensively studied for this problem (Greene 2003, Chapter 22).

10.5.2 Nonparametric Lynden-Bell–Woodroofe estimator

The nonparametric maximum likelihood estimator for randomly truncated data (that is, U following a distribution independent of X), analogous to the Kaplan–Meier estimator for randomly censored data, was derived by astrophysicist Lynden-Bell (1971). Its mathematical properties were studied in a likelihood context by Woodroofe (1985) and as a Markov counting process by Keiding & Gill (1990). We will call this the **Lynden-Bell–Woodroofe (LBW) estimator**. The LBW estimator is the unbiased, consistent, generalized maximum likelihood estimator of the underlying distribution under broad conditions. The estimator is asymptotically normal but convergence may be slow for small samples. An application of the LBW estimator to an astronomical dataset is shown in Figure 10.4.

The LBW estimator for left-truncated data is based on data of the form (x_i, u_i, δ_i) where $x_i \geq u_i$ are observed values, u_i are the sensitivity limits, and δ_i are truncation indicators with 1 for detection and 0 for missing. We assume that the X and U random variables are independent; this assumption is tested below. Following Lawless (2002, Section 3.5), we seek nonparametric estimation of the survival function $S(x) = P(X_i \geq x)$ but can only achieve an estimator for

$$\frac{S(x)}{S(u_{min})} = P(X \geq x | X \geq u_{min}).$$ (10.32)

The maximum likelihood estimator of this function, very similar to the KM estimator in Equation (10.16) for censored data, is

$$S_{LBW} = \prod_{i:x_i < x} \left(1 - \frac{d_i}{n_i}\right)$$ (10.33)

where n_i is the number of points in the set $u_i \leq x \leq x_i$, the product is over the distinct values of x_i, and $d_i = 1$ unless ties are present. Caution is needed if any $n_i = 0$ or $d_i = n_i$. The asymptotically normal Greenwood-type variance is

$$\widehat{Var}_{LBW} = S_{LBW}^2 \sum_{i:x_i < x} \frac{d_i}{n_i(n_i - d_i)}.$$ (10.34)

Figure 10.4 shows an LBW estimator applied to a truncated astronomical dataset, with Greenwood's variance confidence intervals. It is compared to the less reliable $1/V_{max}$ estimator (Schmidt 1968) commonly used in astronomy.

Tsai (1990) gives a valuable test of the assumption that the truncation values are independent of the true values which is essential for the LBW estimator to be a consistent MLE. We consider data of the form (x_i, u_i) where the x_i values are observed only when they lie above the associated truncation limit, $x_i \geq u_i$. Tsai defines a generalized Kendall's τ statistic

$$K_{Tsai} = \frac{1}{2} \sum_{i=1}^{n} \sum_{j=1}^{n} w(x_i, x_j, u_i, u_j) \text{ where}$$ (10.35)

$$w(a,b,c,d) = \begin{cases} 1 & \text{for } (a-b)(c-d) > 0 \text{ and } \max(c,d) \leq \min(a,b) \\ -1 & \text{for } (a-b)(c-d) < 0 \text{ and } \max(c,d) \leq \min(a,b) \\ 0 & \text{otherwise} \end{cases}$$

which should not be significantly different from zero if independent truncation is present. For large datasets, this can be tested with the asymptotically normal test statistic

$$T_{Tsai} = \frac{K_{Tsai}}{\left[\frac{1}{3}\sum_{i=1}^{n}(r_i^2 - 1)\right]^{1/2}} \text{ where}$$

$$r_i = \sum_{j=1}^{n} I(t_j \leq x_i \leq x_j)$$ (10.36)

and I is an indicator function with values 0 and 1. This tests that the **risk set** of r_i values is uniform which is equivalent to a test that the truncation levels are independent of the true values. For small datasets, the significance of K_{Tsai} values should be evaluated numerically by bootstrap or permutation resamples.

A few extensions to bivariate situations related to the LBW estimator have been developed. **Efron & Petrosian (1992) test** for independence between a truncated variable $x_i < u_i$

and a covariate y_i using a vector of normalized rank correlation statistics. A generalized Kendall's τ test can be constructed with this vector. A generalization of Pearson's linear correlation coefficient for truncated data is given by Chen *et al.* (1996).

10.6 Remarks

Astronomers have made surveys which fail to detect the full sample for two centuries, but only in recent decades have they begun using approaches with strong mathematical foundations. In the 1980s, it was recognized that survival methods developed for right-censored problems in other fields are useful for left-censoring common in astronomy. The Kaplan–Meier estimator, two-sample rank tests, and a few other survival methods gained popularity in astronomical studies with upper limits

However, astronomers have not substantially investigated the possible applications of other treatments developed for censored data. The methods outlined here represent only a portion of a large field developed in biometrics, epidemiology, actuarial science, industrial reliability and econometrics that deal with censored data. We encourage astronomers with important statistical problems involving censoring to explore these possibilities.

The situation regarding truncation is unusual: astrophysicist Lynden-Bell developed the unique nonparametric maximum likelihood estimator for randomly truncated univariate data in the 1970s, analogous to the Kaplan–Meier estimator for randomly censored data, yet astronomers have often not recognized this achievement and use less effective methods such as $1/V_{max}$ statistics. Truncation provides less information than censoring, and fewer methods are available for use. Nonetheless, there is substantial potential for improved treatments for truncated data using both nonparametric and parametric techniques.

While the KM and LBW estimators, as nonparametric MLEs of censored and truncated distributions, are powerful tools, the astronomer should use them cautiously. First, the methods assume that the underlying population is homogeneous in the sense that the intrinsic distributions of the measured quantities are the same for the censored (truncated) and detected objects. This may not be true if different classes of astronomical objects are present; for example, in radio continuum surveys, brighter objects are mostly active galactic nuclei like quasars, while fainter objects are mostly starburst galaxies.

Second, although the estimators may maximize a likelihood, the results may be quite inaccurate for small and/or heavily censored values. For example, the median value of the distribution may be unknown, or only very crudely estimated, when the censoring or truncation fraction exceeds 50%. Large samples with heavy censoring or truncation can still be examined, but only the upper portions of the distribution can be accurately estimated.

Third, the KM and LBW estimators are nonparametric in the sense they give valid results for any underlying distribution but only if the censoring or truncation is **random**. That is, the probability that the measurement of an object is censored cannot depend on the value of the censored variable. At first glance, this may seem to be inapplicable to most astronomical problems: we detect the brighter objects in a sample, so the distribution of upper limits always depends on brightness. However, two factors often serve to randomize the censoring distribution: the censored variable may not be correlated with the variable by which the sample was initially identified; and astronomical objects in a sample often lie at greatly different distances so that brighter objects are not always the most luminous. Ideally, the censoring mechanism would be examined for each study to judge whether the censoring is or is not likely to be random; the appearance of the data is not a reliable measure. A frequent (if philosophically distasteful) escape from the difficulty of determining the nature of the censoring in a given eperiment is to define the population of interest to be the observed sample. The KM and LBW estimators then always give a valid redistribution of the upper limits, though the result may not be applicable in wider contexts. Finally, a critical limitation of survival analysis is the assumption that the upper limit values are precisely known from the experimental setup. In other fields of application, one knows exactly when a sick patient dies or when an engine fails. But in astronomical observational situations, censoring and truncation values arise from a probabilistic statement based on independently measured noise levels. For situations with normal (Gaussian) noise, an undetected object's flux may be considered to be below a censored value of three times the root-mean-squared noise level. For situations with Poisson noise, the choice of an upper limit is still widely debated (Mandelkern 2002, Cowan 1998, Kashyap *et al.* 2010). The problem is particularly acute for astronomical datasets containing many data points very close to the detection limit. This fundamental mathematical inconsistency between survival analysis and astronomical nondetections also occurs in ecological applications (Helsel 2005). A closely related deficiency is the absence of heteroscedastic weighting of detections by their measurement error, so that very weak detections are treated mathematically similarly to nondetections.

A promising general approach to these problems is the generalized likelihood for regression developed by Kelly (2007) outlined in Section 7.5.3. He constructs a likelihood that simultaneously treats measurement errors, censoring and truncation in a self-consistent fashion. The likelihood can then be used either for maximum likelihood estimation or Bayesian inference. This approach could be generalized to treat problems other than regression such as density estimation and multivariate analysis. The goal would be to replace the KM and LBW estimators with a single estimator treating measurement errors, censoring and truncation for both univariate and multivariate problems.

10.7 Recommended reading

Helsel, D. R. (2005) *Nondetects and Data Analysis: Statistics for Censored Environmental Data*, Wiley-Interscience, New York
   A brief, readable text that emphasizes left-censoring, the relation between noise and limits, and other problems encountered in astronomy. It covers graphics, maximum

likelihood estimation, interval estimation, two- and k-sample tests, correlation and linear regression.

Klein, J. P. & Moeschberger, M. L. (2005) *Survival Analysis: Techniques for Censored and Truncated Data*, 2nd ed., Springer, Berlin
A comprehensive graduate textbook presenting issues of censoring and truncation and covering many statistical treatments. Methods include survival and hazard function estimation, hypothesis testing, proportional hazards and other regression models and multivariate survival analysis.

Lawless, J. F. (2003) *Statistical Models and Methods for Lifetime Data*, 2nd ed., John Wiley, New York
A comprehensive advanced text covering many aspects of survival analysis including nonparametric estimators, parametric models, parametric and semi-parametric regression models, rank tests, goodness-of-fit tests and multivariate problems. It provides a particularly valuable integration of the LBW and KM estimators for randomly truncated and censored data, respectively.

10.8 R applications

Hundreds of astronomical studies have used the ASURV (Astronomy SURVival analysis) standalone FORTRAN package for treatment of censored data (LaValley *et al.* 1992). This package provides the Kaplan–Meier estimator for left- or right-censored data, four univariate two-sample tests (Gehan, logrank, Peto, and Peto–Prentice tests), three bivariate correlation hypothesis tests (generalized Kendall's τ and Spearman's ρ, Cox's proportional hazard model) and three linear regressions (maximum likelihood line with normal residuals, Buckley–James line with Kaplan–Meier residuals, and the Schmitt–Campbell binned regression line). Some (but not all) of these methods are available in **R** and **CRAN** today as illustrated below, and a great many other methods are provided for censored datasets. **CRAN** provides some additional procedures for truncated datasets such as the Lynden-Bell–Woodroofe estimator for univariate randomly truncated data.

10.8.1 Kaplan–Meier estimator

To illustrate a common type of censoring in astronomical studies, we examine the X-ray luminosity function (XLF) of an optically selected sample of quasars. We select the nearest quasars from the Sloan Digital Sky Survey Data Release 5 quasar sample (Appendix C.8, Schneider *et al.* 2007). The sample has 400 objects, of which 234 are detected and 166 are undetected in the soft X-ray ROSAT All-Sky Survey. Unfortunately, the SDSS catalog does not provide individual X-ray luminosity upper limits, which differ due to both variations in the RASS sky coverage and distances to each object. To estimate luminosity upper limits, we assign random flux limits around $\sim 10^{-12}$ erg s^{-1} cm^{-1} and multiply by $4\pi d^2$ where the distance d is based on measured redshifts. Figure 10.1 shows the resulting distribution of detections (solid circles) and approximate nondetection levels (open triangles).

In the **R** script below, the "ROSAT[ROSAT $< (-8.900)$] $= $ NaN" command changes the nondetection status flag in the original dataset to the "Not a Number" designation recognized by **R**. These objects will later be located using the *is.na* command which gives a vector or matrix of logical "TRUE" entries when NaN entries are present.

```
# Construct sample of X-ray luminosities of low-redshift (z<0.3) Sloan quasars
qso <- read.table("http://astrostatistics.psu.edu/MSMA/datasets/SDSS_QSO.dat",
   head=T, fill=T)

dim(qso) ; names(qso) ; summary(qso)
qso.lowz <- qso[qso[,2]<0.3,]
dim(qso.lowz) ; names(qso.lowz) ; summary(qso.lowz)
attach(qso.lowz)

# X-ray luminosities for detections and (approximately) for upper limits

ROSAT[ROSAT < (-8.900)] <- NaN  # X-ray nondetections
Dist <- (3*10^{10})*z*(3.09*10^{24}) / (73*10^5)
XLum <- 8.3*10^(-12)*10^(ROSAT) * (4*pi*Dist^2)
XLum[is.na(ROSAT)] <- 1.*10^(-12)*runif(1) * (4*pi*Dist[is.na(ROSAT)]^2)

par(mfrow=c(1,2))
plot(log10(XLum[!is.na(ROSAT)]), ylim=c(42.5,45), xlab='',
   ylab='log(Lx) erg/s', pch=20)
plot(log10(XLum[is.na(ROSAT)]), ylim=c(42.5,45), xlab='',
   ylab='log(Lx) erg/s', pch=25)
par(mfrow=c(1,1))
```

The Kaplan–Meier maximum likelihood estimator for the SDSS XLF, shown in the bottom panel of Figure 10.1, is calculated using functions in the *survival* package that is incorporated into **R**. The *Surv* function sets up an appropriate **R** object for input into the *survfit* function which calculates the estimator. The *Surv* object has the censored variable in the first column and the censoring indicator in the second column. Note that most survival codes treat only right-censored data, so we calculate the KM estimator of the negative of the astronomical left-censored variable. The output, which is provided here both as an ASCII table and a plot, gives the value of the estimator at each of the 400 quasar luminosity values, a tie indicator and confidence bounds at a chosen significance level.

The bottom panel of Figure 10.1 shows the KM estimator of the SDSS quasar log X-ray luminosity function with 95% confidence intervals based on a conservative modified

Greenwood's variance. A considerable astronomical research literature is devoted to quantitative evaluation of the quasar luminosity function at various wavebands and redshifts, often involving power-law least-squares fits (e.g. Hopkins *et al.* 2007).

```
# Construct quasar Kaplan-Meier luminosity function and 95% confidence band

library(survival)
Xstatus <- seq(1,1,length.out=400)  # 1 = detected
Xstatus[is.na(ROSAT)] <- 0  # 0 = right-censored
survobj <- Surv(-XLum, Xstatus)
KM.XLF<- survfit(survobj~1, conf.int=0.95, conf.type='plain',
   conf.lower='modified')
KM.output <- summary(survfit(survobj~1))

# Plot Kaplan-Meier estimator with confidence bands

plot(log10(-KM.XLF$time), KM.XLF$surv,  ylim=c(0,1), pch=20, cex=0.5, main='',
   xlab=expression(log*L[x]), ylab='Kaplan-Meier estimator')
lines(log10(-KM.XLF$time), KM.XLF$upper, lty=2)
lines(log10(-KM.XLF$time), KM.XLF$lower, lty=2)
```

We must be careful with limitations of this analysis. From a statistical viewpoint, a comparison of the detected and censoring distributions in Figure 10.1 clearly does not satisfy the requirement of random censoring, so the KM is formally not an MLE. Note however that, even if the individual censored values were approximately random, they are not precisely known because they arise from a probabilistic 3σ-type criterion based on noise, and not from an exact measurement of the survival time as in biometrical or industrial applications. From an astronomical viewpoint, the sample does not fully represent the low-luminosity range of active galactic nuclei because the defining characteristics of a "quasar" imply the nucleus outshines the host galaxy starlight. The true distribution of active galactic nuclear X-ray luminosities should include Seyfert galaxies where the luminosity distribution continues below $\log L_x \sim 43$ erg s^{-1} into the $39 < \log L_x < 42$ erg s^{-1} regime (Ho *et al.* 2001). It is common that the upper ends of astronomical luminosity functions are readily determined while much research is needed to define both the astronomical and statistical properties of the distributions at low luminosities. Despite these limitations, the KM estimator of the sample provides a reasonable treatment of the nondetections in estimating the quasar X-ray luminosity function.

10.8.2 Two-sample tests with censoring

The *survdiff* function in the *survival package* in **R** compares two censored distributions with the logrank test, the Peto–Peto modification of the Gehan–Wilcoxon test, and other tests with weightings based on the Kaplan–Meier estimator. We also use the *surv2.ks*

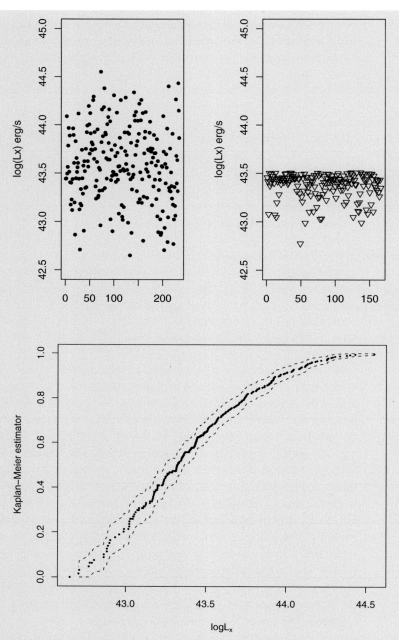

X-ray luminosity function of low-redshift ($z < 0.2$) SDSS quasars. Luminosities of X-ray detected quasars (top left) and approximate upper limits of undetected quasars (top right). Derived Kaplan–Meier estimator with 95% confidence band (bottom).

function in **R**'s *surv2sample* package which performs Kolmogorov–Smirnov, Cramér–von Mises and Anderson–Darling tests on censored samples with significance levels estimated from bootstrap resampling (Kraus 2008).

To illustrate these functionalities, we continue our examination of 400 low-redshift SDSS quasar X-ray luminosity functions (XLF), dividing them into radio-loud (104 objects) and radio-quiet (269 objects) quasars. This classification is based on detection in the FIRST radio survey; 27 SDSS quasars lying outside the FIRST survey field are omitted.

Applying the nonparametric two-sample tests to these two samples, we find that any differences between the radio-loud and radio-quiet XLFs are not significant. The probabilities that the null hypothesis is true (that is, that the two samples are drawn from the same unknown distribution) are $P > 0.1$. Note the value in applying different tests, each of which gives mathematically plausible weightings to the censored data points.

Astronomically, the absence of a difference in the XLFs of radio-loud and radio-quiet optically selected QSOs at first seems at variance with the long-standing statistical correlation between X-ray and radio luminosities in QSOs (Zamorani *et al.* 1981). However, when the low-redshift objects are examined separately from high-redshift strata, other studies have found that the XLFs of the two classes are quite similar (Della Ceca *et al.* 1994). Most of the established correlation is due to differential evolution of the two samples over large redshift ranges and cannot be seen in the low-redshift subsample considered here.

```
# Prepare data for survival two-sample tests
# Construct subsamples for XLFs of radio detected vs. undetected quasars
# Class = 1 are radio detections, Class = 2 are radio nondetections
# X-ray luminosities are obtained from ROSAT counts

radio.obs <- qso.lowz[qso.lowz[,15]>(-1)),]
attach(radio.obs)

radio.class <- seq(1,1,length.out=370)
for(i in 1:370) if(radio.obs[i,15]>0) radio.class[i]=2
ROSAT[ROSAT < (-8.900)] <- NaN
Xstatus <- seq(1,1,length.out=370)
Xstatus[is.na(ROSAT[1:370])] <- 0
Dist <- (3*10^{10})*z*(3.09*10^{24}) / (73*10^5)
XLum <- seq(1,1,length.out=370)
XLum[1:370] <- 8.3*10^(-12)*10^(ROSAT) * (4*pi*Dist^2)
XLum[is.na(ROSAT[1:370])] <- 1.*10^(-12)*runif(1)* (4*pi*Dist[is.na(ROSAT)]^2)

# Apply five two-sample tests for radio detected vs. nondetected subsamples

survdiff(Surv(-XLum,Xstatus) ~ radio.class, rho=0)
survdiff(Surv(-XLum,Xstatus) ~ radio.class, rho=1)
```

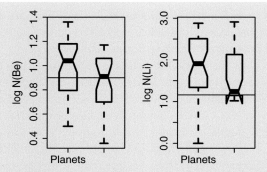

Fig. 10.2 Boxplots of the stellar abundances of beryllium and lithium for stars with and without detected planets. These values are based on the Kaplan–Meier estimator taking nondetections (censoring) into account.

```
install.packages('surv2sample') ; library(surv2sample)
surv2.ks.out <- surv2.ks(Surv(-XLum,Xstatus), radio.class,approx='boot')
surv2.ks.out
```

10.8.3 Bivariate and multivariate problems with censoring

The **CRAN** *NADA* package (Lee 2010) implements methods from the volume by Helsel (2005) for censored environmental data. We apply *NADA* functions to the two-sample, bivariate, doubly censored dataset of photospheric beryllium and lithium abundance measurements in stars with and without host planets presented by Santos *et al.* (2002, C.4). In this dataset, Type = 1 [2] indicates stars with [without] known planets. The censoring indicators are 1 [0] for detected [undetected] abundances.

We start with univariate comparisons of abundances in stars with and without planets using *NADA*'s *cenboxplot* function (Figure 10.2). In *NADA*, the censoring indicator is a vector of logicals where FALSE represents detections. The group indicator is an **R** factor vector created with the *as.factor* command. The *cenboxplot* is similar to the standard **R** boxplot (Section B.6) above a horizontal line showing the highest censored value; the thick bar shows the median, the box hinges show the 25% and 75% quartiles, the notch shows an approximate standard deviation of the median, and the dashed whiskers show a reasonable range of the data excepting outliers. Any structure below the horizontal line takes the distribution of censored points into account using a robust version of the Kaplan–Meier estimator (the **regression on order statistics** method; Helsel 2005, Chapter 6).

In this dataset, the median values of all subsamples lie above the highest censored points, so they can be interpreted in the usual fashion. The presence of planets has no effect on beryllium abundances but may elevate lithium abundances. The significance of this effect is quantified using the *cendiff* function which is *NADA*'s frontend to **R**'s *survdiff*. Here we find any planetary effect on stellar lithium abundances is not statistically significant ($P > 0.1$) using the logrank and Peto–Peto two-sample tests. Santos *et al.* (2002)

arrived at the same result, concluding that the dataset gives no support to the astrophysical scenario where accretion of planets leads to metallicity enhancements in the stellar photosphere.

```
# Read dataset on beryllium and lithium abundances in stars
abun <- read.table('http://astrostatistics.psu.edu/MSMA/datasets/censor_Be.dat',
   header=T)
dim(abun) ; names(abun) ; attach(abun)


# Boxplot of abundances for stars with and without planets

install.packages('NADA') ; library(NADA)
cen_Be <- seq(FALSE, FALSE, length=68) ; cen_Be[Ind_Be==0] <- TRUE
cen_Li <- seq(FALSE, FALSE, length=68) ; cen_Li[Ind_Li==0] <- TRUE
Type_factor <- as.factor(Type)
par(mfrow=c(1,2))
cenboxplot(logN_Be, cen_Be, Type_factor, log=FALSE, ylab='log N(Be)',
   names=c('Planets','No planets'), boxwex=0.5, notch=TRUE, varwidth=TRUE,
   cex.axis=1.5, cex.lab=1.5, lwd=2)
cenboxplot(logN_Li, cen_Li, Type_factor, log=FALSE, ylab='log N(Li)',
names=c('Planets','No planets'), boxwex=0.5, notch=TRUE, varwidth=TRUE,
   cex.axis=1.5, cex.lab=1.5, lwd=2)
par(mfrow=c(1,1))


# Test significance of possible lithium abundance effect

logN_Li1 <- logN_Li[-c(1,23)] ; cen_Li1 <- cen_Li[-c(1,23)]
Type_factor1 <- Type_factor[-c(1,23)] # remove NaN values

cendiff(logN_Li1, cen_Li1, Type_factor1, rho=0)
cendiff(logN_Li1, cen_Li1, Type_factor1,rho=1)
```

Figure 10.3 shows a bivariate scatterplot comparing lithium and beryllium abundances modeled on plots shown by Santos *et al.* (2002). A long series of commands is needed due to the variety of symbols used for different subsamples defined by the *which* function. Here the *arrows* function plots symbols appropriate for upper limits.

```
# Reproduce Santos et al. (2002) plot of stellar beryllium vs. lithium abundance

ind_det1 <- which(Ind_Li==1 & Ind_Be==1 & Type==1) # filled circles
ind_det2 <- which(Ind_Li==1 & Ind_Be==1 & Type==2) # open circles
ind_left1 <- which(Ind_Li==0 & Ind_Be==1 & Type==1)
ind_left2 <- which(Ind_Li==0 & Ind_Be==1 & Type==2)
ind_down1 <- which(Ind_Li==1 & Ind_Be==0 & Type==1)
```

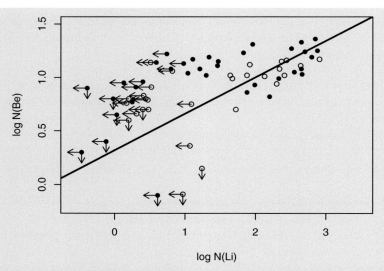

Fig. 10.3 Light-element abundances in the atmospheres of stars with (filled circles) and without (open circles) detected planets (Santos *et al.* 2002) with the Akritas–Thiel–Sen regression line.

```
ind_down2 <- which(Ind_Li==1 & Ind_Be==0 & Type==2)
ind_both1 <- which(Ind_Li==0 & Ind_Be==0 & Type==1)
ind_both2 <- which(Ind_Li==0 & Ind_Be==0 & Type==2)

plot(logN_Li[ind_det1], logN_Be[ind_det1], xlim=c(-0.6,3.5), ylim=c(-0.2,1.5),
   main="", xlab="log N(Li)",
   ylab="log N(Be)", pch=16, lwd=2) # plot detections
points(logN_Li[ind_det2], logN_Be[ind_det2], pch=1)
arrows(logN_Li[ind_left1], logN_Be[ind_left1], logN_Li[ind_left1]-0.2,
   logN_Be[ind_left1],length=0.1) # plot left arrows
arrows(logN_Li[ind_left2], logN_Be[ind_left2], logN_Li[ind_left2]-0.2,
   logN_Be[ind_left2],length=0.1)
points(logN_Li[ind_left1], logN_Be[ind_left1], pch=16)
points(logN_Li[ind_left2], logN_Be[ind_left2], pch=1)
arrows(logN_Li[ind_down1], logN_Be[ind_down1], logN_Li[ind_down1],
   logN_Be[ind_down1]-0.1, length=0.1)
arrows(logN_Li[ind_down2], logN_Be[ind_down2], logN_Li[ind_down2],
   logN_Be[ind_down2]-0.1, length=0.1)
points(logN_Li[ind_down1], logN_Be[ind_down1], pch=16)
points(logN_Li[ind_down2], logN_Be[ind_down2], pch=1)

arrows(logN_Li[ind_both1], logN_Be[ind_both1],
   logN_Li[ind_both1]-0.2, logN_Be[ind_both1], length=0.1) # plot double arrows
arrows(logN_Li[ind_both1], logN_Be[ind_both1], logN_Li[ind_both1],
```

```
    logN_Be[ind_both1]-0.1,length=0.1)
arrows(logN_Li[ind_both2], logN_Be[ind_both2], logN_Li[ind_both2]-0.2,
    logN_Be[ind_both2],length=0.1)
arrows(logN_Li[ind_both2], logN_Be[ind_both2], logN_Li[ind_both2],
    logN_Be[ind_both2]-0.1,length=0.1)
points(logN_Li[ind_both1], logN_Be[ind_both1], pch=16)
points(logN_Li[ind_both2], logN_Be[ind_both2], pch=1)
```

The *cenken* function provides the Akritas–Thiel–Sen (ATS) Kendall τ correlation coefficient and linear regression estimator which we apply to the stellar beryllium–lithium data plotted above. The **R** code below starts with new definitions for the data and censoring vectors required by *cenken*. In this function, "NaN" entries are not automatically eliminated. The output gives Kendall's $\tau = 0.50$ with associated probability $P \ll 10^{-4}$ that no correlation is present. The output ATS slope estimate is $b = 0.34$ (confidence intervals are not provided) and the line is superposed on the scatterplot in Figure 10.3. The *NADA* package also gives **R**'s standard univariate Kaplan–Meier estimator using *cenfit* and univariate two-sample tests using *cendiff*.

```
# Bivariate correlation and regression using Akritas-Thiel-Sen procedure

logN_Li1 <- logN_Li[-c(1,23)] # remove two points with NaN entries
Ind_Li1 <- Ind_Li[-c(1,23)] ;
logN_Be1 <- logN_Be[-c(1,23)]
Ind_Be1 <- Ind_Be[-c(1,23)]
Li_cen <- seq(FALSE, FALSE, length=66) # construct censoring indicator variables
Li_cen[which(Ind_Be1==0)]  <- TRUE
Be_cen=seq(FALSE, FALSE, length=66)
Be_cen[which(Ind_Li1==0)] <- TRUE
cenken_out <- cenken(logN_Be1, Be_cen, logN_Li1, Li_cen)
abline(a=cenken_out$intercept, b=cenken_out$slope, lwd=2)
```

10.8.4 Lynden-Bell–Woodroofe estimator for truncation

We illustrate the basic effects of astronomical truncation and their remediation using the Lynden-Bell–Woodroofe (LBW) estimator applied to a magnitude-limited sample of nearby stars with distances obtained from parallax measurements with the Hipparcos satellite. The sample has 3307 stars from the Hipparcos catalog defined by $V < 10.5$, $0 < RA < 90$ degrees, $\pi < 13.3$ mas (distance < 75 pc) and $\sigma_\pi < 10$ mas. The solid curve in Figure 10.4 shows the luminosity function of the observed sample; the peak around 1 $L_\odot$ is unrealistic and arises from the nondetections of fainter stars at more distant regions of the 75 pc radius volume samples. The distribution of truncation levels from the maximum distance each star could have been detected above $V < 10.5$ is shown by the dashed curve.

```
# Construct a sample of bright nearby Hipparcos stars

hip <- read.table('http://astrostatistics.psu.edu/MSMA/datasets/HIP1.tsv',
   header=T, fill=T)
attach(hip) ; dim(hip); summary(hip)

# Plot luminosity distribution of stars and their truncation limits

AbsMag <- Vmag + 5*log10(Plx/1000) + 5
Lum <- 2.512^(4.84 - AbsMag)
plot(density(log10(Lum)),ylim=c(0,1.7), main='',
   xlab='log L (solar, V band)', lwd=2, cex.lab=1.2)
AbsMaglim <- 10.5 + 5*log10(Plx/1000) + 5
Lumlim <- 2.512^(4.84 - AbsMaglim)
lines(density(log10(Lumlim)), lty=2, lwd=2)
text(0.7, 0.5, 'Hipparcos sample', pos=4, font=2)
text(-0.5, 1.2, 'Truncation limits', pos=4, font=2)
```

Figure 10.4 shows the Lynden-Bell–Woodroofe nonparametric maximum likelihood estimator of the Hipparcos luminosity function with 90% confidence bands estimated through bootstrap resamples. We use the **CRAN** package *DTDA* (doubly truncated data analysis, Moreira *et al.* 2010) based on the algorithm of Efron & Petrosian (1992). Here the apparent decline in the stellar population below a solar mass seen in the detections is replaced by a rise towards low-mass stars. The bootstrap confidence band quantifies the growing uncertainty of the distribution at low luminosity values where the truncation correction becomes very large. This is a computationally intensive calculation, taking several CPU hours on a typical machine for 100 bootstrap resamples.

The estimator agrees reasonably well with the well-established binned luminosity function obtained by Wielen *et al.* (1983) from the Gliese catalog of solar neighborhood stars. Precise agreement is not expected; for example, we have not treated Malmquist bias issues at the $d = 75$ pc boundary (i.e. scattering of stars into and out of the sample due to measurement errors), cleaned the Hipparcos dataset of unusual subpopulations (red giants, white dwarfs, halo stars) or correctly normalized the different estimators. We also compare the LBW estimator to the Schmidt (1968) $1/V_{max}$ luminosity function estimator calculated in $\Delta(\log L_V) = 0.5$ bins (solid curve). We see that the $1/V_{max}$ estimator is considerably less effective in recovering the true distribution when the truncation correction factor is large.

```
# Compute Lynden-Bell-Woodroofe estimator with 90% confidence bands

install.packages('DTDA') ; library(DTDA)
LBW.hip <- efron.petrosian(log10(Lum), log10(Lumlim), boot=T, B=100, alpha=0.1)
summary(LBW.hip)
plot(LBW.hip$time, LBW.hip$survival, pch=20, cex=0.6, main='',
   xlab='log L (solar, V band)', ylab='Density', cex.lab=1.2)
```

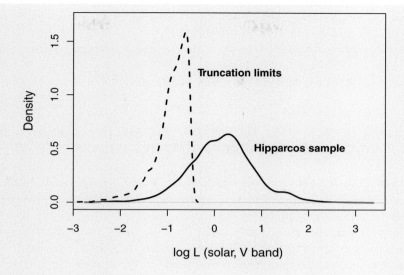

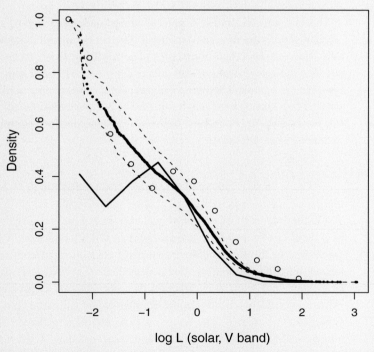

Fig. 10.4   Stellar luminosity function estimated from a sample of 3307 stars from the magnitude-limited (truncated) Hipparcos catalog. *Top:* Distributions of observed stellar luminosities in the V band (solid curve) and of the minimum luminosities for detection of the stars (dashed curve). *Bottom:* Lynden-Bell–Woodroofe estimator of the stellar luminosity function (filled circles with dashed confidence bands) compared to the binned Schmidt $1/V_{max}$ estimator (solid curve) and Wielen's well-established stellar luminosity function (open circles; based on Holtzman *et al.* 1998).

```
upper <- LBW.hip$upper.Sob[-(1000:1013)]
lower <- LBW.hip$lower.Sob[-(1000:1013)]
lines(LBW.hip$time,upper, lty=2) ; lines(LBW.hip$time, lower, lty=2)

# Compare with observed Wielen 1983 local star LF

Wielen.MV <- seq(0, 12, 1)
Wielen.LF <- c(35,126,209,380,676,955,1050,891,1120,1410,2140,2510,4470)
points(log10(2.512^(4.84-Wielen.MV)), Wielen.LF/2500)

# Compare with binned 1/V.max luminosity function (Schmidt 1968)

Vol.max <- (4*pi/3)*(1000/Plx)^3
bin.sum <- function(vol) sum(1/vol)
Lum.bins <- cut(log10(Lum), breaks=seq(-2.5,3.0,0.5), ord=T)
Schmidt.LF <- by(Lum, Lum.bins, bin.sum)
lines(seq(-2.25, 2.75, 0.5), Schmidt.LF/4500, pch=3, lwd=2)
```

The last portion of this script has some interesting **R** usage. The Schmidt $1/V_{max}$ is calculated with the *function* command. The *cut* command divides the ordered luminosities into bins with specified boundaries. The resulting *Lum_bin* vector is an **R** "factor" that encodes the categorization with entries like "(1,1.5]" for a star in the $1.0 < L \leq 1.5$ L$_\odot$ bin. The *by* command is a broadly applicable operation that applies a function (the $1/V_{max}$ statistic *bin_sum*) to an **R** dataframe (the vector of star luminosities *Lum*) using an **R** factor (the bin boundaries *Lum_bin*). The result is a simultaneous computation and binning of the Schmidt $1/V_{max}$ estimator.

10.8.5   Scope of censoring and truncation in R and CRAN

While the treatment of censored and truncated data is limited to basic functionalities in **R**, a number of capabilities are available in **CRAN** packages. The *prodlim* package gives extensions to the Kaplan–Meier estimator such as rapid computation for large samples, nearest-neighbor smoothing, linear dependence on covariates, stratification by covariates or clusters and simulated censored datasets. The *km.ci* package provides various estimates of confidence limits for the Kaplan–Meier estimator in addition to those based on Greenwood's variance. The *Icens* package provides advanced algorithms for computing nonparametric MLEs for censored and truncated data. **CRAN**'s *fitdistrplus* package fits statistical distributions to right-, left- or interval-censored data with goodness-of-fit tests. A wide variety of univariate two-sample tests can be found in **CRAN**'s *surv2sample* package. The function *censboot* in the **boot** package incorporated into **R** provides several procedures for bootstrapping censored data.

Although left-censored astronomical variables can be converted to right-censored by multiplying by -1, some packages specifically permit left-censoring. The *NADA* package described above is oriented towards left-censored datasets. Using the EM algorithm, the package *dblcens* computes the nonparametric maximum likelihood estimator of a univariate dataset where simultaneous left- and right-censoring is permitted.

CRAN has a number of packages for regression involving truncated data and models. *GAMLSS* is a complex of packages giving a sophisticated environment for fitting ordinary or censored datasets with truncated statistical distributions, mixture models, local regressions and user-supplied nonlinear formulas. Models are fitted by maximum likelihood estimation; the Akaike information criterion and other tools assist with model selection. Package *truncreg* is a simpler program for MLE regression of truncated models. The **CRAN** package *lss* gives a probability of correlation and a least-squares-type regression for right-censored data. Elaborate codes are available implementing variants of the Cox proportional hazard model. Two functions using the *survival* package in **R** are particularly useful for the astronomer: *cox.zph* for testing the validity of the proportionality assumption in a given dataset; and *coxph* for testing the significance of correlation with an uncensored covariate. A linear regression with Gaussian residuals for left-censored data is provided by *MCMCtobit* in package *MCMCpack*. The Buckley–James line and other regression tools are implemented in **CRAN**'s *emplik* and *rms* packages.

An annotated list of **CRAN** packages for censoring and truncation can be found at http://cran.r-project.org/web/Survival.html. Dozens are available implementing complex modeling for biometrics, quality control, econometrics and other fields. Their utility for astronomical problems is unexplored. Some implement Bayesian approaches to proportional hazards models, accelerated failure time models and normal mixture models with censoring. Others address censoring in high-dimensional datasets. General formulas for truncating arbitrary probability and cumulative distribution functions are given by Nadarajah & Kotz (2006). They give **R** codes for truncating p.d.f.'s, moments, quantiles and random deviates.

11 Time series analysis

11.1 The astronomical context

Time-domain astronomy is a newly recognized field devoted to the study of variable phenomena in celestial objects. They arise from three basic causes. First, as is evident from observation of the Sun's surface, the rotation of celestial bodies produces periodic variations in their appearance. This effect can be dramatic in cases such as beamed emission from rapidly rotating neutron stars (pulsars).

Second, as is evident from observation of Solar System planets and moons, celestial bodies move about each other in periodic orbits. Orbital motions cause periodic variations in Doppler shifts and, when eclipses are seen, in brightness. One could say that the birth of modern time series analysis dates back to Tycho Brahe's accurate measurement of planetary positions and Johannes Kepler's nonlinear models of their behavior.

Third, though less evident from naked eye observations, intrinsic variations can occur in the luminous output of various bodies due to pulsations, explosions and ejections, and accretion of gas from the environment. The high-energy X-ray and gamma-ray sky is particularly replete with highly variable sources. Classes of variable objects include flares from magnetically active stars, pulsating stars in the instability strip, accretion variations from cataclysmic variable and X-ray binary systems, explosions seen as supernovae and gamma-ray bursts, accretion variations in active galactic nuclei (e.g. Seyfert galaxies and quasi-stellar objects, quasars and blazars), and the hopeful detection of gravitational wave signals. A significant fraction of all empirical astronomical studies concerns variable phenomena; see the review by Feigelson (1997) and the symposium *New Horizons in Time Domain Astronomy* (Griffin *et al.* 2012).

The nature of time series behavior in astronomical data exhibit a great range in properties. Orbital, rotational and pulsational behavior produces periodic variations in position, velocity and/or brightness. Binary-star orbits range from minutes to centuries, neutron star rotations range from milliseconds to minutes, and millions of harmonics have been detected in seismological studies of our Sun's oscillations. Accretion onto compact stars, stellar mass and supermassive black holes produces stochastic variability on time-scales of microseconds to decades. The statistical behavior of the emission is often not Gaussian white noise; $1/f$-type long-term memory processes and other patterns are common. Explosive phenomena produce a bewildering variety of temporal variations that defy simple classification. These include magnetic reconnection events on stellar surfaces, interstellar and interplanetary scintillation of radio sources, relativistic jets in blazars, microlensing events when one star passes in front of another, supernova explosions at the end of a massive star's life, and thermonuclear explosions producing X-ray bursts, and gamma-ray bursts from the violent birth of black holes.

Some astronomical time series are multivariate because the phenomenon is measured (quasi)simultaneously in different wavelength bands. In situations like orbital eclipses or microlensing events, the variations should be simultaneous (though not necessarily with equal amplitude) in all bands. In situations like supernova light curves, blazar synchrotron jets variations, and gamma-ray burst afterglows, the variations may be correlated with some lag-time of astrophysical interest. Major time-domain survey projects such as Pan-STARRS and the Large Synoptic Survey Telescope are underway which will obtain essentially a "movie" of the sky at hundreds of epochs with billions of individual time series measured in several wavelength bands.

The statistical analysis of astronomical time series must confront difficulties associated with data acquisition, in addition to the intrinsic complexity of the emission processes.

1. The measurements of brightness or other variable properties very often are unevenly spaced. The schedule of ground-based visible light observations are regulated by weather, daily solar and monthly lunar cycles. Space-based observations are often regulated by satellite orbits. Many observations are limited by telescope allocation committees which often do not give sufficient access to acquire datasets optimized for later statistical analysis. While standard methods for evenly spaced time series can sometimes be adapted for occasional gaps in the data stream, they cannot be used for datasets that intrinsically consist of measurements at unevenly spaced times. Unfortunately, relatively little attention has been paid to such problems outside of the astronomical community; of the hundreds of volumes written on the analysis of temporal data, only one is specifically devoted to this issue (Parzen 1984). Common approaches are to project the data onto a regularly spaced grid either by binning or by interpolation. Various methods to reconstruct a smooth or evenly spaced time series from irregularly sampled data are discussed by Vio *et al.* (2000). Interpolation procedures may work adequately for smoothly varying behavior but are ineffective for objects exhibiting stochastic short-term autocorrelated variations or periodicities.

2. Individual observations are subject to heteroscedastic measurement errors. The variances of these errors are known from independent measurements of instrumental conditions (such as signal-to-noise ratios, blank sky measurements and instrumental calibrations). Thus, statistical analyses can, in principle, be weighted to account for differences in measurement errors. However, most time series procedures assume homoscedastic errors, and statistical treatments of heteroscedasticity with known variances have not been developed in a time series context.

3. The variations of astronomical sources, and background noise to be subtracted, often show correlation structure. Common deviations are $1/f$-type long-memory processes.

The time series methods described here may be useful for other types of astronomical datasets where one measures a signal along an independent ordered, time-like variable. This includes astronomical spectra where the intensity of light is measured as a function of its wavelength and astronomical images where the intensity is measured as a function of a pixelated spatial location in an image. This connection is rarely made; an interesting example is the adaptation of Bayesian modeling procedures designed for change-point detection in a one-dimensional Poisson time series to source detection in a two-dimensional Poisson image (Scargle 2003).

11.2   Concepts of time series analysis

A broad goal of time series analysis is to represent behavior that may be exhibited over many points in a dataset in a model with a small number of parameters. No single procedure works for all situations. Datasets dominated by periodic behavior may be effectively modeled by Fourier transforms or related types of harmonic analysis. Datasets dominated by deterministic trends may be best modeled by regression. Datasets dominated by stochastic but autocorrelated behavior may be best modeled using autoregressive models. Composite models can be used for complicated time series with combinations of behaviors. In all of these cases, the temporal structure violates the assumption of independence (that is, the data are no longer i.i.d., Section 2.5.2) underlying many statistical methods discussed elsewhere in this volume.

Procedures for all these types of time series modeling are well-developed for evenly spaced data. A large academic library has hundreds of monographs and texts on these methodologies, some in mathematics and statistics, some addressing engineering signal processing, and others oriented towards econometrics. Astronomers tend to use a narrow suite of methods, particularly Fourier analysis, for a wide range of problems. They may also be unfamiliar with the many tools available for model selection and validation.

When the average value of a time series changes with time, one of several types of **trend** is present. A global linear trend has a formulation similar to linear regression treated in Section 7.2,

$$X_t = \alpha + \beta t + \epsilon_t. \tag{11.1}$$

The notation here is common in time series analysis and differs from the notation used elsewhere in the volume: X (or a vector of variables $\mathbf{X}$) is the response variable and time t is the independent variable. The notations X_t and ϵ_t are equivalent to $X(t)$ and $\epsilon(t)$.

More complex trends can be modeled with nonlinear functions or with smoothers and local regression procedures discussed in Chapter 6. Adapting language from signal processing, smoothing to reduce variance from short-term variations is a **low-pass filter**, while examining residuals after fitting a regression model to reduce long-term trends is a **high-pass filter**. Stochastic trends are often treated in the context of **autoregressive models**, such as ARMA models (Section 11.3.3). A **differencing filter**, such as

$$Y_t = X_t - X_{t-1}, \tag{11.2}$$

is commonly used to remove trends of various types (global, periodic or stochastic) to reveal short-term structure. Finally, these procedures can be used in combinations; for example, a likelihood-based **state-space model** can be constructed to fit simultaneously long-term deterministic trends, periodicities, and short-term autocorrelated variations.

Autocorrelation is a basic measure correlated structure in a time series. With a calculation similar to the bivariate linear correlation coefficient presented in Section 3.6.2, the **autocorrelation function** establishes the degree to which values at different separations in time vary together,

$$ACF(k) = \frac{\sum_{t=1}^{n-k}(X_t - \bar{X})(X_{t+k} - \bar{X})}{\sum_{t=1}^{n}(X_t - \bar{X})^2} \tag{11.3}$$

where $\bar{X}$ is the mean of the time series. The integer parameter $k > 0$ here is called the **lag time**, the numerator (properly normalized) is the sample autocovariance function, and the denominator is the sample variance. The plot of $ACF(k)$ for a range of k is often called the **correlogram**. Complicated behavior in the time series may appear as simple behavior in the ACF. Random uncorrelated noise, often called **white noise**, produces near-zero values in the ACF (except for the obvious value $ACF(0) = 1$). Time series with short-term autocorrelation have strong ACF coefficients at small k, while time series with **long-term memory** have ACF signals for a wide range of k. Periodic variations in the time series produce periodic variations in the ACF; these are most efficiently modeled using the Fourier transform. ACFs are not easily interpreted when both trend and stochastic variations are present, so they are often applied to **residual time series** after trend models have been fit and removed.

Stationarity is the property that the temporal behavior, whether stochastic or deterministic, is statistically unchanged by an arbitrary shift in time. An evenly spaced dataset is strictly stationary when the joint distribution of $(X_{t_1}, X_{t_2}, \ldots, X_{t_m})$ and $(X_{s+t_1}, X_{s+t_2}, \ldots, X_{s+t_m})$ is the same for any shift s of time. The concept of a mean value, where the sample mean converges to the population mean, is not relevant to many nonstationary processes. **Weakly stationary** phenomena have constant moments, such as the mean and autocovariance, but may change in other ways. Often the concept of stationarity is applied to the residuals after a model has been fitted; a successful model should produce stationary residuals with negligible autocovariance even if the original time series was nonstationary.

While astronomical systems may occasionally show gradual changes in variability, the clearest cases of **nonstationarity** are systems that change abruptly from one state to another. A remarkable example is the X-ray emission from the accreting Galatic X-ray binary star system GRS 1915+105. Here over a dozen distinct variability states – some dominated by noise, others by quasi-periodicities or explosions – are present (Fender & Belloni 2004). The instant when a time series changes variability characteristics is called a **change-point**.

When periodic phenomena are present, the signal becomes far more concentrated after transformation from the time domain to the frequency domain. This type of analysis has various names: **spectral analysis**, **harmonic analysis** (this term is sometimes restricted to cases where the period is known in advance) or **Fourier analysis** (although some spectral methods do not involve the Fourier transform). Frequency- and time-domain methods are often closely related; for example, the Fourier spectrum can be directly mapped onto the time-domain autocorrelation function, and a periodogram can sometimes be viewed as a partitioning of the variance related to analysis of variance (ANOVA).

Since all data points in a time series contribute to each value of a spectral density function, these methods are based on strong assumptions that the data points are equally weighted (homoscedastic errors) and the temporal behavior does not change during the observation, including changes in mean levels (stationarity). Fourier analysis has additional restrictive assumptions, such as evenly spaced data and sinusoidally shaped variations. Due to these limitations, the results of spectral analysis are often difficult to interpret; for example, the significance of a peak in a periodogram may not be readily computable if a trend is present, the data are not evenly spaced or autoregressive behavior is present.

Finally, we note that most of the methods discussed here assume the temporal behavior is mathematically linear with respect to previous values or ancillary variables. Many types of **nonlinear models** can be treated: time series with variable volatility and other types of heteroscedasticity, piecewise linear models, models with thresholds, autoregressive models with variable coefficients, dependences on high-order autocorrelation (e.g. nonzero bispectrum), or chaotic behavior (Fan & Yao 2003). **Chaos** is a particular form of nonlinear behavior where small changes in initial conditions can, under particular conditions, lead to large changes in outcomes. Chaotic time series have specific characteristics – strange attractors and other organized trajectories in phase space, nonzero Lyapunov exponents, period doubling – which are rarely seen in astronomical datasets (Scargle 1992). Only certain orbits in Solar System dynamics are convincingly chaotic. Nonlinear astronomical time series are more commonly dominated by stochastic behavior rather than deterministic chaos.

11.3   Time-domain analysis of evenly spaced data

11.3.1   Smoothing

In Chapter 6, we discussed a number of important data smoothing operations such as kernel density estimation which transform a series of discrete observations into a continuous function. These can readily be applied to time series data. We outline here a variety of other smoothing methods commonly used to reduce variance in evenly spaced time series.

Perhaps the simplest of these operations is the **central moving average (CMA)** with bandwidth of j time intervals. For even j,

$$\hat{X}_{i,CMA}(j) = \frac{1}{j+1} \sum_{k=-j/2}^{j/2} X_{i+k}. \tag{11.4}$$

Figure 11.3 below shows this and other smoothers applied to an astronomical dataset. This can be modified by averaging only past values rather than past-and-future values, using the robust median rather than the mean, and weighting each value by some predetermined factor such as the measurement error variance.

The **exponentially weighted moving average (EWMA)** is a common smoothing procedure for time series with short-term autocorrelaton

$$\hat{X}_{i,EWMA} = \alpha X_i + (1 - \alpha)X_{i-1} \tag{11.5}$$

where $0 \leq \alpha \leq 1$ and $X_{1,EWMA} = X_1$. The estimator is called exponential because the current value is a weighted average of all previous values with weights decreasing exponentially with time. The value of α can be set by the user, or estimated by least squares to minimize $\sum (\hat{X}_{i,EWMA} - X_i)^2$. The EWMA is closely related to the random walk stochastic process (Section 11.3.3 below), and ARIMA models.

11.3.2 Autocorrelation and cross-correlation

The **autocorrelation function (ACF)** for evenly spaced data shown in Equation (11.3), a fundamental measure of the serial correlation in a time series, is a second-order statistic measuring the ratio of the variance with lag-time k to the sample covariance. Under the null hypothesis that the time series has no correlated structure and the population ACF is zero for all lags (except for ACF(0) which is always unity), the distribution of the sample is asymptotically normal with mean $-1/n$ and variance $1/n$; that is, the distribution of the null case ACF is $ACF(k) = N(-1/n, 1/n)$.

Plots of ACFs (**correlograms**) are frequently displayed with confidence intervals based on this normal approximation to test this null hypothesis that no autocorrelation is present. Comparing a sample ACF to its confidence band gives a simple test for randomness in the correlation structure of a time series. Some ACF patterns are readily interpretable though others are not. Strong values at low k rapidly decreasing at higher k imply a low-order autoregressive process is present (Section 11.3.3). Both monotonic trends and long-memory stochastic processes like $1/f$ noise (Section 11.9) produce slow decreases in ACF values from $ACF(0) = 1$ to large k. Data with a single characteristic time-scale for autocorrelation will produce a wave-like pattern of positive and negative values in the ACF even if phase coherence is not maintained. For strictly periodic sinusoidal variations, these variations follow a cosine curve.

Simple quantities like the sample mean are valid estimates of the population mean for stationary processes, but its uncertainty is not the standard value when autocorrelation is present,

$$\widehat{Var}(\bar{X}_n) = \frac{\sigma^2}{n} \left[1 + 2 \sum_{k=1}^{n-1} (1 - k/n) ACF(k) \right]. \tag{11.6}$$

Qualitatively this can be understood as a decrease in the number of independent measurements due to the dependency of the process. This has a practical application: the comparison of means of two time series (either of different objects or the same object at different times) must take into account this increase in variance. Similar limitations apply to other statistical quantities estimated from autocorrelated time series. The mathematical theory of ergodicity provides the rules of convergence in such cases.

The **partial autocorrelation function (PACF)** at lag k gives the autocorrelation at value k removing the effects of correlations at shorter lags. The general PACF is somewhat complicated to express algebraically; see Wei (2006, Chapter 2) for a derivation. The value of the p-th coefficient $PACF(p)$ is found by successively fitting autoregressive models with order $1, 2, \ldots, p$ and setting the last coefficient of each model to the PACF parameter. For example, a stationary AR(2) model with normal noise will have

$$PACF(2) = \frac{ACF(2) - ACF(1)^2}{1 - ACF(1)^2}.$$ (11.7)

The significance of PACF values can be assessed with respect to the normal approximation for the null hypothesis that no partial correlation is present at the chosen lag.

The PACF thus does not model the autocorrelation of the original time series, but gives insight into the time-scales responsible for the autocorrelated behavior. This can be very useful to the astronomer who often seeks to understand the drivers of a temporal phenomenon. In other fields, it is used as a diagnostic to help in model selection, as in choosing the order of an autoregressive model in the well-established **Box–Jenkins approach** to time series modeling. Astronomers interested in understanding the underlying behavior of an autocorrelated time series are encouraged to view both the ACF and PACF, as often one will have a simpler structure than the other. Figure 11.5 illustrates the ACF and PACF for an astronomical dataset.

Diagnostic **lag k scatter plots** sometimes reveal important structure in an autocorrelated time series (Percival & Walden 1993, Chapter 1). Here one plots all x_t values against x_{t+k} values where k is chosen by the scientist. A random scatter points toward uncorrelated noise, while a linear relationship can point towards stochastic autoregressive behavior. A circular distribution suggests a periodic signal, and clusters of points would suggest a nonstationary time series with distinct classes of variability behavior.

We finally mention a simple measure of autocorrelation in an evenly spaced time series, the **Durbin–Watson statistic** (Greene 2003, Chapter 19). Commonly applied to residuals to assist in econometric regression, the Durbin–Watson statistic is

$$d_{DW} = \frac{\sum_{i=2}^{n}(X_i - X_{i-1})^2}{\sum_{i=1}^{n} X_i^2}$$ (11.8)

which is approximately equal to $2(1 - R)$, where R equals the sample autocorrelation between successive residuals. The statistic ranges from 0 to 4; values around 2 indicate the absence of autocorrelation while values below (above) 2 indicate positive (negative) autocorrelation. For large samples, d_{DW} is asymptotically normal with mean 2 and variance $4/n$. For small samples, critical values of d_{DW} are tabulated but are applicable only to test the alternative hypothesis of an AR(1) process.

11.3.3 Stochastic autoregressive models

The simplest stochastic autocorrelated model is

$$X_i = X_{i-1} + \epsilon_i$$ (11.9)

where the noise term is usually taken to be a homoscedastic normal with zero mean and variance σ^2, $\epsilon_i \sim N(0, \sigma^2)$. This is similar to the classical **random walk** or **drunkard's walk**. In this case, recursive back-substitution shows that

$$X_i = X_0 + \epsilon_1 + \epsilon_2 + \cdots + \epsilon_i.$$ (11.10)

While the mean value remains zero, the ACF starts at unity and decreases slowly with lag-time k,

$$ACF_{RW}(i,k) = \frac{Cov(X_i, X_{i+k})}{\sqrt{Var(X_i)Var(X_{i+k})}} \tag{11.11}$$

$$= \frac{i\sigma^2}{\sqrt{i\sigma^2(i+k)\sigma^2}} \tag{11.12}$$

$$= \frac{1}{\sqrt{1+k/i}}. \tag{11.13}$$

As the differences between adjacent points in a random walk time series is white noise, the ACF of the first-order differences should have no significant values. This fact can be used to test the presence of structure beyond a random walk.

The obvious generalization of the random walk model and the EWMA smoothing procedure is to permit dependencies on more than one past value in the time series with different weights. This is the **autoregressive (AR) model**

$$X_i = \alpha_1 X_{i-1} + \alpha_2 X_{i-2} + \cdots + \alpha_p X_{i-p} + \epsilon_i. \tag{11.14}$$

The AR(1) model has simple analytical properties; the expected mean is zero and the autocorrelation function for lag k is now time-invariant as

$$ACF_{AR(1)}(k) = \alpha_1^k \tag{11.15}$$

for $|\alpha_1| < 1$. Here the ACF decays rapidly if α_1 is near zero, remains high if $\alpha \simeq 1$, and the $PACF(k)$ values average around zero above the lag $k = 1$. Figure 11.7 shows the spectral pattern produced by a high-order AR model fit to an astronomical dataset. Autoregressive models for Poisson distributed event counts have been developed that may be useful for certain astronomical problems.

In the **moving average (MA) model**, the current value of the time series depends on past values of the noise term rather than past values of the variable itself,

$$X_i = \epsilon_i + \beta_1 \epsilon_{i-1} + \cdots + \beta_q \epsilon_{i-q} \tag{11.16}$$

where $\epsilon_i = N(0, \sigma_i^2)$. A wide range of time series can be understood as a combination of AR and MA autoregressive processes. The **ARMA(p,q) model** is

$$X_i = \alpha_1 X_{i-1} + \cdots + \alpha_p X_{i-p} + \epsilon_t + \beta_1 \epsilon_{i-i} + \cdots + \beta_q X_{i-q}. \tag{11.17}$$

Simple cases have analytic properties; for example, assuming homoscedastic normal noise, the ARMA(1,1) model has variance and autocovariance

$$Var_{ARMA(1,1)}(x) = \sigma^2 + \sigma^2 \frac{(\alpha+\beta)^2}{1-\alpha^2}$$

$$ACF_{ARMA(1,1)}(k) = \frac{\alpha^{k-1}(\alpha+\beta)(1+\alpha/\beta)}{1+\alpha\beta+\beta^2}. \tag{11.18}$$

The ACF and PACF can help identify what type of autoregressive behavior is present (Wei 2006, Chapter 6). For an ARMA(p,q) process, the ACF and PACF fall off exponentially at large lag. For a monotonic trend, the ACF remains high for all lags. The ACF shows oscillating behavior when the time series behavior is periodic.

Once we have a model family of autocorrelation behavior, the model coefficients can be estimated by least-squares or maximum likelihood regression methods. Often MLEs are more accurate than least-squares estimators. Error analysis on ARMA parameters can be obtained by some sophisticated mathematical calculations or by bootstrap simulation. Model selection in time series analysis is often based on the **Akaike information criterion** (Percival & Walden 1993, Section 9.9). For a model with p parameters θ_p,

$$AIC(\theta_p) = -2\ln(L(\theta_p)) + 2p \qquad (11.19)$$

where $L(\theta_p)$ is the likelihood for the model under consideration. For AR(p) models, the AIC reduces to the simple form

$$AIC(p) = N\ln(\hat{\sigma}_p^2) + 2p \qquad (11.20)$$

where $\hat{\sigma}_p^2$ is the maximum likelihood estimator of the variance of the ϵ noise term.

The ARMA model is stationary and thus cannot treat time series with systematic trends, either (quasi)periodic or secular. The **autoregressive integrated moving average (ARIMA) model**, proposed by statisticians G. Box and G. Jenkins in 1970, combines ARMA and random walk stochastic processes to treat such problems. The ARIMA(p,d,q) model has p parameters giving the autoregressive dependence of the current value on past measured values, d parameters giving the number of differencing operations needed to account for drifts in mean values ($d = 1$ corresponds to the random walk), and q parameters giving the autoregressive dependence on past noise values.

Another form of nonstationarity is when the variance, rather than the local mean level, of the time series changes during the observation. Commonly called **volatility** in econometric contexts, mathematically it is heteroscedasticity, although with a different origin than the heteroscedasticity arising from measurement errors. Use of the **autoregressive conditional heteroscedastic (ARCH) model** is a major development to treat such problems. Here the variance is assumed to be a stochastic autoregressive process depending on previous values, as in the ARCH(1) model with

$$Var_{ARCH(1)}(\epsilon_i) = \alpha_0 + \alpha_1 Var(\epsilon_{i-1}). \qquad (11.21)$$

The **generalized ARCH (GARCH) model** adds another type of nonlinear dependency on past noise levels. Note that these advanced models must be applied with care; for example, an ARIMA-type model might first be applied to treat long-term trends, followed by applying an ARCH-type model to the residuals for modeling volatility variations. Treatments of nonstationary autoregressive times series in astronomy (including change-points, CUSUM plots, ARCH and state-space models) are discussed by Koen & Lombard (1993). Research on incorporating heteroscedastic measurement errors into ARMA-type modeling has just begun (Tripodis & Buonaccorsi 2009).

It is important to realize that familiar statistical formulae may be very inaccurate for autoregressive processes. Consider, for example, the variance of a time series or the variance of its mean value (Section 3.4.4). For a stationary autocorrelated time series, these variances can be an order of magnitude larger than the values obtained from i.i.d. data.

Furthermore, the sample mean is not asymptotically normally distributed, and convergence to the asymptotic value is slow with a bias that the sample variance is systematically smaller than the true variance (Beran 1989, Chapter 1; Percival 1993).

11.3.4 Regression for deterministic models

The time series above are modeled by stochastic behavior which can be appropriate for astrophysical situations involving accretion, turbulence, rich stellar systems or other phenomena which are so physically complex that the temporal behavior cannot be deterministically predicted. But many astronomical time series – few-body Keplerian orbits, stellar rotation, supernova light curves – can be modeled as outcomes of functional relationships with time and/or covariate time series.

The statistical treatment of time series modeling can be formulated in a similar way to the regression problems discussed in Chapter 7,

$$y(t) = f(t) + g(\mathbf{x}(t)) + \epsilon(t) \qquad (11.22)$$

where $y(t)$ is the observed time series, f represents some functional dependence representing trend, g represents some functional (often assumed linear) dependence on a vector of covariate time series, and ϵ represents a noise term. Unless f and g are both constant, the time series is nonstationary. The noise is often assumed to follow the normal distribution $N(0, \sigma^2)$; if σ varies with time, the time series is called heteroscedastic.

However, there is a critical difference between modeling temporal and other phenomena: time is not a random variable like mass or luminosity, and observations must be taken along a unidirectional time line. The residuals between a time series and its model are themselves a time series, and deviations between the model and data are often correlated in time. In such situations, error analysis emerging from least-squares or maximum likelihood estimation assuming i.i.d. variables will underestimate the true uncertainties of the model parameters. This occurs because, when residuals are autocorrelated, there are fewer effectively independent measurements than the number of data points.

After a model is fitted, it is critical to examine the residual time series for outliers and autocorrelation. The distribution of the **residual ACF** depends on the model in a nontrivial fashion. Useful statistics are generated from the sum of low orders of the residual ACF, such as the test statistic (Chatfield 2004, Section 4.7)

$$Q = n \sum_{k=1}^{K} ACF_{resid}(k)^2. \qquad (11.23)$$

For an ARMA(p, q) model, Q is asymptotically distributed as χ^2 with $(K - p - q)$ degrees of freedom. The **Ljung–Box statistic** is similar to Q. Attention to correlated residuals can be extremely important in astronomy. van Leeuwen (2007) obtains a several-fold improvement in stellar parallactic distances from the Hipparcos astrometric satellite by modeling of the temporal behavior of the model residuals.

A strong advantage of formulating this regression problem from a likelihood approach is that the full capabilities of maximum likelihood estimation and Bayesian analysis become available (Chapters 3 and 7). In MLE, the likelihood can be maximized using the

EM algorithm and parameter uncertainties estimated using the Fisher information matrix. In Bayesian analysis, the parameter ranges and interdependencies can be evaluated using MCMC calculations, including marginalization over parameters of low scientific interest. The appropriate level of model parsimony and complexity of the model can be evaluated with model selection tools such as the Bayesian information criterion (BIC); time series analysts historically have used the closely related Akaike information criterion (AIC).

11.4 Time-domain analysis of unevenly spaced data

The vast majority of time series statistical methodology is designed for data measuring some variables at evenly spaced time intervals. Some limited methodology is available for datasets with intrinsically unevenly spaced observations (Parzen 1984). For example, theorems of convergence and asymptotic normality for moments, autocorrelations and regressions of a time series sampled with various observing sequences have been derived. But few general results emerge when the observing sequence depends on the data, or when either the time series or observing sequence is nonstationary (e.g. with trends or $1/f$-type noise). Econometricians have studied stochastic autoregressive models for irregular sampling, including cross-correlations between multiple time series that are evenly spaced but with different sampling rates (Engle & Russell 1998, Ghysels *et al.* 2007). Parametric modeling of unevenly spaced datasets in the time domain can be pursued within the flexible framework of likelihood-based state-space modeling discussed in Section 11.7.

We discuss here three statistical approaches for time-domain analysis of unevenly spaced data developed by astronomers over four decades. They are widely used, particularly to study the aperiodic variable emission from accreting massive black holes in quasars, BL Lac objects and other active galactic nuclei (e.g. Hufnagel & Bregman 1992).

11.4.1 Discrete correlation function

The discrete correlation function introduced by Edelson & Krolik (1988) is a procedure for computing the autocorrelation function that avoids interpolating the unevenly spaced dataset onto a regular grid. The method can treat both autocorrelation within one time series or cross-correlation between two unevenly spaced time series. Consider two datasets (x_i, t_{xi}) and (z_j, t_{zj}) with $i = 1, 2, \ldots, n$ and $j = 1, 2, \ldots, m$ points respectively. For autocorrelation within a single dataset, let $z = x$. Construct two matrices, the **unbinned discrete correlation function (UDCF)** and its associated time lags,

$$UDCF_{ij} = \frac{(x_i - \bar{x})(z_j - \bar{z})}{\sigma_x \sigma_z} \tag{11.24}$$

$$\Delta t_{ij} = t_j - t_i$$

where σ is the sample standard deviation of each dataset. The UDCF is then grouped into a univariate function of lag-time τ by collecting the $M(\tau)$ data pairs with lags falling within the interval $\tau - \Delta\tau/2 \leq \Delta t_{ij} < \tau + \Delta\tau/2$. The resulting **discrete correlation function** (DCF) and its variance are

$$DCF(\tau) = \frac{1}{M(\tau)} \sum_{k=1}^{M(\tau)} UDCF_{ij} \tag{11.25}$$

$$Var(\tau) = \frac{1}{(M(\tau) - 1)^2} \sum_{k=1}^{M(\tau)} [UDCF_{ij} - DCF(\tau)]^2.$$

Edelson & Krolik provide advice, though without mathematical demonstration, for a number of situations. When homoscedastic measurement errors ϵ contribute significantly to the scatter, the denominator might be replaced by $\sqrt{(\sigma_x^2 - \epsilon_x^2)(\sigma_z^2 - \epsilon_z^2)}$. If a time series has many correlated observations within a small time interval, then the number of points contributed to $M(\tau)$ might be reduced to represent the number of uncorrelated UDCF values in the interval. The DCF can be calculated for negative τ as well as positive τ to reveal possible asymmetries in the correlated behavior. The DCF is undefined for any lag bin that has no associated data pairs.

A similar strategy of grouping lags between data pairs into evenly spaced bins is used in the **slot autocorrelation** estimator for the ACF developed for problems in experimental physics (Benedict *et al.* 2000). For a dataset x_i measured at times t_i, the ACF in a chosen range of lag-times $k\Delta\tau$ is

$$\widehat{ACF}_{slot}(k\Delta\tau) = \frac{\sum_{i=1}^{n} \sum_{j>i} x_i x_j b_k(t_j - t_i)}{\sum_{i=1}^{n} \sum_{j=1}^{n} b_k(t_j - t_i)} \quad \text{where} \tag{11.26}$$

$$b_k(t_j - t_i) = \begin{cases} 1 & \text{for } |(t_j - t_i)/\Delta\tau - k| < 1/2 \\ 0 & \text{otherwise.} \end{cases}$$

Here k are the chosen slots for which the ACF is computed, $\Delta\tau$ is the chosen width of the lag bins, and the Kronecker delta function b selects the data pairs with the appropriate lag-times to enter into the k-th slot. A similar procedure has also been developed for econometrics (Andersson 2007).

There has been some discussion of the accuracy and reliability of peaks in cross- or auto-correlation functions of unevenly spaced data. The centroid may give a more reliable estimate of a correlation lag than the peak, and significance levels can be estimated through resampling and other Monte Carlo procedures (Peterson *et al.* 1998).

In astronomical usage, the choice of bin width $\Delta\tau$ is generally not guided by mathematical considerations. The question is analogous to the choice of bin width in density estimation, as cross-validation to minimize the mean integrated square error (MISE) can be used (Section 6.4.2). Alexander (1997) recommends an equal number of data points contributing to each lag bin, rather than equal lag width. If at least $\sim$11 points are present in each bin, then the bin's distribution is approximately binomial and Fisher's z transform can be applied,

$$z(\tau) = \frac{1}{2} \ln \left(\frac{1 + DCF(\tau)}{1 - DCF(\tau)} \right). \tag{11.27}$$

The $z(\tau)$ values are then normally distributed with known mean and variance, after correction for a small known bias. Based on a few simulations, Alexander argues that $z(\tau)$ has greater sensitivity and smaller variance than the untransformed DCF.

11.4.2   Structure function

The **structure function** (sometimes called the **Kolmogorov structure function**) is a measure of autocorrelation originating in the field of stochastic processes and used in engineering signal processing, geospatial analysis and other applications. The structure function was introduced to astronomy by Simonetti *et al.* (1985) with application to unevenly spaced datasets.

The q-th order structure function is

$$D^q(\tau) = \langle |x(t) - x(t+\tau)|^q \rangle \qquad (11.28)$$

where the angular brackets $\langle \rangle$ indicate an average over the time series and q is the order of the structure function (not necessarily integer). The $q = 2$ structure function is also called the **variogram**, an alternative to the autocorrelation function often used in spatial statistics (Section 12.4.1). The structure function is often more useful than the autocorrelation function when the time series is nonstationary when deterministic trends or stochastic ARIMA-type behavior are present.

When a dataset has a characteristic time-scale of variation τ_c, the structure function is small at shorter τ_c, rises rapidly to a high level around τ_c, and stays at this plateau for longer τ. The shape of the rise can roughly discriminate different noise or trend processes. If a structure function exhibits a power-law dependence on lag, $D^q(\tau) \propto \tau^\alpha$, the time series has a multi-fractal behavior. In this case, the Wiener–Khinchine theorem links the power-law dependence of the $q = 2$ structure function to the power-law dependence of the Fourier power spectrum. Study of the structure function, together with the closely related **singular measures**, can assist in differentiating between white noise, $1/f$ noise, fractional or standard Brownian motion, randomly located steps, smooth deterministic trends and intermittent processes such as turbulence (Davis *et al.* 1994).

The validity and accuracy of structure functions have recently been called into question. Nichols-Pagel *et al.* (2008) find that multi-taper spectral and wavelet estimators have smaller variances than structure function estimators when applied to simulations of turbulent physical systems. Emmanoulopoulos *et al.* (2010) argue from simulation studies that the structure function is unreliable for characterizing autocorrelated time series from unevenly spaced data. Structure function shapes can be very sensitive to the gap structure of the observations and can exhibit features that are not in the time series.

11.5   Spectral analysis of evenly spaced data

Whereas some temporal phenomena can be understood with parsimonious models in the time domain involving deterministic trends or stochastic autoregressive behavior, others are dominated by periodic behavior that is most effectively modeled in the frequency domain. Harmonic analysis studies the variance of a time series as a function of frequency rather than time. Frequency-domain analysis is familiar to the astronomer because many astronomically variable phenomena involve orbits, rotations or pulsations that are exactly or nearly periodic. Time series analysis involving aperiodic phenomena, such as autocorrelated variations commonly seen in human affairs, are best treated in the time domain.

11.5.1 Fourier power spectrum

Classical spectral analysis is based on the Fourier transform and its associated power spectrum. **Fourier analysis** is designed for stationary time series with infinite duration and evenly spaced observations of sinusoidal periodic signals superposed on Gaussian (white) noise. Many astronomical time series deviate from these assumptions with limited-duration observations, gaps or unevenly spaced observations, nonsinusoidal periodic signals, phase incoherence, Poissonian or $1/f$ noise, or nonstationarity. Even when the assumptions of Fourier analysis are approximately met, many judgments must be made. We will briefly review some of the extensive methods in spectral analysis developed to treat these problems. The literature is vast: elementary presentations can be found in Warner (1998), Chatfield (2004) and Cowpertwait & Metcalfe (2009) while more advanced treatments include Priestley (1983) and Percival & Walden (1993).

A time series dominated by sinusoidal periodic variations can be modeled as

$$X_t = \sum_{k=1}^{n} A_i \cos(\omega_k t + \phi_k) + \epsilon_k \tag{11.29}$$

where the parameters A and ϕ are the amplitudes and phases ($0 \leq \phi \leq 2\pi$) at frequencies ω. This time series model is nonstationary except under special conditions. This function is mathematically defined for $n \rightarrow \infty$, but we consider the realistic situation where the data are discrete and finite.

While one could proceed in obtaining estimates of the parameters A_k using regression techniques, it is often more useful to consider the **power spectral density function**, or **power spectrum**, defined to be the Fourier transform of the autocovariance function (Section 11.3.2),

$$\begin{aligned} f(\omega) &= \frac{\sigma_X^2}{\pi} \sum_{k=-\infty}^{\infty} ACF e^{-i\omega k} \\ &= \frac{\sigma_X^2}{\pi} \left[1 + 2 \sum_{k=1}^{\infty} ACF(k) \cos(\omega k) \right] \end{aligned} \tag{11.30}$$

where σ_X^2 is the sample variance. Here $f(\omega)/\sigma_X^2$ can be interpreted as the fraction of the sample variance that is attributable to periodic variations in a narrow frequency interval around ω. The power spectrum can be written in a number of ways – involving trigonometric functions, exponentials and autocovariance functions – and with different normalizations. Note that the power spectrum and the autocorrelation function can be derived from one another and contain the same information.

The power spectrum can be characterized in closed form for some simple temporal models. For a white-noise process, the autocorrelation function is zero and the power spectrum $f(\omega) = \sigma_\epsilon^2/\pi$ is a constant for all frequencies. For a deterministic sinusoidal signal with variable phase, $X_t = \cos(\omega_0 + \phi)$ with ϕ a uniform random variable, the autocorrelation function is $ACF(k) = \cos(\omega_0 k)/(2\sigma_X^2)$ which is periodic and the power spectrum is infinite at ω_0 and zero at other frequencies. For an autoregressive AR(1) model, $x_t = \alpha X_{t-1} + \epsilon_t$, where $ACF(k) = \alpha^{|k|}$, the power spectrum is

$$f(\omega) = \frac{\sigma_\epsilon^2}{\pi[1 - 2\alpha\cos(\omega) + \alpha^2]}. \tag{11.31}$$

When $\alpha > 0$ the spectral density is strong at low frequencies and monotonically declines towards high frequencies. High-order AR or ARMA processes can give more complicated nonmonotonic spectral densities as illustrated in Figure 11.7.

For nontrivial scientific problems, the power spectrum cannot be derived analytically and must be estimated from the time series measurements x_t at evenly spaced increments Δt. This process is often called **spectral analysis**. A common model formulation is the finite Fourier series

$$X_t = a_0 + \sum_{p=1}^{n/2-1} \left[a_p \cos\left(\frac{2\pi pt}{n}\right) + b_p \sin\left(\frac{2\pi pt}{n}\right) \right] + a_{n/2} \cos(\pi t). \tag{11.32}$$

The least-squares parameter expected values are

$$a_0 = \bar{X} = \frac{1}{n} \sum_{t-1}^{n} X_i$$

$$a_p = 2\left[\sum X_t \cos(2\pi pt/n) \right]/n$$

$$b_p = 2\left[\sum X_t \sin(2\pi pt/n) \right]/n$$

$$a_{n/2} = \sum (-1)^t X_t/n \tag{11.33}$$

for the p harmonics $p = 1, 2, \ldots, n/2 - 1$. Spectra are calculated at increments of $2\pi/n, 4\pi/n, \ldots, \pi$ between the low fundamental frequency $2\pi/(n\Delta t)$ and the high Nyquist frequency $\pi/\Delta t$. The spectrum is often plotted as a histogram of the quantity $I(\omega)$

$$I(\omega_p)_S = n\sqrt{a_p^2 + b_p^2}/4\pi \tag{11.34}$$

against frequency ω. This is the **periodogram** developed by A. Schuster in 1898 and very widely used since. An example of a Schuster periodogram for a variable astronomical source is shown in Figure 11.4.

Some astronomical objects, particularly pulsating stars and multi-planetary systems, exhibit multiple periods. In such cases, it is common to seek the dominant periodicity from a periodogram, remove that sinusoid from the time series, seek the next periodicity from the periodogram of the residuals, and iterate until no significant spectral peaks are present. This method is known in signal processing as **iterative pre-whitening**. In astronomy, it is a widely used procedure, often following the **CLEAN algorithm** originally designed for deconvolution in interferometric imaging (Roberts *et al.* 1987).

11.5.2 Improving the periodogram

Schuster's periodogram emerges directly from Fourier analysis, but has statistical limitations. It exhibits a high level of noise, even when the time series has high signal-to-noise, that does not decrease as the observation length increases. In statistical language, it is a biased estimator of the power spectrum $f(\omega)$ in the sense that its variance does not approach zero as n approaches infinity. As mentioned above, the analysis also makes various assumptions (such as infinitely long and evenly spaced data with stationary sinusoidal signals and Gaussian noise) which may not hold for a real astronomical dataset. For example, **spectral leakage** into nearby frequencies occurs if the time series is relatively short with respect to the periods under study and is not an integer multiple of the cycle length. These problems can be considerably alleviated with spectral analysis techniques including detrending, smoothing and tapering.

For nonstationary signals which vary in amplitude in some systematic way, we can **detrend** the dataset with regression or nonparametric smoothers to create a quasi-stationary time series for further harmonic analysis. This reduces strong spectral amplitudes at low frequencies and improves noise throughout the spectrum. High-pass filtering, low-pass filtering and iterative pre-whitening can also be useful to reduce undesired signals or noise in a spectrum.

Smoothing introduces bias but reduces variance in the spectral density estimator. Smoothing can be performed in the frequency domain or in the time domain applied to the original data or the autocorrelation function. A simple smoothing function proposed by Daniell is a moving average which convolves the time series with equal weights across a window of width m. It produces a spectral window with significant side-lobes. Bartlett's window convolves the data with a triangular rather than rectangular kernel and produces smaller side-lobes. Other common choices, such as the Tukey–Hanning, Parzen or Papoulis windows, smoothly decrease the weight with distance from the central frequency. As with all data smoothing operations (Section 6.4.1), a wider window (larger m) reduces variance but increases bias, particularly blurring signals at closely spaced frequencies. Strategies for selecting a window shape and width m are discussed by Percival & Walden (1993, Chapter 6); one seeks to balance the reduction of smoothing window leakage, avoid the introduction of spectral ripples, and maintain sufficient spectral resolution and dynamic range. Figure 11.4 illustrates the effects of two levels of smoothing on an astronomical periodogram.

Tapering, a convolution which reduces the spectral side-lobes and leakage arising from ringing off of the edges of the dataset, treats the limitation that the data do not have infinite duration. The effect of tapering is to reduce the sample size towards the edge of the dataset. Opposite to smoothing, tapering decreases bias due to spectral leakage, but introduces variance as the effective data size is reduced. Here, the time series x_i is replaced by the product $h_i x_i$ where h is a taper function. A large literature exists on the choice of the taper bandwidth and functional form. One common choice is a cosine taper where h_{cos} is applied to the first and last $p\%$ of the dataset,

$$h_{j,cos} = \frac{1}{2}\left[1 - \cos\left(\frac{2\pi j}{|pn| + 1}\right)\right]. \tag{11.35}$$

The 100% cosine taper which affects all data-points is called the Hanning taper. The average of different orthogonal tapers, or **multi-tapering**, is an important procedure permitting, for example, quantitative tradeoff between bias and variance. The discrete prolate spheroidal sequences or Slepian sequences are a favored taper sequence for this purpose.

11.6 Spectral analysis of unevenly spaced data

Most of the methodological work to treat spectral analysis when the observation times are intrinsically unevenly spaced has emerged from astronomy, as these situations do not often occur in common engineering or econometric applications.

Brillinger (1984) gives a broad, mathematical discussion of the spectral analysis of unevenly sampled time series showing that the power spectrum of the observed process will depend both on the underlying process of interest and on its convolution with the observational process. The estimation of the spectrum of the underlying process is a tractable inverse problem if both the observed and observational patterns are stationary random processes. However, the astronomical observational constraints often depend on various extraneous cycles (solar, lunar, satellite orbits), autocorrelated processes (cloud cover) and the vagaries of telescope allocation committees (Section 11.1). The astronomical observing times thus often cannot be considered a stationary random process.

11.6.1 Lomb–Scargle periodogram

Classical Fourier analysis requires time series measurements acquired in evenly spaced intervals, and such datasets are common in many fields of engineering and social science. The uneven spacings common in astronomical time series remove this mathematical underpinning of classical Fourier calculations. Deeming (1975) gives an important and influential discussion of these effects.

A major step was provided by the periodogram developed by N. Lomb (1976) an J. Scargle (1982) which generalizes the Schuster periodogram for unevenly spaced data. The **Lomb–Scargle periodogram (LSP)** can be formulated either as a modified Fourier analysis or as a least-squares regression of the dataset to sine waves with a range of frequencies. Its common expression is

$$P_{LS}(v) = \frac{1}{2\sigma^2}\left[\frac{[\sum_{i=1}^{n} X_i \cos(2\pi v t_i)]^2}{\sum_{i=1}^{n} \cos^2 2\pi v(t_i - \tau(v))} + \frac{[\sum_{i=1}^{n} X_i \sin(2\pi v t_i)]^2}{\sum_{i=1}^{n} \sin^2 2\pi v(t_i - \tau(v))}\right] \qquad (11.36)$$

where σ^2 is the sample variance of the x_i and the parameter τ is defined by

$$\tan(4\pi v \tau) = \frac{\sum_{i=1}^{n} \sin(4\pi v t_i)}{\sum_{i=1}^{n} \cos(4\pi v t_i)}. \qquad (11.37)$$

Press & Rybicki (1989) give an efficient computational algorithm for the LSP based on the fast Fourier transform.

Bretthorst (2003) places the LSP on a broader statistical foundation, showing that it is a solution to Bayes' theorem under reasonable conditions. He constructs a model of sinusoidal variations with time-variable amplitudes at different frequencies superposed by Gaussian noise. Uniform priors are assigned to the range of frequencies and amplitudes, and a Jefferys' prior to the variance. Solving Bayes' theorem analytically gives a generalized periodogram that reduces to the Schuster periodogram for uniformly sample data and to the LSP for arbitrary data spacings. The formalism can also treat nonsinusoidal shapes and multiple periodic signals. All of these are sufficient statistics, and marginal posterior probabilities for each parameter can be computed using Markov chain Monte Carlo procedures.

A number of recent studies have addressed methodological issues concerning the LSP. Reegen (2007) derives the LSP from a modified discrete Fourier transform with rotation in Fourier space, emphasizing methods for estimating reliable false alarm probabilities. Zechmeister & Kürster (2009) propose a generalized LSP based on weighted regression in the time domain with a floating mean, emphasizing application to nonsinusoidal behavior typical of exoplanets with eccentric orbits. Vio *et al.* (2010) rederive the LSP with a matrix algebraic formalism, allowing tests for periodic signals in the presence of nonstationary noise. Baluev (2008) and Sturrock & Scargle (2009) give further strategies for reliable assessment of peaks in the LSP, the latter based on Bayesian inference. Townsend (2010) presents an algorithm for efficient calculation of the LSP on graphics processing units, and another computational approach is proposed by Palmer (2009).

While the LSP is popular in the astronomical community for spectral analysis of unevenly spaced datasets, other approaches have been developed for applications in engineering and physics. Marquardt & Acuff (1984) present a **direct quadratic spectrum estimator (DQSE)**

$$Z_{DQSE}(\omega) = \frac{1}{T} \sum_{i=1}^{m} \sum_{j>i} D(t_j - t_i) F(t_i, t_j) \cos(2\pi\omega(t_j - t_i)) x(t_i) x(t_j) \qquad (11.38)$$

where T is the observation duration, D is the Tukey–Hanning lag window, and F is a data spacing factor: $F = 1$ for equally spaced data and $F = 1/\lambda^2$ for random Poisson sampling with intensity λ. For arbitrary data spacings,

$$F(t_i, t_j) = \frac{1}{2}(t_{i+1} - t_{i-1})(t_{j+1} - t_{j-1}). \qquad (11.39)$$

A limit on F is introduced when the data spacing is very large.

The laboratory study of fluid turbulence using a laser Doppler anemometer gives velocity measurements at erratic times when tracer particles flow through the measurement device. This gives an unevenly sampled time series of a strongly autocorrelated, $1/f$-type (long-memory) process. This problem has led to the independent development of estimators of autocorrelation functions and power spectra (Benedict *et al.* 2000). The power spectral density is then derived from the **slot autocorrelation function** (Section 11.3.2) with smoothing or windowing to reduce the variance. The calculation can be computationally expensive.

Benedict *et al.* (2000) compare the performance of several spectral estimators for simulated datasets with $1/f$-type noise. They find that power spectra based on the slot correlation estimator with certain sophisticated smoothing options have excellent performance. The DQSE has low bias but shows high variance while the Lomb–Scargle estimator systematically overestimates the power at high frequencies. Rivoira & Fleury (2004) present an iterative procedure for estimating the number of lags and the window function that smooths the ACF in a fashion that balances bias and variance to minimize the mean integrated square error (MISE, Section 6.4.1). Their estimators give a lower MISE than the smoothed slot estimator and their computation is more efficient.

11.6.2 Non-Fourier periodograms

Astronomers have developed and extensively used a variety of methods for periodicity searches in unevenly spaced datasets and nonsinusoidal signals where Fourier methods have difficulties. The most common strategy involves **folding the data modulo a trial period**, computing a statistic on the folded time series (now a function of phase rather than time), and plotting the statistic for all independent frequencies. These methods have many advantages. They measure the strength of signals that are strictly periodic, but not necessarily sinusoidal in shape. This is particularly appropriate for variations due to orbital or rotational eclipses where the periodic signal often has a characteristic flat-bottom **U**-shape that can occupy an arbitrary phase fraction of each period. They are also relatively insensitive to the duration and uneven spacing of the dataset, and some methods readily permit heteroscedastic weighting from measurement errors. As with any harmonic analysis, however, adjudicating the significance of a peak in the periodogram can be difficult, particularly as the observation spacings interact with a periodic or autocorrelated signal to producing spurious aliased structure in the periodograms.

The simplest statistic is the **minimum string length** of the dataset folded modulo a trial period (Dworetsky 1983). This is the sum of the length of lines connecting values of the time series as the phase runs from zero to 2π. When an incorrect period is chosen, the signal is scattered among many phases and the line length is large, while for the correct period, the signal is collected into a small phase range and the line length is reduced. As distances rather than squared distances are considered, this is an L_1 statistic that is robust against non-Gaussianity and outliers (Chapter 5). This statistic is similar to two commonly used procedures in statistics: least absolute deviation (LAD) regression (Section 7.3.4) and the first element of the autocorrelation function of residuals from a regression function (Box & Pierce 1970). But in these cases, the line lengths are computed vertically for each individual point rather than diagonally between adjacent points.

Three related methods are based on least-squares (L_2) statistics of the time series data folded modulo trial periods, but then grouped into bins. They are very widely used in variable-star research. The **Lafler & Kinman (1965) statistic** for each trial period is

$$\theta_{LK}^2 = \frac{\sum_{j=1}^m (\bar{\phi}_j - \bar{\phi}_{j+1})^2}{\sum_{j=1}^m (\phi_j - \bar{\phi})^2} \tag{11.40}$$

where $\bar{\phi}_j$ the average value within the j-th phase bin, $\bar{\phi}$ is the global mean for the full sample, and m is the number of phase bins. In econometrics, this statistic (with slightly different normalization) is known as the Durbin–Watson statistic commonly used for analysis of correlation in time series. J. von Neumann (1941) calculated its distribution for normally distributed variables.

The **Stellingwerf (1978) statistic** is the ratio of the sum of inter-bin variances to the sample variance,

$$\theta_S^2 = \frac{\sum_{j=1}^{m} \sum_{j=1}^{n_j} (\phi_{ij} - \bar{\phi}_j)^2}{(n-m) \sum_{i=1}^{n} (\phi_{ij} - \bar{\bar{\phi}})^2} \tag{11.41}$$

where n_j is the number of data points in the j-th phase bin. The mean values can be weighted by heteroscedastic measurement errors. This is known as the **phase dispersion minimization** periodicity search method.

The **ANOVA (analysis of variance) statistic** presented by Schwarzenberg-Czerny (1989) is the ratio of the sum of inter-bin variances to the sum of intra-bin variances,

$$\theta_{AoV}^2 = \frac{(n-m) \sum_{j=1}^{m} n_j (\bar{\phi}_j - \bar{\phi}_{ij})^2}{(m-1) \sum_{j=1}^{m} \sum_{j=1}^{n_j} (\phi_{ij} - \bar{\phi}_j)^2}. \tag{11.42}$$

This ANOVA statistic (assuming normal homoscedastic errors and sufficiently large n_j values) follows an F distribution with $(m-1)$ and $(n-m)$ degrees of freedom,

$$E[\theta_{AoV}^2] = \frac{n-m}{n-r-2}$$
$$Var[\theta_{AoV}^2] = \frac{2(n-m)^2(n-3)}{(m-1)(n-m-2)^2(n-m-4)}. \tag{11.43}$$

The ANOVA statistic may be preferable over the Stellingwerf statistic because the numerator and denominator are independent of each other. Note that the ANOVA statistic takes no cognizance of the shape of the variation in the binned phase time series, while the Lafler–Kinman (or Durbin–Watson) statistic is sensitive to smooth variations across the binned phase time series. They are not equivalent, and both can be used to give periodograms sensitive to different characteristics of the putative periodicity.

As with any histogram, there is no mathematical guidance regarding the choice of phase origin or bin boundaries; evenly spaced bins are commonly adopted, but this is not required (Section 6.3). The emerging periodogram may be sensitive to these choices. This limitation is addressed by the **Gregory-Loredo Bayesian periodogram** designed for nonsinusoidal periodic variations (Gregory & Loredo 1992). After folding the data modulo trial periods, an ensemble of grouping options is considered with different bin zero-points and widths. A likelihood is constructed including arbitrary levels for each bin level. The Bayesian odds ratio for periodic vs. constant emission is calculated in a high-dimensional parameter space, and the results are marginalized over uninteresting binning options to maximize the significance of the periodicity.

Schwarzerberg-Czerny (1999) has compared the performances of these period-finding procedures for a variety of simulated datasets. No single method outperforms others for all situations. It is important, if possible, to match the method to the complexity of the model. For periodic behavior with fine-scale structure, methods with narrow phase bins (or the unbinned string length statistic) are most sensitive while for broad structures that fill the phased time series, Lomb–Scargle or ANOVA procedures are better. Gregory & Loredo's method that accounts for the model structure is an attractive approach.

The search for exoplanets is an important new application of periodicity searching and has triggered a flood of methodological studies. Here, a planet periodically passes in front of a star and produces a partial eclipse seen as a slight (typically 0.1–0.01%) diminution in the stellar brightness. The variation has a characteristic U-shape with a flat bottom. Data acquisition may be either evenly or unevenly spaced, and the data weightings may be either homo- or heteroscedastic. Proposed techniques include a least-squares **box-fitting algorithm** (Kovács *et al.* 2002) and a simplified version of the Gregory–Loredo approach using a χ^2-type statistic (Aigrain & Irwin 2004).

The problem is particularly challenging because the signal is very weak, and uninteresting sources of variability (instrumental effects, night-time conditions and stellar magnetic activity) may dominate. These effects can have $1/f$-type characteristics which can be very difficult to treat (Section 11.9). Approaches for treating these problems include principal components analysis (Tamuz *et al.* 2005), a gauge filter from image processing (Guis & Barge 2005), detrending and low-pass filtering (Renner *et al.* 2008), and application of covariance-corrected significance levels (Pont *et al.* 2006).

Statistician P. Hall (2008) has discussed some mathematical issues of periodicity searches using epoch-folded time series. He suggests, for example, that each folded time series be smoothed with the Nadaraya–Watson kernel or another local regression technique (Section 6.6.1). Another approach involves regression with sinusoidal functions; this can treat multiple and nonstationary periodicities.

11.6.3 Statistical significance of periodogram peaks

The significance of a single peak in a Schuster periodogram from classical Fourier analysis can be formally calculated using a χ^2 distribution assuming homoscedastic normal (white) noise with known variance, no periodic signal, and an infinite, evenly spaced dataset. However, the calculation is quite different if the variance is not known in advance but must be estimated from the data; this is the case seen in astronomical data. It is also often difficult to establish the degrees of freedom of the problem when the periodogram has been modified by smoothing and/or tapering. The significance level is also incorrect if the time series has any trend or autocorrelation, or if multiple periodicities are present. Even if conditions are appropriate for this significance test, the probability applies to a single frequency and not to the entire spectrum, so that the significance must be corrected by a large factor based on the false alarm rate of testing many potential periods. The difficulties of evaluating the significance of spectral peaks in Fourier analysis are discussed by Percival & Walden (1993).

For the Lomb–Scargle periodogram with unevenly spaced data, Horne & Baliunas (1986) argue that the significance of a single peak can be expressed as the sum of gamma functions which simplify to an exponential distribution with density

$$f(z) = \frac{1}{\sigma^2} e^{-z/\sigma^2}, \qquad (11.44)$$

where σ is the total variance of the data including any signals. When searching for the existence of an unknown periodicity over N_p independent periods, the probability that some spectral peak is higher than z_0 is

$$P(Z > z_0) = 1 - (1 - e^{-z_0})^{N_p}. \qquad (11.45)$$

Although this result is widely used in astronomy to evaluate the significance of features in the LSP, it has a number of difficulties. First, like the exponential distribution for the standard Fourier periodogram, it assumes that the time series consists only of a single periodic signal in homoscedastic white (Gaussian) noise, $\epsilon \sim N(0, \sigma_0^2)$, and that σ_0 is known in advance. But if the variance is estimated from the data, which is almost always the case, the F distribution must be used rather than the exponential distribution, and the resulting probabilities are substantially different (Koen 1990). The exponential distribution of the LSP also requires that $\sum_{j=1}^n \cos(wt_j) \sin(wt_j) = 0$, a condition that depends on the spacing of the observation times. If autocorrelated and quasi-periodic behavior, either noise or astrophysical in origin, is present, there is no analytic estimator for the statistical significance of a peak in the Lomb–Scargle periodogram (or any other periodogram for unevenly spaced data, Section 11.6 below). The problem of significance estimation in Fourier spectra is exacerbated here because the signals can interact nonlinearly with gaps in the unevenly spaced datasets, giving spurious peaks and strong aliasing.

As analytical approaches are ineffective under these circumstances, numerical methods are essential for investigating the significance of a periodogram peak. First, periodograms of permutations of the data among the fixed observation times can be examined to establish the probability distribution of the periodogram in the presence of uncorrelated noise. Second, periodic test signals can be superposed on permuted datasets to study any peak frequency displacements, broadening and alias structures. Third, autocorrelated but aperiodic noise can be superposed on the permuted datasets to examine whether spurious spectral peaks readily appear. Reegen (2007) and others recommend other procedures for estimating LSP significance levels (Section 11.6.1).

11.6.4 Spectral analysis of event data

The methods described so far consider time series of the form (x_i, t_i) where a real-valued intensity is measured at unevenly spaced times. A related problem arises in high-energy astrophysics − X-ray, gamma-ray and neutrino astronomy − where individual events are detected. If the source is emitting at a constant flux, then the arrival times follow a Poisson distribution with an exponential distribution of intervals between adjacent events. A Poissonian noise component may also be present. X-ray and gamma-ray sources are often variable and, particularly if they arise from isolated or accreting rotating neutron stars, will exhibit periodicities. Mathematically, these would be classified as **nonstationary Poisson processes** with periodic components. Some methods for spectral analysis have been developed to treat this problem.

The most commonly used periodogram for event data is based on the Rayleigh probability distribution function (Leahy *et al.* 1983, Protheroe 1987),

$$f(x, \sigma) = \frac{x}{\sigma^2} e^{-x^2/(2\sigma^2)} \qquad (11.46)$$

which can be generated from the square root of the quadratic sum of two normally distributed random variables. This statistic can be used as a test for uniformity in angle around a circle.

It is a special case ($m = 1$) of the $\boldsymbol{Z_m^2}$ **statistic** which can be written

$$Z_m^2 = \frac{2}{n} \sum_{k=1}^{m} \left[\left(\sum_{i=1}^{n} \sin(2\pi k\phi_i) \right)^2 + \left(\sum_{i=1}^{n} \cos(2\pi k\phi_i) \right)^2 \right] \qquad (11.47)$$

where ϕ_i are the data points in the interval 0–2π after folding with a trial period. The Rayleigh test is sensitive to a broad, sinusoidal signal while higher orders of Z_m^2 are sensitive to periodic structures with smaller duty cycles. Both are preferable to the traditional use of Pearson's χ^2 statistic that requires grouping of the folded events into bins of arbitrary origin and width. At high count rates, the periodograms based on the Rayleigh statistic approach those based on the Fourier series. The use of these statistics for periodicity searches with event data in gamma-ray astronomy is reviewed by Buccheri (1992).

The $\boldsymbol{H}$ **test** based on the Z_m^2 statistic was developed by de Jager *et al.* (1989) to find the optimal order m that minimizes the mean integrated square error (MISE; see Section 6.4.1) for the given dataset. This criterion gives

$$H = \max(Z_m^2 - 4m - 4) \qquad (11.48)$$

where the maximization is calculated over choices of m. The H test has excellent performance for both narrow and broad pulses in simulated data. It is commonly used today by researchers searching for high-energy pulsars using ground-based (e.g. MAGIC, HESS, HEGRA) and space-based (e.g. AGILE, Fermi) gamma-ray telescopes.

A periodogram based on the **Kuiper statistic** has occasionally been used instead of the Rayleigh statistic on event data folded with trial periods (Paltani 2004). This statistic is similar to the Kolmogorov–Smirnov statistic, measuring the supremum distance between the empirical distribution function of folded event arrival times and a model based on the null hypothesis of no variation. The Kuiper test has better performance on simulated data with narrow pulses superposed on noise, while the Rayleigh statistic shows higher sensitivity on simulated data with broad pulses.

Swanepoel *et al.* (1996) propose a periodogram for event data based on a simple non-parametric statistic applied to epoch folded event data. The statistic is the minimum phase interval that contains a specified number of events. The statistic is unbiased and the bootstrap is used to estimate confidence levels.

11.6.5 Computational issues

The range of frequencies for calculating a spectrum is not obvious for unevenly spaced or event data. The low-frequency cutoff can be scaled inversely to the observation length so that a few periods will be present. The high-frequency cutoff for evenly spaced data is the **Nyquist frequency** $\nu_N = 1/(2\Delta t)$ where Δt is some time interval between observations. Periodic structure in the time series at higher frequencies will be aliased into the calculated spectrum. However, the Nyquist frequency is defined for unevenly spaced data, and the estimated spectrum has reduced sensitivity to periodicities at some range of high frequencies.

When data spacings are very uneven and the time series duration is long, the folded phase diagrams can differ considerably for even very small changes of trial period. It thus may be necessary to compute the power spectrum, using the Lomb–Scargle or other method, at many frequencies above the Nyquist frequency limit appropriate for evenly spaced data (Pelt 2009).

It is well-known that a completely random sequence of observing times removes spurious aliased spectral peaks when the period of the variation is near the observation interval Δt (Shapiro & Silverman 1960). But even small deviations from random sequences can introduce aliases. A particularly insidious problem arises when the underlying time series with strong autocorrelation, but no strict periodicity, is observed at nonrandom unevenly spaced observation times. Then large, spurious peaks in the spectrum can appear, giving the illusion of strong periodic behavior. The possible presence of spurious peaks can be tested with permutations and simulations of the dataset, but a definitive conclusion requires new observations to test predictions. The problem of reliable inference of periodicity from unevenly spaced time series has beset the field of variable and binary-star research for decades (e.g. Morbey & Griffin 1987). NASA's Kepler mission is now providing denser and more evenly spaced light curves for $\sim$150,000 stars (Appendix C.15).

11.7 State-space modeling and the Kalman filter

Regression and stochastic parametric models (Sections 11.3.3 and 11.3.4) are often solved by likelihood methods. A general likelihood-based formulation of time series models called **state-space modeling** was developed in the mid-twentieth century for engineering applications where signals, often from complex dynamic mechanical systems, must be monitored for changes and adjustment of the system. The **Kalman filter** is an efficient algorithm for updating the likelihood as new data arise; here the term "filter" has meaning closer to "smoothing" and "forecasting" than the alterations of time series like a "high-pass filter".

We model the development of a multivariate time series $\mathbf{X}_t$ in two equations, somewhat analogously to the errors-in-variable models discussed in Section 7.5:

$$\mathbf{X}_t = \mathbf{h}_t \mathbf{S_t}(\theta) + \epsilon_t$$
$$\mathbf{S_t}(\theta) = \mathbf{G}_t \mathbf{S}_{t-1}(\theta) + \eta_t. \qquad (11.49)$$

In the measurement equation, $\mathbf{X}$ is a vector of observed values of a collection of variables at time t. $\mathbf{S}$ is the **state vector** that depends on the parameters θ, and $\mathbf{h}$ is the matrix that maps the true state space into the observed space. In the state equation, $\mathbf{G}$ is the **transition matrix** that defines how the system changes in time. The ϵ and η terms are uncorrelated and stationary normal errors. The system has a short-term memory.

A very simple state-space model would be a univariate system with both h and G as constant. This reproduces the random walk model of Section 11.3.3. A more complex nonstationary ARIMA-type linear growth model can be written as

$$X_t = Y_t + \epsilon_t$$
$$Y_t = Y_{t-1} + \beta_{t-1} + \eta_t$$
$$\beta_t = \beta_{t-1} + \zeta_t \qquad (11.50)$$

where β represents the random walk component. A linear regression model relating X to some covariate Z commonly written as $X_t = \alpha_t + \beta_t Z_t + \epsilon_t$ can be formulated as the state-space model (11.49) where θ is the two-element vector $\theta^T = [\alpha_t, \beta_t]$, $\mathbf{h}_t^T = [1, z_t]$, $\mathbf{G}_{11} = \mathbf{G}_{22} = 1$, and η with zero variance.

The advantage of a state-space formulation for time series modeling is its capability of treating complex models with multiple input time series, linear and nonlinear deterministic dependencies, periodicities, autoregressive stochastic processes, known and unknown noise terms with heteroscedasticity, and so forth. The goals of the calculation might be interpolation and extrapolation of observed time series, estimating confidence bands around time series, detecting change-points in system behavior, or estimating the θ parameters to understand the underlying dynamical process.

The Kalman filter is a recursive procedure for updating the state vector $\mathbf{S}_{t-1}$ by incorporating the new data $\mathbf{X}_t$ under a least-squares or maximum likelihood criterion. It uses the current estimate of a fixed transition matrix $\mathbf{G}$ and the variance–covariance matrix of $\mathbf{S}$ to construct a gain matrix for correcting $\mathbf{S}_{t-1}$ to $\mathbf{S}_t$. In a least-squares context, the update equations were easily computable even before computers as they are linear and do not depend on past states of the system. Today, MLE and Bayesian approaches are commonly applied using numerical techniques such as the EM algorithm and Markov chain Monte Carlo. The Kalman filter is a simple example of **dynamic Bayesian networks**, and Equations (11.49) can be reformulated in terms of **hidden Markov models**. State-space models can be recovered from sparse or noisy data using the Kalman filter with techniques from **compressive sensing**.

Methodology for state-space modeling has been developed for 50 years and has great flexibility in treating sophisticated problems (Durbin & Koopman 2001). The methods can either identify or incorporate changes in behavior (nonstationarity). Hierarchical models can link variables observed at the telescope with intrinsic state variables determined by astrophysical theory. State-space methods can treat models which are nonlinear and noise which is non-Gaussian (including Poissonian noise). Bayesian methods can give parameter posterior distribution functions using importance sampling techniques. Model goodness-of-fit can be evaluated using the Akaike and Bayesian information criteria.

The volume by Kitagawa & Gersch (1996) illustrates the power of state-space methods for problems encountered in geosciences and astronomy such as seismological signals of earthquakes, annual temperature variations, Poissonian measurements of rapid X-ray variations in black-hole binary systems, and quasi-periodic variations in sunspot number. They adopt a quasi-Bayesian approach to state-space analysis involving **smoothing priors**. Several of their analysis procedures may be useful in astronomy. In particular, they formulate a likelihood for a state-space model of a continuous process observed at unevenly spaced intervals (Kitagawa & Gersch 1996, Sections 5.3 and 7.3). The likelihood can be used either for estimation of model parameters, or for smoothing the dataset into a continuous curve.

State-space models are rarely used in astronomy today, but may be valuable for complicated problems with dense evenly spaced data streams such as X-ray timing studies of accreting black holes (Vio *et al.* 2005). Consider, for example, the search for gravitational wave signals from a laser interferometric observatory like LIGO. Here one has a stream of primary signals representing positions of the fringes, ancillary signals representing conditions of the instrument, sinusoidal variations of instrumental origin (LIGO's violin modes), and stochastic noise which may be correlated with instrument conditions. The state-space model would then have terms representing all known effects, and test data would be used to establish the state vector. Likelihood theory could then be used to search for change points (such as chirps from a binary star in-spiral) or astrophysical signals (such as periodicities from rotating neutron stars). Analogous treatments could be made for study of X-ray accretion disk systems, radio pulsars, asteroseismology and exoplanetary, transit detection. State-space models and Kalman filters can also be applied to problems with $1/f$-type long-term memory behavior (Section 11.9).

11.8 Nonstationary time series

As outlined in Section 11.1, astronomical sources exhibit a great variety of nonstationary variability where the mean levels change with time. These include explosions and transients (X-ray bursters, supernovae, gamma-ray bursts, eruptive variable stars, magnetic flaring on stellar surfaces), high and low states (accretion systems like cataclysmic variables and X-ray binaries) and episodic ejections (BL Lac objects, Herbig–Haro objects). There are also many statistical approaches to the analysis of nonstationary time series. Several strategies commonly used in astronomical research, or with promising prospects, are outlined here.

1. Regression for deterministic behavior   If a regression model can justifiably be formulated − polynomial, autoregressive or nonlinear based on astrophysical theory − then best-fit parameters can be estimated by least-squares, maximum likelihood or Bayesian methods. Either standard regression methods (Section 11.3.4) or the sophisticated state-space approach (Section 11.7) can be applied. The model may lead to astrophysical insights, or be used as a phenomenological fit to remove long-term trends to allow analysis of the stationary residual time series.

2. Bayesian blocks and bump hunting   This is a semi-parametric regression method designed for signal characterization and change-point detection in nonstationary astronomical Poisson time series (Scargle 1998, Dobigeon *et al.* 2007). Here the data consist of a sequence of events, typically photon arrival times from an X-ray or gamma-ray telescope. This procedure constructs a likelihood assuming a sequence of constant flux levels with discontinuous jumps at an arbitrary number of change-points at arbitrary times. The segmentation is conducted by using a hierarchical Bayesian approach with Gibbs sampling to jointly estimate the flux levels and change-point times. The method avoids the limitations of common methods relying on grouping events into bins and then applying methods that rely on Gaussian statistics. Bayesian blocks are widely used to study variable X-ray sources. A similar model can be applied to Gaussian data.

3. Change-point analysis   When the behavior of a nonstationary time series changes abruptly, the boundary between the two behaviors is called a change-point. A host of time-domain methods is available to detect change-points (Brodsky & Darkhovsky 1993,

Chen & Gupta 2000). When likelihood methods are used to establish the model of the initial state, the likelihood ratio test can be used to find changes in mean, variance, trend, autocorrelation or other model characteristics. Nonparametric tests (Section 5.3.1), and the engineer's **CUSUM** (cumulative sum) control chart, are also widely used to find structural changes in time series. Change-point analysis tools are currently rarely used in astronomy.

4. Wavelet analysis   The wavelet transform, like the Fourier transform, translates the information of a time series into another space through convolution with an orthonormal basis function (Percival & Walden 2000, Mallat 2009). But unlike the Fourier transform, the wavelet transform is well-suited to nonstationary phenomena, particularly if the variations occur on a range of temporal scales or over small portions of the time series. The wavelet transform of X_t at scale s and time u is

$$W_X(u, s) = \frac{1}{\sqrt{s}} \int_{-\infty}^{\infty} X_t g\left(\frac{t - u}{s}\right) dt \qquad (11.51)$$

where g is a simple basis function (such as the Gaussian, Mexican hat, Daubechies, or cubic spline wavelet function) which rises and falls from zero with some shape. Selection of wavelet coefficients and performing an inverse transform can filter the time series to enhance features at certain scales. This can reduce high-frequency noise, low-frequency trends, or provide efficient data compression (such as the JPEG 2000 standard). Wavelet analysis provides a flexible and powerful toolbox for analyzing nonstationary signals with nonlinear trends, $1/f$-type noise (Section 11.9), change-points, transient signals or signals that are most evident in time-frequency displays. Figure 11.9 illustrates a discrete wavelet transform applied to an astronomical dataset.

Wavelets have been used in hundreds of astronomical studies over 20 years for the study of gamma-ray bursts, solar phenomena, multiperiodic variable stars and a wide variety of spatial analyses. Foster (1996) investigates the performance of wavelet analysis in the analysis of unevenly spaced time series of variable stars, and finds that standard wavelet transforms can give unreliable results. He recommends a wavelet weighted by the local density and variance.

5. Singular spectrum analysis   Singular spectrum analysis (SSA) is an approach to time series analysis with roots in multivariate statistics that has become popular in geosciences (including meteorology, oceanography and climatology), dynamical physics and signal processing applications (Elsner & Tsionis 1996). It is a nonparametric, matrix-oriented procedure designed to differentiate and extract broad classes of structures such as trends, oscillatory behavior and noise. SSA first obtains the eigenvectors of the **trajectory matrix** with the lag-covariance $\sum_{t=1}^{n-|i-j|} X_t X_{t+|i-j|}$ as the (i, j)-th element. The time series is then projected onto these orthogonal functions to give temporal principal components. SSA is useful for smoothing time series with complex behavior, extracting nonsinusoidal oscillating signals, interpolating gaps in the time series and locating structural changes in nonstationary time series. In astronomy, it has been used in a few studies of variable solar phenomena.

6. Gamma test   The gamma statistic has recently been proposed as a nonlinear curve-fitting technique for temporal and spatial signals based on a minimum-variance criterion using neighboring values within a specified window (Jones 2004). Local variances are estimated by fitting a simple model to the local structure, calculating the variance for a range of neighbors around each point, and extrapolating these values to zero distance. This is repeated for each data point in the time series. The result is a filtering of the dataset that suppresses noise and enhances localized signals. In astronomical applications, it has been shown to be effective in revealing extended, low-surface-bright features in autocorrelated noise (Boyce 2003). The gamma statistic computation is unusually efficient with $O(n \log n)$ operations and thus may be appropriate for megadatasets.

7. Hidden Markov models   Hidden Markov models (HMMs) are a sophisticated and flexible mathematical framework that can simultaneously model stationary stochastic processes (such as red noise or periodicities), predefined structures of various shapes and durations (such as eclipses or outbursts) and their temporal relationships (such as change-points between variability states). As with state-space models (Section 11.7), HMMs are regression techniques where the variables which determine the state of the system are not directly observable, but are hidden behind some stochastic behavior. In Fraser's (2009) phrasing: "Trying to understand a hidden Markov model from its observed time series is like trying to figure out the workings of a noisy machine from looking at the shadows its moving parts cast on a wall, with the proviso that the shadows are cast by a randomly-flickering candle." These models can either be considered alone or in combination with artificial neural networks, state-space models, Kalman filters, Bayesian methods, decision trees and other methods within a likelihood context.

HMMs have been widely applied in speech recognition, image processing, bioinformatics and finance. The developments and applications of HMMs are extensive with hundreds of monographs available; Fraser (2009) provides an accessible introduction. HMMs have not yet been applied to astronomical problems, but are planned for automated classification of millions of spectra to be obtained with the Chinese LAMOST survey telescope (Chen 2006). HMMs could be effective in treating signal extraction problems such as exoplanetary transits and supernova events in the billions of photometric light curves expected from the LSST and other survey telescopes.

11.9 $1/f$ noise or long-memory processes

While many time series in the social sciences are dominated by short-term correlation behavior that is reproduced by ARMA-type models discussed in Section 11.3.3, time series in astrophysical and other physical systems are often dominated by long-term autoregressive behavior (Press 1978). It is common that the variance increases with characteristic time-scale, and the power spectrum can be fit by a power-law function at low frequencies. The physicist and astronomer calls this **$1/f$ noise**, **flicker noise** or **red noise** from an inverse correlation of power against frequency f in a Fourier power spectrum. The statistician calls time series of this type **long-memory** (also called **persistent**, **long range** and **fractional**) noise processes, in contrast to **short-memory** processes that can be understood with low-order ARMA models. A few results from the statistical and econometric literature are summarized here; details are presented in the volumes by Beran (1989), Robinson (2003) and Palma (2007). Some early results were derived by B. Mandelbrot in his well-known studies of fractal processes.

Long-memory processes can arise in several ways (Beran 1989, Chapter 1). One possibility is the combination of many short-memory subprocesses. The sound of a bubbling brook has a $1/f$-type noise structure because it aggregates the events associated with different interactions between the moving water and obstructions in the flow. In astronomy, the observed emission may appear to arise from an unresolved source, but in reality is the aggregation of variations at different locations on (say) the surface of a star or accretion disk. Another possibility is that a single process with self-similar hierarchical variation is involved, perhaps regulated by stochastic differential equations. A turbulent interstellar cloud may have this property, although the variations we study are spatial not temporal. A third possible source of long-memory variations are instrumental and observational problems: calibration drifts, changes in background conditions during a night or satellite orbit, or interactions between stages in a complex data reduction procedure.

A long-memory process is often modeled with an autocorrelation function that decays as a power law at large time lags $k \rightarrow \infty$,

$$ACF_{LM}(k) \propto k^{2d-1} \tag{11.52}$$

for $0 < d < 1/2$. **Hurst's self-similarity parameter** $H = d+1/2$ is sometimes used. In the context of ARMA-type models, this is called a **fractional ARIMA process (FARIMA or ARFIMA)**. For example, the variability of a FARIMA(0,d,0) model is intermediate between those of a stationary AR(1) and a nonstationary random walk model. The corresponding Fourier spectral density function at low frequencies is

$$f_{LM}(\nu) \propto |\nu|^{-2d}. \tag{11.53}$$

The $1/f$-noise case corresponds to $d = 1/2$ which is marginally nonstationary.

Long-memory processes in general can be modeled using likelihood methods. These include maximizing the likelihood with the **Durbin–Levinson algorithm**, solving the state-space model with the Kalman filter, deriving Bayesian posterior distributions for assumed priors and loss functions, and constructing the periodogram using Whittle's approximate likelihood. The **Whittle (1962) likelihood** is a modified MLE applied in the Fourier domain designed for autoregressive processes. When the variance of the noise variations changes with time, then heteroscedastic models are needed. These likelihood calculations can be technically difficult, asymptotic properties of the statistics can be complicated, and small-sample behavior may be unknown.

For the problem of estimating the slope d in a Gaussian fractional noise process, and discriminating long-memory processes ($d > 0$ in Equation 11.52) from a random walk, three solutions have been widely used.

First, an estimate for the slope d can be obtained by least-squares regression on the periodogram (Robinson 1995). Given the periodogram from the discrete Fourier transform

$$I(\nu) = \left| \frac{\sum_{t=1}^{n} X_t e^{it\nu}}{\sqrt{2\pi n}} \right|^2 \tag{11.54}$$

and defining

$$a_j = \ln(4 \sin^2(\nu_j/2)), \tag{11.55}$$

the least-squares slope estimator is

$$\hat{d}_LS(l) = \frac{\sum_{j=l+1}^{m}(a_j - \bar{a})\ln(I(\nu_j))}{\sum_{j=l+1}^{m}(a_j - \bar{a})^2} \tag{11.56}$$

where $0 \le l < m < n$ and $\bar{a}$ is the mean of the time series. This method is often called the **Geweke–Porter-Hudak (GPH) estimator** (Geweke & Porter-Hudak 1983). However, its mathematical properties (e.g. asymptotic bias and variance) are not strong. In particular, it is difficult to decide the precise frequency range for computing the linear regression; recommendations for the optimal number of periodogram values include $m \simeq n^{1/2}$ and $m \simeq n^{4/5}$ where n is the number of data points in the time series (Hurvich *et al.* 1998). The regression can be performed on a smoothed or multi-tapered periodogram; a similar method of regression on the binned periodogram was independently developed by astronomers Papadakis and Lawrence (1993). Other modifications to the GPH estimator have also been suggested (Andersson 2002).

Second, a least-squares time-domain procedure for estimating the slope d, called **detrended fluctuation analysis**, originally developed to study DNA sequences, has become popular in a variety of fields (Peng *et al.* 1994). The time series x_t is grouped into k bins each with $m = n/k$ data points. A linear regression is performed within each bin and the residual variance is computed:

$$\sigma_k^2 = \frac{1}{m}\sum_{i=1}^{m}(X_t - \hat{\alpha}_k - \hat{\beta}_k t)^2. \tag{11.57}$$

Note that the regression lines will be discontinuous at the bin boundaries. The average of these variances, $F = (1/k)\sum_{j=1}^{k}\sigma_k^2$, is related to the slope of the long-memory process d as $F(k) \propto k^{d+1/2}$ for a Gaussian fractional noise process. The slope d can then be estimated by another linear regression between $\log F$ and $\log k$. The method is described by Palma (2007, Section 4.5.4) and is applied to an astronomical dataset in Figure 11.8.

Third, an estimator for d has been developed based on the **discrete wavelet transform** of the time series. A vector of squared wavelet amplitudes summed over different time-scales is distributed as χ^2 with dependency on d. The slope is then estimated from a linear regression involving this vector and the original time series. In general, wavelet analysis of nonstationary $1/f$-type processes can be effective.

In general, statistical evaluation of $1/f$-type (fractional ARMA) processes encounters many pathologies (Beran 1989, Chapters 8 and 10; Robinson 2003, Chapter 10). While the sample mean of a fractional ARMA process is unbiased, its convergence to the true mean can be very slow and standard confidence intervals are incorrect. The standard sample variance is biased, and some long-memory processes have infinite variance. Both the spectral density at zero frequency and the area under the ACF over all lags can be infinite. Caution is also needed when seeking relationships between a long-memory random variable and covariates, as these time series show excursions from the mean over various time-scales and short samples often show spurious trends. Standard statistics for hypothesis testing, like the likelihood ratio test, no longer follow a χ^2 distribution. Nonparametric e.d.f.-based goodness-of-fit tests, like the Kolmogorov–Smirnov test, can be very inaccurate: even true models may be rejected due to long-memory variations. While some of these problems arise with all autocorrelated processes, they are particularly exacerbated with long-memory processes.

Clearly, simple statistical evaluation of $1/f$-type time series is fraught with difficulties. For time series with mild variations, or with very long durations, cautious use of standard methods may be tractable. Variable transformation, such as considering the differenced time series $x_i - x_{i-1}$ or the inverse of the periodogram, sometimes helps. A variety of sophisticated mathematical approaches to these problems have been developed over the past several decades with associated theorems and convergence calculations (Beran 1989, Robinson 2003, Palma 2007). Research on long-memory processes is rapidly developing with applications in genetics, finance, human physiology, Internet studies, geology, meteorology, ecology and other fields. There are many new methods available for astronomers to consider when treating $1/f$-type processes in astronomical time series.

11.10 Multivariate time series

As mentioned in Section 11.1, astronomical time series are sometimes measured in several wavelength bands which may show similar, lagged or different variations depending on the astrophysical situation. This is the domain of multivariate time series analysis which is well-developed for problems encountered often in fields like econometrics and meteorology. For example, temporal variations in the prices of fertilizer, corn and pork may be correlated with time lags associated with the agricultural process of growing grain for livestock consumption. Analogously, the peak emission from supernova explosions or gamma-ray burst afterglows may evolve from short to long wavelengths as the expanding fireball cools or becomes transparent to different wavelengths of light.

For a bivariate stochastic temporal process measured in evenly spaced observations of the form (X_i^1, X_i^2) for $i = 1, \ldots, n$, the sample second-order moments are the two individual autocorrelation functions and the **cross-correlation function**

$$CCF(k) = \frac{1}{n} \frac{\sum_{i=1}^{n-k}(X_t^1 - \bar{X}^1)(X_t^2 - \bar{X}^2)}{\sqrt{Var(X^1)Var(X^2)}} \tag{11.58}$$

for lags $k \geq 0$. Here the numerator is called the cross-covariance function and the denominator is the product of the two sample standard deviations. This estimator of the true cross-correlation function is consistent and asymptotically unbiased, but its variance may be large and its distribution difficult to derive as it convolves the variations of the individual time series. The CCF is thus often calculated on time series residuals after periodic, trend or autocorrelated signals are removed by pre-whitening, regression or ARMA-type modeling. If the CCF shows a single strong peak at some lag k, then this may reflect similar structure in the two time series with a delay. However, the CCF may show structure for other reasons, so a careful check of this interpretation is needed.

The bivariate cross-correlation function can be readily generalized to multivariate time series using a correlation matrix function where the (i, j)-th element of the matrix is the bivariate CCF for variables x^i and x^j. In econometrics and other fields, one often seeks explanatory and predictive relationships between a single dependent variable and a suite of covariates, all of which vary in time and may be correlated with each other with different lag-times. A large literature exists to address these dynamic regression, **distributed lag** or **vector autoregressive (VAR)** models. But astronomers can rarely acquire evenly spaced time series data in many variables for a single object, so this approach will not be generally useful in astronomy.

11.11 Remarks

Time series analysis is a vast field with a myriad methods that might be applied. In some cases, the typical astronomical analysis of time series data is well-motivated. When sinusoidal periodicities are present or suspected, Fourier analysis is an excellent approach. Astronomers have led innovations in spectral analysis for finding periodicities in unevenly space data and nonsinusoidal signals. Many of these methods are reasonable, and the interpretation by users is often simplistic. The Lomb–Scargle periodogram (perhaps the most highly cited astrostatistical method in the astronomical literature) has a strong mathematical foundation, but its performance is sometimes imperfect and the statistical significance of its peaks is difficult to establish. When secular trends are present, then regression is a correct procedure; indeed, here the independent variable t is fixed, and this model is more appropriate to time series data than to regression between two random variables (Chapter 7).

Astronomers have actively generated non-Fourier periodograms to treat common problems of unevenly spaced data points, nonsinusoidal recurrent patterns, and heteroscedastic measurement errors, and Poisson rather than Gaussian processes. There is ample need for statisticians to evaluate these methods, establishing their mathematical properties and limitations.

However, astronomers have habitually used spectral analysis, rather than time-domain modeling, to analyze aperiodic and stochastic phenomena. Although all information in the original time series is contained in the Fourier coefficients (as it is in the autocorrelation function), the periodogram does not concentrate the signal well for aperiodic autoregressive signals. For $1/f$-type noise, for example, a FARIMA model might capture the essential behavior in a handful of parameters which requires dozens of components of a Fourier spectrum.

It is important to recognize that when any kind of autocorrelated structure is present — periodicities, secular trends or aperiodic autocorrelated variations — then standard statistics may not apply; in particular, variances and standard deviations are often underestimated. This is relevant to many astronomical problems. Consider, for example, the estimation of the equivalent width of a faint spectral line in a univariate spectrum dominated by continuum emission. Here the wavelength is a fixed time-like variable so time series methods apply.

The common procedure is to evaluate the variance of the signal around the putative line, compute the signal-to-noise ratio at the wavelength of interest, and evaluate its significance from the normal distribution. But the continuum signal often has autocorrelation due either to structure in the source continuum emission or to instrumental effects. In such cases, the assumption of i.i.d. underlying the standard computation of the variance is wrong and the equivalent width significance can be overestimated. The problem is particularly exacerbated when the autocorrelation function does not converge as in some $1/f$-type noise behaviors, the statistician's long-memory processes.

One reason astronomers often prefer Fourier to time-domain analysis is the astrophysical interpretation of the parameters. There is often a lack of a clear relationship between the derived model parameters of an ARMA-like model and the scientific goals of the study. In economic or meteorological studies, the meaning of the parameter values of an FARIMA or ARCH model is not as important as obtaining a successful prediction of future values. But the astronomer has less interest in predicting the future X-ray luminosity of an accretion disk than in understanding the physical processes underlying the variations. Astrophysical theories today are rarely designed to predict autoregressive behavior. One unusual success in linking advanced time series model parameters to a physical process is the interpretation of the variations of the X-ray source Sco X-1 which exhibits red noise and quasi-periodic oscillations on time-scales from milliseconds to hours. Using Fourier and wavelet decompositions, the empirical time series model has been linked to a physical model known as the **dripping rail** (Scargle *et al.* 1993). We encourage future astrophysical theoretical studies to make predictions of autoregressive properties to match the statistical benefits of time-domain time series modeling of non-periodic phenomena.

Astronomers can benefit from other statistical approaches to time series analysis, particularly for nonstationary phenomena. Wavelet analysis and Bayesian blocks are examples of methods gaining popularity for treating multi-resolution features that are present only at certain times. State-space and hidden Markov modeling are sophisticated and flexible approaches having considerable potential for understanding complex astronomical time series problems.

The statistical findings summarized here are not restricted to time series analysis, but will be relevant to any astronomical situation where the data are a function of a single ordered time-like variable. This includes astronomical spectra where intensity is a function of wavelength, and images where astronomical spectra are a function of pixelated position on a detector. We encourage experimentation with time series methodology for the analysis of astronomical spectra and images.

11.12 Recommended reading

Due to the wide range of time series methodologies, we list below only broad-scope texts. Recommended topical books include Warner (1998, elementary) and Percival & Walden (1993, advanced) for spectral analysis, Kitagawa & Gersch (1996) and Durbin & Koopman (2001) for state-space methods, Fan & Yao (2003) for nonlinear models, Mallat (2009) and

Percival & Walden (2000) for wavelet analysis, and Beran (1989) and Palma (2007) for long-memory processes ($1/f$-type noise).

Chatfield, C. (2004) *The Analysis of Time Series: An Introduction*, 6th ed., Chapman & Hall/CRC, Boca Raton

An excellent elementary presentation of time series analysis. Coverage includes stationary stochastic processes, autocorrelation, autoregressive modeling, spectral analysis, state-space models, multivariate modeling and brief introductions to advanced topics.

Cowpertwait, P. S. P. & Metcalfe, A. V. (2009) *Introductory Time Series with R*, Springer, Berlin

A short and clear elementary treatment of times series analysis with **R** applications covering autocorrelation, forecasting, stochastic models, regression, stationary and non-stationary models, long-memory processes, spectral analysis, multivariate models and state-space models.

Shumway, R. H. & Stoffer, D. S. (2006) *Time Series Analysis and Its Applications with R Examples*, 2nd ed., Springer, Berlin

A well-written intermediate-level text with useful descriptions of **R** implementation. Topics include exploratory data analysis, regression, ARMA-type models, spectral analysis and filtering, long-memory and heteroscedastic processes, state-space models and cluster analysis.

Wei, W. W. S. (2006) *Time Series Analysis: Univariate and Multivariate Methods*, 2nd ed., Pearson, Harlow

A more detailed text on time series. Coverage includes autocorrelation, ARMA-type models, regression, model selection, Fourier and spectral analysis, vector time series, state-space models and the Kalman filter, long-memory and nonlinear processes, aggregation and sampling.

11.13 R applications

Excellent tutorials on time series analysis using **R** are provided by Shumway & Stoffer (2006) and Cowpertwait & Metcalfe (2009). Elementary methods are reviewed by Crawley (2007).

We use the variability of Galactic X-ray source GX 5-1 to illustrate many of the time series capabilities of **R**. GX 5-1 is an X-ray binary-star system with gas from a normal companion accreting (falling) onto a neutron star. Highly variable X-rays are produced in the inner accretion disk, often showing stochastic red noise and quasi-periodic oscillations of uncertain origin. The dataset, described in Appendix C.12, consists of 65,536 measurements of photon counts in equally spaced 1/128-second bins obtained with the Japanese Ginga satellite during the 1980s. Although strictly a Poisson process, the count rates are sufficiently high that they can be considered to be normally distributed.

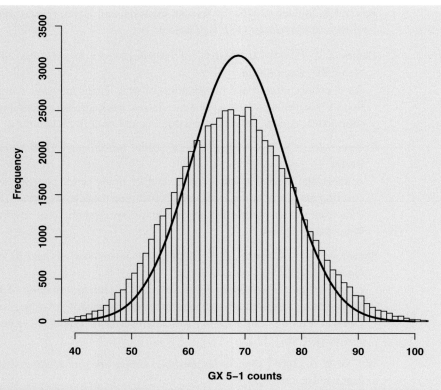

Fig. 11.1   Histogram comparing the distribution of X-ray counts in 1/128-second time increments to a normal distribution.

Using methods that are rarely considered in astronomy, but which are standard in many other applications of time series analysis, we have arrived at some unusual conclusions with respect to standard astronomical interpretations of rapid X-ray variations in accretion binary systems (van der Klis 2000). The structure can be attributed to high-order autoregressive behavior plus a very-low-frequency component. While red noise is evident, we find its interpretation as a simple power-law $1/f^\alpha$ component to be inaccurate. While a quasi-periodic component is evident, we find it is entirely attributable to autoregressive behavior with no strict periodicities involved.

11.13.1 Exploratory time series analysis

We start with visual exploration of the dataset in a variety of displays. The histogram of counts in each bin shows that the variance is 24% larger than expected from a white noise process, and asymmetry about the mean is present (Figure 11.1). Examination of the time series in its raw form (Figure 11.2) and after various smoothing operations (Figure 11.3) does not reveal obvious nonstationary structure to account for the extra variance, although

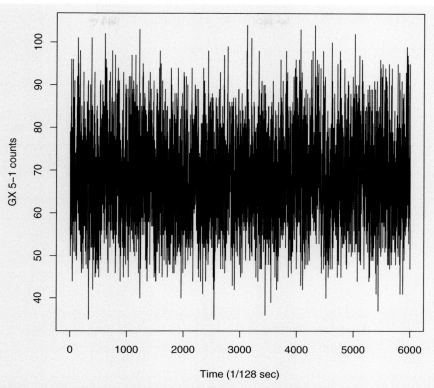

Fig. 11.2   Raw time series of the X-ray emission from the Galactic X-ray binary system GX 5-1 measured with the Ginga satellite observatory. This plot show the first few percent of the dataset.

some of the smoothers give a hint of autocorrelated variations with a characteristic time-scale around 20–50 seconds. GX 5-1 thus appears to exhibit stochastic, possibly autocor-related and quasi-periodic, variability that can now be investigated in detail.

In the **R** code below, we start by reading this ASCII dataset using the *scan* function rather than *read.table*, as we want the values to be placed into a single vector rather than into a tabular matrix. The counts and time stamps are collected into an object with **R** class "time series" using the *ts* function; this allows a wide variety of time series functions to operate on the dataset. The normal function superposed on the histogram uses the function *curve* because *hist* does not permit superposition using *lines*. In Figure 11.3, we display five smoothers offset vertically. Two are based on the GX.ts time series object and use the *filter* and *kernapply* functions. Three operate on the original counts and time stamp vectors and use kernel density estimation, a nearest-neighbor super-smoother and the *lowess* local regression fit described in Section 6.6.2. In three cases, we chose a bandwidth of seven time bins, but for the super-smoother we use the cross-validation bandwidth default and for *lowess* we choose a span of 0.05. Examination of different smoothers and bandwidths is encouraged during the exploratory stage of a time series analysis.

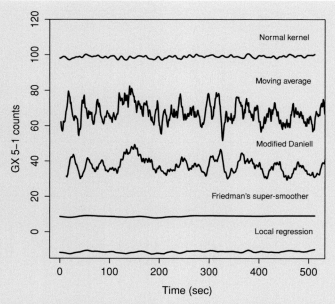

Fig. 11.3 The full GX 5-1 time series subject to various smoothing procedures: normal kernel density estimator, moving average estimator, modified Daniell (boxcar) smoother, Friedman's cross-validation super-smoother, and Cleveland's local regression model. Except for the moving average smoother, the curves are displaced vertically for clarity.

```
# Read in GX 5-1 data and create time series

GX.dat <- scan("http://astrostatistics.psu.edu/MSMA/datasets/GX.dat")
GX.time <- seq(from=0, to=512, length.out=65536)
GX.ts <-  ts(GX.dat, GX.time) ; GX.ts.offset <- ts(GX.dat-30, GX.time)

# Compare histogram of counts to normal distribution

hist(GX.dat, breaks=100, xlim=c(40,100), ylim=c(0,3500), xlab='GX 5-1 counts',
   font=2, font.lab=2, main='')
curve(dnorm(x,mean=mean(GX.dat), sd=sqrt(mean(GX.dat)))*65536, lwd=3, add=T)
sd(GX.dat) / sqrt(mean(GX.dat))  # result is 1.24

# Examine raw and smoothed time series

plot.ts(GX.ts[1:6000], ylab='GX 5-1 counts', xlab='Time (1/128 sec)',
   cex.lab=1.3, cex.axis=1.3)
plot(GX.time,GX.dat, ylim=c(-10,115), xlab='Time (sec)', ylab='GX 5-1 counts',
   cex.lab=1.3, cex.axis=1.3, type='n')
lines(ksmooth(GX.time, GX.dat+30, 'normal', bandwidth=7), lwd=2)
```

```
text(450, 110, 'Normal kernel')
lines(filter(GX.ts, sides=2, rep(1,7)/7), lwd=2)
text(450, 85, 'Moving average')
lines(kernapply(GX.ts.offset, kernel('modified.daniell', 7)), lwd=2)
text(450, 50, 'Modified Daniell')
lines(supsmu(GX.time, GX.dat-60), lwd=2)
text(400, 20, 'Friedman's super-smoother')
lines(lowess(GX.time, GX.dat-80, 0.05), lwd=2)
text(450, 0, 'Local regression')
```

11.13.2 Spectral analysis

Many astronomical time series analyses emphasize harmonic or spectral analysis. The **R** script below starts with a manual construction of the Schuster periodogram based on the script provided by Shumway & Stoffer (2006, Chapter 2), and then uses the automated function *spec.pgram* (Figure 11.4, top panel). Note that entries in the raw periodogram can be very uncertain; for this spectrum, the 500th spectral value is 119 with 95% range [32,4702] assuming an asymptotic χ^2 distribution.

We then proceed to experiment with various smoothers and tapers to reduce the variance of the raw periodogram. A taper has little effect in this case; they are more important for shorter datasets. Smoothing, however, has a dramatic effect on the periodogram appearance and much is learned about the excess variance of the time series (Figure 11.4). It appears to rise from two distinct processes: a noise component below ∼0.05 rising towards lower frequencies, and a strong but broadened spectral peak around 0.17–0.23. These are the red noise and quasi-periodic oscillations addressed in hundreds of studies of accretion binary-star systems like GX 5-1 (van der Klis 2000).

```
# Raw periodogram

f <- 0:32768/65536
I <- (4/65536) * abs(fft(GX.ts) / sqrt(65536))^2
plot(f[2:60000], I[2:60000], type="l", ylab="Power", xlab="Frequency")

Pergram <- spec.pgram(GX.ts,log='no',main='')
summary(Pergram)
Pergram$spec[500] # value of 500th point
2*Pergram$spec[500] / qchisq(c(0.025,0.975),2)

# Raw and smoothed periodogram

par(mfrow=c(3,1))
spec.pgram(GX.ts, log='no', main='', sub='')
spec.pgram(GX.ts, spans=50, log='no', main='', sub='')
spec.pgram(GX.ts, spans=500, log='no', main='', sub='')
```

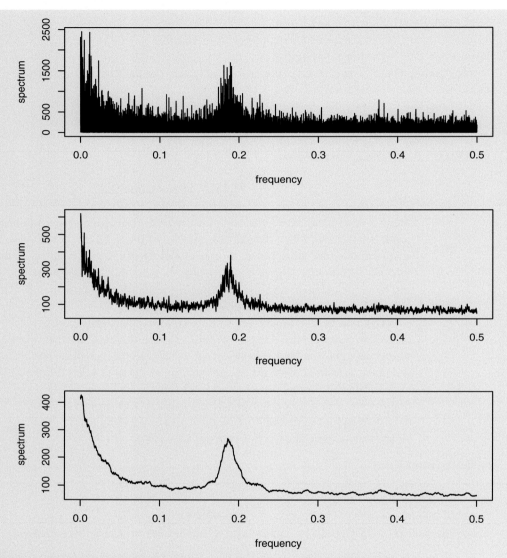

Fig. 11.4 Raw and smoothed (modified Daniell with $w = 50$ and 500 bandwidths) periodograms for GX 5-1.

11.13.3 Modeling as an autoregressive process

The red noise and (quasi-)periodic structure in the GX 5-1 data are clearly seen in plots of the (partial) autocorrelation function (Figure 11.5). The PACF is more informative: the strongest predictor of a current value are the values separated by 4–6 time increments, and oscillations have periods around 5–6 increments. The envelope of significant autocorrelation extends to lags of 20–30 increments.

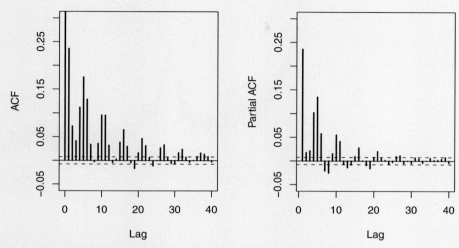

Fig. 11.5   Autocorrelation and partial autocorrelation functions for GX 5-1 time series.

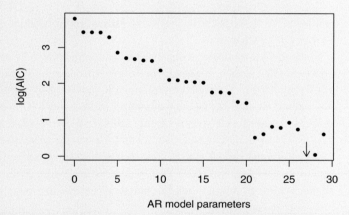

Fig. 11.6   Values of the Akaike information criterion as a function of model complexity for autoregressive AR(p) models of the GX 5-1 time series.

The presence of autocorrelation and (from the smoothed time series) absence of changes in the average level motivates an AR (rather than ARMA or ARIMA) model. The *ar* function in **R** selects a model using the minimum Akaike information criterion. Several computational options are provided; we choose ordinary least squares. AR(27) is the best model, although any model with order $p > 22$ is adequate (Figure 11.6). The largest positive coefficients are at lags $= 1$, 4, 5 and 6 and negative coefficients at lags 10 and 11, again indicating oscillation with period around five increments. Parameter uncertainties can be based on asymptotic theory in the function *ar* or by bootstrap resampling of the model

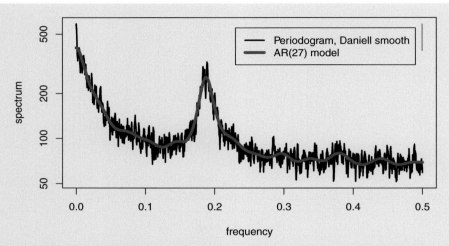

Fig. 11.7 Smoothed periodogram of the GX 5-1 time series and its AR(27) model. For a color version of this figure please see the color plate section.

residuals. An **R** script for the bootstrap resampling procedure is presented by Shumway & Stoffer (2006, Chapter 3).

Since the AR model is stochastic, we cannot meaningfully compare its time series to the observed GX 5-1 data. However, the Fourier spectrum of the model can be compared to the GX 5-1 spectrum. The function *ar.spec* in **R** runs *ar*, chooses the model with the minimum AIC, and computes the Fourier transform of the model. This program found that AR(27) is the best model, as it gives a slightly stronger spectral signal around frequency 0.19 than the AR(12) model. Figure 11.7 shows that the spectral properties of GX 5-1 are remarkably well-reproduced by the model: the steep red noise signal at frequencies below 0.05, the shallow red noise around 0.05−0.1, the height and width of the QPO peak around 0.19, the slight asymmetry around the peak, the weak harmonic around 0.38, and the spectral noise level at high frequency.

```
# Autocorrelation functions of the GX 5-1 time series

par(mfrow=c(1,2))
acf(GX.ts, 40, main='', ci.col='black', ylim=c(-0.05,0.3), lwd=2)
pacf(GX.ts, 40, main='', ci.col='black', ylim=c(-0.05,0.3), lwd=2)


# Autoregressive modeling

ARmod <- ar(GX.ts, method='ols')
ARmod$order # model selection based on AIC
ARmod$ar # best-fit parameter values
ARmod$asy.se.coef$ar # parameter standard errors
```

```
plot(0:29, log10(ARmod$aic[1:30]), xlab='AR model parameters',
    ylab='log(AIC)', pch=20)
arrows(27, 0.4, 27, 0.0, length=0.1)


# Spectrum of AR model

ARspec <- spec.ar(GX.ts, plot=F)
GXspec <- spec.pgram(GX.ts, span=101, main='', sub='', lwd=2)
lines(ARspec$freq, ARspec$spec, col='red', lwd=4)
legend(0.23, 550, c('Periodogram, Daniell smooth', 'AR(27) model'),
    lty=c(1,1), lwd=c(2,4), col=c('black','red'))
```

11.13.4 Modeling as a long-memory process

The only spectral feature in Figure 11.7 not accounted for by the AR model is the excess signal at the very lowest frequency. This indicates that the GX 5-1 behavior has an additional long-term memory component. As discussed in Section 11.9, we can treat long-memory processes in both the time domain and frequency domain. We can generalize the AR model above to nonstationary ARIMA and FARIMA models. The *arima* function in **R** calculates the likelihood in a state-space formulation using the Kalman filter (Section 11.7). However, in this case with 27 AR parameters, it is computationally expensive to calculate a range of ARIMA models with additional differencing and moving averaging components. Trials with fewer AR parameters suggests that models such as ARIMA(12,1,1) provide little benefit over the corresponding AR model.

Several estimators of the long-memory parameter d are provided by **CRAN** packages based on both spectral and time-domain techniques. Recall from Section 11.9 that $d = \alpha/2$ and the Hurst parameter is $H = d + 1/2$, where the spectral power is $P \propto (1/f)^{\alpha}$. The function *fdGPH* in **CRAN** package *fracdiff* (Fraley 2009) calculates the Geweke–Porter-Hudek estimator from linear regression on the raw log periodogram; this method gives $d = 0.10 \pm 0.04$. The function *fdSperio* uses Reisen's (1994) regression on the log periodogram smoothed with a lag Parzen window and obtains the same value with higher precision, $d = 0.10 \pm 0.01$. These values indicate a relatively weak long-memory process with noise power scaling roughly as $1/f^{0.2}$. Similar calculations are provided by function *hurstSpec* in package *spectral*.

In the time domain, a FARIMA model calculation obtained from the **CRAN** package *fracdiff* gives a best-fit value of $d = 0.06$. The **CRAN** package *fractal* has several functions which compute the long-memory Hurst parameter from the time-domain data (Constantine & Percival 2010). These results are noticeably inconsistent. We obtain $H = 0.03$ using detrended fluctuation analysis; 0.12 from an arc-hyperbolic-sine transformation of the autocorrelation function; and 0.66–1.00 from various blockings of the time series. Clearly, it is difficult to establish a consistent value for the $1/f^{\alpha}$-noise spectral shape. The *fractal* package also has useful capabilities for simulating several types of fractal time series for

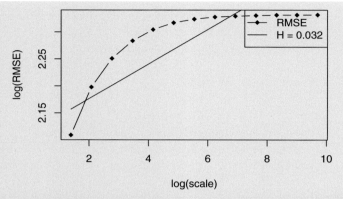

Fig. 11.8   Log-log plot of the root mean square error against scale from the detrended fluctuation analysis of the GX 5-1 time series. The slope of the linear fit gives the long-range memory Hurst parameter.

comparison with a real dataset: a $1/f^{\alpha}$ process; a fractionally differenced process and fractional Gaussian noise; and fractional Brownian motion.

The likely cause of these discrepant results is that the red noise in GX 5-1 is not truly fractal, such as a $1/f^{\alpha}$ process where the scaling of the variance on different scales would be a power law. A plot from the detrended fluctuation analysis shows a poor fit of a power-law model to the data (Figure 11.8), so that different estimation procedures of the fractal dimension arrive at different results.

```
# Spectral estimates of the long-range memory parameter d

install.packages('fracdiff') ; library(fracdiff)
d.FARIMA <- fracdiff(GX.ts, nar=27, nma=1, ar=ARmod$ar) ; d.FARIMA$d
d.GPH <- fdGPH(GX.ts)
d.Reisen <- fdSperio(GX.ts)


# Time domain estimates of the long-range memory parameter H=d+1/2

install.packages('fractal') ; library(fractal)
d.DFA <- DFA(GX.ts) ;  d.DFA
plot(d.DFA)
d.ACVF <- hurstACVF(GX.ts) ; d.ACVF
d.block <- hurstBlock(GX.ts)  ; d.block
```

11.13.5 Wavelet analysis

In examination of the GX 5-1 time series, we did not see any short-lived structures such as rapid increases or decreases in emission. However, it is valuable to make a wavelet decomposition of the time series to assist in visualization of the dataset at different time-scales. Here we use **CRAN**'s package *waveslim* (Whitcher 2010) to construct the discrete

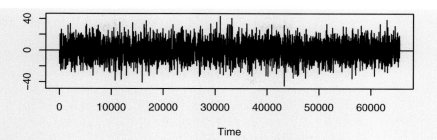

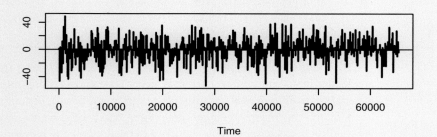

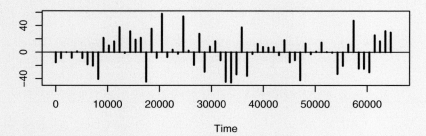

Fig. 11.9   Wavelet coefficients during the GX 5-1 time series at three time resolutions.

wavelet transform using Mallat's pyramid algorithm and Daubechies' orthonormal basis functions at ten different temporal scales. This package implements a variety of wavelet methods for one-, two-, and three-dimensional data as described in the volume by Gencay *et al.* (2001). Figure 11.9 shows the wavelet coefficients for the time series at three scales. As expected, no obvious structure is seen.

```
# Discrete wavelet transform

install.packages('waveslim') ; library(waveslim)
GX.wav <-  dwt(GX.ts,n.levels=10)
par(mfrow=c(3,1))
plot.ts(up.sample(GX.wav[[4]],2^4),type='h',axes=F,ylab='') ; abline(h=0)
plot.ts(up.sample(GX.wav[[7]],2^7),type='h',axes=F,ylab='',lwd=2) ; abline(h=0)
plot.ts(up.sample(GX.wav[[10]],2^{10}),type='h',axes=F,ylab='',lwd=2) ; abline(h=0)
par(mfrow=c(1,1))
```

11.13.6   Scope of time series analysis in R and CRAN

R supports the class *ts* for evenly spaced time series data. Basic analysis includes autocorrelation functions (*acf* and *pacf*) and linear filters with the function *filter*. Modeling includes univariate (*arima*) and multivariate (*VAR*) autoregressive models, maximum likelihood structural regression models (*StructTS*) and local regression models (*stl*).

CRAN's *zoo* package (with its extension *xts*) creates the *zoo* class for ordered observations such as irregularly spaced time series (Zeileis & Grothendieck 2005). Editing, merging, plotting, summaries, computing lags and windows, aggregation into bins, interpolation between values, treatment of *NA* values, conversion from other **R** classes, and other infrastructure is provided, but support for duplicate observation times is limited. At the present time, there are no time series analysis applications.

CRAN has dozens of time series packages for evenly spaced data; an overview is provided at http://cran.r-project.org/web/views/TimeSeries.html. While many cover similar material primarily oriented towards econometrics and finance, a number of packages may be useful to astronomers. If outliers are present, and need to be downweighted or removed, the *robfilter* package provides a variety of iterated filters and regressions based on robust statistics (Section 7.3.4). The *RTisean* package provides a number of functions for high-pass filtering, noise reduction, polynomial modeling and more. Several very capable packages for wavelet analysis are available including *wavelets*, *wavethresh* and the *wmtsa* package accompanying the monograph by Percival & Walden (2000). The package *sde* implements methods for stochastic differential equations discussed in the monograph by Iacus (2008). Bayesian approaches are implemented in *bspec* for autocorrelation and spectral analysis, *ArDec* for autoregressive decomposition, and *MSBVAR* for dynamic vector autoregressive modeling.

An **R** script for the Lomb–Scargle periodogram, designed for genetics research problems, is available outside of **CRAN** at http://research.stowers-institute.org/efg/2005/LombScargle.

12 Spatial point processes

12.1 The astronomical context

Spatial data consists of data points in p dimensions, usually $p = 2$ or 3 dimensions, which can be interpreted as spatial variables. The variables might give locations in astronomical units or megaparsecs, location in right ascension and declination, or pixel locations on an image. Sometimes nonspatial variables are treated as spatial analogs; for example, stellar distance moduli based on photometry or galaxy redshifts based on spectra are common proxies for radial distances that are merged with sky locations to give approximate three-dimensional locations.

The methods of spatial point processes are not restricted to spatial variables. They can be applied to any distribution of astronomical data in low dimensions: the orbital distributions of asteroids in the Kuiper Belt; mass segregation in stellar clusters; velocity distributions across a triaxial galaxy or within a turbulent giant molecular cloud; elemental abundance variations across the disk of a spiral galaxy; plasma temperatures within a supernova remnant; gravitational potential variations measured from embedded plasma or lensing distortions of background galaxy shapes; and so forth.

The most intensive study of spatial point processes in astronomy has involved the distribution of galaxies in the two-dimensional sky and in three-dimensional space. One approach, pioneered by Abell (1958), is to locate individual concentrations or "clusters" of galaxies. The principal difficulty is the overlapping of foreground and background galaxies on a cluster, diluting its prominence in two-dimensional projections. The greatest progress is made when spectroscopic redshifts are obtained that, due to Hubble's law of universal expansion, allows the third dimension of galaxy distances to be estimated with reasonable accuracy. Redshift surveys, along with X-ray studies of the copious hot intergalactic gas in rich clusters, led to the discovery that clusters are often not Gaussian in structure. Elongated, filamentary, double, hierarchical, unequilibrated and honeycomb-like structures are common. However, galaxy redshifts are expensive to measure, and less accurate photometric redshift estimates are available for hundreds of millions of galaxies.

A wide variety of statistical methods to characterize this complicated galaxy spatial point process using both spectroscopic and photometric distance estimators is reviewed in the volume by Martínez & Saar (2002). Recent progress is often based on the giant Sloan Digital Sky Survey that imaged hundreds of millions of galaxies and spectroscopically measured a million galaxy redshifts. Studies of clustering structure using the Sloan Survey dataset have used Wiener filtering (Kitaura *et al.* 2009), normal mixture models

(Einasto *et al.* 2010; Hao *et al.* 2010), topological measures (Choi *et al.* 2010), modified single-linkage clustering (Tago *et al.* 2010) and two-point correlation functions (Kazin *et al.* 2010). The astrophysical origin of this structure is broadly understood in the context of the formation of large-scale structure in the expanding Universe. But detailed statistical comparisons between data and cosmological model predictions are difficult to achieve.

Another important class of problems with spatial point data involve the locations of photons or particles in a detector. Photon-counting detectors are particularly common in high-energy astrophysics. Huge investments have been made to collect such datasets from the Chandra X-ray Observatory, Fermi Gamma-Ray Space Telescope, IceCube Neutrino Observatory, Pierre Auger Cosmic Ray Observatory, and similar instruments. Here the spatial variables are positions in the two-dimensional detector; a common additional variable is the energy of the photon or particle.

Point processes on a circle or sphere, rather than an infinite Euclidean space, also appear in astronomy. Polarization, fluid flows and stellar proper motions all have directionality with orientation variables restricted to the range $0 < \theta < 2\pi$. All-sky surveys produce distributions of objects across the celestial sphere. Specialized statistics for spatial variables on a circle or sphere are needed for these situations.

Spatial point processes are related to other branches of applied statistics. If the observation locations fill a fixed grid, such as pixels in a solid state detector, then the problems approach those of image analysis. A spatial point process can be considered a distribution of points in an arbitrary *p*-space, where no attention is paid to the identification of some variables as spatial and others as physical properties. Thus methods of multivariate analysis, such as principal components analysis, are sometimes applied to spatial data, and multivariate classification methods are often close kin to spatial clustering methods. Nonetheless, primarily due to its prevalence in Earth sciences from ecology and forestry to geography and mining, the analysis of spatial point processes has its own niche of methodology and nomenclature which can benefit a variety of challenges faced in astronomical research.

12.2 Concepts of spatial point processes

We start with some basic definitions (Stoyan 2006). The **point processes** considered here are sets of irregular patterns of points in a *p*-dimensional space. Most methods discussed here are developed for two- or three-dimensional spaces. Spatio-temporal methods can be applied to 2+1 dimensions where the third dimension is a fixed time-like variable. Point processes are **stationary** if their properties are invariant under spatial translation, and they are **isotropic** if their properties are invariant under rotation.

The **intensity** of a stationary point process is λ where

$$E[N] = \lambda v_p \tag{12.1}$$

is the expected number of points in a unit of volume v_p of *p*-space. In a **Poisson point process**, the distribution of the counts in such volumes follows a Poisson distribution,

$N \sim Pois(\lambda v_p)$ (Section 4.2). A **stationary** Poisson point process where λ is constant across the space produces a pattern commonly called **complete spatial randomness** (CSR). A sprinkle of sand grains on a table or the distribution of background events in a detector are examples of CSR. A more general Poisson point process is **inhomogeneous** where $\lambda(\mathbf{x})$ varies with location in the space; in two dimensions, $\mathbf{x}$ is a vector of (x, y) locations. Many other point process models have been developed, some emphasizing isolated clusters, others emphasizing textures, and yet others emphasizing anisotropic structures where the behavior is not invariant under rotations.

Often, the data points have associated nonspatial variables. In astronomy, these may be fluxes or luminosities in different bands, surface temperature, mass, radial velocity, proper motion and classificatory properties. These are called **marked point processes**. The **mark** can be derived from ancillary observations or from the spatial distribution itself, such as the distance to the nearest neighbor.

Spatial modeling has been developed principally for application in Earth sciences such as geology, geography, soil science, mining and petroleum engineering, and oceanography. The data may be pixelated images from remote sensing, bore or bathymetry measurements, measurements of rainfall, snow cover or soil type. The basic data structure is a spatially changing variable, or vector of variables, at sampling points that may be either irregularly spaced or in a regular grid. The model may be exploratory, constructing smoothed estimators from discrete observations, or explicative, seeking relationships between spatial patterns of different properties.

The spatial statistics described here each seek to quantify one of several related properties: Moran's I, Geary's c and the variogram γ treat **spatial autocorrelation** or variance (Section 12.4); kriging and other **interpolation** procedures map continuous spatial behavior (Section 12.5); and F, G, L, K and two-point correlation functions measure **spatial clustering** or aggregation across the full sample space (Section 12.6). The measures of autocorrelation and clustering use only the location of the points, while the interpolators can map both spatial densities and mark variables.

While any of these statistics can be calculated for a spatial dataset, some phenomena are most efficiently described as a clustering process, while others are best modeled as an autocorrelated random field or as a map with changing means and variances. The choice of approach is made by the scientist and may depend on the underlying model of the physical process. For example, clustering views the sample as aggregations of points based on stochastic dependencies, while mapping views the sample as representatives of a continuous density distribution.

Note that the global statistics of spatial autocorrelation, clustering and variance have important limitations for scientific inference. First, different spatial processes can lead to identical values of I or c and identical shapes of K or a variogram. Thus we cannot derive a definitive spatial model from these functions alone. This is well-known in galaxy clustering where the power-law angular two-point correlation function with power-low slope -1.8 was found to be compatible with many cosmological models. Second, the distributions of the statistics often depend on the spatial coordinate system, or equivalently, on the distance metric chosen. This matter is particularly important when the variables have different physical units, as in galaxy clustering studies when the two sky spatial variables are in degrees and the distance dimension is scaled to redshift. Converting these variables to consistent units (e.g. megaparsecs) requires assumptions regarding the Hubble expansion and localized velocity dispersions or flows. Third, these functions are often sensitive to the choice of area due to edge effects. Fourth, most of these functions assume that spatial autocorrelations are isotropic. This is clearly a serious limitation for galaxy clustering or molecular cloud studies where filamentary structures are pervasive.

12.3 Tests of uniformity

Unless strong structure is obviously present, an analysis of a spatial point process often begins with a test of the hypothesis that the distribution is consistent with CSR. Various quantities characterizing the spatial point process of a distribution can be compared with analytic predictions or simulations of CSR distribution. Several such tests for uniformity are reviewed by Diggle (2003) and Fortin & Dale (2005).

The e.d.f. of the **nearest-neighbor distribution** $G(r)$ can be constructed and compared with CSR expectations. For large n and ignoring edge effects, the CSR distribution in a two-dimensional area A containing n points is

$$G_{CSR}(r) = 1 - e^{-\pi nr^2/A}. \tag{12.2}$$

Simulations are needed when edge effects are important and for small-n samples. A similar result applies to the function F which measures the distances from a set of random locations in the space to the nearest object. The distributions G and F are identical for a homogeneous Poisson process, but differ when any form of point aggregation or regularity is present. Averages and ratios of G, F and similar statistics give less biased and more scale-free measures when CSR does not hold (Section 12.6.1).

Some CSR tests are based on sums of nearest-neighbor distances of a point process (Fortin & Dale 2005, Chapter 2). Consider, for example, the statistic

$$Q = \frac{\pi \lambda}{n} \sum_{i=1}^{n} W_{i1}^2 \tag{12.3}$$

where λ is the average surface density of points and W_{i1} is the Euclidean distance between the i-th point and its nearest neighbor (indicated by the subscript "1"). Q is asymptotically normal with $N(1, 1/n)$. A recent comparison of nearest-neighbor statistics found that a modified Pollard's P performs well using the first five nearest neighbors,

$$P(j) = \frac{12j^2 n \left[n \ln\left(\sum_{i=1}^{n} Y_{ij}^2/n\right) - \sum_{i=1}^{n} \ln(Y_{ij}^2) \right]}{(6jn + n + 1)(n - 1)} \tag{12.4}$$

where $j = 1, 2, 3, 4$ and 5 and Y_{ij} are the distances between i randomly placed locations and j represents the j-th nearest neighbor. If CSR holds, $(n - 1)P(j)$ follows a χ^2 distribution with $(n - 1)$ degrees of freedom. For clustered distributions, $P(j) > 1$.

The major difficulty with CSR tests based on nearest-neighbor and related statistics is bias due to edge effects. This problem is repeatedly encountered in spatial statistics. One approach is to use a reduced sample, omitting or reweighting points and areas which are closer to a survey edge than to their neighbor. Various procedures for such edge correction are in use. Another approach is to construct composite statistics that are less vulnerable to edge effects, such as the Lieshout–Baddeley J function discussed in Section 12.6.

12.4 Spatial autocorrelation

12.4.1 Global measures of spatial autocorrelation

The unbiased maximum-likelihood estimator of the intensity λ of a stationary Poisson point process is the intuitive value

$$\hat{\lambda} = \frac{n}{v_p} \tag{12.5}$$

where n is the number of points and v_p is the unit volume in p-space.

Moran's I and **Geary's contiguity ratio** c, developed in the 1940s and 1950s, are important indices of spatial autocorrelation that are extensions of Pearson's coefficient of bivariate correlation. They are localized, weighted correlation coefficients designed to detect departures from spatial randomness. The correlation is often inversely weighted with distance according to some kernel w. Here the spatial data are first grouped into m identical spatial bins, and m_j gives the count of objects in the j-th bin for $j = 1, 2, \ldots, n$. Then

$$
\begin{aligned}
I(d) &= \frac{1}{W(d)} \frac{\sum_{i=1 \neq j}^{n} \sum_{j=1 \neq i}^{n} w_{ij}(d)(m_i - \bar{m})(m_j - \bar{m})}{\frac{1}{n}\sqrt{\sum_{i=1}^{n}(m_i - \bar{m})^2}} \\
c(d) &= \frac{1}{2W(d)} \frac{\sum_{i=1 \neq j}^{n} \sum_{j=1 \neq i}^{n} w_{ij}(d)(m_i - m_j)^2}{\frac{1}{n-1}\sqrt{\sum_{i=1}^{n}(m_i - \bar{m})^2}},
\end{aligned}
\tag{12.6}
$$

where $\bar{m}$ is the global mean counts in the bins, $w_{ij}(d)$ is a matrix of the kernel function and $W(d)$ is the sum of w_{ij} elements. Plots of $I(d)$ against distance scale d are called **correlograms**. Fortin & Dale (2005) illustrate the behavior of correlograms for a variety of spatial patterns. Random clustering produces positive values on scales m less than the cluster sizes. Periodically spaced clusters, or samples with just a few distinct clusters, will produce oscillations around zero corresponding to the distance between clusters. Correlograms can also be extended to treat the spatial interactions of two or more populations.

The expected values of Moran's I and Geary's c for normal noise without clusters or autocorrelation are very simple,

$$E(I) = \frac{-1}{n-1} \to 0 \text{ for large } n$$
$$E(c) = 1. \tag{12.7}$$

When spatial autocorrelation is strong, Moran's I approaches 1 and Geary's c approaches 0. The significance level of I or c can be estimated by normal approximations, but tests on the full correlogram require care due to dependence between the values.

A vector indicating the strength of autocorrelation at different distance scales gives more information than a scalar measure like I or c. Astrophysical interest sometimes focuses on the dispersion, or variance, of a mapped quantity. For example, in a turbulent molecular cloud, spatial fluctuations in brightness and velocity on different physical scales reveal the dynamical state. This problem arises very often in Earth sciences, and the **variogram** is widely used to map the variance on different scales across the survey area. The term **semi-variogram** is often used because half of the average squared difference between pairs of values is plotted against the separation distance of the points,

$$\gamma(d) = \frac{1}{2m(d)} \sum_{j=1}^{m(d)} [z(\mathbf{x}_j) - z(\mathbf{x}_j + \mathrm{d})]^2. \tag{12.8}$$

Here the sum is over $m(d)$ pairs of points which are at a distance d from each other, and z is the mark variable measured at vector location $\mathbf{x}$. Anisotropic versions of the variogram have been developed for elongated patterns.

From a mathematical viewpoint, the variogram describes the second-moment properties of a spatial process $S(\mathbf{x})$,

$$V(\mathbf{x}_i, \mathbf{x}_j) = \frac{1}{2} Var[S(\mathbf{x}_i) - S(\mathbf{x}_j)]. \tag{12.9}$$

In the special case of a stationary, isotropic, Gaussian (i.e. the joint distribution of S is multivariate normal) spatial process, the variogram simplifies to the sum of two terms,

$$V(d) = \sigma^2[1 - \rho(d)] \text{ where}$$
$$\rho = \frac{Cov[S(\mathbf{x}_i), S(\mathbf{x}_j)]}{\sigma^2} \tag{12.10}$$

is the **spatial covariance function**, d is the distance between data points at $\mathbf{x}_i$ and $\mathbf{x}_j$, and σ^2 is the constant variance of the process.

Among geostatisticians, the σ^2 offset at zero distance in a variogram is called the **nugget effect**; it is attributable to variations smaller than the minimum separation of measured points. The correlation often increases to a higher value; the rise is called the **sill** and the distance at which the variogram asymptotes to a constant value is called the **range**. The sill is the total variance or asymptotic value of $\gamma(d)$ as $d \to \infty$; if γ does not converge, then the process is nonstationary and a trend should be removed. If the sill varies with direction, the process is anisotropic.

Variogram curves are typically modeled by simple functions. In a circular model, the variance grows linearly with separation to a fixed distance. Parabolic, exponential,

Gaussian and power-law models are also used, and multiple components may be present. Models are typically fit by weighted least-squares or maximum likelihood estimation. The variogram is a powerful tool in the geosciences due to the prevalence of Tobler's first law of geography: "Everything is related to everything else, but near things are more related than distant things." Study of astronomical point processes that follow this pattern will benefit from these geostatistical techniques. But some astronomical point processes, such as the anisotropic and multiscale characteristics of galaxy clustering, will be too complex to be fully characterized by a global variogram. A variogram derived from an astronomical dataset with a functional model is shown in Figure 12.2 below.

While these statistics give scalar and vector measures of the global autocorrelation, more sophisticated parametric models of stationary autoregressive images and spatial processes have been developed. An important class involves **Markov random fields (MRFs)** which require that the local density depends only on neighboring values and is independent of any other variables. Source-free background regions of astronomical images are often simple MRFs; the values are noise that is spatially autocorrelated by the point spread function determined by the telescope optics. The rippled background in images derived from interferometry might also be modeled as an MRF. A large field of image analysis has emerged based on modeling MRFs, particularly when the underlying statistics are Gaussian (Winkler 2002). The goal of characterizing the image in a few parameters may involve Bayesian inference methods such as maximum a postiori (MAP) estimation and Markov chain Monte Carlo (MCMC) calculations. Simulated annealing, genetic algorithms and multi-resolution methods can assist in covering large regions of model parameter space. In astronomical image analysis, the smoothness prior may be the telescope point spread function. Models can be used for image denoising, texture analysis, source and edge detection, feature recognition and deconvolution. Robust methods can treat outliers such as cosmic-ray effects in CCD images. MRF methods can be applied both locally and globally in an image.

12.4.2 Local measures of spatial autocorrelation

In various problems of geography, epidemiology and astronomy, the spatial autocorrelation function is nonstationary; that is, its characteristics vary across the surveyed region. This certainly applies to the galaxy distribution where the mean densities are inhomogeneous on all scales below $\sim$100 megaparsecs. For these situations, Anselin (1995) proposed a class of **local indicators of spatial association** (LISA) including maps of local values of Moran's I and Geary's c. The **local Moran's I**, for example, averages the spatial variance among the k-nearest neighbors of the i-th point. As with the global Moran's I in Equation (12.6), the scientist chooses in advance k, the number of neighbors for local averaging, and w_{ik}, the weight matrix such as a Gaussian kernel or inverse power law extending out to some chosen distance. In localized regions of high density, the k neighbors are at small distances from the i-th point and are heavily weighted by w_{ik}, giving high values of I_i. In regions of low density, the neighbors are more distant with low weights, giving low values of I_i. Under CSR assumptions, the expected value of the local Moran's I_i is a weighted version of Equation (12.7).

The local Moran's I_i is often used as an exploratory mapping tool for nonstationary point processes. While superficially resembling k-nearest-neighbor (k-nn) density estimation outlined in Section 6.5.2, it differs in that positive values of I_i occur when local values are either all above or all below the global mean. Negative values occur when the variations are on small scales such that the nearest k values are both above and below the mean. It is tricky to use this statistic for inference regarding the significance of nonstationarity, as adjustments are needed for the large number of significance tests performed and for violations of CSR assumptions.

Another LISA statistic is **Getis–Ord G^*** (Ord and Getis 2001) which measures the ratio of local averages to the global average of a point process. Maps of G^* are essentially a form of k-nn density estimation where the user has control over three parameters regulating the local averages: the number k of nearest neighbors in local averages; the weighting matrix w_{ik} which serves the role of the density kernel; and the scale d over which the G^* map is computed. G^* maps can have high fidelity in reproducing clustering patterns in spatial point processes but, as with local Moran's I_i, care must be taken in assessing the statistical significance of local variations in the presence of global spatial autocorrelation. A discussion of global and local mapping statistics is given by Fortin & Dale (2005).

For nonstationary point processes such as hierarchical galaxy clustering, an adaptive intensity λ in Equation (12.5) can be estimated locally at vector location $\mathbf{x}$. Illian *et al.* (2008, Chapter 4), for example, recommend a local average of the inverse of areas of Voronoi tiles; this procedure is applied to a galaxy dataset in Figure 12.6.

A variety of standard statistical procedures can be applied to spatial processes and images. For example, a t test can assess the correlation between two spatial processes, Pearson's r correlation coefficient can measure the correlation between two variables in a single spatial process, and a linear regression can find the slope across an image. But when spatial autocorrelation is present, the significance levels of the standard tests are altered because the effective sample size is smaller than the number of observations. Essentially, the assumption of independence in i.i.d. that underlies most hypothesis testing and model fitting has been violated.

12.5 Spatial interpolation

If a spatial point process is an observable representation of an underlying continuous distribution, then interpolation is an effective approach to estimating this distribution. The underlying quantity of interest could be the density of objects, where the spatial positions suffice for interpolation. But often we seek the spatial variation of a mark variable where interpolation is made between measurements of mass, temperature, abundance, gravitational potential, or other astrophysical variables made at distinct locations. A variety of interpolation methods was discussed in Chapter 6. Some are local regression where a functional form (such as a low-order polynomial) is fitted to local data values. Nonparametric kernel smoothing and k-nn methods can be applied. Here we discuss similar methods that typically arise in geostatistical studies of spatial data.

When Tobler's law applies, it makes sense that the value at unmeasured locations is principally determined by the closest measurements. A common model is **inverse distance weighted (IDW) interpolation** where the weighting of measures scales inversely as a power law of the distance to the location under consideration. The value z at vector location $\mathbf{x_0}$ is assigned a weighted average of measured values according to

$$\hat{z}(\mathbf{x}_0) = \frac{\sum_{i=1}^{n} z(\mathbf{x}_i) d_i^{-\alpha}}{\sum_{i=1}^{n} d_i^{-\alpha}} \qquad (12.11)$$

where $d_i = \sqrt{\mathbf{x}_i - \mathbf{x}_0}$ is the vectorial distance from the i-th measurement to $\mathbf{x}_0$ and $\alpha > 0$. The index α is usually chosen between 0 and 3 with smaller values giving smoother interpolations. An IDW interpolation of a mark variable in a galaxy spatial distribution is shown in Figure 12.7.

Other interpolation methods are based on weighted averages of local values derived from triangular or polygonal **tessellations** of the dataset. These methods include triangulated irregular network and natural neighbor interpolation. These estimators are appropriate when sharp discontinuities in the underlying variable are present which would be blurred by functional interpolations like IDW. When, in contrast, the user knows that the underlying distribution does not have rapid changes, then spline fits can be useful. They allow specification of constraints on the curvature, or tension, of the local curves fitted to the data.

The most common method for interpolation of marked spatial data in the geosciences is **kriging**, developed by geostatistician G. Matheron in the 1960s based on earlier work by mining engineer D. Krige. Kriging is a suite of related linear least-squares, minimum variance methods for density estimation in two or three dimensions with the goal of localizing features of inhomogeneous structures rather than globally characterizing homogeneous distributions. The continuous surfaces of the mark variable are derived from discrete data, either evenly or unevenly spaced, by interpolation using the variogram described in Section 12.4.1. For a stationary Gaussian spatial process, kriging gives the minimum mean square error (MSE) predictor for the unobserved continuous distribution.

In **ordinary kriging**, we seek the weightings λ_i at each of $i = 1, 2, \ldots, n$ points in the dataset that give the unbiased estimator of the continuous response variable z from the measurements z_i as a function of the location vector $\mathbf{x}$,

$$\hat{z}(\mathbf{x}) = \sum_{i=1}^{n} \lambda_i z(\mathbf{x}_i). \qquad (12.12)$$

Here $\mathbf{x}$ is the locational vector, (x, y) if two dimensions are involved, and z is the measured physical quantity (mass, gravitational potential, velocity, etc.). The weights, normalized so $\sum_i \lambda = 1$, are derived from the variogram γ emphasizing distances within the range of each data point. The weights are chosen to minimize the variance

$$\widehat{\sigma^2}(\mathbf{x}) = E[(\hat{Z}(\mathbf{x}) - Z(\mathbf{x}))^2]$$
$$= 2 \sum_{i=1}^{n} \lambda_i \gamma(\mathbf{x}_i - \mathbf{x}_0) - \sum_{i=1}^{n} \sum_{j=1}^{n} \lambda_i \lambda_j \gamma(\mathbf{x}_i - \mathbf{x}_j) \qquad (12.13)$$

where γ is the variogram in Equation (12.8) and $\mathbf{x}_0$ is some point chosen to be the origin. The λ coefficients are traditionally found using the method of Lagrangian multipliers, and are now found using maximum likelihood estimation or stochastic simulations.

These techniques allows estimation of the localized errors in the interpolated map. High variance can arise both from a sparsity of local measured points or choice of a poorly fitting variogram model. Applying the procedure can benefit from some tricks such as selection of a range to give sufficient measurements for interpolation but not too large to require inversion of a large matrix. Like a spline fit but unlike kernel density estimation, the kriged values at the measurement locations are exactly equal to the measured values. A kriging estimator and its variogram for a galaxy clustering sample is shown in Figure 12.8.

Many variants of kriging are in common use. **Simple kriging** is used if the mean is known prior to the observations. If the dataset is large, adjacent points can be merged in **block kriging.** If two correlated variables are measured at each location, then in **cokriging** they are both used for mapping using the cross-variogram. **Universal kriging** first removes trends in the data (often with polynomial surfaces) and then computes the kriging fit to the variogram of the residuals. If a secondary variable can map these trends, then kriging with an **external drift** can be used. **Nonlinear kriging** is used when the trends are not linear in the variables, such as sinusoidal, exponential or Gaussian functions. If the autocorrelation structure changes across the map so that a global variogram cannot be reliably applied, then local variograms can be computed in moving windows. **Multitype kriging** can treat categorical covariates, and multiple interacting marked variables can be treated using **multivariate kriging**. Monographs in geostatistics provide details on these procedures.

12.6   Global functions of clustering

12.6.1   Cumulative second-moment measures

Our discussion of data smoothing in Chapter 6 showed that the distribution of distances to neighboring points gives a measure of local density. Applying this idea to spatial point processes leads to a class of important global measures of clustering:

$$G(d) = \frac{\#[d_{NN}(\mathbf{s_i}) < d]}{n}$$

$$F(d) = \frac{\#[d_{NN}(\mathbf{p_i}) < d]}{n} \tag{12.14}$$

are the cumulative distribution of **nearest-neighbor (nn) distances** computed either from the data point locations ($\mathbf{s}_i$ for G) or a set of randomly chosen locations ($\mathbf{p}_i$ for F). F is often called the **empty space function** because it is centered around empty locations. Generally, for weakly clustered processes F is more sensitive than G, and conversely for strongly clustered processes.

Ripley's K function is the most widely used global measure of clustering. It is typically defined as the average number of points within distance d of observed points divided by the process intensity λ,

$$K(d) = \frac{1}{\hat{\lambda}n} \sum_{i=1}^{n} \#[S \text{ in } C(\mathbf{s}_i, d)] \qquad (12.15)$$

where C denotes a circle of radius d centered on the data point location $\mathbf{s}_i$. For a two-dimensional homogeneous Poisson process, $K(d)$ increases quadratically with d and $\lambda = n/a$ where a is the total area of the survey. The denominator can be replaced by $\hat{\lambda}^2$. For graphical convenience and to stabilize the variance, **Besag's L^* function** is commonly used in place of K where

$$L^*(d) = \sqrt{\frac{K(d)}{\pi}} - d \qquad (12.16)$$

in two dimensions, and $L^* \propto K^{1/p}$ in p dimensions. K and L^* contain the same information, but graphs of L^* appear more horizontal and, for a homogeneous Poisson process, have variances that are constant with d. Figure 12.4 shows the K and L^* functions for a galaxy clustering application.

Under the assumption of CSR, a homogeneous Poisson process with intensity λ, these three spatial autocorrelation functions have expected values

$$\begin{aligned} E[G(d)] &= E[F(d)] = 1 - e^{-\lambda \pi d^2} \\ E[K(d)] &= \pi d^2 \\ E[L^*(d)] &= 0. \end{aligned} \qquad (12.17)$$

$K(d)$ is asymptotically normal with variance

$$Var[K(d)] = \frac{2\pi d^2}{\hat{\lambda}^2 A} \qquad (12.18)$$

as the area $A \to \infty$, but this approximation is rarely accurate. Chetwynd & Diggle (1998) give a more complicated approximation to the variance of $K(d)$ for survey areas of arbitrary shape, again assuming CSR. For large samples, a model-free estimate of the variance can be obtained by computing $K_i(d)$ in i sub-areas and examining the variance with respect to the average of these $K_i(d)$ functions.

Scientists are cautioned that the F, G, K and L^* global measures of clustering can be strongly biased by edge effects. Every contribution to these functions is affected by edges at distances d comparable to the scale of the survey. The distances to neighbors will be overestimated, or equivalently, the number of points within a given distance will be underestimated, when edges are reached. Several edge-correction procedures for K and the other global statistics are in common use. For example, a weight w_i can be introduced inside the sum of Equation (12.15) representing the reciprocal of the fraction of the circumference of the circle of radius d around the i-th point that lies outside the survey area. Once edge-correction is applied, the variance of K can no longer be estimated analytically, and bootstrap resampling must be used to estimate significance levels and confidence intervals. Diggle (2003) gives a good discussion of various estimates of second-order functions, the variances and edge-correction strategies.

The **Lieshout–Baddeley J function** compares the inter-event distances G to distances from a fixed point F,

$$J(r) = \frac{1 - G(r)}{1 - F(r)}. \tag{12.19}$$

J has be shown to be highly resistant to edge effects while still sensitive to clustering structure (Lieshout & Baddeley 1996). Note that J can be viewed as the ratio of two survival functions but itself is not a c.d.f. or survival function as it is not bounded by the interval [0,1]. For CSR, $J(r) = 1$ for all r, although the converse is not always true. Values of $J(r) < 1$ indicate clustering and $J(r) > 1$ indicate spatial repulsion or a lattice-like distribution. $J(r)$ becomes constant beyond the scale of interaction. It seems to inherit the advantages of both F and G for weak and strong clustering processes, respectively. The J function for a galaxy clustering sample is shown in Figure 12.5.

In addition to measuring the clustering strength and scales of a single population, **clustering interactions** of two or more populations in the same survey can be compared using these global statistics (Fortin & Dale 2005, Chapter 2). For example, the counts of a single population in Equation (12.15) can be replaced by counts of one population around points of the other population. The statistic

$$L_{12}^*(d) = \sqrt{\frac{n_2 K_{12}(d) + n_1 K_{21}(d)}{\pi (n_1 + n_2)}} - d \tag{12.20}$$

will then be positive on scales where the two populations are segregated, negative where they are aggregated, and zero where then have no interaction. This procedure might be used, for example, to study the distribution of elliptical and spiral galaxies in rich clusters. Other extensions to the global clustering functions can reveal clustering dependence on mark variables.

Though not as well developed as the isotropic statistics outlined above, some of the statistics discussed here can be formulated for anisotropic spatial point processes. These appear in astronomy in gravitationally bound systems that have not yet reached virial equilibrium, most importantly in the filaments and voids appearing in the large-scale structure and the distribution of subgalactic clumps during galaxy formation. Measures include orientation distribution functions derived from Ripley's K and other second-order statistics, the nearest neighbor orientation function, the empty space function, and certain morphological functions. These are described Illian *et al.* (2008, Chapter 4).

12.6.2 Two-point correlation function

Rather than using K, J or related global statistics to quantify spatial autocorrelation, astronomers have used the **two-point correlation function** in hundreds of studies since the seminal work by Totsuji & Kihara (1969) and Peebles (1973) on galaxy clustering. The two-point correlation function counts the number of objects in annuli around each point rather than counts within circles around each point, and is thus related to the differential of the K function.

The astronomers' two-point correlation function is derived from the joint probability P_{12} that two objects in a spatial point process lie in infinitesimal spheres of volume (area in two dimensions) dV_1 and dV_2 around two vector locations $\mathbf{x}_1$ and $\mathbf{x}_2$ (Martínez & Saar 2002, Chapter 3),

$$dP_{12} = \lambda_2(\mathbf{x}_1, \mathbf{x}_2)dV_1 dV_2 \tag{12.21}$$

where λ_2 is the second-order product density of the process. The astronomers' two-point correlation function $\xi(\mathbf{d})$ is a function of vectorial distance $\mathbf{d}$ between $\mathbf{x}_1$ and $\mathbf{x}_2$ formed by treating λ_2 as the sum of two components, one a uniform process with volume density $\bar{\rho}$ and the other a correlated process,

$$dP_{12} = \bar{\rho}^2[1 + \xi(\mathbf{d})]dV_1 dV_2. \tag{12.22}$$

The two-point correlation function ξ thus measures the covariance of the nonuniform structure of the point distribution. In most studies, the process is assumed to be stationary and isotropic within the survey under investigation, so the two-point correlation function becomes a function of scalar distance,

$$\xi(d) = \frac{\lambda_2(d)}{\bar{\rho}^2} - 1. \tag{12.23}$$

In statistics, the measure $\xi(d) + 1$ is called the **pair correlation function**. It is the differential of Ripley's K function discussed in Section 12.6.1; ξ is computed in annuli specified by the scientist while K is computed for all distances d without binning.

Astronomers have developed several estimators of the differential $\xi(d)$ function which, as with discussions of the cumulative K and related functions, seek to reduce the bias associated with edge effects (Martínez & Saar 2002). They combine the concepts of the G and F functions which measure points around measured points and around random locations, respectively, although using Monte Carlo simulations of CSR distributions. The simulations can incorporate complicated selection effects of the observations such as magnitude limits convolved with the galaxy luminosity function, the redshifting of galaxy spectra, and intrinsic galaxy color evolution. The 1974 Peebles–Hauser estimator considers the ratio of the number of point pairs around observed data points, $DD(d)$, to the number of point pairs in a CSR simulated dataset, $RR(d)$,

$$\hat{\xi}_{PH} = \left(\frac{n_{CSR}}{n}\right)^2 \frac{DD(d)}{RR(d)} - 1 \tag{12.24}$$

where $n_{CSR} \gg n$ is the number of points in the simulated sample and n is the number of points in the observed sample. Replacing n_{CSR}/n^2 by $n_{CSR}(n_{CSR} - 1)/n(n - 1)$ makes the estimator unbiased for a Poisson process. The counts are obtained in annuli $(d, d + \Delta d)$ around each location, and edge effects are treated by placing the CSR distribution into the same survey area as the observations.

Several estimators of the two-point correlation function were introduced as alternatives to ξ_{PH} with improved performance: the 1983 Davis–Peebles $\hat{\xi}_{DP}$, 1993 Hamilton $\hat{\xi}_H$, and

1993 Landy–Szalay $\hat{\xi}_{LS}$ estimators:

$$\hat{\xi}_{DP} = \frac{n_{CSR}}{n}\frac{DD(d)}{DR(d)} - 1$$

$$\hat{\xi}_{H} = \frac{DD(d)\,RR(d)}{DR(d)^2} - 1$$

$$\hat{\xi}_{LS} = 1 + \left(\frac{n_{CSR}}{n}\right)^2 \frac{DD(d)}{RR(d)} - 2\frac{n_{CSR}}{n}\frac{DR(d)}{RR(d}$$

(12.25)

where $DR(d)$ is the number of pairs between the observed and simulated CSR distributions in the $(d, d + \Delta d)$ annulus. They are designed to improve treatment of edge effects and to reduce bias from inaccuracies in the estimates of the mean density $\bar{\rho}$ and large-scale autocorrelation. The estimators can be extended to higher orders such as the three-point correlation function. Martínez & Saar (2002) and others evaluate the performance of these estimators and their variances, discussing their relationship to measures of spatial clustering developed by statisticians.

The variance of the two-point correlation estimators is often estimated with a normal approximation,

$$\widehat{Var}[\xi(d)] = \frac{[1 + \xi(d)]^2}{DD(d)}.$$

(12.26)

Terms involving the simulated CSR datasets are small because $n_{CSR} \gg n$. However, it is recognized that this estimator does not take the spatial autocorrelation into account. Straightforward bootstrap variances are also biased by the prevalence of multiple points with zero separations. A **block bootstrap** approach can be more effective if the sample is sufficiently large, where local structure within a block is maintained. A promising new approach has been developed by Loh (2008) where each point is assigned a mark variable associated with its local correlation function, and these marks are efficiently bootstrapped to give variances to the global two-point correlation function.

As mentioned above, the pair correlation function of spatial analysis,

$$g(d) = \xi(d) + 1,$$

(12.27)

is related to the first derivative of the K function, K', in p dimensions as (Illian *et al.* 2008, Chapter 4)

$$g(d) = \frac{K'(d)}{pv_p r^{p-1}}$$

$$= \frac{K'(d)}{2\pi d} \quad \text{for} \quad p = 2.$$

(12.28)

Thus, the relationship of $\xi(d)$ and $g(d)$ to $K(d)$ is similar to that of a probability density function $f(x)$ to its corresponding distribution function $F(x)$. To compute $g(d)$, Illian *et al.* (2008) recommend that the point process first be smoothed with a kernel density estimator and then divided by an edge-corrected estimator of $\widehat{\lambda^2}$ recommended by Hamilton and Landy & Szalay. Other researchers apply spline smoothing to the function $Z(d) = K(d)/4\pi d^2$), set $Z(0) = 1$, and numerically compute its derivative.

Although the two-point correlation function contains the same information as the K function, it is not widely used by statisticians or scientists in other fields. This is largely due to difficulties in its computation as a differential rather than cumulative estimator. For all methods, choices must be made regarding the binning to form the pair correlation function and, as discussed in Section 6.3, statisticians are hesitant to construct binned estimators without mathematical guidance on how the grouping is best achieved. The scientist must choose the bin origin, width (if constant) or width function (if variable), and center or centroid of the bin as an average distance. The K function, like the cumulative e.d.f. discussed in Section 5.3.1, is free from these issues; it is directly computed from the point process data without the need to make decisions on binning. The cumulative estimators are more stable for small samples.

Despite its infrequent use outside of astronomy, Illian *et al.* (2008) state in their comprehensive monograph on spatial point processes that "[we] recommend the pair correlation function $g(r)$ as the best, most informative second-order summary characteristic, even though its statistical estimation is relatively complicated ... due to the serious issues of bandwidth choice and estimation for small r. ... While it contains the same statistical information as the K- or L-function, it offers the information in a way that is easier to understand, in particular for beginners."

12.7 Model-based spatial analysis

Parametric models of spatial point processes have a structure similar to models discussed in Chapters 7 and 11 on regression and time series. The hypothesis of CSR can be written as a mean plus simple global noise term following a Poisson distribution,

$$X = \mu + \epsilon \tag{12.29}$$

where $\epsilon = Pois(\lambda)$. This spatial pattern has no autocorrelation, clustering or trend.

For non-CSR spatial point processes, as noted in Section 12.2, we can take two rather different approaches to modeling. One is based on a concept of independent interacting points producing clusters or other density variations, and the other is based on a concept of discrete representations of an underlying continuous process. Many implementations of these approaches have been developed with techniques for parameter estimation (using second-order functions, maximum likelihood estimation or Bayesian methods) and goodness-of-fit testing (Diggle 2003, Chapters 5–7; Illian *et al.* 2008, Chapters 6 and 7). We first review examples from astronomical treatments of galaxy clustering, and then a few examples from statistics of point processes in general.

12.7.1 Models for galaxy clustering

Galaxy clustering is often viewed as a manifestation of a continuous process which arises from cosmological growth of structure in the expanding Universe. Perhaps the first parametric model, without cosmological interpretation, was Edwin Hubble's demonstration in the 1930s that galaxy counts in spatial cells follow a lognormal distribution. More complex models emerged as the astrophysical theory of the growth of large-scale structure developed. In a seminal study of the self-similar growth of gravitating structure, Press & Schechter (1974) derived the number distribution $n(M, R)$ of galaxy clustering following

$$n(M, R) = aM^{-1-\alpha} e^{-b(M^{1-\alpha}/R)^2} \tag{12.30}$$

where the mass M scales with the number of galaxies, R is the region radius over which the galaxies are counted, $1/3 \leq \alpha \leq 1/2$, and a and b are scalar parameters.

Saslaw (2000) similarly derived a quasi-equilibrium distribution of galaxies where the probability of finding n galaxies in volume V is

$$P(n, V) = \frac{\bar{n}V(1-b)}{n!} [\bar{n}V(1-b) + nb]^{n-1} e^{-\bar{n}V(1-b)+nb} \tag{12.31}$$

where $\bar{n}$ is the expected value of n and b is a scalar parameter arising from gravo-thermodynamic theory.

Bardeen *et al.* (1986) adopted a model based on Gaussian random fields where galaxies form at peaks of dark matter distributions in the expanding Universe. They derived amplitude distributions and n-point spatial correlation functions of density peaks in general three-dimensional Gaussian fields representing galaxy clustering. At locations of high dark matter densities ν, the number distribution of galaxies is predicted to follow

$$n(\nu) = a(\nu^2 - 1)e^{-\nu^2/2}, \quad \nu > 1. \tag{12.32}$$

The alternative approach of individual interacting particles was first applied to galaxy clustering by the astrostatistician team of Jerzy Neyman and Elizabeth Scott in the 1950s. They modeled galaxy counts as the result of two Poisson processes, one placing cluster seeds in space and a second constructing "daughter" clusters around the seeds. These models are now considered to be artificial, as realistic simulations of the origins of galaxy clustering are now possible with large-scale N-body simulations of gravitational attraction between dark matter galactic halos in an expanding universe (Springel *et al.* 2005). Such simulations can be mined to study the origins of galaxy clustering, luminosity distributions and internal structures.

Extensions of the **Neyman–Scott double-Poisson model** for galaxy clusters are commonly used in spatial statistics, though less so in astronomy. **A Matérn cluster process**, for example, constructs clusters with power-law density profiles around the Poisson-distributed seeds. **Cox processes** are flexible model families involving two stochastic processes, one to establish an inhomogeneous intensity function across the space and a second to place individual objects proportional to the local density. The interaction between particles can often be expressed as a function of pairwise distance; these are called **Gibbs processes** (in physics) or **Markov point processes** (in statistics).

Finally, we note that a wide range of specialized parametric modeling and nonparametric procedures have been developed for studies of spatial point processes in cosmology. Some methods use Fourier analysis or topological analysis of galaxy redshift surveys to characterize clustering, including the bispectrum to study non-Gaussianity. Other methods seek to map dark matter fields through indirect means including reconstruction of "peculiar" velocity fields and spatial shear fields of background galaxies from gravitational lensing by foreground structures. These are discussed in Chapters 8−10 of Martínez & Saar (2002). Some methods of finding individual clusters in any multidimensional point process are discussed in Section 9.3.

12.7.2 Models in geostatistics

A wide range of spatial interaction models can be formulated as a regression problem by generalizing Equation (12.29). For example, the mean density of objects μ can be spatially varying in a fashion linearly dependent on a vector of mark variables $M(\mathbf{x})$

$$X(\mathbf{x}) \;=\; \mu(\mathbf{x}) + \epsilon \quad \text{where}$$
$$\mu(\mathbf{x}) \;=\; \beta \mathbf{M}(\mathbf{x}). \tag{12.33}$$

The noise ϵ can follow a Poisson distribution, zero-mean Gaussian distribution or a Markov random field. The formulation, parameter fitting, validation, simulation and interpretation of clustered and autocorrelated spatial point process models can be a complex enterprise. Interested readers are referred to Illian *et al.* (2008, Chapters 6 and 7), Möller & Waagepetersen (2003) and Anselin & Rey (2010).

Some relatively simple classes of point process regression models are widely used in geostatistics. In **geographically weighted regression (GWR)**, maps of several variables are available and one seeks to explain one map in terms of the others with a linear regression that takes spatial autocorrelation into account. The response random variable $Y(\mathbf{x})$ at vector location $\mathbf{x}$ depends both on the value of covariates $\mathbf{X}(\mathbf{x})$ and on neighboring values of Y,

$$Y(\mathbf{x}) = \mu(\mathbf{x}) + \beta(\mathbf{x})\mathbf{X}(\mathbf{x}) + \epsilon \tag{12.34}$$

where $\epsilon = N(0, \sigma^2)$ is the usual normal noise term (Section 7.2). If the spatial dependency of μ and β can be expressed as a localized kernel matrix w_{ij}, then a least-squares solution is found by iterative minimization of the quantity

$$\min_{\hat{\mu}(\mathbf{x}_i), \hat{\beta}(\mathbf{x}_i)} \sum_j \left[y_j - \hat{y}_j(\hat{\mu}(\mathbf{x}_i), \hat{\beta}(\mathbf{x}_i)) \right]^2 w_{ij} \tag{12.35}$$

at each location $(\mathbf{x})$. Common forms of the kernel w_{ij} include the normal function and $[1 - d_{ij}^2/h^2]^{-2}$ where d_{ij} is the distance to k nearby data points. The choice of kernel width h can be based on cross-validation (Section 6.4.2) or a penalized likelihood like the Bayesian information criterion (Section 3.7.3). The k-th-nearest-neighbor (k-nn) kernel is also adaptive to the local density. Note that the goal of GWR is to map explanatory regression coefficients, whereas the goal of kriging is to accurately map the measured quantity. GWR techniques are presented by Fotheringham *et al.* (2002).

Another approach to modeling multivariate spatial autocorrelation assumes the regression coefficients are global rather than local, but the response variable is also dependent on its neighboring values. We can write this **spatial autoregressive (SAR)** model in a hierarchical fashion analogous to our treatment of measurement errors in Section 7.5,

$$Y(\mathbf{x}) = \mu + \delta(\mathbf{x})$$
$$\delta(\mathbf{x}) = \beta \mathbf{W}(d)\delta(\mathbf{x}) + \epsilon \tag{12.36}$$

where μ is a mean value, $\delta(\mathbf{x})$ are spatially correlated residuals, $-1 < \beta < 1$ gives the strength of the spatial autocorrelation, $\epsilon = N(0, \sigma^2)$ is a white noise error term, and $\mathbf{W}(d)$ is a normalized weight matrix dependent on the distance d from the data point at $\mathbf{x}$. $\mathbf{W}(d)$ plays the role of a kernel in density estimation (Section 6.4). Common options include a uniform distribution $\mathbf{W} = 1$ out to some fixed distance d_0, a Pareto distribution $\mathbf{W} \propto d^{-\alpha}$ with $1 \leq \alpha \leq 3$, a normal distribution $\mathbf{W} \propto N(0, d_0)$, or a restriction to k nearest neighbors. Astronomers might insert the telescope point spread function for $\mathbf{W}$ in the case of autocorrelated images. The variance–covariance matrix for SAR models is then

$$\mathbf{\Omega} = \sigma^2[(\mathbf{I} - \beta\mathbf{W})'(\mathbf{I} - \beta\mathbf{W})]^{-1} \tag{12.37}$$

where $\mathbf{I}$ is the identity matrix.

These SAR models are spatial analogs of temporal autoregressive models discussed in Section 11.3.3 but, unlike temporal models where current values depend only on past behavior, spatial values can depend on behavior in any direction around the location in question. Recall that whenever any non-CSR structure is present, the effective sample size is less than the observed sample size of n points.

12.8   Graphical networks and tessellations

While the radial distribution of nearest neighbors is considered in various methods discussed above, the p-dimensional distribution of neighbors provides more detailed information about the local autocorrelation and clustering structure of a point process. The visualizations of these connections between nearby points are important elements of **graph theory**. The **Delaunay triangulation** joins all neighboring triplets of points with triangles. The network of line segments that connects all points without closed loops having the smallest total length is called the **minimal spanning tree (MST)**.

Dirichlet (or Voronoi or Thiessen) tessellations are based on line bisectors perpendicular to the Delaunay line segments, partitioning the space into polygons around the data points. The locations within each polygon are closer to an associated data point than any other data point. Analysis of spatial point processes can be based on the nearest-neighbor, MST, Delaunay triangulation or Dirichlet tessellation. The mathematics and applications of these graph networks and their applications to spatial point processes are reviewed at an elementary level by Fortin & Dale (2005, Chapter 2), examined with respect to galaxy clustering by Martínez & Saar (2002), and presented in detail by Okabe et al. (1999). Bayesian estimation methods applied to tessellation and related methods for cluster analysis are discussed by Lawson & Denison (2002).

MSTs and Dirichlet tessellations can address a variety of issues concerning spatial point processes. Progressive pruning of the long branches of the MST gives a well-defined procedure for defining a hierarchy of spatial clusters which can then be analyzed for scientific purposes. This procedure is identical to the nonparametric single-linkage hierarchical agglomerative clustering algorithm discussed in Section 9.3.1, which is colloquially known in the astronomical community as the **friends-of-friends** algorithm. The distribution of MST branch lengths or Dirichlet tile areas gives insight into the clustering structure that is complementary to that provided by the second-order statistics like K and J. Density estimators or interpolators of the point distribution can be constructed because the local density of points is related to the inverse of the area of local tiles. This is illustrated with a galaxy clustering sample in Figure 12.6.

These graph networks have been used in studies of astronomical point processes to a moderate extent. Ebeling & Wiedenmann (1993) developed an influential application to the location of photons in a detector. Combining Voronoi tessellation and percolation, they discriminate uniform Poisson background noise from clusters of photons associated with faint X-ray or gamma-ray sources. The method is particularly effective in finding low-surface-brightness extended structures in noise, and has been used as an adaptive smoothing algorithm for NASA's Chandra X-ray Observatory.

12.9 Points on a circle or sphere

Statistical methods have been developed over several decades for points on a circle (Jammalamadaka & SenGupta 2001) and on a sphere (Fisher *et al.* 1987). Directional data are valuable in astronomy for study of orientations in the sky, distributions on the celestial sphere, and vectorial quantities such as polarization angles or dynamical flows.

Moments of a univariate distribution of angles on a circle, such as the mean and variance, are not generally useful because values near 0 and 2π are closer than values near 0 and π. Angular values must be summed vectorially. Assuming unit lengths, the average direction $\bar{\theta}$ of a sample of angles $\theta_1, \theta_2, \ldots, \theta_n$ is the solution of

$$\cos(\bar{\theta}) = C/R \text{ and } \sin(\bar{\theta}) = S/R \text{ where}$$
$$C = \sum_{i=1}^{n} \cos\theta_i, \ S = \sum_{i=1}^{n} \sin\theta_i, \text{ and}$$
$$R^2 = C^2 + S^2. \tag{12.38}$$

The quantity

$$\bar{R} = R/n, \tag{12.39}$$

called the **mean resultant length**, approaches 1 when the data are clustered towards one direction and approaches 0 when they are distributed in a balanced (though not necessarily uniform) fashion. The **sample circular variance** and standard deviation are

$$Var = 1 - \bar{R}$$
$$s.d. = \sqrt{-2\ln(1 - Var)}. \tag{12.40}$$

Note that the circular standard deviation is not the square root of the variance as it is with functions on the real line. A more useful measure of spread is the **circular dispersion**

$$\delta = \frac{1 - \sum_{i=1}^{n} \cos 2(\theta_i - \bar{\theta})/n}{2\bar{R}^2}. \tag{12.41}$$

From a parametric viewpoint, aside from the uniform distribution $f_U = 1/2\pi$, circular data are often modeled by ordinary probability density functions g that are wrapped around a circle of unit radius. The **circular normal** distribution introduced by R. von Mises is the most commonly used model for unimodal deviations from a uniform distribution. The **von Mises distribution** is

$$f_{VM}(\theta) = \frac{1}{2\pi I_0(\kappa)} e^{\kappa \cos(\theta - \mu)} \quad \text{where}$$

$$I_0 = \sum_{r=0}^{\infty} \left(\frac{\kappa}{2}\right)^{2r} \left(\frac{1}{r!}\right)^2, \tag{12.42}$$

where μ is the mean direction, κ is the **concentration parameter**, and I_0 is the modified Bessel function. The von Mises distribution is circular for $\kappa = 0$ (uniform distribution of angles), has an egg-like shape for intermediate $2 \leq \kappa \leq 5$ (moderately concentrated distribution of angles), and a very elongated shape for large κ (narrow range of angles). The maximum likelihood sample mean and mode coincide at angle μ.

The maximum likelihood estimator for von Mises parameters is $\hat{\mu} = \bar{\theta}$ and $\hat{\kappa}$ is the solution to the equation

$$A(\kappa) \equiv I_1(\kappa)/I_0(\kappa) = \bar{R}. \tag{12.43}$$

While these Bessel functions should be evaluated numerically, an approximation for large κ gives $A(\kappa) \simeq 1 - 1/(2\kappa) - 1/(8\kappa^2)$ while for small κ, $A(\kappa) \simeq \kappa/2 - \kappa^3/16$. The uncertainties of these estimators can be estimated from asymptotic normal approximations, the Fisher information matrix or parametric bootstrap resampling.

Jammalamadaka & SenGupta (2001) give these and other procedures for the von Mises distribution including goodness-of-fit tests, estimation of mixture models, tests for pre-selected μ_0 or κ_0 values, two- and k-sample tests, bivariate correlation, circular-linear and circular-circular bivariate regression. The mathematics for the von Mises distribution is more straightforward than for the wrapped normal distribution or other unimodal circular distributions. For example, if an angular variable θ follows the von Mises distribution, $\theta \sim vM(\mu, \kappa)$, then the quantity $\beta = \sqrt{\kappa}(\theta - \mu)$ follows the standard normal distribution, $\beta \sim N(0, 1)$. For Bayesian estimation of the von Mises distribution, the μ and κ parameters do not have simple conjugate priors, but some results have been obtained.

The generalization of the von Mises distribution for spherical data is known as the **Fisher p.d.f.** It is rotationally symmetric around the mean direction and, as in Equation (12.42), has a concentration parameter κ measuring the spread of angles around this mean direction. The **Watson distribution** is closely related to the Fisher distribution but has two antipodal elongations; it is used to model bipolar point distributions on a sphere. The **Kent distribution**, the analog of the bivariate normal distribution on the surface of a sphere, is appropriate for elliptical point patterns with arbitrary orientation and spreads.

Other spherical distributions are effective for point distributions that concentrate in a small circular region or girdle a great circle or other curve.

For both circular and spherical data, a number of nonparametric tests are available to evaluate angular distributions. These can be used either when the distributions appear smooth or when they show complicated multimodality. Rank tests are used to evaluate whether two distributions are significantly different, or whether they have significantly different mean directions. A test against the hypothesis of a uniform distribution of angles, $f_U = 1/(2\pi)$, was developed by Lord **Rayleigh** in 1919, rejecting uniformity when the resultant length is too long. It was later improved by **Kuiper** for improved sensitivity against bimodal distributions. **Watson's U^2 test** is a circular version of the nonparametric Cramér–von Mises test, and others like **Bingham's test** and **Giné's G_n test** have been developed. Rank and linear correlation coefficients have been developed to evaluate possible correlation between an angular variable and a second angular or linear variable. These and other applications (goodness-of-fit tests, robust methods, density estimation, simulation and bootstrap) for directional data are reviewed by Jupp & Mardia (1989). Some additional procedures for spherical data are available. Astronomers treating distributions from all-sky surveys might consider several nonparametric kernel density estimators for spherical data presented by Hall *et al.* (1987).

12.10 Remarks

Many astronomers study the spatial distributions of galaxies in space, stars in the Galaxy, photons on an image, and object properties in p-space. But their analyses have generally been restricted to a few statistical methods such as (adaptive) kernel smoothing or the friends-of-friends algorithm to localized individual features and the two-point correlation function as a global measure of clustering. While these are not weak techniques, they represent only a fraction of the repertoire developed in the field of spatial analysis and commonly used in the Earth sciences.

Discriminating the concepts of spatial patterning, and consequently choosing which spatial statistics should address particular objectives, is sometimes confusing (Fortin & Dale 2005). The topology of a point process can be studied through network statistics such as minimal spanning trees, Voronoi tessellation and nearest-neighbor statistics (Martínez & Saar 2002). Detection and characterization of spatial structure is facilitated by global variance statistics like Moran's I and Geary's c, variograms, second-order functions like K or the two-point correlation function, and transformation methods like Fourier analysis or wavelets. Statistical inference involving hypothesis testing and parameter estimation can be pursued using variograms, second-order functions, autoregressive and random field modeling. Interpolation can be performed using kernel density estimation, inverse distance weighting, spline fits and kriging. Wavelet applications to spatial point processes are not common in the Earth sciences, but are promising for astronomical problems involving both spatial distributions of galaxies and photons in sparsely filled images. Wavelet and related (such as curvelet and ridgelet) methods are presented with astronomical applications in the monographs by Starck & Murtagh (2006) and Starck *et al.* (2010).

Astronomers have long experience in studying galaxy clustering, most commonly using the two-point correlation function that is the differential form of Ripley's K function, the most commonly used tool for clustering analysis of spatial point processes in other fields. A large literature in astronomy has developed these binned differential clustering measures in parallel to the large literature in spatial statistics based on unbinned cumulative clustering measures. The monographs by Martínez & Saar (2002) and Illian *et al.* (2008) provide important links between these approaches.

We encourage astronomers to examine the J and L^* functions which have high sensitivity to spatial clustering and low sensitivity to edge effects, in studies of spatial clustering. Important opportunities for the analysis of photon-counting astronomical images are also not being adequately exploited. For example, the extensive methodology based on Gaussian and Markov random fields may be promising for treating spatial autocorrelation arising from an independently known telescopic point spread function.

12.11   Recommended reading

Diggle, P. J. (2003) *Statistical Analysis of Spatial Point Patterns*, 2nd ed., Arnold, London
   A slim volume presenting many useful methods in spatial analysis. Coverage includes tests of CSR, nearest-neighbor and second-order correlation functions, inhomogeneous and clustered models, parameter estimation and nonparametric methods.

Fortin, M.-J. & Dale, M. (2005) *Spatial Analysis: A Guide for Ecologists*, Cambridge University Press
   A readable, less mathematical introduction to spatial analysis methods with clear summaries of methods. Issues treated include spatial autocorrelation and stationarity, nearest-neighbor and second-order statistics, circumcircle methods, global autocorrelation statistics, interpolation and kriging, spatial clustering and partitioning and spatio-temporal analysis.

Fotheringham, A. S. & Rogerson, P. A., eds. (2009) *The SAGE Handbook of Spatial Analysis*, SAGE, London
   An up-to-date collection of informative and engaging review articles on spatial analysis from the perspective of geographers. Topics include data visualization, geographic information systems, data mining, geostatistics, clustering, spatial autocorrelation, interpolation, spatial regression, Bayesian analysis and neural networks.

Illian, J., Penttinen, A., Stoyan, H. & Stoyan, D. (2008) *Statistical Analysis and Modelling of Spatial Point Patterns*, John Wiley, New York
   A comprehensive and advanced monograph on modern methods of spatial analysis. Topics include the homogeneous and inhomogeneous Poisson point process, finite and stationary point processes, nearest-neighbor and empty space distributions, second- and higher-order functions, tessellations, marked and multivariate point processes, cluster and random field modeling, parameter estimation and goodness-of-fit testing.

Martínez, V. J. & Saar, E. (2002) *Statistics of the Galaxy Distribution*, Chapman & Hall/CRC, Boca Raton

An excellent and authoritative presentation of both statistical and astronomical perspectives on galaxy clustering as a spatial point process. Statistical topics include: the pair correlation, void probability, *K* and *J* functions; Neyman–Scott and Cox models; Voronoi tessellations; multi-fractal analysis; Gaussian random fields; Fourier analysis; Wiener filtering; topological measures; Minkowski functionals and shape statistics; minimal spanning tree; wavelets; and cluster finding algorithms.

12.12 **R** applications

We illustrate **R**'s graphical and analytical capabilities for spatial point processes using a galaxy redshift survey of the Shapley Supercluster, the richest concentration of galaxies in the nearby Universe (Drinkwater *et al.* 2004). The dataset has 4215 galaxies over ∼100 square degrees with five variables: two spatial dimensions, right ascension and declination; optical magnitude which is an inverse logarithmic measure of a galaxy's brightness; radial velocity (equivalent to redshift) which is a distorted measure of the third dimension, distance; and the measurement error of the radial velocity. We can consider this to be a marked spatial point process with three spatial dimensions and two ancillary variables.

The three spatial variables − right ascension, declination and velocity − are plotted in Figure 12.1 using three different visualization tools. First, the full set of 4215 galaxies is plotted in three dimensions using the *rgl* package in core **R**. The *rgl* package is a real-time rendering device driver that can generate three-dimensional graphical shapes and terrains, lighting, background environments, translucency, and so forth. We use its *rgl.points* function after standardizing the variables using the *scale* function. Second, a region of low galaxy density without major clusters is shown using **R**'s standard *plot* function. This is a two-dimensional plot where the third dimension, galaxy velocity, is used to scale the symbol sizes of individual points. Third, a region of high galaxy density is displayed with the **CRAN** package *scatterplot3d* (Ligges & Mächler 2003).

```
# Construct three galaxy redshift datasets, plot using spatstat

shap <- read.table('http://astrostatistics.psu.edu/MSMA/datasets/Shapley_galaxy.dat',
   header=T, fill=T)
attach(shap) ; dim(shap) ; summary(shap)
shap.hi <- shap[(R.A. < 205) & (R.A. > 200) & (Dec. > -34) & (Dec. < -29) ,]
shap.lo <- shap[(R.A. < 214) & (R.A. > 209) & (Dec. > -34) & (Dec. < -27) ,]
shap.clus <- shap[(R.A. <204.5) & (R.A. > 200.4) & (Dec. > -32.5) & (Dec. < -31.0)
   & (Vel > 11000) & (Vel < 18000),]

# Plot in 3-dimensions using rgl, plot and scatterplot3d
```

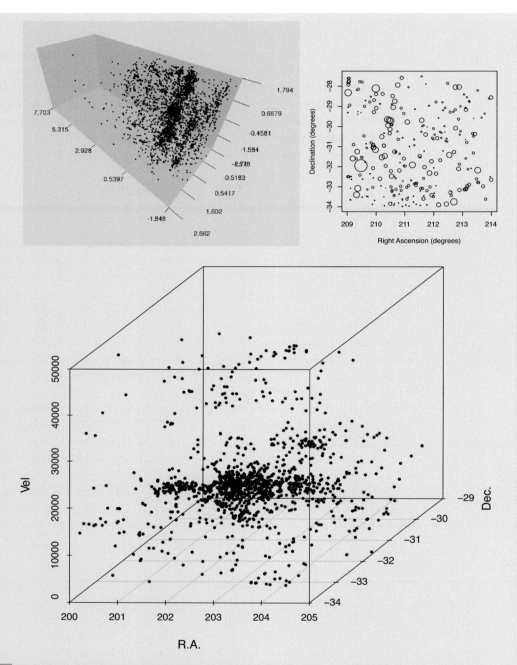

Fig. 12.1 Three-dimensional views of the Shapley Supercluster galaxy redshift survey: *Top left*: Full survey produced with **R**'s *rgl* package; *top right*: low-density region produced with **R**'s *plot* function with symbol sizes scaled to galaxy distance; *bottom*: high-density region produced with **CRAN**'s *scatterplot3d* package.

```
install.packages('rgl') ; library(rgl)
rgl.open()
rgl.points(scale(shap[,1]), scale(shap[,2]), scale(shap[,4]))
rgl.bbox()
rgl.snapshot('Shapley.png')
rgl.close()

plot(shap.lo[,1], shap.lo[,2], cex=(scale(shap.lo[,4])+1.5)/2,
   xlab='Right Ascension (degrees)', ylab='Declination (degrees)')

install.packages('scatterplot3d') ; library(scatterplot3d)
scatterplot3d(shap.hi[,c(1,2,4)], pch=20, cex.symbols=0.7, type='p', angl=40,
   zlim=c(0,50000))
```

12.12.1 Characterization of autocorrelation

The **CRAN** package *spdep* provides global and local calculations of Moran's I, Geary's c, and Getis–Ord G^* statistics as well as other functionalities such as the minimal spanning tree (Bivand *et al.* 2010). It also fits spatial autoregressive models by maximum likelihood estimation. The procedures require preparation of nearest-neighbor lists for each point, and a choice of weightings. Here we choose simple binary weighting, while other choices involve sums of links to each point. Crawley (2007, Chapter 24) provides an **R** script for computing spatial nearest neighbors, and for removing objects where the nearest-neighbor distance may be overestimated due to proximity to the edges of the survey.

Applying *spdep* procedures to the right ascension distribution of the Shapley Supercluster low-density region, we find very strong evidence for clustering with $I = 0.988$ and $c = 0.005$. Using the *moran.test* function, the Moran hypothesis test for autocorrelation give $P \ll 10^{-4}$ as the expected value is 0.00 ± 0.08. The significance is confirmed by bootstrap resamples using the Monte Carlo simulation function *moran.mc*. Similar results are found using the Geary c statistic and the declination distribution. Function *EBImoran.mc* gives a Monte Carlo test for Moran's I using an empirical Bayes techniques. Note that Moran's I and similar statistics can be applied to the residuals of a spatial model to see whether the model fully accounts for the correlated structure.

```
# Preparation of nearest neighbor lists for spdep analysis

install.packages('spdep') ; library(spdep)
shap.lo.mat <- as.matrix(shap.lo[,1:2])
nn.shap.lo <- knearneigh(shap.lo.mat, k=1) ; str(nn.shap.lo)
nb.shap.lo <- knn2nb(nn.shap.lo) ; plot.nb(nb.shap.lo, shap.lo[,1:2])
nb.wt.shap.lo <- nb2listw(nb.shap.lo,style='B') ; summary(nb.wt.shap.lo)

# Application of Moran's I and Geary's C statistics
```

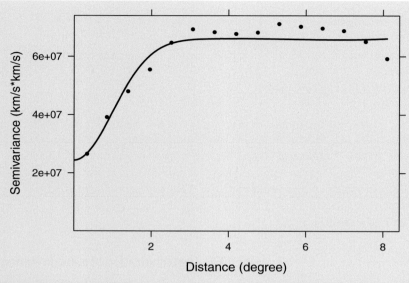

Fig. 12.2   The geostatistical variogram giving a summary and an unrealistically simple model of the velocity structure as a function of angular separation of galaxies in the Shapley Supercluster sample. This analysis used **CRAN**'s *gstat* package.

```
moran(shap.lo.mat[,1], nb.wt.shap.lo, n=length(nb.shap.lo),
  S0=Szero(nb.wt.shap.lo))
moran.test(shap.lo.mat[,1], nb.wt.shap.lo)
moran.mc(shap.lo.mat[,1], nb.wt.shap.lo, nsim=10000)
geary(shap.lo.mat[,1], nb.wt.shap.lo, n=length(nb.shap.lo),
  n1=length(nb.shap.lo)-1, S0=Szero(nb.wt.shap.lo))
geary.test(shap.lo.mat[,1], nb.wt.shap.lo)
```

12.12.2 Variogram analysis

We now examine the autocorrelation in more detail using the variogram which shows the variance in the point process as a function of spatial scale. Again, we consider the Shapley Supercluster redshift data, treating the galaxy radial velocity as the mark variable with two spatial dimensions, right ascension and declination. Our analysis is based on the **CRAN** packages *geoR* (Ribeiro & Diggle 2001) described in the monograph by Diggle & Ribiero (2007, http://www.leg.ufpr.br/geoR/) and *gstat* (Pebesma 2004; http://www.gstat.org).

We first compute the variogram for the Shapley Supercluster velocities using *gstat*'s *variogram* function. Similar capabilities are provided by *variofit* in the *geoR* package. The result is shown as filled circles in Figure 12.2. If we assume a simplistic stationary, isotropic, Gaussian model, we can generate various variogram models using the function *vgm* and perform a least-squares fit using *fit.variogram*. We find that a Gaussian model can fit the variogram reasonably well (the curve in Figure 12.2). This model can be roughly interpreted as follows. The nugget 2.4×10^7 (km s^{-1})^2 (semi-variance γ at zero distance) indicates a characteristic velocity dispersion of $\sqrt{2\gamma}$ or 7000 km s^{-1} at a fixed sky location. The sill of 4.2×10^7 (km s^{-1})^2 indicates an increase of 9100 km s^{-1} velocity dispersion due to correlations on scales up to a range of $1.4°$ assuming Gaussian structures.

However, while this simple variogram analysis may be helpful for understanding simpler astronomical datasets with spatial autocorrelation, it may not be meaningful for complicated galaxy redshift datasets. The galaxy clustering is probably only stationary on scales larger than shown here and, with its elongated structures in both sky and velocity dimensions, is definitely not isotropic. In addition, velocity variances arise from two different effects: uninteresting chance superpositions of galaxies at different distances of the Hubble flow, and physically interesting velocity dispersions within rich clusters.

The limitations of a simple random field model is illustrated by analysis with the *geoR* package. Some preparation of the datasets is needed for *geoR* analysis. The procedures do not permit multiple points at the same spatial locations so the **R** functions *which* and *duplicated* are used to find and remove a few galaxies in close pairs. The three dimensions of interest are then converted into a *geodata* class object using *as.geodata*. We then perform a maximum likelihood fit to a family of Gaussian random fields using the function *likfit* using default parameter options. In this case, the fit (not shown) is poor. The *likfit* optimization can be computationally intensive.

```
# Variogram analysis: gstat

install.packages('gstat') ; library(gstat)
shap.variog <- variogram(Vel~1, locations=~R.A.+Dec., data=shap)
variog.mod1 <- vgm(7e+07, "Gau", 3.0,2e+07)
variog.fit <- fit.variogram(shap.variog, variog.mod1) ; variog.fit
plot(shap.variog, model <- variog.fit, col='black', pch=20,
    xlab='Distance (degree)', ylab="Semivariance (km/s*km/s)", lwd=2)


# Variogram analysis: geoR

install.packages('geoR') ; library(geoR)
shap1 <- shap[-c(which(duplicated(shap[,1:2]))),]
shap.geo <- as.geodata(shap1, coords.col=1:2, data.col=4)
points.geodata(shap.geo, cex.min=0.2, cex.max=1.0, pt.div='quart', col='gray')
plot.geodata(shap.geo, breaks=30)

shap.vario <- variog(shap.geo, uvec=seq(0, 10, by=0.5))
plot(shap.vario, lwd=2, cex.lab=1.3, cex.axis=1.3, lty=1)
shap.GRF1 <- likfit(shap.geo, ini.cov.pars=c(4e7,0.2))
summary.likGRF(shap.GRF1)
lines.variomodel(shap.GRF1, lwd=2, lty=2)
```

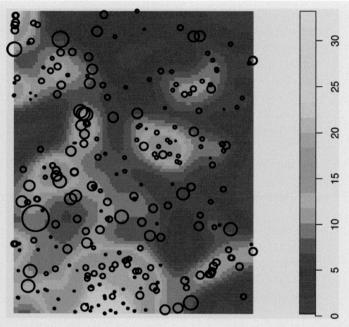

Fig. 12.3 The low-density region of the Shapley Supercluster galaxy redshift survey plotted with **CRAN**'s *spatstat* package. The color image shows a kernel density estimator of the galaxy distribution, and the circles show the individual galaxies with symbol size scaled to their velocities. For a color version of this figure please see the color plate section.

12.12.3   Characterization of clustering

The **CRAN** package *spatstat* provides a large suite of analysis tools and informative plots on spatial uniformity and clustering (Baddeley 2008, Baddeley & Turner 2005). Only a small portion is illustrated here. We apply these to the low galaxy density regions of the Shapley Supercluster. Some preparation for *spatstat* analysis is needed. A window with the ranges in the (x, y) plane must be specified. The window containing the spatial point process need not be rectangular; polygons, binary masks, tessellations, and convex hulls of point processes can be used. *spatstat* provides a variety of tools for manipulating and characterizing windows; here we compute the window centroid and area. The data must then be converted to an object of the planar point pattern (ppp) **R** class. The *summary* function in *spatstat* gives useful information about the point process under study. The functions *pairdist*, *nndist* and *distmap* give the pairwise distances between points, the nearest-neighbor distances, and an image of the empty space distances to the nearest point, respectively. These are used in computing the K, G and F distributions plotted below.

We use the *plot* function in *spatstat* in two ways; first, we show a color filled contour map of the two-dimensional galaxy distribution smoothed with a Gaussian kernel with $0.3°$ width; and second we add the points showing individual galaxy locations with symbol size scaled to galaxy velocities as the mark variable. The result is shown in Figure 12.3. The *spatstat* package provides tools for creating marked variables from ancillary data or other point patterns.

The *spatstat* calculations of the K and other global functions of the point process pay particular attention to the role of edge effects. For small datasets like the Shapley low-density region considered here, these effects can be significant. For Ripley's K function, the *Kest* function in *spatstat* provides three edge corrections algorithms; we show Ripley's isotropic correction here. We plot the L^* function so that models assuming complete spatial randomness (CSR) are horizontal. A collection of the four summary functions (K, F, G and J) can be plotted using *allstats*.

The K and L functions for the low-density Shapley Supercluster region is shown in Figure 12.4. We quickly see the importance of edge correction: we would infer from the K and L estimators without edge correction that the galaxy distribution is nearly identical to a CSR process, but the galaxy K and L estimators with edge correction shows a growing spatial correlation at larger distances.

We use *spatstat*'s *envelope* function to evaluate the significance of this effect. (An error may occur because another function called *envelope* appears in the *boot* package; it can be removed temporarily with the command detach('package:boot'). The simulation shows the range of K function shapes from an ensemble of simulations of CSR processes with the observed number of points in the same window shape with edge correction applied. **R**'s *polygon* function is used to plot the envelope of simulations as a gray region on the plot. From examination of Figure 12.4, we conclude that, on scales larger than $0.7°$ where the observed K and L estimators lies outside the envelope of CSR simulations, the low-density region of the Shapley Supercluster shows significant clustering. Similar analysis of larger samples with stronger clustering show deviations from CSR at smaller angular distances.

```
# Preparation for spatstat analysis

install.packages('spatstat') ; library(spatstat)
shap.lo.win <- owin(range(shap.lo[,1]), range(shap.lo[,2]))
centroid.owin(shap.lo.win) ; area.owin(shap.lo.win)
shap.lo.ppp <- as.ppp(shap.lo[,c(1,2,4)], shap.lo.win) # planar point pattern
summary(shap.lo.ppp)
plot(density(shap.lo.ppp,0.3), col=topo.colors(20), main='', xlab='R.A.',
   ylab='Dec.')
plot(shap.lo.ppp, lwd=2, add=T)


# K function for the Shapley low density region

shap.lo.K <- Kest(shap.lo.ppp, correction='isotropic')
shap.lo.K.bias <- Kest(shap.lo.ppp, correction='none')
plot.fv(shap.lo.K, lwd=2, col='black',  main='', xlab='r (degrees)', legend=F)
plot.fv(shap.lo.K.bias, add=T, lty=3, lwd=2, col='black', legend=F)


# Draw envelope of 100 simulations of CSR process
```

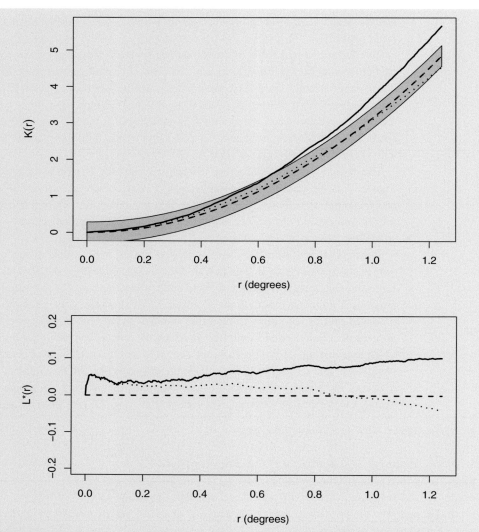

Fig. 12.4 Ripley's K function (top) and L^* function (bottom) for the low-density region of the Shapley Supercluster calculated with **CRAN**'s *spatstat* package. Curves show the estimators with edge correction (solid); estimator without edge correction (dotted); and theoretical function assuming complete spatial randomness (dashed). The gray envelope in the top panel gives the 99% range of K for simulated samples assuming complete spatial randomness.

```
shap.lo.K.env <- envelope(shap.lo.ppp, fun=Kest, nsim=100, global=T)
xx <- c(0, shap.lo.K.env$r, rev(shap.lo.K.env$r), 0)
yy <- c(c(0, shap.lo.K.env$lo), rev(c(0,shap.lo.K.env$hi)))
polygon(xx, yy, col='gray')
plot.fv(shap.lo.K, lwd=2, col='black', main='', add=T, legend=F)
plot.fv(shap.lo.K.bias, add=T, lty=3, lwd=2, col='black', legend=F)
```

```
# Similar plot for the L* function

shap.lo.L <- Lest(shap.lo.ppp, correction='isotropic')
shap.lo.L.bias <- Lest(shap.lo.ppp, correction='none')
plot(shap.lo.L$r, (shap.lo.L$iso - shap.lo.L$r), lwd=2, col='black',
   main='', xlab='r (degrees)', ylab='L*(r)', ty='l', ylim=c(-0.2,0.2))
lines(shap.lo.L$r, (shap.lo.L$theo - shap.lo.L$r), lwd=2, lty=2)
lines(shap.lo.L$r, (shap.lo.L.bias$un - shap.lo.L$r), lwd=2, lty=3)
```

We now compute two other global statistics of clustering for the low-density region of the Shapley Supercluster shown in Figure 12.3. The top panel of Figure 12.5 shows that the J function falls substantially below unity, the expected value for a CSR process, for angles as close as $\sim 0.05°$. The J statistic appears more sensitive than the K statistic in this regime. Two edge-corrections to the J function are shown, one based on the Kaplan–Meier estimator of the F and G functions, and another on a reduced sample that avoids edges.

The bottom panel of Figure 12.5 shows the pair correlation function. We plot it on a log-log scale as commonly done for the galaxy two-point correlation function, and compare the result to a $w(\theta) \propto \theta^{-0.78}$ power law obtained from an all-sky survey of galaxies at similar magnitude and redshift (Maller *et al.* 2005). A CSR process without clustering would give a flat distribution. The pair correlation function in *spatstat* is calculated following the procedure recommended by Stoyan & Stoyan (1994). It uses an Epanechnikov smoothing kernel with rule-of-thumb bandwidth scaling inversely with the spatially average Poisson intensity $\lambda^{-1/2}$ and applies Ripley's isotropic edge correction. Note that the pair correlation function and the Lieshout–Baddeley J function both show a steep fall in correlation at small distances within $\sim 0.02°$ followed by a flattening in the correlations around $\sim 0.05-0.2°$. The J estimator also shows a second steepening around $0.2-2°$. The J function here shows more noise because it has not been subject to smoothing.

```
# Baddeley J function for the Shapley low-density region

plot(Jest(shap.lo.ppp), lwd=2, col='black', cex.lab=1.3, cex.axis=1.3, main='',
   xlab='r (degrees)', legend=F)


# Two-point correlation function

shap.lo.pcf <- pcf(shap.lo.ppp)
plot(shap.lo.pcf, xlim=c(0.0,0.2))
plot(log10(shap.lo.pcf$r[2:512]), log10(shap.lo.pcf$trans[2:512]), type='l',
   lwd=2, xlab='log r (degrees)', ylab='log pair correlation fn')
lines(c(-1,0), c(0.78+0.48,0.48), lwd=2, lty=2)
lines(c(-2,0), c(0,0), lwd=2, lty=3)
```

All of the above analysis provides global statistical measures of spatial clustering, but does not actually find individual clusters. Such methods are discussed in Section 9.3.

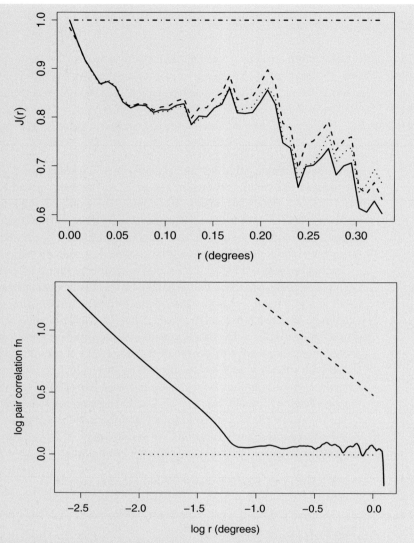

Fig. 12.5 Two global clustering functions for the low-density region of the Shapley Supercluster. *Top*: Baddeley J function: J estimator with two edge corrections (solid and dotted); J estimator without edge correction (dashed); and theoretical J function assuming complete spatial randomness (dot-dashed). *Bottom*: Pair (two-point) correlation function: PCF estimator with edge correction (solid); power law found for similar 2MASS galaxies (dashed, Maller *et al.* 2005); and theoretical PCF assuming CSR (dotted).

12.12.4 Tessellations

While tessellations are sometimes used to analyze geostatistical or other spatial point procsses, they represent an approach based on mathematical graph theory. Here we apply functions in the **CRAN** package *spatstat* described by Baddeley (2008) to the low-density

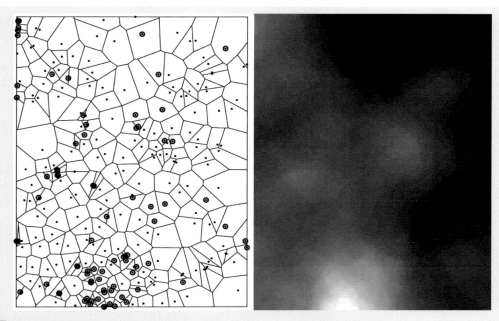

Fig. 12.6 Dirichlet (Voronoi) tessellation of the low-density Shapley region: *Left*: Plot of tessellation tiles with galaxies shown as dots. Those associated with small-area tiles are marked with circles. *Right*: Image of an adaptive density estimator based on surface densities scaled to local tile areas. The gray scale shows a factor of 8 range in surface density.

portion of the Shapley Supercluster two-dimensional galaxy locations. They are closely related to functions in package *deldir* (Delaunay–Dirichlet).

The *spatstat* package provides several calculations relating to tessellations. The *dirichlet* function computes the Dirichlet (or Voronoi) tessellation of a point process giving polygonal tiles with edges equally spaced from adjacent points. These polygons can be converted into windows for additional *spatstat* analysis, and an adaptive density estimator is provided based on the Dirichlet tiling of the point pattern. The *Delaunay* function computes the Delaunay triangulation of a point process, a network of lines connecting adjacent points. Line segments, rather than infinitesimal points, can be analyzed. Many calculations can be made using the mark variable; for example, tests of CSR can be made in tiles defined by values of the mark variable. Other functions in *spatstat* allow subdividing tiles, weighting points by their tile areas, and locating tiles which touch on window boundaries.

We start with calculating the Dirichlet tessellation of the galaxy positions from a class **planar point pattern** constructed above, storing the result in a structured list of polygonal edges. These can be directly plotted (Figure 12.6, left). The function *tiles* converts the format into *spatstat*'s window format, from which *area.owin* extracts the tile areas.

While not a widely used method, the tessellation areas can be used to identify clusters of closely spaced objects, including groups as small as binary objects. We first examine the histogram of tile areas; most have areas < 0.15 square degrees, with a few isolated galaxies having areas up to ∼0.6 square degrees. We arbitrarily choose to display galaxies with tiles ≤ 0.06 square degrees to represent tight galaxy groupings. **R**'s *cut* function breaks the distribution into ≤ 0.06 and > 0.06 square degree subsamples, allowing the clustered galaxies to be plotted with distinctive symbols in Figure 12.6, left. The format "(0,0.06]" is a string associated with a *factor* which **R** uses to classify objects into categories.

The *spatstat* package includes an adaptive density estimator derived from the Dirichlet tessellation. The function *adaptive.density* randomly removes a fraction of the dataset (here we choose 10%) to make a crude tessellation. The density of galaxies in the remaining 90% of the dataset is computed for each tile, and the result is placed into a pixelated image of the region. This process is repeated (here we choose 100 times) and the images are averaged. The result is a smoothed image with more detail in high-density areas where tile areas are small, and less detail in low-density areas where tile areas are large (Figure 12.6, right).

The two maps of galaxy grouping based on tessellation areas in Figure 12.6 can be compared to the simpler uniform kernel density estimator map shown in Figure 12.3.

```
# Compute Dirichlet (Voronoi) tessellation

shap.lo.dir <- dirichlet(shap.lo.ppp)
summary(shap.lo.dir) ; plot(shap.lo.dir, main='')
shap.lo.tile <- tiles(shap.lo.dir) ; str(shap.lo.tile)
shap.lo.area <- list(lapply(shap.lo.tile, area.owin)) ; str(shap.lo.area)

# Select small area tiles as clusters

hist(as.numeric(shap.lo.area[[1]]), breaks=30)
shap.lo.clus <- cut(as.numeric(shap.lo.area[[1]]), breaks=c(0,0.06,1))
plot(shap.lo.dir, main='')
points(shap.lo.ppp, pch=20, cex=0.5)
points(shap.lo.ppp[shap.lo.clus=='(0,0.06]'], pch=1,lwd=2)
```

12.12.5 Spatial interpolation

Here we illustrate spatial interpolation using two methods common in the Earth sciences described in Section 12.5: inverse distance weighted (IDW) interpolation, and ordinary kriging. We apply these methods to the full Shapley Supercluster dataset, and seek a smooth map of the mean velocity field as a function of sky location. In this case, due to overlapping clusters, this mean velocity field does not give direct astrophysical insight into galaxy clustering, but it does illustrate the use of estimating smooth spatial distributions of mark variables from unevenly sampled point data.

IDW interpolation is implemented by the function *idw* in **CRAN** package *gstat*. The smooth estimator is calculated on a lattice which is constructed using **R**'s *seq* function to make evenly spaced univariate grid points on each axis, and *expand.grid* which produces the two-dimensional lattice. The function *spplot* is a spatial plotting program to display the

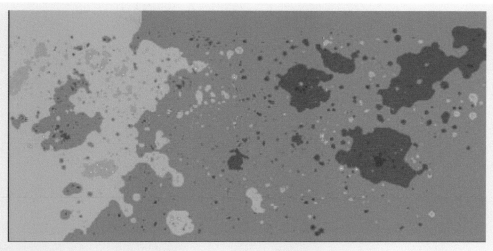

Fig. 12.7 Spatial interpolation of the velocity distribution in the full Shapley Supercluster region based on the inverse distance weighted interpolation estimator with index $\alpha = 1.5$. For a color version of this figure please see the color plate section.

IDW estimator; we use the "topo.colors" color scheme. The resulting map is shown in Figure 12.7. The map shows mean velocities ranging from $\sim$10,000 km s^{-1} (blue) to $\sim$40,000 km s^{-1} (yellow). The map highlights the smaller foreground and background clusters rather than the major components of the Shapley Supercluster with velocities $\sim$10–20,000 km s^{-1}.

As kriging is very commonly used in the Earth sciences, it is treated in a number of **CRAN** packages. Here we use *geoR* where the procedures are described in detail by Diggle & Ribeiro (2007). The procedure starts with a variogram fit similar to that conducted above using the *geoR* functions *variog* and *variofit*. We set up a grid for the kriging predictor function using *pred_grid*; it has a similar function to *expand.grid* in base **R**. The function *krige.control* sets many parameters of the calculation which is performed by *krige.conv*. This conventional kriging function has options for simple, ordinary, universal and extended trend kriging. *geoR* also provides the function *krige.bayes* that computes kriging solutions by Bayesian inference. Both analytic solutions based on conjugate priors and numerical solutions based on Markov chain Monte Carlo computations are treated. The solutions are often similar to the standard kriging solutions, but more can be learned about the distributions of parameter values. This method is described in Chapter 7 of Diggle & Ribeiro (2007).

The kriging estimator for the eastern portion of the Shapley Supercluster is shown in Figure 12.8. The first panel shows the velocity estimator, and the second panel shows its variance. We superpose the positions of the galaxies, with smaller symbols representing low-velocity galaxies and larger symbols representing high-velocity galaxies. The variance map gives important insights into variations in the estimator's accuracy. The estimator is

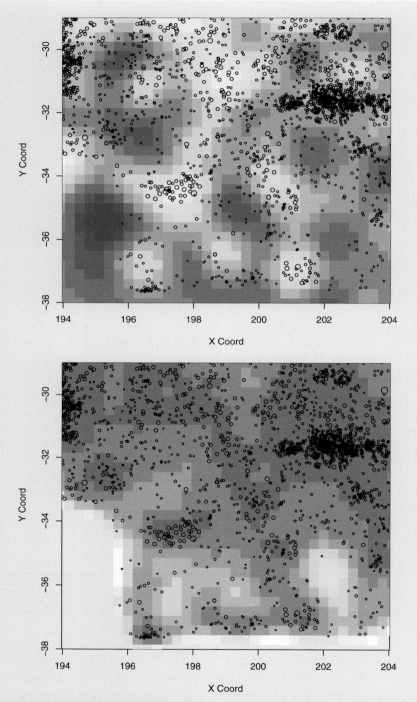

Fig. 12.8 Spatial interpolation of the Shapley Supercluster velocity distribution using ordinary kriging. *Top*: Estimated mean velocity distribution shown as a color map with galaxy locations superposed. *Bottom*: Variance of the kriging estimator shown as a color map. For a color version of this figure please see the color plate section.

most accurate around the rich clusters, but many of the estimator features in low-density regions have high variance and are unreliable.

We note that these interpolations of the spatial distribution of this particular set of galaxy velocities are not very interesting from an astronomical point of view, as this dataset has unrelated galaxies at a wide range of distances superposed onto nearby sky positions. However, in other datasets where the points represent a more homogeneous structure, these interpolations would be scientifically valuable.

```
# IDW spatial interpolation using the gstat package

xrange <- c(193,216) ; yrange <- c(-38,-26.5)
shap.grid <- expand.grid(x=seq(from=xrange[1], to=xrange[2], by=0.05),
y <- seq(from=yrange[1], to=yrange[2], by=0.05))
names(shap.grid) <- c("R.A.","Dec.")
gridded(shap.grid) <- ~R.A.+Dec.
plot(shap.grid) ; points(shap[,1:2], pch=20)

shap.idw <- idw(shap[,4]~1, locations=~R.A.+Dec., shap[,1:2], shap.grid,
    idp=1.5)
sp.theme(regions = list(col = topo.colors(100)), set=T)
spplot(shap.idw)


# Ordinary kriging using the geoR package

shap.vario <- variog(shap.geo, uvec=seq(0, 10, by=0.2))
plot(shap.vario, lwd=2, lty=1)
shap.variofit <- variofit(shap.vario, cov.model='gaussian')
lines(shap.variofit, lty=2)

shap.grid <- pred_grid(c(193,217), c(-38,-29), by=0.3)
KC <- krige.control(obj.model=shap.variofit)
shap.okrig <- krige.conv(shap.geo, loc=shap.grid, krig=KC)

image(shap.okrig, xlim=c(195,203), ylim=c(-38,-27))
points(shap.geo, cex.min=0.3, cex.max=1.5, add=T)
image(shap.okrig, loc=shap.grid, val=sqrt(shap.okrig$krige.var),
    xlim=c(195,203), ylim=c(-38,-27), zlim=c(3800,5000))
points(shap.geo, cex.min=0.3, cex.max=1.5, add=T)
```

12.12.6 Spatial regression and modeling

The *spatstat* and *geoR* packages provide sophisticated capabilities for fitting a variety of spatial models to an observed point process, as described by Baddeley (2008) and Diggle & Ribeiro (2007, Chapters 3 and 4). These include broad families of models such as inhomogeneous Poisson processes with spatially varying intensities and nonstationary, anisotropic Gaussian random fields for elongated structures. Models can involve just spatial locations or can be functionally linked to mark variables. Models can be simulated and fitted to datasets using likelihood methods with validation based on the χ^2 test after grouping into spatial quadrats or by examining the residual field. Model selection can be based on the Akaike information criterion. Note that these codes cannot be readily applied to large datasets, as their matrix operations have order $O(N^3)$.

We have tested one of these models to reproduce Ripley's K function for the Shapley Supercluster dataset. The *Kinhom* function in the *spatstat* package examines a family of nonstationary point process where the Poisson intensity λ is obtained locally within a fixed distance of each galaxy. The best-fit inhomogeneous Poisson model fits the observed K distribution very well from $\sim 0.02°$ to $1.2°$.

12.12.7 Circular and spherical statistics

CRAN's *CircStats* package (Lund & Agostinelli 2009) provides a suite of methods for univariate orientation data with $0° < \theta < 360°$. A guide to *CircStats* usage is given by Jammalamadaka & SenGupta (2001). The package *circular* has some similar functionalities.

We choose as a directional database the sample of 2007 stars with distances around $40-45$ parsecs obtained from the Hipparcos satellite astrometric catalog where we examine the direction of the stellar proper motions. Several steps are needed to prepare the univariate circular dataset from the Hipparcos catalog. First, we clean the data by removing stars with very large parallax uncertainty ($\sigma_{plx} > 5$ mas yr^{-1}) and those with incomplete data using the **R** function *na.omit* to remove objects with *NA* (Not Available) data entries. Second, we remove a small asymmetry in the proper motions be subtracting the median of the motions in the right ascension and declination directions. This is not an optimal procedure; it would be more accurate to remove the solar reflex motion for each star individually. Third, we convert each proper motion vector to a directional angle with respect to $(\mu_{R.A.}, \mu_{Dec}) = (0, 0)$ using **R**'s arctangent function *atan*.

The plot of proper motions of the remaining 1921 stars in Figure 12.9 (top) shows two distinctive characteristics: a $180°$ asymmetry that is the well-known velocity ellipsoid of stellar kinematics in the Galaxy; and a compact clump of stars around $(\mu_{R.A.}, \mu_{Dec}) \simeq (110, 20)$ representing the Hyades star cluster. Figure 12.9 (bottom) shows circular plots made with the very similar *circular* and *CircStats* **CRAN** packages, one binned and the other smoothed with a Gaussian kernel. Both the excess around $\theta = 80°$ from Hyades stars and the broad north–south asymmetry are evident.

We proceed with the statistical evaluation of the univariate directional distribution of the 1921 stars. The *circ.summary* and *circ.disp* functions give the mean and standard deviation of the directions, $167° \pm 56°$. Several tests for uniformity are performed: the Kuiper test (*kuiper.test*), the Rayleigh test (*r.test*), Rao's spacing test (*rao.test*), and the Watson test (*watson.test)* for both a uniform and a von Mises distribution. Probabilities are mostly

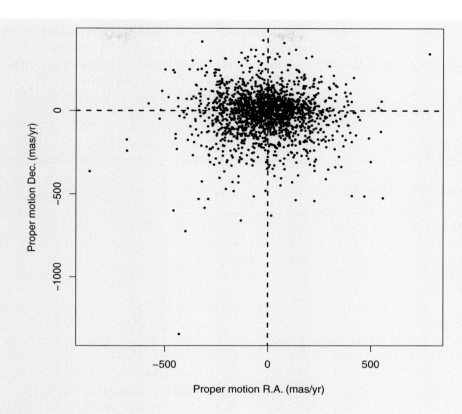

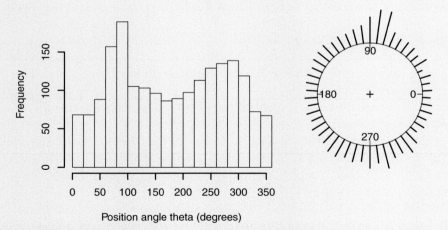

Fig. 12.9 Hipparcos proper motions of stars at distances 40–45 pc, including Hyades cluster stars. *Top*: Proper motions in right ascension and declination, after removal of median values. *Bottom left*: Histograms of proper motion position angles. *Bottom right*: Circle plot showing the distribution of position angles θ for Hipparcos star motions.

$P \simeq 1-5\%$ that the distribution is uniform. A maximum likelihood estimator of the κ concentration parameter of a von Mises distribution with bias correction is calculated with bootstrap confidence intervals using *vm.ml* and *vm.bootstrap.ci*. The MLE parameters values with 95% confidence intervals are $\mu = 2.9^{3.9}_{2.1}$ and $\kappa = 0.07^{0.13}_{0.01}$.

```
# Preparation of circular datasets from Hipparcos proper motion catalog

hip <- read.table("http://astrostatistics.psu.edu/MSMA/datasets/HIP.dat",
   header=T, fill=T)
dim(hip) ; summary(hip)
hist(hip[,8], breaks=50)
hip1<- hip[hip[,8]<5,] ; hip2 = na.omit(hip1) ; hip3 = hip2
hip3[,6] <- hip2[,6] - median(hip2[,6])
hip3[,7] <- hip2[,7] - median(hip2[,7])
attach(hip3) ; dim(hip3)
nstar <- length(hip3[,6])


const=360. / (2*pi)
theta <- numeric(length=nstar)
for (i in 1:nstar) {
if(pmRA[i]>=0 & pmDE[i]>=0) theta[i] = atan(pmRA[i] / pmDE[i]) * const
if(pmDE[i]<0) theta[i] = 180. + atan(pmRA[i] / pmDE[i]) * const
if(pmRA[i]<0 & pmDE[i]>=0)   theta[i] = 360. + atan(pmRA[i] / pmDE[i]) * const
}
hist(theta, breaks=20, lwd=2, xlab='Position angle theta (degrees)', main='')

# Proper motion and circular plots

install.packages('CircStats') ; library(CircStats)
circ.summary(theta)
circ.plot(theta, cex=0.3, pch=20, stack=T, bins=50, shrink=2)
theta=theta / const
plot(pmRA, pmDE, pch=20, cex=0.6, xlab='Proper motion R.A. (mas/yr)',
   ylab='Proper motion Dec. (mas/yr)', main='')
abline(0, 0, lty=2, lwd=2) ; abline(0, 100000, lty=2, lwd=2)
est.kappa(theta, bias=T)

# Basic statistics, tests for uniformity, and von Mises tests

install.packages('circular') ; library(circular)
circ.summary(theta)
sqrt(circ.disp(theta)[4])
```

```
kuiper(theta) ; r.test(theta); rao.spacing(theta)
watson(theta) ; watson(theta,dist='vm')

vm.ml(theta,bias=T)
vm_boot <- vm.bootstrap.ci(theta,bias=T)
vm_boot$mu.ci ; vm_boot$kappa.ci
```

12.12.8 Scope of spatial analysis in **R** and **CRAN**

The *sp* library for spatial data built into **R** is described by Bivand *et al.* (2008). It is particularly designed for graphing combinations of images, points, lines and polygons in effective fashions. The **R** package *sp* and **CRAN** package *rgdal* provide a large infrastructure for spatial classes and coordinate reference systems, including the all-sky Aitoff projection often used in astronomy.

Lists of **CRAN** packages for spatial analysis are given at http://cran.r-project.org/web/views/Spatial.html and by Baddeley (2008). The package *spatial* provides a number of basic procedures associated with the classic book by Ripley (1988). *splancs* (Rowlingson & Diggle 1993) is another older package with many useful functions: simulations of complete spatial random distributions; nearest-neighbor distributions; F, G and K global autocorrelation functions; three-dimensional K functions; statistics for comparing K functions of different datasets; two- and three-dimensional kernel smoothing; and an interactive viewer of three-dimensional point distributions. The package *maptools* gives advanced capabilities of displaying polygons and other shapes. The *spatgraphs* and *SGCS* packages implement a variety of tools in graph theory. The *spatialsegregation* package specializes in tools to quantify the segregation and mingling of classes of points.

The packages *geoR* and *geoRglm* implement an approach to model-based geostatistics described by Diggle & Ribeiro (2007). *gstat* also emphasizes geostatistical procdures (Pebesma 2004). *spBayes* gives Bayesian analysis capabilities for spatial models (Finlay *et al.* 2007). *fields* provides kriging for very large datasets, in addition to other interpolation methods. *RandomFields* models and simulates stochastic spatial patterns. *spgwr* implements geographically weighted regression methods described by Fotheringham *et al.* (2002).

Several packages treat circular and spherical data in addition to the *circular* and *CircStats* packages for univariate directional data introduced above. The *CircSpatial* package provides advanced geostatistical analysis for circular random variables associated with spatial variables including circular-spatial variograms and kriging, simulation of circular random fields, residual analysis, and visualization tools. The *geosphere* package implements spherical geometry including great circles, polygons on a sphere, and bearings. *ssplines* and *skmeans* provide narrow-scope programs to compute a smoothing spline fit and a k-means partitioning algorithm for spherical data. The *CircStats* package provides functions beyond those illustrated here, including random number generators for the von Mises and other circular distributions, a test for change points, von Mises mixture models, Watson's two-sample test, and bivariate correlation coefficients and regression.

Outside of **R**, two notable geostatistical software packages for spatial analysis are available in the public domain. These have the advantage of interactive graphics, but lack **R**'s comprehensive statistical analysis environment. The Stanford Geostatistical Modeling Software, *SGeMS* (http://sgems.sourceforge.net) package includes a variety of variogram, kriging and stochastic simulation tools in addition to image manipulation, display and statistics capabilities. Written in C++ with links to Python, it is described by Remy *et al.* (2009). *GeoVISTA* (http://www.geovista.psu.edu) is a large interactive software environment for spatial data analysis and visualization. Based on Java, *GeoVista* provides scatterplot and biplot matrices, sophisticated three-dimensional graphics, parallel coordinate plots, multivariate classification with hierarchical clustering, minimal spanning trees and self-organizing maps.

Appendix A **Notation and acronyms**

A

Notation

$\#$ . . . . number of (Section 2.2)

$\widehat{}$ . . . . estimated quantity, often an MLE (Section 3.2)

$A \cup B, A \cap B$ . . . . union, intersection of two sets (Section 2.3)

β . . . . model parameter (Section 7.2)

C_p . . . . Mallow's statistic for multivariate model validation (Section 8.4.1)

D_M . . . . Mahalanobis distance in multivariate analysis (Section 8.3)

d_{DW} . . . . Durbin–Watson statistic (Section 11.3.2)

$E[X]$ . . . . expected value, or mean, of X (Section 2.5)

ϵ . . . . random error, often distributed as $N(0, 1)$ (Section 7.2)

f . . . . probability density function (p.d.f.) (Section 2.5)

F . . . . cumulative distribution function (c.d.f.) (Section 2.5)

g . . . . arbitrary function (Section 2.5)

Γ . . . . gamma distribution (Section 4.5)

h . . . . bandwidth in density estimation (Section 6.3)

Λ . . . . Wilk's multivariate generalization of the likelihood ratio (Section 8.3)

$K(F, G, J)$ . . . . Ripley's K and related functions of spatial clustering (Section 12.6)

k-nn . . . . k nearest neighbors (Section 6.5.2)

$\lim\limits_{n \to \infty}$ . . . . limit as n increases without bound (Section 2.5.1)

$M(\theta)$ . . . . model with parameters θ (Section 3.8)

Med . . . . median (Section 5.3.2)

μ . . . . population mean (Section 2.5)

n . . . . number of objects in a sample (Section 2.7)

$N(\mu, \sigma^2)$ . . . . normal distribution with mean μ and variance σ^2 (Section 4.3)

$\xi(d)$ $(\hat{\xi}_{DP}, \hat{\xi}_H, \hat{\xi}_{LS}, \hat{\xi}_{PH})$ . . . . two-point spatial correlation function (and its estimators)
(Section 12.6.2)

$P(A)$ . . . . probability of A (Section 2.4)

$P(A|B)$ . . . . probability of A given B (Section 2.4)

P_{LS} . . . . Lomb–Scargle periodogram for unevenly spaced data (Section 11.6.1)

$r(\rho)$ . . . . Pearson's sample (population) correlation coefficient (Section 8.2.1)

R_a^2 . . . . adjusted coefficient of determination in regression (Section 7.7.1)

ρ_S . . . . Spearman's rho rank corrrelation statistic (Section 5.6)

$s^2(\sigma^2)$ . . . . univariate sample (population) variance (Section 4.3)

$\mathbf{S}^2 \, (\boldsymbol{\Sigma}^2)$ . . . . sample (population) covariance matrix of a multivariate random variable
(Section 8.2)

$S(x)$ .... survival function (Section 10.2)

T^2 .... Hotelling's multivariate generalization of the t^2 statistic (Section 8.3)

$\theta^2_{AoV}, \theta^2_{LK}, \theta^2_S$ .... variance-based statistics for periodicity searches (Section 11.6.2)

τ_K .... Kendall's tau rank corrrelation statistic (Section 5.6)

$Var(X)$ .... variance of X (Section 2.5)

x_i .... i-th observation (Section 2.10)

x_{ij} .... i-th observation of the j-th variable (Section 8.2)

$\bar{x}$ .... sample mean (Section 2.10)

x_{std} .... sample standardized variable (Section 9.2.2)

$x \in A$ .... for values of x lying within A (Section 2.3)

X .... univariate random variable (r.v.) (Section 2.5)

X^2 .... weighted summed squared residuals related to χ^2 (Section 7.4)

$\mathbf{X}$ ..... multivariate variable, vector of univariate r.v.'s $\mathbf{X} = (X_1, X_2, \ldots, X_p)$ (Section 2.5)

$Y = g(X)$ .... random variable with functional dependence on X (Section 2.5)

$z_{1-\alpha/2}$ .... $100(1 - \alpha)\%$ quantile of the normal distribution (e.g. $\alpha = 0.05$) (Section 4.1)

$\phi(X; \mu, \sigma^2)$ .... p.d.f. of the normal distribution (Section 4.3)

$\Phi(X; \mu, \sigma^2)$ .... c.d.f. of the normal distribution (Section 4.3)

χ^2 .... χ^2 distribution (Section 4.3)

Acronyms

ACF .... autocorrelation function (Section 11.2)

AD .... Anderson–Darling statistic/test (Section 5.3.1)

AGN .... active galactic nuclei (Section 3.8)

AIC .... Akaike information criterion (Section 3.7.3)

ANN .... automated neural networks for classification (Section 9.6.5)

AR .... autoregressive model for time series (Section 11.3.3)

ARMA .... autoregressive moving average model (Section 11.3.3)

ARIMA, ARFIMA, GARCH ...... generalizations of the ARMA model (Section 11.3.3)

ASH .... averaged shifted histograms (Section 6.3)

BF .... Bayes factor (Section 3.8.4)

BIC .... Bayesian information criterion (Section 3.7.3)

CART .... classification and regression trees (Section 9.6.3)

c.d.f. .... cumulative distribution (Section 5.3.1)

CLT .... Central Limit Theorem (Section 2.10)

CSR .... complete spatial randomness (Section 12.2)

CUSUM .... cumulative sum control chart (Section 11.6.2)

CV .... cross-validation (Section 6.4.2)

CvM .... Cramér–von Mises statistic/test (Section 5.3.1)

DCF (UDCF) .... (unbinned) discrete correlation function (Section 11.4.1)

e.d.f. .... empirical distribution function (Section 5.3.1)

EM .... expectation-maximization algorithm (Section 3.4.5)

GLM .... generalized linear modeling (Section 7.6)

HL .... Hodges–Lehmann shift statistic (Section 5.4.2)
HMM .... hidden Markov models (Section 11.8)
HPD .... highest posterior density in Bayesian inference (Section 3.8.2)
i.i.d. .... independent and identically distributed random variable (Section 2.5.2)
IQR .... interquartile range (Section 5.3.3)
KL .... Kullback–Liebler distance (Section 5.3.1)
KM .... Kaplan–Meier estimator in survival analysis (Section 10.3.2)
KS .... Kolmogorov–Smirnov statistic/test (Section 5.3.1)
KW .... Kruskal–Wallis statistic (Section 5.4.2)
LBW .... Lynden-Bell–Woodroofe estimator for truncation (Section 10.5.2)
LDA .... linear discriminant analysis (Section 9.6.1)
LISA .... local indicators of spatial association (Section 12.4.2)
LM .... long-memory time series process (Section 11.9)
LOWESS (LOESS) .... Cleveland's (robust) local regression (Section 6.6.2)
LSP .... Lomb–Scargle periodogram (Section 11.6.1)
ΛCDM .... consensus astrophysical cosmological model (Section 3.8)
MAD (MADN) .... (normalized) median absolute deviation (Section 5.3.3)
MANOVA .... multivariate analysis of variance (Section 8.3)
MAP .... maximum a posteriori estimate in Bayesian inference (Section 3.8.2)
MARS .... multivariate adaptive regression splines (Section 8.4.5)
MCMC .... Markov chain Monte Carlo simulations (Section 3.8.6)
MCLUST .... model-based clustering algorithm (Section 9.5)
ME ....measurement error regression model (Section 7.5)
MLE .... maximum likelihood estimation (Section 3.3)
MSE (MISE) .... mean (integrated) square error (Section 6.4)
MST .... minimal spanning tree for point processes (Section 12.8)
MVN .... multivariate normal distribution (Section 8.2.2)
MVUE .... minimum variance unbiased estimator (Section 3.3)
MWW .... Mann–Whitney–Wilcoxon statistic (Section 5.4.2)
OLS .... ordinary least-squares regression (Section 7.3.1)
P-P plot .... probability-probability plot (Section 4.3)
PACF .... partial autocorrelation function (Section 11.3.2)
PCA .... principal components analysis (Section 8.4.2)
p.d.f. .... probability distribution function (Section 4)
p.m.f. ....probability mass function (Section 4)
Q-Q plot .... quantile-quantile plot (Section 2.6)
RSS ....sum of squared residuals in regression (Section 7.6)
SSA .... singular spectrum analysis (Section 11.8)
TS .... Thiel–Sen median slope regression line (Section 7.3.4)

Appendix B Getting started with R

B.1 History and scope of R/CRAN

The development of **R** (R Development Core Team, 2010) as an independent public-domain statistical computing environment was started in the early 1990s by two statisticians at the University of Auckland, Ross Ihaka and Robert Gentleman. They decided to mimic the **S** system developed at AT&T during the 1980s by John Chambers and colleagues. By the late 1990s, **R** development was expanded to a larger core group, and the Comprehensive **R** Archive Network (**CRAN**) was created for specialized packages. The group established itself as a non-profit *R Foundation* based in Vienna, Austria, and began releasing the code biannually as a GNU General Public License software product (Ihaka & Gentleman 1996).

   R grew dramatically, both in content and in widespread usage, during the 2000s. **CRAN** increased exponentially with $\sim$100 packages in 2001, $\sim$600 in 2005, $\sim$2500 in 2010, and $\sim$3,300 by early 2012. The user population is uncertain but was estimated to be $\sim$2 million people in 2010.

   R consists of a collection of software for infrastructure analysis, and about 25 important packages providing a variety of important data analysis, applied mathematics, statistics, graphics and utilities packages. The **CRAN** add-on packages are mostly supplied by users, sometimes individual experts and sometimes significant user communities in biology, chemistry, economics, geology and other fields. Tables B.1 and B.2 give a sense of the breadth of methodology in **R** as well as **CRAN** packages (up to mid-2010). The **CRAN** entries here are restricted to methods that might interest astronomers and, in a number of cases, several packages treat a single problem. For example, about 15 packages provide functionalities related to wavelet analysis. Table B.3 lists some topics with **R/CRAN** scripts associated with recent texts and monographs. Table B.4 lists the major links between **R** and other languages, software systems, formats and communication protocols. Interfaces in some cases are bidirectional; for example, **R** code can call Python subroutines and Python code can call **R** functions. Many of these interfaces were produced by the Omegahat Project for Statistical Computing (http://www.omegahat.org/).

B.2 Session environment

R can be downloaded from http://r-project.org. Compiled binaries with immediate self-installation are available for Windows and MacOS as well as code for several

Table B.1 Selected statistical functionalities in **R**.

arithmetic (scalar/vector/array)	maximum likelihood estimation
bootstrap resampling	multivariate analysis #
convolution	multivariate classification #
correlation coefficients	multivariate cluster analyses #
distributions #	neural networks
exploratory data analysis	nonlinear regression
fast Fourier transform	nonparametric statistics
generalized linear modeling	optimization
graphics #	projection pursuit regression
Kalman filtering	smoothing and cross-validation
linear and matrix algebra	spatial point processes
linear and quadratic programming	statistical tests #
linear modeling, robust regression	survival analysis
local regression	time series analysis #

# indicates extensive capabilities

flavors of Linux. In Windows and MacOS, the **R** console is obtained by double-clicking the icon; in Linux, it is obtained by typing "R" on the terminal.

Data and **R** products from a current **R** session are stored in a file called *.Rdata*. A number of utilities are available for establishing the location and current contents of the *.Rdata* file. The function *getwd* gives the working directory and *setwd* changes the directory. Note that parentheses are needed for most **R** functions; the commands are written "function(arguments)" as in *getwd()*. The functions *ls* and *objects* give the names of **R** objects in the current environment; *get* returns a specified object, and *rm* deletes it. Objects tend to proliferate during complicated interactive data analysis, and can be overwritten. It is advisable not to reuse object names for different purposes, and to remove unused objects. The command "rm(list=ls( ))", deleting all objects in the environment, should be used very cautiously. The contents of the **R** objects in the current workspace can be saved in file *.Rdata* upon departure from the console using the *quit* function or manually closing the console window. Note that *.Rdata* can quickly become very large if big datasets are analyzed; it can be deleted if the contents are not needed in the future.

R is so large and complex that even experienced users constantly refer to the help files that accompany every function and operation, written *help(function)* or *?function*. A broader search of the contents of all available help files can be made with the *help.search('character string')* or *??'character string'* commands. The help files have a standard structure: brief description, usage and list of arguments, details on the procedure, references, related help files, and examples. As with the scripts in this volume, the examples are fully self-contained **R** scripts that can be cut and pasted directly into the **R** console to illustrate the function.

A history of the 512 most recent **R** commands is stored in the file *.Rhistory*. The history file can be searched, stored and retrieved using the *history*, *savehistory* and *loadhistory* functions. The keyboard arrows allow the user to easily revise and repeat recent commands. **R** also allows direct interface with the host computer's file system with functions like *list.files*, *file.create* and *files.choose*.

Table B.2 Selected statistical functionalities in **CRAN**.

adaptive quadrature	multimodality test
ARIMA modeling	multivariate normal partitioning
Bayesian computation	multivariate outlier detection
boosting	multivariate Shapiro–Wilks test
bootstrap modeling	multivariate time series
classification and regression trees	neural networks
combinatorics	nonlinear least squares
convex clustering and convex hulls	nonlinear time series
conditional inference	nonparametric multiple comparisons
Delaunay triangulation	omnibus tests for normality
elliptical confidence regions	orientation data
energy statistical tests	outlier detection
extreme value distribution	parallel coordinates plots
fixed point clusters	partial least squares
genetic algorithms	periodic autoregression analysis
geostatistical modeling	Poisson–gamma additive models
GUIs (Rcmdr, SciViews)	polychoric and polyserial correlations
heteroscedastic t regression	principal component regression
hidden Markov models	principal curve fits
hierarchical partitioning and clustering	projection pursuit
independent component analysis	proportional hazards modeling
interpolation	quantile regression
irregular time series	quasi-variances
kernel-based machine learning	random fields
kernel smoothing	random forest classification
k-nearest-neighbor tree classifier	regressions (MARS, BRUTO)
Kolmogorov–Zurbenko adaptive filtering	ridge regression
least-angle and lasso regression	robust regression
likelihood ratios and shape analysis	segmented regression break points
linear programming (simplex)	Self-Organizing Maps
local regression density estimators	space-time ecology
logistic regression and spline fits	spatial analysis and kriging
map projections	structural regression with splines
Markov chain Monte Carlo	three-dimensional visualization
Markov multistate models	truncated data
Matlab emulator and tesselations	two-stage least-squares regression
mixture discriminant analysis	variogram diagnostics
mixture models	wavelet toolbox
model-based clustering	weighted likelihood robust inference
multidimensional analysis	

The function *sessionInfo* gives the **R** version and currently attached packages. Although **R** is publicly available through a GNU General Public License, it is copyrighted and should be referenced in publications that use its functionalities. The reference is given by the *citation* function; the author of **R** is the *R Development Core Team*. Each user-supplied

Table B.3   **R** scripts associated with textbooks.

Bayesian statistics	image analysis
bootstrapping	kernel smoothing
circular statistics	linear regression
contingency tables	relative distribution methods
data analysis	smoothing
econometrics	survival analysis
engineering statistics	time-frequency analysis
generalized additive models	

Table B.4   **R** interfaces.

Languages: BUGS, C, C++, FFI, Fortran, Java, JavaScript, Python, Perl, XLisp, Ruby
Headers: XML, KML
I/O file structures: ASCII, binary, bitmap, ftp, gzip, FITS, HDF5, netCDF
        MIM, Oracle, SAS, S-Plus, SPSS, Systat, Stata, URL, .wav, OpenOffice
Web formats: URL, cgi, HTML, Netscape, SOAP
Statistics packages: GRASS, Matlab, XGobi
Spreadsheets: Excel, Gnumeric
Graphics: Grace, Gtk, OpenGL, Tcl/Tk, Gnome, Java Swing
Databases: MySQL, SQL, SQLite
Science/math libraries: GSL, Isoda, LAPACK
Parallel processing: PVM
Interprocess communication: CORBA
Network connections: sockets, DCOM
Text processors: LaTeX

package installed from the **CRAN** archive has a reference obtained from *citation(package.name)*.

B.3   R object classes

An important concept in **R** is the **class**, an attribute of **R** objects central to later operations and analysis. Some classes denote the item type: *numeric*, *character*, *Date* or *logical*. Numeric objects can be *integer*, *real*, *complex* or *double* (precision). The *vector* class is used for one-dimensional objects, while the *array* and *matrix* classes are used for two-dimensional objects. The *list* class can have a hierarchical structure with embedded vectors, matrices and other substructures.

Some classes have more elaborate definitions: a *factor* object encodes a vector into categories, and a *table* object builds a contingency table of counts for each combination of factors. A *ts*, or time series, object gives values as a function of an evenly spaced time-like variable. A *stepfun* object has jumps at specified locations. Each **R** class has associated utilities. For example, the function *is.matrix* tests if its argument is a matrix, returning a logical TRUE/FALSE, while the function *as.matrix* attempts to coerce its argument into a matrix.

Often a major function in **R** or **CRAN** will create a specialized class for further analysis. The *hclust* function for hierarchical cluster analysis of multivariate data creates an *hclust* object giving the merging sequence and other technical information. Sometimes these classes have a great deal of content in complicated hierarchical formats. The function *str(x)* reveals the detailed internal structure of any **R** object, and is particularly useful for objects created by advanced functions. The *plot.hclust*, *print.hclust* and *summary.hclust* functions (and similarly for other classes) are specialized versions of **R**'s generic *plot*, *print* and *summary* functions to present the results of the hierarchical clustering in useful fashions that do not require detailed knowledge of the internal structure of the object class. The *plot(x)* command, for example, will automatically operate differently if *x* is an *hclust* or a *ts* object, revealing the cluster tree or the time series variations, respectively. Examine *methods(print)* to see some important classes available in **R**; many more are defined in **CRAN** packages.

The **dataframe** is a particularly important class in **R**. It can be viewed as an elaboration on an *array* or *matrix* where the columns are associated with variables' names. However, unlike a matrix, the columns need not be vectors of a single class; different columns can consist of real numbers, integers, logical (TRUE/FALSE) or factor (category) variables. The *data.frame* structure is often required for **R** modeling functions.

Another special object is *NA* representing "Not Available". This is a logical constant with values TRUE or FALSE indicating that a value is missing in a vector, dataframe, or other object. For numeric variables, *NaN* may appear representing "Not a Number". The user often must pay particular attention to *NA* and *NaN* entries as different functions require different instructions on their treatment. A variety of utilities is available for treating missing values: *is.na*, *na.omit*, *na.pass*, and so forth.

B.4   Basic operations on classes

Binary operations are readily performed for scalar and vector objects. Arithmetic operations are $+$, $-$, $*$, $/$, and $\hat{}$ for addition, subtraction, multiplication, division and exponentiation. Common mathematical operations include *abs*, *sqrt*, *round*, *exp* and *log10* which provide absolute value, square root, rounding (truncation and other options are available), exponentiation and common (base 10) logarithms. Note that the function *log* gives natural (base *e*) logarithms. Trigonometric operations include *sin*, *asin*, *sinh*, *asinh* and similarly for the cosine and tangent.

R includes a variety of comparison and logical operations. A value is assigned to a name using the "$<-$" assignment operation. This can be replaced with the more familiar "$=$" operator in many, but not all, circumstances. The relation between two objects can be made using the operators $x < y$ ($x <= y$), $x > y$ ($x >= y$), $x == y$, $x != y$. These generate logical returns if *x* is less than (less than or equal), greater than (greater than or equal),

equal, or not equal to y. Logical operators for comparing two TRUE/FALSE objects include "|" for OR, "&" for AND and "!" for NOT. The *which* function returns the array indices of a logical vector or array containing TRUE entries.

The functions *if* (or the *if . . . else* combination) and *for* can be used for controlling flow in **R** analysis. The argument of the *if* function is a logical scalar that, if TRUE, allows the following expression to be applied. The *for* function performs a loop over a specified sequence of a variable. The *if* and *for* functions can be used in combination with *while*, *break*, *next* and *else* to give more control over the operational flow of a script.

A number of vector operations is commonly used in **R**-based data analysis. Any vector can be characterized with the *length* function, but a number of other functions are generally used with numeric vectors:

1. The "[[" and "[" operators are used to select a single or multiple elements; for example, $x[1:10]$ and $x[[3]]$ extract the first ten and the third element of the vector x, respectively. $x[1:10,-1]$ deletes the first column of a matrix. For an **R** *list* with more complex structure, the \$ operator locates vectors or arrays within a hierarchy. Thus, the expression $x\$second[[7]]$ extracts the seventh element of the vector called *second* within the list x.
2. More complex filters of vector elements can be constructed. The statement $x[expression]$ will extract elements satisfying chosen constraints. These can be recursive; for example, $x[x < 5]$ or $x[which(x < 5)]$ will give values of the vector x satisfying the constraint that the element value is below 5. The *subset* function extracts values meeting logical conditions. The *sample* function draws random subsamples from a vector; when the argument *replace=TRUE* is specified, this gives a bootstrap resample of the vector.
3. A vector can be divided based on specified levels or criteria using the *factor*, *cut* and *split* functions. These can be used to bin the data into groups.
4. Simple vectors can be constructed using the c function to concatenate existing quantities, the *rep* function to produce repeated entries, and *seq* to produce ordered sequences.
5. Arithmetic operations on numeric vectors include *sum* and *prod* giving scalar sums and products of the elements. The functions *cummax*, *cummin*, *cumsum* and *cumprod* give vectors with the cumulative maximum, minimum, sum and product of the original values. Simple statistics of numeric vectors are provided by the *min*, *max*, *median*, *mean*, *mad* (median absolute deviation), *sd* (standard deviation), *var* (variance) and *quantile* functions. The *fivenum* and *summary* functions combine several of these statistics. Other useful functions include *sort* and *order* and *rank*. More complex operations can be applied to all elements of a vector using the *apply* function, including *aggregate* that applies a function to specified subsets of a vector.

Very similar capabilities are available for two-dimensional arrays, matrices, lists and dataframes. Elements can be specified like $x[1:100,5]$ for the first hundred elements of the fifth column, or $x[1:10,]$ for all columns of the first ten rows. New objects are constructed using the *array*, *matrix*, *list* or *data.frame* functions. The size of the object is found with the *dim*, *nrow* and *ncol* functions. New rows and columns can be added with the *rbind* and *cbind* functions. Arithmetic is provided by *colMeans* and *colSums*, and similarly for rows. The function *expand.grid* creates a two-dimensional dataframe based on two supplied vectors. Matrix mathematical operations include t for matrix transpose, %*% for matrix multiplication, *diag* for extracting diagonal elements, *det* for calculating the determinant, and *solve* for inverting a matrix or solving a system of linear equations. The functions *attach* (and *detach*) associate variable names to vector columns in lists and dataframes during the **R** session. Dataframes can be combined using *merge*.

B.5 Input/output

Throughout this volume, the most common format of data input is an ASCII table ingested with **R**'s *read.table* function. Variants for ASCII input include *read.csv* or *read.delim* when entries are separated by commas or tabs, or the *scan* function. Binary data can be obtained with the *readBin* function, and *download.file* obtains files from a URL on the Internet. Large datasets can be compressed onto the disk and reentered into **R** using the *save* and *load* functions. Several file compression formats are supported.

Each of these input functions has a corresponding output function, such as *write.table* and *writeBin*. The *write*, *cat* and *print* functions provide flexible outputs for data in various **R** classes; *print* affords pretty formatting using *format*. External connections can be created with *open*, and the *sink*, *pipe* and *capture.output* functions output data to these connections. The *dump* and *dput* functions produce ASCII representations of **R** objects. The *file.create* and related functions provide a low-level interface to the host computer's file system. The simplest output for an object is to display its contents in the **R** console by stating the object's name. The *show*, *head*, *tail* and *print* functions give similar displays to the console.

For several decades, the astronomical community has developed the Flexible Image Transport System (FITS) standard for communication of astronomical images, spectra, tables and other data structures between observatories and software systems (Wells *et al.* 1981, Hanisch *et al.* 2001). Enormous effort has been applied to precisely define hundreds of keywords in FITS headers and develop associated software (http://fits.gsfc.nasa.gov). Recently, FITS headers have been converted to XML format within the VOTables structure created for the emerging International Virtual Observatory (Graham *et al.* 2008). Vast amounts of astronomical data are publicly available in FITS or VOTable formats through observatory archives and Virtual Observatory portals.

Two FITS input/output scripts are available today for **R**. First, the **CRAN** package *FITSio* was developed by Andrew Harris at the University of Maryland. It reads and writes FITS files, parsing FITS headers, converting FITS images into **R** arrays and FITS binary tables into **R** dataframes. Five functions are provided: *readFITS* to input a single image or binary table, *readFITSheader* to parse the header, *readFITSarray* and *readFITSbinarytable* read an image and table into an **R** array, and *readFrameFromFITS* reads a table into an **R** dataframe. At present, only one output option is currently available, *writeFITSim* to produce a single FITS image file.

The second program, *R-FITS*, is not currently available in **CRAN**; it runs under Linux and can be downloaded from http://hea-www.harvard.edu/AstroStat/rfits. Developed by Peter Freeman from Carnegie Mellon University, *R-FITS* is an **R** wrapper for NASA's official *cfitsio* subroutine library in C and FORTRAN that has extensive ability to parse FITS headers and extract data of various types (http://heasarc.gsfc.nasa.gov/docs/software/fitsio/).

R and **CRAN** provide an enormous range of graphical output capabilities outlined in the graphics Task Views at http://cran.r-project.org/web/views/Graphics.html. The *plot* function is ubiquitous and, when applied to the output of many functions, produces well-designed specialized graphics. **CRAN**'s *ggplot2* package, described in the volume by Wickham (2009), provides a user-friendly suite of enhancements to standard **R** graphics. *RSvgDevice* is a device driver for the widespread Scalable Vector Graphics language. The **CRAN** package *cairoDevice* gives access to the Cairo vector graphics library (http://cairographics.org/) designed to work with **RGTK2**, an interface to the *Gimp* Tool Kit. *JGR* gives a Java GUI for R, and *JavaGD* routes **R** graphics commands to a Java Graphics Device. Interactive graphics are provided by the *iPlots* and *rggobi* packages. A particularly useful function for exploratory multivariate analysis in *iPlots* is *ipcp* providing interactive parallel coordinates plots. The **R** Graphical Manual (http://www.oga-lab.net/RCM2) and the **R** Graph Gallery (http://addictedtor.free.fr/graphique) display a wide range of graphics with associated **R** scripts. A broad overview of visualization for statistical data analysis is provided by Chen *et al.* (2008).

B.6   A sample R session

We illustrate some of the uses of elementary **R** functions outlined above, followed by an analysis of a hypothetical dataset based on concepts presented in Chapters 4–7. The semicolon ";", like the carriage return, serves as a divider between commands. Throughout the volume, we use it to produce a more compact presentation for closely related functions. Comments are placed to the right of the hash symbol "#". We encourage inexperienced users to examine help files frequently. Note that **R** is case-sensitive: for example, the command *sessioninfo()* produces an error while *sessionInfo()* gives the desired information.

We start by downloading **R** from http://www.r-project.org for a Windows, MacOS or Linux operating system. Start **R** by double-clicking the icon or typing "R" into the terminal. In Section I of the **R** session below, we obtain some basic information about the session environment. The working directory can be changed using *setwd*. Note how most **R** function commands include "( )" brackets, even if no arguments are needed by the function. We create a simple vector *a.test* using the *c* function that concatenates values. This new vector is then displayed in several ways.

In Section II, we create a dataset of 500 (x_i, y_i) data pairs with a peculiar nonlinear relationship between the two variables with heteroscedastic errors (i.e. the scatter depends on the variable values). Several useful **R** functions appear here: the *seq* function to produce a regular sequence of values between specified limits; the *sample* function draws random values from a dataset; and the *rnorm* function gives normal (Gaussian) random deviates. Note that arguments to functions need not be constants, as in the standard deviation of the scatter. Parentheses are used liberally to avoid ambiguities regarding sequence of operations.

The *set.seed* function gives optional control over the random numbers used in the resampling or other stochastic operations. Here we fix the random number sequence to allow the calculation to be repeated. The function also allows a choice between several sophisticated random number generators; the default is the Mersenne–Twister with a period of $2^{19937} - 1$ and a 624-dimensional seed (Matsumoto & Nishimura 1998).

```
# I. Query the session environment and make a vector of three numbers

getwd()
setwd('/Users/myself/newdir')
sessionInfo() ; citation()
a.test <- c(33, 44, 55)
ls() ; class(a.test) ; str(a.test)
a.test ; write(file='output', a.test)


# II. Create a 500x2 bivariate dataset where the X values are evenly
# distributed between 0 and 5 and the Y values have a nonlinear
# dependence on X with heteroscedastic Gaussian scatter

help(seq) ; help(sample) ; help(rnorm)
set.seed(1)
x <- sample(seq(0.01, 3, length.out=500))
y <- 0.5*x + 0.3^(x^2) + rnorm(500, mean=0, sd=(0.05*(1+x^2)))
```

Section III below presents operations commonly called **exploratory data analysis**, an approach particularly promoted by statistician John Tukey (1977). Here we consider the distributions of the X and Y variables separately. The *summary* function presents the minimum, quartiles, mean and maximum of the distribution on the console. They can also be obtained with **R**'s *fivenum* function. We can save information appearing on the console in a disk file. Here we first determine that the structure of the *summary* output is an **R** *table* class, so it is reasonable to record it into a file using *write.table*. Note the use of the **rbind** function that binds the rows of multiple tables or dataframes into a single object.

The univariate summary information is then plotted using the *boxplot* function. The boxplot, or box-and-whiskers plot, is a visualization of Tukey's **five-number summary** of a distribution with **hinges**, **whiskers**, **notches** and **outliers**. The box width can be scaled to the sample size. Though rarely used in astronomy, it is a compact and visually effective display of robust measures of location and spread, appropriate for both small and moderately large univariate datasets.

In our example here, the X distribution is symmetrical (by construction) while the Y distribution is skewed with outliers. The plots are preceded by the *mfrow* argument of the *par* function that sets a layout for a two-panel figure. Many options for plotting displays are controlled through *par*, and users are encouraged to become familiar with its arguments. When the user is pleased with a figure constructed interactively in **R**, it can be written to a disk file using the *postscript* function, or similar function for JPEG

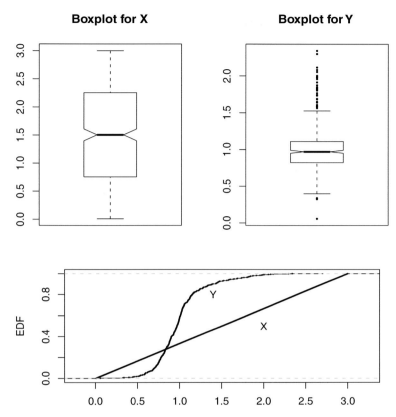

Univariate plots of a hypothetical two-dimensional dataset. *Top:* Boxplots. *Bottom:* Empirical distribution functions

or other graphical formats; see *help(Devices)* for a listing. The Postscript figure is written to disk when the *dev.off* function is given after the plotting commands are completed. A convenient alternative for producing a Postscript figure identical to the latest figure shown is the function *dev.copy2eps*.

We then plot the univariate empirical distribution functions (e.d.f.) using **R**'s *ecdf* function. Internal annotations to the figure are added using the *text* function. The XY boxplots and the two e.d.f.s are shown in the panels of Figure B.1.

```
# III. Examine, summarize and boxplot univariate distributions
summary(x) ; summary(y)
str(summary(x))
write.table(rbind(summary(x),summary(y)),file='summary.txt')

help(par) ; help(boxplot)
par(mfrow=c(1,2))  # set up two-panel figure
boxplot(x,  notch=T, main='Boxplot for X')
boxplot(y,  notch=T, pch=20, cex=0.5, main='Boxplot for Y')
```

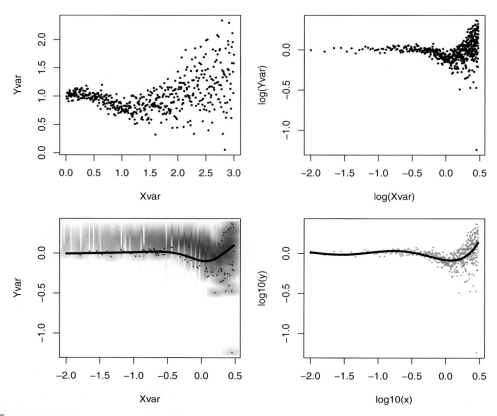

Fig. B.2   *Top:* Scatterplots, original and after logarithmic transformation, of a hypothetical two-dimensional dataset. *Bottom:* Bivariate nonparametric spline and least-squares fit to a fifth-order polynomial.

```
par(mfrow=c(1,1))
dev.copy2eps(file='box.eps')

plot(ecdf(x),pch=20,cex=0.3,main='',ylab='EDF',xlab='')
plot(ecdf(y),pch=20,cex=0.3,add=T)
text(2.0,0.5,"X") ; text(1.4,0.8,"Y")
dev.copy2eps(file='edf.eps')
```

In Section IV, we treat the two variables as a multivariate dataset. The first step is to bind the vectors by columns into a matrix with *cbind*, and coerce the result from a "matrix" class to a "dataframe" class with *as.data.frame*. The *str* function reveals the structure of the dataset at each stage. We assign variable names to the columns using *names* and make the dataframe readily available for further analysis using the *attach* function. Section V produces standard bivariate scatterplots of the original variables, and after logarithmic transformation. The *cor.test* function gives parametric and nonparametric correlation coefficients with associated probabilities. These are shown as the top panels of Figure B.2.

```
# IV. Put X and Y into a dataframe

help(cbind) ; help(as.data.frame) ; help(names)
xy <- cbind(x, y) ; str(xy)
xy <- as.data.frame(xy)
names(xy) <- c('Xvar', 'Yvar') ; str(xy)
attach(xy) ; ls()

# V. Bivariate plots and correlation tests

help(plot)
par(mfrow=c(2,2))
plot(xy, pch=20, cex=0.5)
plot(log10(xy), pch=20, cex=0.5, xlab='log(Xvar)', ylab='log(Yvar)')

length(x[x>2.5])
cor.test(x[x>2.5],y[x>2.5], method='pearson')
cor.test(x[x>2.5],y[x>2.5], method='kendall')
```

In the final sections of this **R** session, we illustrate use of two more sophisticated functions in **R**. Section VI gives the function *smoothScatter*, residing in the *graphics* library that is bundled into **R**, which produces a two-dimensional kernel density estimator and plots it with a color palette with 100 points in low-density regions (Figure B.2, lower left). We have superposed a cubic B-spline fit to the data with the degree of smoothing set by generalized cross-validation. The density estimator allows a flexible choice of kernel bandwidth, and the spline fit offers many choices of knots and smoothing parameter.

Section VII uses **R**'s *lm*, or linear modeling, function to fit a quintic polynomial function by unweighted least-squares to the logarithmic bivariate distribution. We create separate variables for the terms of the polynomial, and we place them into an **R** formula with the syntax "$y \sim x$". The structure of the fit result shows that the *lm*-class object is a complex list with a dozen components: best-fit coefficients, fitted values, residuals, details of the QR decomposition, and more. These results are accessed using the "$" syntax, as in *yfit$fit*. The *plot* function gives several preset graphics for evaluating the quality of the fit, but we choose to plot the fitted values superposed on the scatterplot of the data using the *lines* function (Figure B.2, lower right). Note the use of the *sort* and *order* functions; the former gives the sorted data values of the X variable while the latter gives the row numbers of the sorted values.

```
# VI. Kernel density estimator with spline fit

smoothScatter(log10(xy), lwd=4, pch=20, cex=0.2,
    colramp = colorRampPalette(c("white",gray(20:5/20))))
lines(smooth.spline(log10(xy)), lwd=3)
```

```
# VII. Least-squares polynomial regression
logx <- log10(x) ; logx2 <- (log10(x))^2 ; logx3 <- (log10(x))^3
logx4 <- (log10(x))^4 ; logx5 <- (log10(x))^5
yfit <- lm(log10(y) ~ logx + logx2 + logx3 + logx4 + logx5)
str(yfit)
plot(log10(x), log10(y), pch=20, col=grey(0.5),cex=0.3)
lines(sort(log10(x)), yfit$fit[order(log10(x))], lwd=3)
dev.copy2eps(file="smooth.eps")
```

B.7 Interfaces to other programs and languages

Considerable effort has been made to connect **R** to other programs, languages and analysis systems. External **R** scripts can be easily run in the console using the *source* function, but more effort is needed to run programs in other languages. As listed in Table B.4, **R** connects to BUGS (Bayesian inference Using Gibbs Sampling), C, C++, FORTRAN, Java, JavaScript, Matlab, Python, Perl, XLisp and Ruby. In some cases, the interface is bidirectional allowing **R** functions to be called from foreign programs and foreign programs to be called from **R** scripts. These include **CRAN**'s *RMatlab* to and from Matlab, *rJava* to Java and *RNCBI* from Java, *Rcpp* to and from C++, and *rJython* to Python and *rpy* from Python. **R** can also be used within the Python-based Sage mathematics software system. The **CRAN** package *inline* facilitates the integration of C, C++ and FORTRAN code within **R** scripts. Furthermore, the *Rserve* package is a socket server for TCP/IP that allows binary requests to be sent to **R** from other languages or environments.

R offers several functions to run compiled C or FORTRAN code within an **R** script: *.C* and *.Fortran* call routines independent of **R** classes, while *.Call* and *.External* use **R**-dependent code in the external code. The functions have the syntax *.C('external.name', input.arguments, PACKAGE = 'external.dll')*. Care must be taken that the argument types are compatible; for example, an **R** *numeric* class object will usually correspond to a "double precision real" number in C and FORTRAN. This can be implemented using **R** utilities, as in *.C('external.name', as.numeric(arg1), as.integer(arg2), PACKAGE = 'external.dll')*. The *.C* and *.Fortran* functions return an **R** list similar to the input arguments with changes made by the C or FORTRAN code. The reader is encouraged to read Chambers (2008, Chapters 11 and 12) for a presentation of interfaces between **R** and C, FORTRAN, and other languages (C++, Perl, Python, Java, Oracle, MySql, Tcl/Tk).

B.8 Computational efficiency

Some operations in **R** are as efficient as C or FORTRAN implementations, while others are much slower. Below we consider the timing of a simulated random walk time series (Section 11.3.3) with one million points. The elapsed times, including CPU and system overhead, were obtained using **R**'s *system.time* function on a MacOS computer.

```
# Benchmarking R
# Some computationally efficient R functions
set.seed(10)
w <- rnorm(1000000)            #  0.1 sec
wsum <- cumsum(w)           #  0.01 sec
wsort <- sort(w)                  #  0.2 sec
wacf <- acf(w)                    #  1.3 sec
wfft <- fft(w)                      #  0.3 sec

# Some computational inefficient functions
x <- numeric(1000000)
for (i in 2:1000000) x[i] = x[i-1] + w[i]        #  5 sec
xsum = function(i) { x[i] <<- x[i-1] +w[i] }
xsumout <- lapply(2:1000000, xsum)     #  9 sec

# Output operations
quartz() ; plot(x,pch=20,cex=0.2)                 #  18 sec
X11(); plot(x, pch=20)                               #  3 sec
postscript(file='random_walk.ps',height=2000,width=2000)
plot(x,pch=20,cex=0.2) ; dev.off()         #  3 sec
write(x)                                                 #  11 sec
save(x)                                                  #  1 sec
install.packages('FITSio') ; library(FITSio)
writeFITSim(file='random_walk.fits',x)       #  1 sec
```

We see that a wide variety of built-in **R** functions are very efficient, including random number generators, vector summations, sorting, autocorrelation functions and fast Fourier transforms. Output operations are reasonable for a million points with some devices (e.g. X11 and Postscript plots, binary *save* and FITS files) and slower than other devices (e.g. MacOS Quartz plot, ASCII *write*). But the scalar *for*-loop is tens of times less efficient. Ihaka & Lang (2008) used a similar test of 10,000 summations of an $n = 100,000$ vector to compare the performance of several languages. They found that Java and C are seven times faster than **R**'s summation function. Python is 2.5 times faster than **R** as interpreted code but Python's summation function is four times slower than **R**'s.

For many functionalities, the computational efficiency of **R** is comparable to that of IDL, a commercial data and image analysis program commonly used in astronomy. In the benchmark above, based on a million random numbers, we find that IDL is 2−3 times faster in random number generation and fast Fourier transforms, similarly rapid in summing a vector, and ~2 times slower in sorting a vector. The summation by loop, however, was much faster than the **R** *for* function, performing a million sums in 0.2 seconds.

It is clearly advantageous to write computationally intensive algorithms in C or FORTRAN and call them from **R**. The **R** function *Rprof* is useful for improving the efficiency of **R** scripts; it profiles the system time taken at each step of a script. Its results can be plotted using **CRAN**'s *profr* package. In addition to inefficient operation for some computationally intensive scalar problems, **R** has been criticized for multiple copying operations of datasets during complicated calculations such as linear modeling.

Work in several directions is underway to alleviate these efficiency problems, particularly for megadatasets. First, **R** compilation is now (2012) being converted to byte code, similar to Java and Python, rather than the current parse tree method. This is leading to a several-fold speedup for many loops and scalar calculations. Second, the foundations of **R** may be rewritten in the Lisp language, projected to give a 10^2-fold speedup for scalar operations (Ihaka & Lang 2008). Third, large-scale vectorized computations can be implemented on parallel computers and new technology chips (Rossini *et al.* 2007). A growing number of **CRAN** packages now provide distributed processing on computing clusters and graphics processing units. Fourth, several **CRAN** packages provide access to disk-based datasets that are too large for memory. Fifth, a subset of **R** has been implemented in a new high-throughput language called **P** (http://ptechnologies.org). Annotated links to many of these efforts can be found at http://cran.r-project.org/web/views/HighPerformanceComputing.html. In addition, a private company provides specialized **R** codes and services for terabyte-class datasets and production processing (http://www.revolutionanalytics.com).

Related to computational efficiency, one can inquire into the coding efficiency of **R** in comparison to other data analysis languages. Press & Teukolsky (1997), authors of the distinguished volume *Numerical Recipes*, compare the coding efficiency of older languages like FORTRAN with newer high-level languages like *IDL* and *Mathematica*. They consider a simple data analysis task in astronomy: find upper quartile values of magnitudes in a sample of galaxies preselected to have velocities between 100 and 200 km s^{-1}. We compare these code snippets to an **R** script.

```
# Code comparison (Press & Teukolsky 1997)
# Fortran 77
n=0
do j=1,ndat
if(vels(j).gt.100..and.vels(j).le.200.) then
n=n+1
temp(n)=mags(j)
endif
enddo
call sort(n,temp)
answer=temp((n*0.75)

# IDL
temp = mags(where(vels le 200. and vels gt 100., n))
answer=temp((sort(temp))(ceil(n*0.75)))

# R
answer = quantile(mags[vels > 100. & vels <= 200.], probs=0.75)
```

R syntax is similar to *IDL* syntax but is more compact: *IDL*'s *where* command is implicit in **R**, and the statistical function *quantile* replaces data manipulation with *sort* and *ceil*.

B.9 Learning more about R

R/CRAN is a very large system and significant effort is needed to explore its capabilities and to learn how to adapt it to specific research needs. We can recommend several tools for extending knowledge of **R** and **CRAN** capabilities and usage:

1. Several official **R** manuals produced by the R Development Core Team are available at http://r-project.org. They provide *An Introduction to R* (~100 pages) as a learning tool, and *The R Reference Index* (~3000 pages) as a searchable collection of the help files for **R** functions. Other manuals are designed for programmers and specialists. On-line **R** tutorials oriented towards astronomers have been prepared by Alastair Sanderson (http://www.sr.bham.ac.uk/~ajrs/R), Penn State's Center for Astrostatistics (http://astrostatistics.psu.edu/datasets) and the Indian Institute for Astrophysics (http://www.iiap.res.in/astrostat/).

2. Several short "cheat sheets" giving brief reminders for common **R** functions and formats. We can recommend the *R Reference Card* by Tom Short (http://cran.r-project.org/doc/contrib/Short-refcard.pdf).

3. To find a specific statistical method within the **CRAN** collection, a global Web search (e.g. Google) with "CRAN package.name" or "CRAN functionalities" is often effective. A similar capability is provided by *Rseek* at http://rseek.org. The **R** Graphical Manual gives searchable lists of tens of thousands of **CRAN** functions and images at http://www.oga-lab.net/RGM2.

4. About 30 *CRAN Task Views*, maintained by volunteer experts, provide annotated lists of **CRAN** packages with functionalities in broad areas of applied statistics. Task Views are available for Bayesian inference, graphics, high-performance computing, machine learning, multivariate statistics, robust statistics, spatial analysis, survival analysis and time series analysis. Other Task Views are oriented towards particular user communities such as chemometrics, econometrics, genetics and medical imaging.

5. Over 100 books have been published in English with "**R**" in the title, and more have **R** scripts in the text. They range from elementary textbooks to specialized monographs. The Springer-Verlag publishers are in the process of publishing a large collection of short didactic volumes in their "*Use R!*" series. Selections of these books are listed in the Recommended reading at the end of this and other chapters.

6. *The* **R** *Journal*, formerly *The* **R** *Newsletter*, is an on-line refereed journal published biannually. It describes important add-on packages and applications, providing hints for users and programmers, listing new features and **CRAN** packages, and announcing **R**-related conferences. Articles on **CRAN** packages also appear in the on-line *Journal*

of Statistical Software (http://www.jstatsoft.org). Annual *useR!* conferences are held and the *Bioconductor* community holds additional conferences.

B.10 Recommended reading

Adler, J. (2010) *R in a Nutshell*, O'Reilly, Sebastopol CA
A comprehensive reference book with extensive **R** scripts, plots and instructions. Statistical methods covered include probability distributions, hypothesis tests, regression, classification, machine learning, time series and biostatistics. The volume includes an introductory tutorial, **R** structure and syntax, graphics, and a complete annotated list of functions in base **R** and its built-in packages.

Chambers, J. M. (2008) *Software for Data Analysis: Programming with R*, Springer, Berlin
A volume by the principal author of the **S** statistical software system, highly recommended for scientists involved in serious **R** programming. Topics include basic **R** programming, creating a **CRAN** package, computing with text, introducing new **R** classes and methods, and interfaces with other languages.

Dalgaard, P. (2008) *Introductory Statistics with R*, 2nd ed., Springer, Berlin
A clear elementary presentation by a member of the **R** core team. Topics include essentials of the **R** language, the **R** environment, data handling, probability distributions, description statistics, graphics, hypothesis tests, regression, ANOVA, contingency tables, multiple regression, linear models, logistic regression, Poisson regression, nonlinear regression and survival analysis.

Jones, O., Maillardet, R. & Robinson, A. (2009) *Introduction to Scientific Programming and Simulation Using R*, CRC Press, Boca Raton
An unusual volume providing methods for scientific and stochastic modeling rather than statistical analysis using **R** with tutorials which are helpful for learning **R** programming. The book presents basic programming with **R**, graphics, probability and random variables, numerical computations (e.g. root finding, integration, optimization), Monte Carlo and simulation methods, importance sampling and related methods. Accompanied by the **CRAN** package *spuRs*.

Venables, W. N. & Ripley, B. D. (2002) *Modern Applied Statistics with S*, 4th ed., Springer
A distinguished and well-written volume with a wealth of intermediate-level guidance for statistical analysis using **R**. Topics include basics of the **S** and **R** languages, graphics, classical statistics, robust estimation, density estimation, bootstrap methods, linear modeling, generalized linear models, nonlinear and projection pursuit regression, tree-based partitioning, exploratory multivariate analysis, supervised classification, neural networks, support vector machines, survival analysis, time series analysis and spatial analysis. This work is the basis of **R**'s *MASS* library.

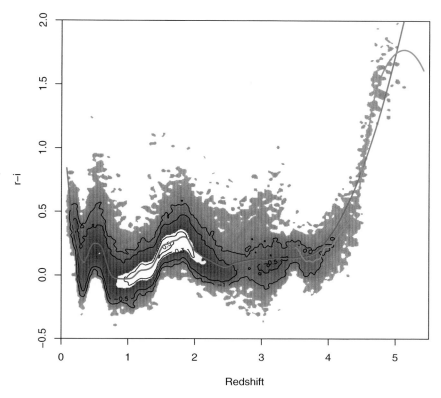

Fig. 6.5 Bivariate distribution of $r - i$ color index and redshift of SDSS quasars. Gray-scale and contours are calculated using averaged shifted histograms. Red and green curves are calculated using LOESS local regression. See text for details.

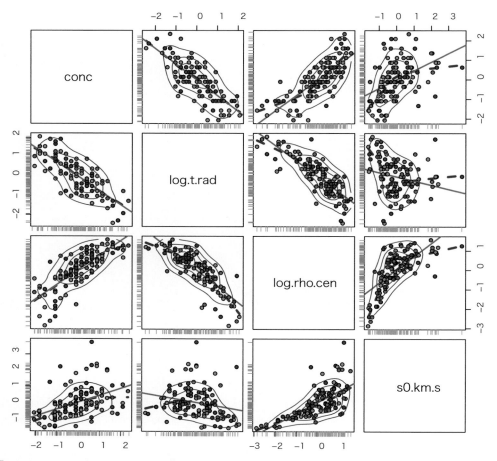

Fig. 8.4 An elaborated pair plot of Galactic globular cluster dynamical variables with colored points, linear regression (red lines), local regression (dashed blue curves), contours of a kernel density estimator, and rug plots along the axes.

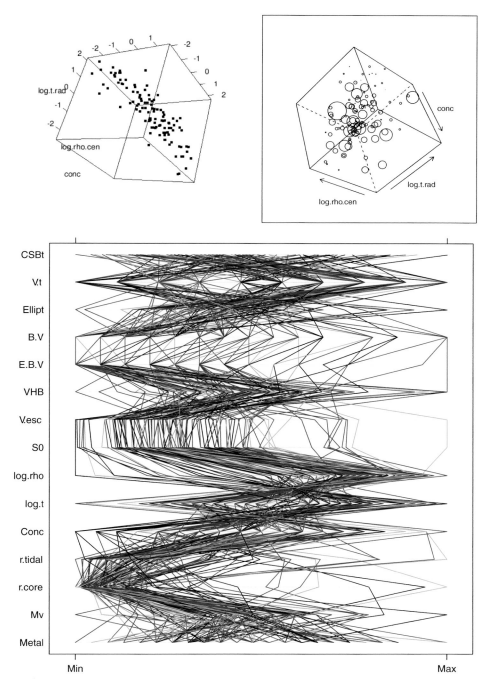

Fig. 8.7 Various multivariate displays of globular cluster dynamical variables: (a) snapshot from the *rgl* three-dimensional real-time rendering device; (b) "cloud" scatterplot of three variables with symbol size scaled to a fourth variable; and (c) 1+1 dimensional parallel coordinate plot of all variables.

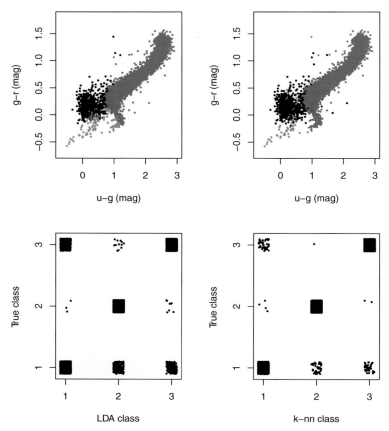

Fig. 9.8 Two supervised classifications of the SDSS test dataset: linear discriminant analysis (left) and k-nearest-neighbor classification with $k = 4$ (right). The top panels show the classification in the $(u - g) - (g - r)$ color–color diagram for the test dataset, and the bottom panels show the misclassifications of the training dataset. We follow the plotting scheme of Figure C.2 in the top panels where black symbols represent quasars, red symbols represent main-sequence and red-giant stars, and green symbols represent white dwarfs.

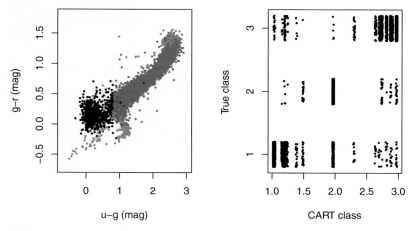

Fig. 9.10 CART classification of the SDSS point source dataset. *Left:* $(u - g) - (g - r)$ color–color diagram of the test dataset with colors as in Figure 9.8. *Right:* Validation of classifier on the training dataset.

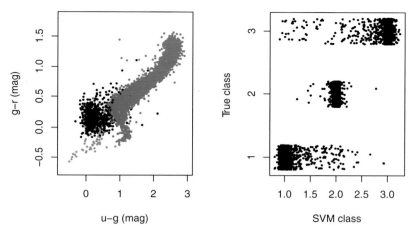

Fig. 9.12 Support vector machine classification of the SDSS point source dataset. *Left:* $(u - g) - (g - r)$ color–color diagram of the test dataset with colors as in Figure 9.8. *Right:* Validation of classifier on the training dataset.

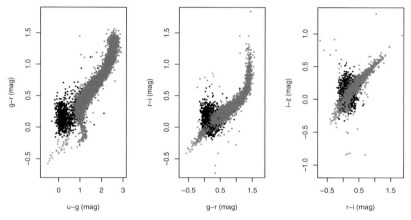

Fig. 9.13 Final SVM classification of SDSS test sources shown in the $(u - g) - (g - r)$, $(g - r) - (r - i)$, and $(r - i) - (i - z)$ color–color diagrams.

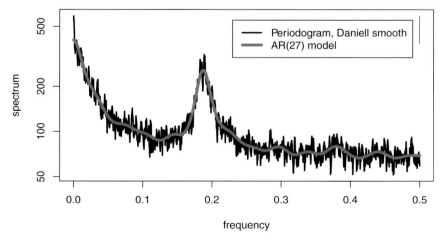

Fig. 11.7 Smoothed periodogram of the GX 5-1 time series and its AR(27) model.

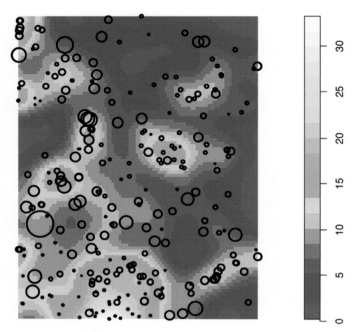

Fig. 12.3 The low-density region of the Shapley Supercluster galaxy redshift survey plotted with **CRAN**'s *spatstat* package. The color image shows a kernel density estimator of the galaxy distribution, and the circles show the individual galaxies with symbol size scaled to their velocities.

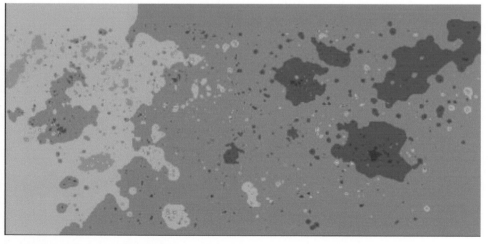

Fig. 12.7 Spatial interpolation of the velocity distribution in the full Shapley Supercluster region based on the inverse distance weighted interpolation estimator with index $\alpha = 1.5$.

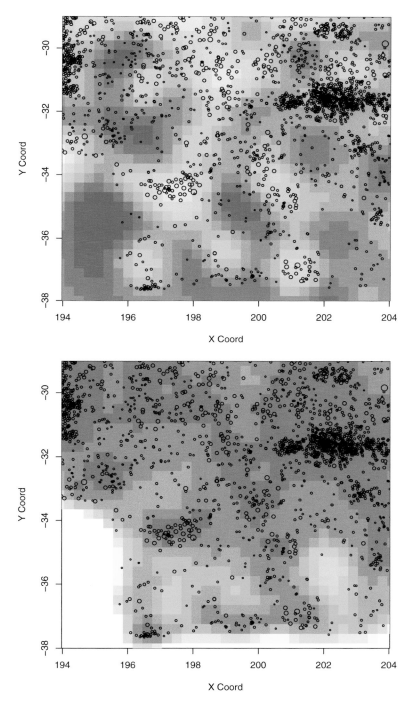

Fig. 12.8 Spatial interpolation of the Shapley Supercluster velocity distribution using ordinary kriging. *Top*: Estimated mean velocity distribution shown as a color map with galaxy locations superposed. *Bottom*: Variance of the kriging estimator shown as a color map.

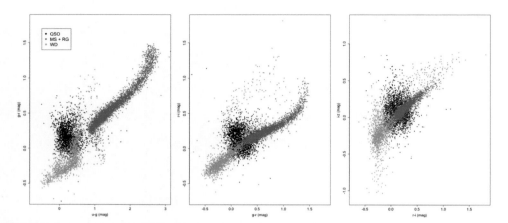

Fig. C.2 Optical photometric color–color diagrams for three training sets of the Sloan Digital Sky Survey. Black symbols show quasars, red symbols show main-sequence and red-giant stars, and green symbols show white dwarfs. The population ratios are arbitrary in this training set.

Appendix C **Astronomical datasets**

As demonstrated throughout this volume, astronomical statistical problems are remarkably varied, and no single dataset can exemplify the range of methodological issues raised in modern research. Despite the range and challenges of astronomical data analysis, few astronomical datasets appear in statistical texts or studies. The Zurich (or Wolff) sunspot counts over $\sim$200 years showing the 11 year cycle of solar activity is most commonly seen (Section C.13).

We present 20 datasets in two classes drawn from contemporary research. Thirteen datasets are used for **R** applications in this volume; they are listed in Table C.1 and described in Sections C.1–C.13. The full datasets are available on-line at http://astrostatistics.psu.edu/MSMA formatted for immediate use in **R**. Six additional datasets that, as of this writing, are dynamically changing due to continuing observations are listed in Table C.2 and described in Sections C.14–C.19. Most of these are time series of variable phenomena in the sky.

Tables C.1–C.2 provide a brief title and summary of statistical issues treated in each dataset. Here N_d is the number of datasets, n is the number of datapoints, and p is the dimensionality. In the sections below, for each dataset we introduce the scientific issues, describe and tabulate a portion of the dataset, and outline appropriate statistical exercises.

The datasets presented here can be used for classroom exercises involving a wide range of statistical analyses. Some problems are straightforward, others are challenging but within the scope of **R** and **CRAN**, and yet others await advances in astrostatistical methodology and can be used for research purposes. More challenging exercises are marked with (†).

Many additional datasets can be extracted from public astronomical databases and archives. Easily accessed databases include the Vizier catalog service curated by the Centre des Données Astronomique de Strasbourg with $\sim$9000 published tables (http://vizier.u-strasbg.fr/viz-bin/VizieR) and the NASA/IPAC Extragalactic Database with properties of $\sim$500,000 galaxies and related objects (http://ned.ipac.caltech.edu/). Vast datasets that require more expertise to extract and understand are available from NASA's many archive research centers and satellite mission science centers (http://archive.stsci.edu/sites.html). All major observatories, both ground-based and space-based, have public online data archives. Finally, the International Virtual Observatory (http://www.ivoa.net/ and http://www.usvao.org/) has developed tools for accessing and analyzing distributed data archives at many observatories (http://www.euro-vo.org/pub/fc/software.html). Training on Virtual Observatory tools is available at http://www.us-vo.org/summer-school/. All of these resources can be used for education and research in astrostatistics.

One unusual characteristic of several datasets is the use of the historical astronomical unit of "magnitudes" to measure the observed brightness (**apparent magnitude** denoted

Table C.1 Astronomical datasets used in the text.

#	Dataset	Problem	N_d	n	p	Sec
1	Asteroids	Inference, density estimation	1	26	1	C.1
2	Protostar populations	Contingency table	2	small	2	C.2
3	Globular cluster magnitudes	Two sample, inference	2	81,360	1	C.3
4	Stellar abundances	Two sample, censoring	2	29,39	2	C.4
5	Galaxy clustering	Spatial point process, clustering	1	4215	5	C.5
6	Hipparcos stars	Mixture models, regression	1	2719	8	C.6
7	Globular cluster properties	Multivariate analysis, regression	1	147	19	C.7
8	Sloan quasars	Multivariate analysis	1	77,429	15	C.8
9	Sloan point sources	Classification, data mining	1	17,000	5	C.9
10	Sloan training sets	Classification, data mining	3	3000, 3000, 2000	5	C.9
11	Galaxy photometry	Clustering, mixture model	1	572	2	C.10
12	Elliptical galaxy profiles	Regression, model selection	3	52, 57, 40	2	C.11
13	X-ray source variability	Stochastic time series	1	65,536	1	C.12

Table C.2 Dynamic astronomical datasets.

#	Dataset	Problem	N_d	n	p	Sec
14	Sunspot numbers	Time series, periodicities	1	>3000	1	C.13
15	Exoplanet orbits	Multivariate analysis, clustering	several	>500	2–20	C.14
16	Kepler light curves	Time series	~150,000	~10^2–10^3	1	C.15
17	Sloan Digital Sky Survey	Multivariate analysis, data mining	several	up to 10^8	2–10	C.16
18	Fermi light curves	Time series, censoring	~75	10^1–10^2	1	C.17
19	Swift gamma-ray bursts	Multivariate analysis	1	>600	2–20	C.18

m_{band}) or intrinsic luminosity (**absolute magnitude** denoted M_{band}) of objects in visible or infrared spectral bands. Magnitudes are inverted logarithmic measures; for example, the absolute magnitude of a star in the visual V band is $M_V = M_\odot - 2.5 \log_{10}(L_V/L_\odot)$ where $M_\odot$ is the absolute magnitude and $L_\odot$ is the luminosity of the Sun. Thus an increase of five magnitudes corresponds to a decrease in luminosity by a factor of 100; a quasar with $M_i = -29$ is 100 times more luminous than one with $M_i = -24$ in the near-infrared i band.

C.1   Asteroids

Statistical issues: univariate estimation, measurement errors, multimodality

Asteroids are small rocky bodies orbiting the Sun. We consider here the billions of asteroids in the main Asteroid Belt between the orbits of Mars and Jupiter. They are remnants of the formation of the Solar System 4.5 billion years ago, and their properties give crucial clues to the processes of planetesimal growth and the dynamics of the Solar System. Spectra and

Table C.3	Asteroid densities.	
Asteroid	Density	Error
1_Ceres	2.12	0.04
2_Pallas	2.71	0.11
4_Vesta	3.44	0.12
10_Hygiea	2.76	1.20

reflectivity show several types of surfaces, but the internal structure of asteroids is more difficult to establish. A crucial clue is provided by the density; for example, asteroids made of solid rock will have densities around 3–5 g cm^{-3}; porous rock asteroids will have lower densities. Note that an asteroid of solid water ice will have density 1 g cm^{-3}. To calculate asteroid densities, one needs measurements of the radius (obtainable from telescope data) and the mass that can only be measured by examining the orbit of some smaller body, a man-made probe or a natural companion asteroid.

Due to the difficulties in obtaining asteroid masses, densities have been measured for only a handful of asteroids. Each density has an estimated measurement error depending on the accuracy of the orbital information of the companion, and other problems. Evidence for bimodality has been noted; most have densities consistent with solid rock, but a few (particularly asteroid Mathilde) are either porous or dominated by ice. Asteroid densities can also be associated with laboratory measurements of meteorite densities (Britt & Consolmagno 2003), as meteorites are likely pieces of asteroids that have collided with the Earth.

Densities for 22 asteroids with detailed discussion of asteroid composition and porosity are presented by Britt *et al.* (2002, data online at http://www.psi.edu/pds/resource/density.html accessed in 2010). Five additional measurements were reported by Behrend *et al.* (2006) and Marchis *et al.* (2006). The value for Antiope was dropped due to the absence of a measurement error. The aggregated dataset called *asteroid_dens.dat* has 26 rows and three columns:

Asteroid Common name
Dens Bulk density in units of g cm^{-3}
Err Standard deviation uncertainty of the density

Four lines of the database are shown in Table C.3. The distribution of asteroid densities is shown in Figure 6.3.

Statistical exercises

1. Estimate the distribution of asteroid densities, with and without weighting by the measurement errors, using a variety of density estimation techniques. Estimate the moments of the distribution.
2. Test for bimodality using both parametric normal mixture models, nonparametric density estimators and dip test with bootstrap resampling. Perform these tests with and without

Table C.4	Protostellar jets.		
	Jet		
Multiplicity	No	Yes	Total
Single	9	5	14
Multiple	2	5	7
Total	11	10	21

weighting or deconvolution of measurement errors. Is there evidence for two classes of asteroids, solid rock and porous? (†)

C.2 Protostar populations

Statistical issues: contingency tables, hypothesis tests

The youngest stars are found embedded in and distributed around molecular clouds where they form from gravitational collapse of interstellar gas. The ages of young stars are associated with the infrared spectral energy distributions of their circumstellar disks, and are classified from youngest to oldest as: Class 0 (age $\sim 10^{4-5}$ yr, massive accreting disk and envelope); Class I (age $\sim 10^5$ yr, massive accreting disk); Class I/II or "flat" spectrum; Class II (age $\sim 10^6$ yr, accreting disk); Class II–III "transition" (nonaccreting outer disk); and Class III (age 10^{6-7} yr, no disk). The Class II systems, known as T Tauri stars, were first discovered in the 1940s while the Class 0–I systems were discovered only when infrared and millimeter telescopes were developed some decades later. Class III systems are mostly located by their X-ray emission. Astronomers also found that the younger protostars often eject high-velocity, collimated jets of gas perpendicular to the protoplanetary disks. These Herbig–Haro objects and bipolar flows are a byproduct of the accretion process and may have important effects on the surrounding molecular cloud. A good popular presentation of star formation, planet formation and the astrophysics of young stars is given by Bally & Reipurth (2006).

The first dataset presented here (Table C.4) is a 2×2 contingency table concerning a possible relationship between the multiplicity of protostars and the presence of a radio continuum jet (Reipurth *et al.* 2004). The astrophysical question is whether close gravitational interactions in binary- or triple-star systems is necessary to drive the jets, or whether the outflows emerge spontaneously from isolated protostars.

The second dataset (Table C.5) is a $r \times c$ contingency table with four samples divided into five classes. These are census results from recent multiwavelength surveys of nearby star-formation regions using NASA's infrared Spitzer Space Telescope and Chandra X-ray Observatory as well as ground-based telescopes. If the census is sufficiently complete, several science questions can be addressed: the star-formation efficiency can be estimated

Table C.5 Protostellar disks.

Cloud	Evolutionary class					Reference
	0–0/I	I–flat	II	Trans	III	
Serpens	21	16	61	17	22	Winston *et al.* 2007
Chamaeleon I	1	14	90	4	95	Luhman *et al.* 2008
Taurus	2	42	179	5	126	Luhman *et al.* 2010
η Cha	0	1	2	5	10	Sicilia-Aguilar *et al.* 2009

from the ratio of the total mass of young stars to the gaseous cloud mass; the stellar initial mass function can be estimated from the distribution of stellar masses; the history of star formation in the cloud can be evaluated from the distribution of young stellar ages; and the evolution of protoplanetary disks (a critical element in the complex processes of planet formation) can be traced. The observations are difficult and different research groups sometimes disagree on classification criteria. Here we have grouped together Class 0 and 0/I objects, Class I and flat spectrum objects. "Trans" refers to Class II/III or transition disks.

Inference based on these contingency tables is presented in Section 5.9.4.

Statistical exercises

1. Compare the χ^2 test and Fisher exact test to evaluate whether multiple protostars are more likely to produce jets than single protostars.
2. Use the protostellar jet data to investigate various approaches to the binomial proportion problem (Section 4.1.1).
3. Evaluate the age sequence of the four star-forming clouds from the ratios of younger to older stars. Establish whether the age differences are statistically significant.
4. Estimate the fraction of transition disks to total disks (Class 0–I–II), and evaluate whether this fraction is constant among the star-forming clouds. This is an important constraint for planet formation theory.

C.3 Globular cluster magnitudes

Statistical issues: univariate two-sample tests, parametric estimation, truncation

A globular cluster (GC) is a collection of $10^4 - 10^6$ ancient stars concentrated into a dense spherical structure structurally distinct from the field population of stars (Ashman & Zepf 1998). Historically, studies of individual stars within GCs were crucial for unravelling the evolution of stars, particularly their red-giant phases, and the structure of our Milky Way Galaxy (MWG). Understanding the collective global properties of GCs in the MWG is now

Table C.6 Globular cluster magnitudes.			
Milky Way Galaxy		**M 31**	
MWG_GC	K	M31_GC	K
47_Tuc	−11.790	303-026	17.537
NGC_362	−10.694	DAO023	15.845
NGC_1261	−9.452	305-D02	15.449
Eridanus_4	−5.140	306-029	12.710

a challenge, as they probably are a small remnant of a once-larger population of clusters and small satellite galaxies, most of which have been disrupted. Tracing the populations of GCs in nearby galaxies can give clues regarding the different evolution of galaxies including galactic star-formation histories (evidenced through GC metallicities), cannabilism (a large galaxy will acquire GCs from smaller merging galaxies), and galactic structure (the GC spatial distribution should reflect the gravitational potential of the galaxy).

The distribution of GC luminosities (i.e. the collective brightness of all the stars in a GC) is known as the globular cluster luminosity function (GCLF). For unknown reasons, the shape of the GCLF appears to be approximately lognormal in shape with a universal (though with some dependence on metallicity) mean and standard deviation. The peak of the GCLF has thus been used as one of several methods to measure the distances to galaxies. It is difficult to obtain complete samples of GCs in other galaxies as surveys are typically magnitude limited; that is, truncated due to flux limits at the telescope.

Here we consider the univariate brightness distributions of 81 GCs in our MWG and 360 GCs in the nearest large spiral galaxy, M 31 or the Andromeda Galaxy obtained from Nantais *et al.* (2006). A few rows of the dataset are shown in Table C.6. We use the near-infrared K-band magnitudes which are less vulnerable to interstellar absorption than visual-band magnitudes. Measurement errors are small and are ignored here. In the MWG dataset *GlobClus_MWG.dat*, we give the absolute magnitude which is an inverted logarithmic unit for intrinsic luminosity. In the M 31 dataset *GlobClus_M31.dat*, we give the apparent magnitude which is offset from the absolute magnitude by the distance modulus, measured to be 24.44 ± 0.1 for M 31. In both cases, K-band photometry is obtained from the 2MASS All-Sky Survey and is available for only a (biased) subset of the full MWG and M 31 GC populations.

The distributions of the Milky Way Galaxy and M 31 globular cluster magnitudes are displayed and discussed in Figure 4.2.

Statistical exercises

1. Fit a normal model to the MWG and M 31 samples and test whether the model is valid. Estimate the offset between the sample magnitudes.
2. Apply parametric and nonparametric two-sample tests to establish whether the MWG and M 31 GC luminosity functions are consistent with each other.
3. Calculate the distance modulus of M 31 and compare with the value 24.41±0.1 obtained with Cepheid variable distances.

4. Test whether the M 31 sample is consistent with a normal distribution (see Table C.6). First assume the sample is complete and unbiased. Second, truncate the sample for magnitudes $K > 15.5$ (the completeness limit of the 2MASS survey) and estimate the parameters of the normal model using maximum likelihood or moment techniques (Cohen 1991; Secker & Harris 1993). (†)

C.4 Stellar abundances

Statistical issues: bivariate censoring, survival analysis, correlation

One of the most remarkable recent discoveries in contemporary astronomy has been exoplanets orbiting normal stars near the Sun. A great deal of research has focused on whether the host star properties have any relationship to the presence or absence of a planetary system. The main finding has been that the probability of finding a planet is a steeply rising function of the star's metal content. But it was unclear whether this arises from the metallicity at birth or from later accretion of planetary bodies. Here we present a censored dataset from Santos *et al.* (2002) where the authors seek differences in the abundances of the light elements beryllium (Be) and lithium (Li) in stars that do and do not host planets. These elements are expected to be depleted by internal stellar burning, so that excess Be and Li should be present only in the planet accretion scenario of metal enrichment.

The dataset consists of 39 stars known to host planets and 29 stars in a control sample of stars without planets. Due to internal stellar processes, Be abundances are correlated with stellar mass which is traced by stellar surface temperature (the effective temperature or T_{eff}). Be and Li abundances are seen to be interdependent in a complicated fashion, but little difference is seen between the planet-hosting and control samples. Regression lines of the detections only show a slight elevation in Be abundance for planet-hosting stars, but this difference evaporates when a Buckley–James regression line is considered that includes the effects of censoring.

The columns of the dataset *censor_Be.dat* (Table C.7) are:

Star Star name (HD = Henry Draper catalog)
Sample 1 indicates planet-hosting stars and 2 is the control sample
T$_{eff}$ Effective stellar surface temperature (in kelvin)
I(Be) Censoring indicator of beryllium measurement: 1 indicates detected, 0 indicates censored
logN(Be) Beryllium abundance with respect to the Sun's abundance
σ(Be) Measurement standard deviation of log N(Be)
I(Li) Censoring indicator of lithium measurement
logN(Li) Lithium abundance with respect to the Sun's abundance

The lithium and beryllium abundances in this stellar sample are discussed in Section 10.8.3.

Table C.7 Stellar abundances with and without planets.

Star	Sample	T_{eff}	I(Be)	logN(Be)	σ(Be)	I(Li)	logN(Li)
BD-103166	1	5320	1	0.50	NaN	1	NaN
HD_6434	1	5835	1	1.08	0.10	0	0.80
HD_9826	1	6212	1	1.05	0.13	1	2.55
HD_10647	1	6143	1	1.19	0.10	1	2.80
HD_10697	1	5641	1	1.31	0.13	1	1.96

Statistical exercises

1. Construct the univariate Kaplan–Meier distributions of Be and Li abundances for the planet-hosting and control samples. Apply survival analysis two-sample tests.
2. Apply bivariate correlation tests and linear regressions to the Be vs. Li plot for both samples. Note that the generalized Kendall's τ developed by Brown et al. (1974), implemented in the standalone *ASURV* package, permits censoring in both variables.
3. Perform linear regressions considering heteroscedastic weighting and/or censoring for the Be abundances. (†)
4. Extend the two-sample test to the trivariate case involving T_{eff}, $\log N_{Be}$ and $\log N_{Li}$. Akritas & Siebert (1996) developed a trivariate partial correlation coefficient for multiply censored data based on Kendall's τ. Write an R wrapper for the standalone code available at StatCodes (http://astrostatistics.psu.edu/statcodes). (†)

C.5 Galaxy clustering

Statistical issues: spatial point processes, multivariate clustering, Poisson processes, measurement errors

The distribution of galaxies in space is strongly clustered due to their mutual gravitational attraction. The Milky Way Galaxy resides in its Local Group which lies on the outskirts of the Virgo Cluster of galaxies, which in turn is part of the Local Supercluster. Similar structures of galaxies are seen at greater distances, and collectively the phenomenon is known as the large-scale structure (LSS) of the Universe. The clustering is hierarchical, nonlinear, and anisotropic. The latter property is manifested as galaxies concentrating in huge flattened, curved superclusters surrounding voids, resembling a collection of soap bubbles. The basic characteristics of the LSS are now understood astrophysically as arising from the gravitational attraction of matter in the Universe expanding from the Big Bang ~14 billion years ago. The observed three-dimensional patterns are well-reproduced by simulations with attractive cold dark matter and repulsive dark energy in addition to attractive baryonic

Table C.8 Galaxy redshifts in the Shapley Concentration.				
R.A.	Dec.	Mag	Vel	SigVel
193.02958	−32.84556	15.23	15056	81
193.04042	−28.54083	17.22	16995	32
193.04042	−28.22556	17.29	21211	81
193.05417	−28.33889	18.20	29812	37
193.05542	−29.84056	12.55	2930	38

(ordinary) matter. The properties of baryonic and dark components needed to explain LSS agree very well with those needed to explain the fluctuations of the Cosmic Microwave Background and other results from observational cosmology.

Despite this fundamental understanding, there is considerable interest in understanding the details of galaxy clustering, such as the growth of rich galaxy clusters and their collection into supercluster-scale structures. The richest nearby supercluster of interacting galaxy clusters is called the Shapley Concentration. It includes several clusters from the Abell catalog of rich galaxy clusters seen in the optical band, and a complex and massive hot gaseous medium seen in the X-ray band. Optical measurement of galaxy redshifts provides crucial information but represents an uncertain convolution of the galaxy distance and gravitational effects of the clusters in which they reside. The distance effect comes from the universal expansion from the Big Bang, where the recessional velocity (galaxy redshift) follows Hubble's law $v = H_0 d$, where v is the velocity in km s^{-1}, d is the galaxy distance from us in Mpc (million parsecs, 1 pc = 3 light years), and H_0 is Hubble's constant known to be about 73 km s^{-1} Mpc.

The dataset arises from a galaxy redshift survey by Drinkwater *et al.* (2004) where redshifts (recessional velocities in km s^{-1}) are measured for 4215 galaxies in the Shapley Concentration. The dataset (Table C.8) has the following columns:

R.A. Right ascension (coordinate in the sky similar to longitude on Earth, in degrees)
Dec. Declination (coordinate similar to latitude, in degrees)
Mag Galaxy magnitude
Vel Velocity of the galaxy moving away from Earth (in km s^{-1}, after various corrections are applied)
SigVel Standard deviation of the velocity measurement

Visualization and statistical analysis of this galaxy redshift survey is presented in Section 12.12.

Statistical exercises

1. Apply and compare a variety of multivariate clustering algorithms to this dataset. Astronomers often use single-linkage nonparametric hierarchical agglomeration or the friends-of-friends algorithm.

2. Apply maximum likelihood mixture models for assuming individual clusters are multi-variate normals. Examine penalized likelihoods like the Bayesian information criterion for model selection.

3. Apply a wavelet decomposition, with both standard and anisotropic bases (e.g. curvelets, ridgelets) to elucidate the anisotropic hierarchical structure. (†)

4. Estimate the void probability distribution or otherwise quantify the regions with low galaxy densities. (†)

5. Examine the effects of nonstandard metrics in the clustering calculation: spherical vs. Euclidean geometry and mass-weighted points (log-mass scales with log-luminosity which scales with $-2.5*\text{Mag}$). (†)

6. Evaluate the "fingers of God" effect in the Shapley redshift distribution. This is a distortion in the velocity dispersion of the densest regions due to the gravitational potential of dark matter and the intracluster gaseous medium. (†)

C.6   Hipparcos stars

Statistical issues: multivariate clustering, mixture models, regression, spatial point processes

A hundred years ago, the key to understanding the properties and evolution of stars was a plot of the brightness of stars against their color. In observed units this is the color–magnitude diagram and, if distances are known, the Hertzsprung–Russell (HR) diagram, first studied around 1910. It plots the luminosity of stars along the ordinate, and the color (or equivalently, the surface temperature or spectral type) of stars along the abscissa. When a random selection of bright stars is plotted, one sees the main sequence of hydrogen-burning stars and the evolved red-giant stars burning helium and other elements. The main sequence is most clearly seen when the stars in a coeval (that is, formed together), codistant open cluster of stars are plotted. The nearest open cluster is the Hyades cluster and its HR diagram has been carefully studied for many years.

The main difficulty in establishing the HR diagram for a sample of stars is determining their distances. For nearby stars, this can be done with extremely precise measurements of their parallax, tiny positional motions every year as the Earth orbits the Sun. In the 1990s, the European Space Agency launched a satellite called Hipparcos that measured stellar parallaxes with more precision than previously achieved for $\sim$100,000 stars. The Hipparcos catalog is thus often used for HR diagram studies. Bright Hyades members have a mean parallax of 22 mas in the Hipparcos database corresponding to a distance of 45 parsecs (pc). A study of the Hyades cluster using Hipparcos data has been made by Perryman *et al.* (1998).

We have extracted a subset of 2719 Hipparcos stars which include many Hyades members with the selection criterion that the parallax lies between 20 and 25 mas corresponding to distances between 40 and 50 pc. This dataset (Table C.9) has the following columns:

Table C.9 Hipparcos star properties.

HIP	Vmag	RA	DE	Plx	pmRA	pmDE	e_Plx	B-V
2	9.27	0.003797	−19.498837	21.90	181.21	−0.93	3.10	0.999
38	8.65	0.111047	−79.061831	23.84	162.30	−62.40	0.78	0.778
47	10.78	0.135192	−56.835248	24.45	−44.21	−145.90	1.97	1.150
54	10.57	0.151656	17.968956	20.97	367.14	−19.49	1.71	1.030
74	9.93	0.221873	35.752722	24.22	157.73	−40.31	1.36	1.068

HIP Hipparcos star number

Vmag Visual-band magnitude

RA Right ascension (degrees), positional coordinate in the sky equivalent to longitude on the Earth

DE Declination (degrees), positional coordinate in the sky equivalent to latitude on the Earth

Plx Parallactic angle (mas = milliarcseconds). 1000/Plx gives the distance in parsecs (pc)

pmRA Proper motion in RA (mas yr^{-1}). RA component of the motion of the star across the sky

pmDE Proper motion in DE (mas yr^{-1}). DE component of the motion of the star across the sky

e_Plx Measurement error in Plx (mas)

B-V Color index (mag)

The HR diagram can be plotted by plotting log L vs. $B - V$ where (roughly) the log-luminosity in units of solar luminosity is constructed: $\log_{10}(L) = (15 - Vmag - 5\log_{10}(Plx))/2.5$. It is difficult to select Hyades members from field stars in a reliable and complete fashion. Their sky positions are centered around RA = 67 degrees and DE = +16 degrees, but they also share converging proper motions with vector components pmRA and pmDE.

This sample of Hipparcos stars is used to estimated the stellar luminosity function in Section 10.8.4.

Statistical exercises

1. Find Hyades cluster members, and possibly Hyades supercluster members, by nonparametric and parametric multivariate clustering. Compare the sample with Perryman *et al.* (1998), and reproduce some of their other results.
2. Isolate the Hyades main sequence and fit with nonparametric local regression techniques.
3. Locate outliers far from the main sequence associated with white dwarfs, red giants and Galactic halo stars.
4. Using bootstrap and other techniques, estimate probabilities of individual star membership in the Hyades cluster. (†)
5. Use the heteroscedastic measurement error values e_Plx to weight the multivariate clustering and membership probabilities. (†)

Table C.10 Galactic globular cluster properties.									
Name	Gal.long	Gal.lat	R.sol	R.GC	Metal	Mv	r.core	r.tidal	Conc
NGC_104	305.90	−44.89	4.6	8.1	−0.8	−9.6	0.5	60.3	2.1
NGC_288	151.15	−89.38	8.2	12.1	−1.4	−6.6	4.0	37.0	0.9
NGC_362	301.53	−46.25	8.7	9.9	−1.4	−8.4	0.5	25.9	1.7
NGC_1261	270.54	−52.13	16.1	18.3	−1.2	−7.8	1.9	37.1	1.3
Pal_1	130.07	19.02	13.7	20.3	−1.0	−2.5	0.6	19.9	NA

Name	log.t	log.rho	S0	V.esc	VHB	E.B-V	B-V	Ellipt	V.t	CSB
NGC_104	7.9	5.0	13.2	56.8	14.1	0.0	1.0	3.8	3.0	5.6
NGC_288	9.0	2.0	2.8	10.0	15.3	0.0	0.9	8.1	1.0	11.0
NGC_362	7.8	4.8	10.3	41.8	15.4	0.0	0.9	6.4	6.0	6.1
NGC_1261	8.6	3.2	5.5	20.7	16.6	0.0	0.9	8.2	8.0	8.7
Pal_1	7.1	2.4	0.7	2.9	16.8	0.2	1.1	13.6	2.0	12.4

C.7 Globular cluster properties

Statistical issues: multivariate analysis, regression

Globular clusters have provided important clues to a number of fields of astrophysics. Here we examine the global properties of well-studied clusters in the Milky Way Galaxy — location in the Galaxy, structural and dynamical properties, photometric properties — to address issues concerning the evolution of the Galaxy. Globular clusters are sometimes seen in the halo of the Galaxy far from the Galactic Center. These clusters tend to have low metallicities indicative of formation when the Galaxy formed $\geq$ 10 billion years ago. Other clusters are concentrated in the Galactic Plane (low Galactic latitude) with higher metallicities indicative of formation at later times. Yet other clusters likely were accreted from the cannibalism of small satellite galaxies. Clusters also have a range of internal structures. Some are strongly centrally concentrated with small core radii, while others have a more diffuse structure. Stars in the outer regions of the cluster may have been stripped away by tidal interactions with the Galaxy.

We provide a multivariate dataset drawn from the catalog of Webbink (1985) with a variety of measured and inferred properties showing a variety of interrelationships. The dataset (Table C.10) has columns:

Name Common name
Gal.long Galactic longitude (degrees)
Gal.lat Galactic latitude (degrees)
R.sol Distance from Sun (kpc)
R.GC Distance from Galactic Center (kpc)
Metal Log metallicity with respect to solar metallicity

Mv Absolute magnitude
r.core Core radius (pc)
r.tidal Tidal radius (pc)
Conc Core concentration parameter
log.t Log central relaxation timescale (yr)
log.rho Log central density ($M_\odot/pc^{-3}$)
S0 Central velocity dispersion (km s^{-1})
V.esc Central escape velocity (km s^{-1})
VHB Level of the horizontal branch (mag)
E.B-V Color excess (mag)
B-V Color index (mag)
Ellipt Ellipticity
V.t Integrated V magnitude (mag)
CSB Central surface brightness (mag per square arcsec)

This multivariate dataset is extensively studied in Section 8.8.

Statistical exercises

1. Perform regressions for the cluster structure variables with cluster luminosity (Mv) as the response variable.
2. Seek relationships between the locational, photometric and structural variables, perhaps using canonical correlation analysis.
3. Divide clusters into halo and Galactic Plane subpopulations using multivariate clustering methods. Compare the subsamples using two-sample tests, MANOVA and other tools.
4. Identify collinear variables and apply regression techniques (ridge regression, lasso) designed for collinearity. (†)

C.8 SDSS quasars

Statistical issues: multivariate analysis, regression, censoring

Most or all large galaxies have a supermassive black hole (SMBH, 10^6–10^9 M$_\odot$ solar masses) in their nuclei. Gas from the interstellar medium or a disrupted star may fall onto the SMBH through an accretion disk (to "accrete" means to "fall onto"). This accretion disk can become exceedingly hot and can eject a jet of material at relativistic (near the speed of light) velocities. The disk and jet radiate light across the electromagnetic spectrum (radio, infrared, visible, ultraviolet, X-ray, gamma-ray) with great efficiency. In most galaxies today, like our own Milky Way Galaxy, the SMBH is starved of gas and little light is produced. In other galaxies, like Seyfert galaxies or radio galaxies, the light is very strong, particularly in spectral bands other than the visible band where the stars of the host galaxies are bright. In rare cases, the light from the accreting SMBH exceeds the starlight in all spectral bands by enormous factors. These are the quasars or quasi-stellar objects (QSOs), the most luminous

Table C.11 SDSS quasars.							
SDSS	z	u_mag	sig_u_mag	g_mag	sig_g_mag	r_mag	sig_r_mag
000006.53+003055.2	1.8227	20.389	0.066	20.468	0.034	20.332	0.037
000008.13+001634.6	1.8365	20.233	0.054	20.200	0.024	19.945	0.032
000009.26+151754.5	1.1986	19.921	0.042	19.811	0.036	19.386	0.017
000009.38+135618.4	2.2400	19.218	0.026	18.893	0.022	18.445	0.018
000009.42−102751.9	1.8442	19.249	0.036	19.029	0.027	18.980	0.021
000011.41+145545.6	0.4596	19.637	0.030	19.466	0.024	19.362	0.022
000011.96+000225.3	0.4790	18.237	0.028	17.971	0.020	18.025	0.019

i_mag	sig_i_mag	z_mag	sig_z_mag	FIRST	ROSAT	Mp
20.099	0.041	20.053	0.121	0.000	−9.000	−25.100
19.491	0.032	19.191	0.068	0.000	−9.000	−25.738
19.165	0.023	19.323	0.069	−1.000	−9.000	−25.085
18.331	0.024	18.110	0.033	−1.000	−9.000	−27.419
18.791	0.018	18.751	0.047	0.000	−9.000	−26.459
19.193	0.025	19.005	0.047	−1.000	−9.000	−22.728
17.956	0.014	17.911	0.029	0.000	−1.660	−24.046

objects in the Universe that can be seen at high redshifts (i.e. great distance from us). Collectively, Seyfert galaxies, radio galaxies, quasars and similar objects are called active galactic nuclei (AGN).

The spectrum, variability and (when resolved in telescopes) structure of quasars are studied in detail to understand the complex processes of accreting SMBHs and their environs. But an important subfield relies on wide-field surveys to obtain statistically meaningful samples of quasars and characterize their bulk properties such as brightness in various spectral bands, redshifts and luminosities. Such survey and photometric (brightness) studies lead to classification of quasar subtypes (e.g. Type I, Type II, radio-loud/radio-quiet, BAL, BL Lacs, Lyman-alpha dropouts), to measurement of the quasar luminosity function (distribution of luminosities), and to cosmic evolution studies (how the population changes with redshift).

The Sloan Digital Sky Survey (SDSS) (York *et al.* 2000) has performed the largest wide-field photometric and spectroscopic survey leading to a well-defined sample of tens of thousands of quasars with very precise photometric measurements in five visible bands and accurate redshifts (Schneider *et al.* 2007). Some of these also have radio, infrared and X-ray detections from other surveys.

The SDSS team has produced a catalog of 77,429 quasars from its Fifth Data Release, 95% of them previously unknown (Schneider *et al.* 2007). The dataset *SDSS_QSO.dat* presented here (Table C.11) has 60,120 quasars with 15 properties:

SDSS Sloan Digital Sky Survey designation based on right ascension and declination
z Redshift (scales with distance)

u_mag and sig_u_mag Brightness in the *u* (ultraviolet) band in magnitudes with standard deviation (mag). The heteroscedastic measurement errors for each magnitude are determined by the SDSS team from knowledge of the observing conditions, detector background and other technical considerations.

g_mag and sig_g_mag Brightness in the *g* (green) band with standard deviation (mag)

r_mag and sig_r_mag Brightness in the *r* (red) band with standard deviation (mag)

i_mag and sig_i_mag Brightness in the *i* (further red) band with standard deviation (mag)

z_mag and sig_z_mag Brightness in the *z* (further red) band

FIRST Brightness in the radio band, in magnitudes scaled from the flux density measured in the NRAO FIRST survey at 20m. 0 indicates the quasar is undetected by FIRST, while −1 indicates it was not observed by FIRST.

ROSAT Brightness in the X-ray band, in log(Count rate) from the ROSAT All-Sky Survey (RASS) in the 0.2–2.4 keV band. −9 indicates not detected by RASS.

$\mathbf{M_P}$ Absolute magnitude in the *i* band.

Aspects of this large multivariate database are displayed in Sections 6.9 and are analyzed in Sections 7.10 and 10.8.

Statistical exercises

1. In paired *ugriz* photometry plots, find outliers and compare them with measurement errors. Note the survey structure in the i_mag distribution: quasar identifications are complete below $i = 19$ and incomplete for fainter magnitudes.
2. Seek redshift dependencies (i.e. cosmic evolution) in the X-ray/optical and X-ray/radio brightness ratios.
3. Seek photometric predictors for radio-loudness and for high redshift ($z > 4$–5). These may appear as outliers in a $(u - g)$ vs. $(g - r)$ color–color plot.
4. Study the relationships between X-ray, radio and optical emission using survival analysis.
5. Update the database to the most recent SDSS quasar survey (e.g. Schneider *et al.* 2010), and examine the effects of a doubling of the sample size.
6. Seek a photometric predictor for redshift. This might involve a regression of redshift as a function of *ugriz* magnitudes and colors, or application of data mining methods such as neural networks or k-nearest-neighbor classifiers. It is important to quantify the precision of the predictor as a function of photometric properties. This is an active area of research (Weinstein *et al.* 2004; Yèche *et al.* 2010, Wu & Jia 2010). (†)

C.9 SDSS point sources

Statistical issues: multivariate classification, mixture models, regression

The first Sloan Digital Sky Survey (SDSS) provides a catalog of ∼230 million objects in the visible band covering one-fourth of the celestial sphere. The SDSS photometric catalog gives brightnesses in five photometric bands: *u* (ultraviolet), *g* (green), *r* (red), *i* and *z*

Table C.12 SDSS point source test dataset.						
u_mag	g_mag	r_mag	i_mag	z_mag	ra	dec
19.39	17.06	16.10	15.75	15.56	180.0148538	20.90261459
17.98	17.07	16.76	16.66	16.60	180.0130685	24.47397840
19.17	18.01	17.59	17.44	17.36	180.0084428	23.83174719
20.74	18.58	17.64	17.31	17.14	180.0069980	20.84820751
20.71	19.75	19.35	19.22	19.17	180.0012797	20.79797715

(very-near-infrared) bands. About a million spectra have also been obtained. The survey is described by York *et al.* (2000) and data products can be obtained at http://sdss.org. Although it uses a rather modest 2.5-meter diameter telescope, the SDSS has been one of the most important and productive projects in modern astronomy, generating thousands of publications.

For the test set for classification, we collect 17,000 sources from Data Release 7 of the SDSS survey lying in a random $5° \times 5°$ square on the sky, which are preselected to have point-like morphologies. This omits spatially resolved galaxies. We have applied a simple filter on magnitudes $15 < \text{mag} < 21$ to avoid saturation of bright sources and uncertain measurements on faint sources, and require that internal estimates of magnitude errors be < 0.05 mag. This dataset, *SDSS_17K.dat*, shown in Table C.12 has 12 columns:

u_mag, g_mag, r_mag, i_mag and z_mag Brightness in five spectral bands (in magnitudes)

ra and dec Right ascension and declination (in degrees)

The dataset was acquired with SQL commands similar to:

```
# SDSS SQL queries from http://cas.sdss.org/astro/en/tools/search/sql.asp
# Test set of SDSS point sources
# SDSS_test.csv

SELECT  TOP 17000 psfmag_u, psfmag_g, psfmag_r, psfmag_i, psfmag_z, ra,dec
FROM Star WHERE
ra between 180 and 185 and dec between +20. and +25.
and (psfmag_u>15.) and (psfmag_u<21.) and (psfmag_g>15.) and (psfmag_g<21.)
and (psfmag_r>15.) and (psfmag_r<21.) and (psfmag_i>15.) and (psfmag_i<21.)
and (psfmag_z>15.) and (psfmag_z<21.)
and psfmagerr_u<0.05 and psfmagerr_g<0.05 and psfmagerr_r<0.05
and psfmagerr_i<0.05 and psfmagerr_z<0.05
```

As objects of the same class can lie at a wide range of distances from Earth, the apparent magnitudes by themselves are not useful indicators of source class. The distance differences are removed by considering ratios of brightnesses in the five bands; as magnitudes are logarithmic units of brightness, these ratios are given by color indices like $(u - g)$ and

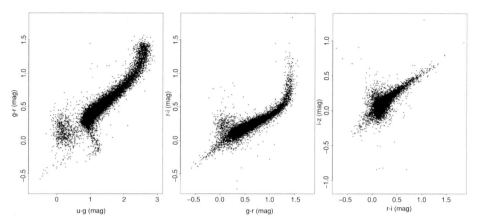

Fig. C.1   Optical photometric color–color diagrams for the test dataset of 17,000 Sloan Digital Sky Survey point sources: $(u - g) - (g - r)$, $(g - r) - (r - i)$, and $(r - i) - (i - z)$ color–color diagrams.

$(r - i)$. The five photometric variables thus provide four color indices, and our classification effort here is performed in this four-dimensional space. The four variables are commonly displayed in three bivariate color–color scatter diagrams; these are shown for the SDSS test dataset in Figure C.1.

The SDSS point sources have been well characterized with spectroscopic followup, providing excellent training sets. The largest population are main-sequence stars with red giants at $g - r \sim 1.3$ (Ivezić *et al.* 2002). Low-redshift quasars form a roughly spherical cluster at $u - g < 0.6$ (Schneider *et al.* 2010) but are mixed with white dwarfs in some color–color diagrams (Eisenstein *et al.* 2006). Rarer high-redshift quasars are spread over a large region below the main sequence. Real substructure is present; for example, the curved distribution from $(u - g, g - r) = (0.5, 0.2)$ to $(0.1, -0.4)$ is a subclass of white dwarf stars. Rare bluer A-F stars, including RR Lyrae variables, constitute a small elongated group extending from the main-sequence locus around $(1.0, 0.2)$ to $(1.2, -0.2)$. Some classes of point sources, such as high-redshift quasars and brown dwarfs, are entirely missing from this plot as they appear only in the redder bands of the SDSS survey. The color–color diagram is considerably more complex when spatially extended galaxies are added.

Our SDSS training sets are constructed as follows. We obtain a sample of 5000 main-sequence and red-giant stars from a sky location separate from the test dataset with filters $u - g > 0.6$ and $15 < \mathrm{mag} < 18$ using the following SQL commands:

```
# Training set of 5,000 SDSS stars (main sequence and red giants)
# SDSS_stars.csv

SELECT  TOP 5000 psfmag_u, psfmag_g, psfmag_r, psfmag_i, psfmag_z, ra,dec
FROM Star WHERE
```

Table C.13 SDSS star training set.						
u_mag	g_mag	r_mag	i_mag	z_mag	ra	dec
20.533161	17.968031	16.827314	16.373095	16.091063	190.0012612	21.91061999
18.894476	17.674576	17.208895	17.033516	16.97294	190.0023299	21.87464109
18.290901	17.022612	16.625498	16.431305	16.399673	190.0122402	21.70299891
19.784182	17.820679	17.003235	16.721306	16.541569	190.0129122	23.66791335
18.874607	17.638632	17.120625	16.907972	16.849592	190.0129511	24.79750668

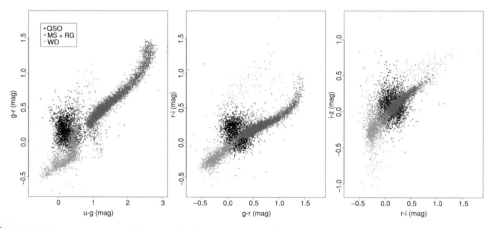

Fig. C.2   Optical photometric color–color diagrams for three training sets of the Sloan Digital Sky Survey. Black symbols show quasars, red symbols show main-sequence and red-giant stars, and green symbols show white dwarfs. The population ratios are arbitrary in this training set. For a color version of this figure please see the color plate section.

```
ra between 190 and 195 and dec between +20. and +25.
and (psfmag_g>15.) and (psfmag_g<18.) and (psfmag_r>15.) and (psfmag_r<18.)
and (psfmag_i>15.) and (psfmag_i<18.) and (psfmag_z>15.) and (psfmag_z<18.)
and (psfmagerr_g<0.05) and (psfmagerr_r<0.05) and (psfmagerr_i<0.05)
and (psfmagerr_z<0.05)
and (psfmag_u - psfmag_g > 0.6)
```

A training set of 10,090 white dwarfs from Eisenstein *et al.* (2006) is downloaded from the Vizier Catalogue Service (http://vizier.u-strasbg.fr/, catalog J/ApJS/167/40). A training set of 2000 spectroscopically confirmed quasars is selected from the SDSS DR 4 dataset described in Section C.8. These three SDSS training datasets are culled and combined using the following **R** script. The resulting color–color scatterplots are shown in Figures C.1 and C.2 for the test and training datasets, respectively.

The quasar training dataset, *SDSS_QSO.dat*, is the same as that used in Section C.8. The main-sequence (and red-giant) training set from Ivezić *et al.* (2002) is shown in Table C.13 with seven columns:

Table C.14	SDSS white-dwarf training set.						
Sp_Class	u_mag	g_mag	r_mag	i_mag	z_mag	ra	dec
DA_auto	19.186	18.835	18.884	19.052	19.407	0.028147	−0.781616
DA	19.421	19.565	20	20.277	21.152	0.029836	−9.727753
DQ	19.413	19.091	19.052	19.12	19.339	0.048182	−8.835651
DA_auto	20.423	20.041	19.925	20.03	20.012	0.051346	−0.845157
DA	19.285	18.891	18.81	18.886	18.958	0.093902	−10.86171

u_mag, g_mag, r_mag, i_mag and z_mag Brightness in five bands (in magnitudes)
ra and dec Sky coordinates (in degrees)

The white-dwarf training set from Eisenstein *et al.* (2006) is shown in Table C.14 with
eight columns:

Sp_Class White dwarf spectral class
u_mag, g_mag, r_mag, i_mag and z_mag Brightness in five bands (in magnitudes)
ra and dec Sky coordinates (in degrees)

The following **R** script has two purposes: to document the construction of the test and
training datasets for SDSS point source classification; and to illustrate common operations
needed to manipulate published astronomical datasets for **R** analysis. For each of the three
training sets, we construct color indices, such as $u - g$ and $r - i$, from the SDSS photometric
data. For the quasar training set, we filter the published sources to remove objects with very
faint magnitudes and inaccurate photometry using **R**'s *which* function. In the white-dwarf
dataset, we remove objects with missing photometry with the *na.omit* function. New dataframes are constructed using *cbind* to add new columns, *data.frame* to place the resulting
array into an **R** dataframe, and *names* to add variable labels. The main-sequence star, quasar
and white-dwarf datasets are then concatenated into a single training set using the *rbind*
function.

```
# SDSS point sources test dataset, N=17,000 (mag<21, point sources, hi-qual)

SDSS <- read.csv('http://astrostatistics.psu.edu/MSMA/datasets/SDSS_test.csv', h=T)
dim(SDSS) ; summary(SDSS)
SDSS_test <- data.frame(cbind((SDSS[,1]-SDSS[,2]), (SDSS[,2]-SDSS[,3]),
(SDSS[,3]-SDSS[,4]), (SDSS[,4]-SDSS[,5])))
names(SDSS_test) <- c('u_g', 'g_r', 'r_i', 'i_z')
str(SDSS_test)

par(mfrow=c(1,3))
plot(SDSS_test[,1], SDSS_test[,2], xlim=c(-0.7,3), ylim=c(-0.7,1.8), pch=20,
cex=0.6, cex.lab=1.5, cex.axis=1.5, main='', xlab='u-g (mag)', ylab='g-r (mag)')
plot(SDSS_test[,2], SDSS_test[,3], xlim=c(-0.7,1.8), ylim=c(-0.7,1.8), pch=20,
cex=0.6, cex.lab=1.5, cex.axis=1.5, main='', xlab='g-r (mag)', ylab='r-i (mag)')
```

```
plot(SDSS_test[,3], SDSS_test[,4], xlim=c(-0.7,1.8), ylim=c(-1.1,1.3), pch=20,
cex=0.6, cex.lab=1.5, cex.axis=1.5, main='', xlab='r-i (mag)', ylab='i-z (mag)')
par(mfrow=c(1,1))


# Quasar training set, N=2000 (Class 1)

qso1 <- read.table('http://astrostatistics.psu.edu/MSMA/datasets/SDSS_QSO.dat', h=T)
dim(qso1) ; summary(qso1)
bad_phot_qso <- which(qso1[,c(3,5,7,9,11)] > 21.0 | qso1[,3]==0)
qso2 <- qso1[1:2000,-bad_phot_qso,]
qso3 <- cbind((qso2[,3]-qso2[,5]), (qso2[,5]-qso2[,7]), (qso2[,7]-qso2[,9]), (qso2[,9]-qso2[,11]))
qso_train <- data.frame(cbind(qso3, rep(1, length(qso3[,1]))))
names(qso_train) <- c('u_g', 'g_r', 'r_i', 'i_z', 'Class')
dim(qso_train) ; summary(qso_train)


# Star training set, N=5000 (Class 2)

temp2 <- read.csv('http://astrostatistics.psu.edu/MSMA/datasets/SDSS_stars.csv', h=T)
dim(temp2) ; summary(temp2)
star <- cbind((temp2[,1]-temp2[,2]), (temp2[,2]-temp2[,3]), (temp2[,3]-temp2 [,4]),
(temp2[,4]-temp2[,5]))
star_train <- data.frame(cbind(star, rep(2, length(star[,1]))))
names(star_train) <- c('u_g','g_r','r_i','i_z','Class')
dim(star_train) ; summary(star_train)


# White dwarf training set, N=2000 (Class 3)

temp3 <- read.csv('http://astrostatistics.psu.edu/MSMA/datasets/SDSS_wd.csv', h=T)
dim(temp3) ; summary(temp3)
temp3 <- na.omit(temp3)
wd <- cbind((temp3[1:2000,2]-temp3[1:2000,3]), (temp3[1:2000,3]-temp3[1:2000, 4]),
(temp3[1:2000,4]-temp3[1:2000,5]), (temp3[1:2000,5]-temp3[1:2000,6]))
wd_train <- data.frame(cbind(wd, rep(3, length(wd[,1]))))
names(wd_train) <- c('u_g', 'g_r', 'r_i', 'i_z', 'Class')
dim(wd_train) ; summary(wd_train)


# Combine and plot the training set (9000 objects)

SDSS_train <- data.frame(rbind(qso_train, star_train, wd_train))
names(SDSS_train) <- c('u_g', 'g_r', 'r_i', 'i_z', 'Class')
str(SDSS_train)


par(mfrow=c(1,3))
plot(SDSS_train[,1], SDSS_train[,2], xlim=c(-0.7,3), ylim=c(-0.7,1.8), pch=20,
```

```
    col=SDSS_train[,5], cex=0.6, cex.lab=1.6, cex.axis=1.6, main='', xlab='u-g (mag)',
        ylab='g-r (mag)')
legend(-0.5, 1.7, c('QSO','MS + RG','WD'), pch=20, col=c('black','red','green'),
cex=1.6)
plot(SDSS_train[,2], SDSS_train[,3], xlim=c(-0.7,1.8), ylim=c(-0.7,1.8), pch=20,
col=SDSS_train[,5], cex=0.6, cex.lab=1.6, cex.axis=1.6, main='', xlab='g-r (mag)',
ylab='r-i (mag)')
plot(SDSS_train[,3], SDSS_train[,4], xlim=c(-0.7,1.8), ylim=c(-1.1,1.3), pch=20,
col=SDSS_train[,5], cex=0.6, cex.lab=1.6, cex.axis=1.6, main='', xlab='r-i (mag)',
ylab='i-z (mag)')
par(mfrow=c(1,1))
```

The resulting **R** dataframes for the test and training sets have five variables:

$u - g$ color based on ultraviolet and green magnitudes
$g - r$ color based on the green and red magnitudes
$r - i$ color based on two red magnitudes
$i - z$ color based on two far-red magnitudes

Class in the training dataset only: 1 = main-sequence stars; 2 = quasars; 3 = white dwarfs

Multivariate classification of SDSS point sources using this training set is presented in Section 9.9.

Statistical exercises

1. Characterize the main-sequence, quasar, and white-dwarf training sets using parametric multivariate normal models. Apply MANOVA tests to show how their locations in four-dimensional color space differ. Using quantile-quantile plots, show that the MVN assumption is poor for these distributions.
2. Model the distributions of the training sets shown in Figure C.2 in two, three and four dimensions. Use nonparametric kernel density estimators, nonlinear regression and normal mixture models.
3. Display the four-dimensional distributions of the test and training sets using a variety of statistics (e.g. three-dimensional perspective plots; parallel coordinates plots) and dynamic (e.g. **CRAN**'s *Rggobi* package) visualization tools.
4. Model the four-dimensional distribution of the training sets using nonlinear multiple regression, lasso regression and MARS. (†)

C.10 Galaxy photometry

Statistical issues: density estimation, multivariate clustering, mixture models

In the visible spectral band, and adjacent ultraviolet and near-infrared bands, the light from galaxies arises from billions of individual stars. Careful modeling of galaxy spectra, both at high resolution showing atomic lines and at low resolution showing broad color trends, give insights into the stellar populations and histories of individual galaxies. Spiral and irregular galaxies with active star formation will emit more blue and ultraviolet light, while elliptical galaxies without current star formation are redder. More distant galaxies are redshifted by the expansion of the Universe, so that the observed spectra are redder than the emitted spectra.

The COMBO-17 project, Classifying Objects by Medium-Band Observations in 17 Filters, obtained photometry of $\sim$50,000 galaxies in 17 bands in $\sim$1 square degree of the southern sky (Wolf *et al.* 2003). The spectral bands range from 400 nanometers in the far blue to 900 nanometers in the far red. The red sequence is clearly differentiated from the bluer galaxies out to redshifts $z \sim 1$. With redshifts estimated from the photometric shapes, this dataset has measured the rapid decline in cosmic star-formation rate, the evolution of galaxy luminosity functions, the dynamic evolution of luminous elliptical galaxies, and other aspects of cosmic evolution (e.g. Bell *et al.* 2004, Faber *et al.* 2007).

The COMBO-17 dataset has $\sim$65 variables, but many of these measures are nearly redundant, as magnitudes in adjacent bands are necessarily close. The variables thus have a great deal of collinearity and the underlying structure is quite simple. A large COMBO-17 dataset can be obtained from the Vizier Web service, Table II/253A. We have extracted a simple two-dimensional subset for 572 galaxies at low redshift, $z < 0.3$, with one magnitude and one color index; the distribution is shown in Figure 9.1. The dataset *COMBO17lowz.dat* has columns:

M_B Blue band absolute magnitude
$M_{280} - M_B$ Color index between the ultraviolet 280 nm band and the blue band

This dataset is plotted in Figure 9.1 and analyzed in Section 9.9.

Statistical exercises

1. Using nonparametric clustering techniques, locate and characterize the blue spiral galaxies, the green valley, the red sequence of elliptical galaxies, and bright cluster galaxies (red with high luminosity).
2. Model these classes with normal mixture models.

C.11   Elliptical galaxy profiles

Statistical issues: nonlinear regression, model selection

Elliptical galaxies, first identified photographically by Edwin Hubble, appear to have a monotonically decreasing surface brightness from their centers to their outer regions. Hubble (1930) modeled the brightness distribution as $I \propto (1 + r/r_c)^2$ where the brightness is constant over a core radius and declined as $1/r^2$ in the outer regions. This formula can be physically interpreted as an isothermal sphere; that is, the stars are in virial equilibrium with constant velocity dispersion. Later studies showed that the core is small and the shape is more complicated. A recent formulation that appears to fit a wide range of elliptical galaxy brightness profiles is (Sérsic 1968)

$$\log_{10} I(r) = \log_{10} I_e + -b_n[(r/r_e)^{1/n} - 1] \tag{C.1}$$

where $b_n \simeq 0.868n - 0.142$, $0.5 < n < 16.5$, and the observed surface-brightness profile is measured in units of

$$\mu(r) \propto -2.5 \log_{10} I(r) \text{ mag arcsec}^{-2}. \tag{C.2}$$

Unfortunately, no straightforward structural or dynamical explanation has been found to explain this distribution.

We have obtained three datasets from the study of elliptical galaxy surface photometry in the nearby Virgo Cluster by Kormendy *et al.* (2009). They seek departures from the Sérsic profile as diagnostics of galaxy-formation processes; excess light components are attributed to star formation from the accretion of gas-rich smaller galaxies. This is known as the wet merger model for elliptical galaxy formation. The three Virgo galaxies are: NGC 4406 (=Messier 86), NGC 4472 (=Messier 49), and the smaller elliptical NGC 4551. Each dataset has two columns:

radius Radius from the galaxy center in arcseconds
surf_mag Surface brightness at that radius in V-band magnitudes per square arcsecond

The galaxy radial profiles are shown in Figure 7.5 and modeled in Section 7.10.5.

Statistical exercises

1. Fit a variety of model families to these elliptical galaxy profiles: Hubble's isothermal sphere, King's truncated sphere, de Vaucouleurs $r^{1/4}$ model, Sérsic's $r^{1/n}$ model, and Hernquist's sech(r/r_c) (1990) model. Perform residual analysis for the best-fit parameters for each model. Estimate model selection criteria for both nested (Sérsic and de Vaucouleurs models) and nonnested model families. (†)

C.12 X-ray source variability

Statistical issues: time series analysis, Fourier analysis, autoregressive models, long-memory processes

The sky at X-ray wavelengths, observable only from space-borne telescopes, is dominated by a small number of very bright X-ray sources in the Galactic Plane. These are X-ray binary-star systems where a massive companion in a stellar binary has evolved and exploded in a supernova, leaving behind a neutron star or black hole. Gas from the companion star then spirals onto the compact object, forming an accretion disk that emits copiously in the X-ray band.

These accretion flows are not smooth, and a wide range of fascinating temporal behavior in the X-ray emission is produced. One or more strictly periodic components may be present due to rotation of the neutron star and/or orbital eclipses. Another common component is red noise, high-amplitude aperiodic but correlated variations at low frequencies. Sometimes quasi-periodic oscillations (QPO) are seen as strong broadened peaks in the Fourier periodograms of the X-ray emission. Episodes of episodic flaring events are also seen. Perhaps the most remarkable of these systems is the black-hole binary system GRS 1915+105 with over a dozen distinct variability states (Belloni *et al.* 2000, Harikrishnan *et al.* 2011).

Correlations are often seen between the intensities, spectra and temporal behavior of these systems. A number of different physical explanations for these phenomena have been proposed, involving, for example, misalignments between the accretion disk and black-hole spin axis, wave instabilities in the disk flow, or disk precession caused by general relativistic effects near the collapsed object (Ingram & Done 2010).

One of the best-studied cases of X-ray binaries with QPO variations is GX 5-1, an accreting neutron star system (Jonker *et al.* 2002). We provide here a dataset *GX.dat* consisting of 65,536 contiguous measurements of the source brightness obtained by Norris *et al.* (1990) using the Japanese Ginga satellite. The dataset should be read row-by-row. Each measurement gives the number of X-ray photons detected during intervals of 1/128 second duration covering 512 seconds. This is one of a variety of astronomical time series datasets presented by Hertz & Feigelson (1995) and available online at http://xweb.nrl.navy.mil/timeseries/timeseries.html.

This time series is shown in Figure 11.2 and is extensively analyzed in Section 11.13.

Statistical exercises

1. Characterize the dataset using wavelet transforms in a manner similar to the study of Sco X-1 by Scargle *et al.* (1993). (†)
2. Apply a state-space model and solve sequentially using the Kalman filter. Consider both autoregressive and variable-phase periodic models. (†)

C.13 Sunspot numbers

Statistical issues: time series analysis, periodicities

Ancient Chinese astronomers, and independently Galileo Galilei in 1610 with his telescope, discovered that the Sun has small dark spots that emerge on one side and disappear on the other side as the Sun rotates with a period around 30 days. In 1908, American spectroscopist George Hale found that sunspots are the consequence of concentrations of strong magnetic fields on the solar surface. Since then, a large field of solar physics has grown to understand the complicated morphologies, evolution, causes and consequences of these solar active regions. In particular, the solar fields can undergo a violent reconnection event, where a sudden and powerful solar flare occurs, producing energetic particles and plasmas that are not seen over most of the solar surface. These flares influence the space weather environment of Earth-orbiting satellites, the Earth's magnetosphere, the upper layers of the Earth's atmospheres, and occasionally terrestrial electrical systems.

The number of sunspots visible on the Sun, usually averaged over a month or a year, shows a high-amplitude nearly periodic variation with a period of around 11 years (Figure C.3). This is actually a 22-year cycle as the polarity of the magnetic fields reverse every 11 years. Numerous time series analyses of the solar sunspot numbers and solar flares have been conducted, most often for the purpose of short-term prediction of terrestrial consequences. Periodicities have been claimed on time-scales ranging from the ~30 day rotation period to hundreds of years. The most well-accepted period, in addition to the 11-year cycle, is at 155 days (Rieger *et al.* 1984, Lean 1990). Relations between cycle length and amplitude have also been reported (Solanki *et al.* 2002).

Continuous time series of monthly sunspot numbers from 1749 to 2011 with >3100 entries can be obtained from NASA's Solar Physics group at Marshall Space Flight Center http://solarscience.msfc.nasa.gov/SunspotCycle.shtml.

Statistical exercises

1. Construct and improve (by smoothing and tapering) the Fourier periodogram of the sunspot dataset.
2. Seek secondary periodicities such as the reported 155-day effect. (†)
3. Seek relationships between the periods, heights and shapes of adjacent 11-year cycles. (†)

C.14 Exoplanet orbits

Statistical issues: time series analysis, periodicities, multivariate analysis

Until 1995, astronomers had direct evidence for only one planetary system orbiting a normal star, our Solar System. Protostars were known to have dusty protoplanetary disks consistent with the formation mechanism developed for our Solar System, but the planets themselves were elusive because they are so small and faint. However, extremely accurate (parts in a million) spectroscopy of bright, nearby solar-type stars has recently revealed small periodic Doppler (radial velocity) shifts indicating the star is wobbling back-and-forth as some unseen object orbits. The first star showing a Doppler wobble that was widely attributed to a substellar (i.e. planetary) body was 51 Pegasi, reported by Swiss astronomers M. Mayor and D. Queloz in 1995. Knowing the period and velocity amplitude of the wobble, and the mass of the primary star, standard theory based on Newton's laws is applied to infer the mass and orbit of the orbiting object. Typical inferred masses are around the mass of Jupiter – thus the discovery of exoplanets had emerged.

Today (early 2011) over 400 planets have been found around stars using the radial velocity method, over 100 by the tiny diminution of light as a planet passes in front of the

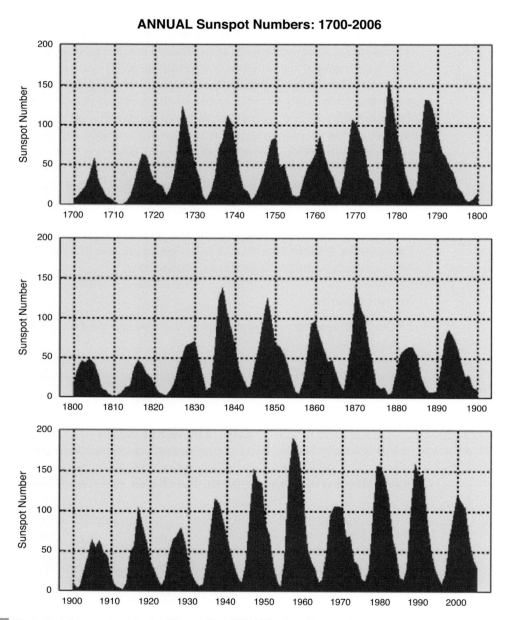

Fig. C.3 Time series of sunspot numbers over 300 years (from NOAA/NGDC, http://www.ngdc.noaa.gov/stp/ solar/ssndada.html).

star (transit method), and over 40 by other methods. NASA's Kepler mission (Section C.15) is now reporting thousands of probable transit planets. Planetary masses range from a few Earth-masses to over 10 Jupiter-masses, orbital periods range from 1 day to a decade, and eccentricities range from zero to 93%.

The findings have opened up many new astrophysical investigations concerning the origin and evolution of planetary systems. Clearly, planetary systems show enormous diversity, and the structure of our Solar System is not typical. Multivariate analyses and classifications have been attempted on the datasets, but only a few results of astrophysical interest have yet emerged. Note that scientific interpretation requires detailed knowledge of the selection effects of the surveys entering the databases. Many observational programs are underway, including space-based observatories, to detect and characterize more exoplanets, and the field is rapidly growing.

Online datasets of exoplanetary source lists and properties are provided by three organizations: The Extrasolar Planets Encyclopedia (http://exoplanet.eu/), the Exoplanet Data Explorer (http://exoplanets.org), and the New Worlds Atlas (http://planetquest.jpl.nasa.gov/). Figure C.4 show two important bivariate plots relating properties of the planets and their orbits.

The estimation of orbital properties and mass of orbiting planets from radial velocity data is statistically challenging. The time series is sparse, unevenly spaced, and with heteroscedastic measurement errors of known variance. The periodic behavior is nonsinusoidal if the orbital eccentricity is nonzero. Harmonic periodicity search methods include (Section 11.5): simple Fourier analysis, the Lomb–Scargle periodogram for unevenly spaced data, maximum likelihood estimation and Bayesian inference. The estimation of the number of planets orbiting a star exhibiting radial velocity variations is a particularly difficult problem in nonlinear model selection. While there is no online repository for radial velocity measurements, datasets for selected systems can be found in the references given in the Web sites listed above.

Statistical exercises

1. Apply a variety of multivariate clustering and analysis methods to the multivariate datasets of extrasolar planetary properties.
2. Estimate the planetary mass–radius relationship, including confidence bands, using kernel smoothing, nonparametric regression and other density estimation methods.
3. Obtain some radial velocity datasets and estimate the number of orbiting planets. A particularly important and difficult case is presented by Vogt *et al.* (2010). (†)

C.15 Kepler stellar light curves

Statistical issues: time series analysis, periodicities, autoregressive models

As outlined in Section C.14, ground-based spectroscopic and photometric observations have succeeded in detecting over 500 planets, despite the small sizes and masses of the planets compared to the host stars. The transit method, involving detection of periodic tiny decreases in star brightness as a planet passes in front of the stellar disk, can benefit from space-based observations where fluctuations in the Earth's atmosphere do not limit photometric

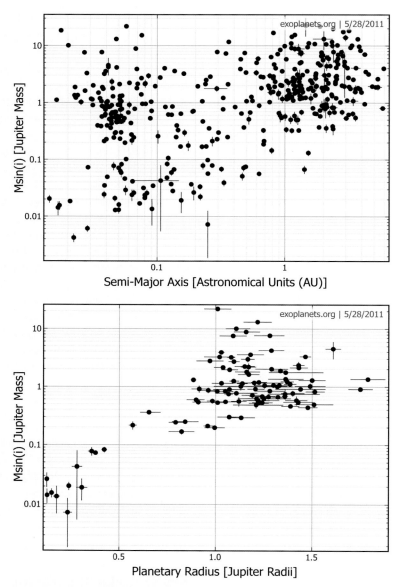

Fig. C.4 Selected bivariate plots from the multivariate databases of extrasolar planets obtained from the Exoplanet Data Explorer (http://exoplanets.org). *Top:* Planet mass plotted against planet orbit size. *Bottom:* Planet mass plotted against planet radius (for transiting planets only).

precision. Among the several satellites launched to date for precise stellar photometric measurements, NASA's Kepler mission is producing the largest sample. Launched in 2009, the Kepler telescope is monitoring the brightnesses of ~160,000 stars with 30-minute temporal sampling with typical photometric precision of ~10^{-5} over several years. The time

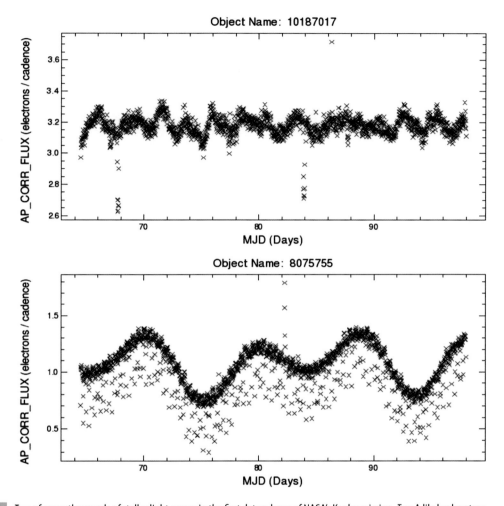

Fig. C.5   Two of many thousands of stellar light curves in the first data release of NASA's Kepler mission. *Top:* A likely planetary transit superposed on autocorrelated variations perhaps due to stellar magnetic activity. *Bottom:* Smooth variations may be due to an eclipsing contact binary, and brief dips to transits by a third star or planet. Plots obtained from the NASA/IPAC/NExSci Star and Exoplanet Database (http://nsted.ipac.caltech.edu).

series is nearly continuous, interrupted only occasionally due to science data downloading and spacecraft anomalies.

The extraordinary number and accuracy of Kepler light curves provide a unique resource for both stellar and planetary research. Stellar variability phenomena include acoustic oscillations on timescales of minutes, flares and other types of magnetic activity on timescales of hours to days, global pulsations on timescales of days, binary and multiple star eclipses, and transit dips due to planets. Two sample light curves from the first data release of the mission are shown in Figure C.5. Light curves for the full Kepler sample can be downloaded from the NASA/IPAC/NExSci Star and Exoplanet Database (NSTED, http://nsted.ipac.caltech.edu).

Statistical exercises

1. Apply autoregressive modeling to several stars exhibiting correlated but aperiodic brightness variations. Examine residuals, likelihood-based criteria, and nonparametric tests for model validation. Discuss possible astrophysical interpretations of the fitted models.
2. Examine several stars exhibiting periodic variations and discuss the differences between the periodograms provided by the NSTED Periodogram Service. Apply Monte Carlo simulation techniques to test the significance of periodogram peaks. (†)
3. Apply nonlinear multiparameter models to eclipsing binaries based on the work of Wilson & Devinney (1971). (†)

C.16 Sloan Digital Sky Survey

Statistical issues: data mining, multivariate analysis, multivariate classification, spatial point processes, parametric modeling

The Sloan Digital Sky Survey (SDSS) has been an extraordinarily productive sky survey in visible light. Using a specially built modest-sized telescope at Apache Peak Observatory (APO) in New Mexico, over one-third of the celestial sphere has been imaged using a wide-field camera in five spectral bands generating accurate photometric measurements of hundreds of millions of objects. In addition, spectra have been obtained for over a million objects using a multiplexed fiber-coupled spectrograph.

While the SDSS is not the most sensitive or highest-resolution telescope, the size, uniformity and reliability of its data products have led to several thousand publications generating nearly 150,000 citations since 1996. Some of the principal science results of the survey are outlined at http://www.sdss.org/signature.html; two important findings based on large statistical samples are shown in Figure C.6. A small portion of the SDSS data products was extracted for statistical analysis in Sections C.8 and C.9. Here we outline the scope of the project through 2014:

Sloan Legacy Survey A single epoch of five-band photometry of 230 million objects and spectra of 930,000 galaxies, 120,000 quasars and 225,000 stars. Important statistical issues include: characterization of the three-dimensional distribution of galaxies (Section 12.12); discovery of rare classes such as brown dwarfs and high-redshift quasars; classification of stars, galaxies and quasars (Chapter 9); understanding physical processes of accreting black holes and the distribution of intergalactic matter from quasar spectra.

Sloan Extension for Galactic Understanding and Exploration A spectroscopic survey of 240,000 stars in our Milky Way Galaxy to elucidate the kinematics, populations

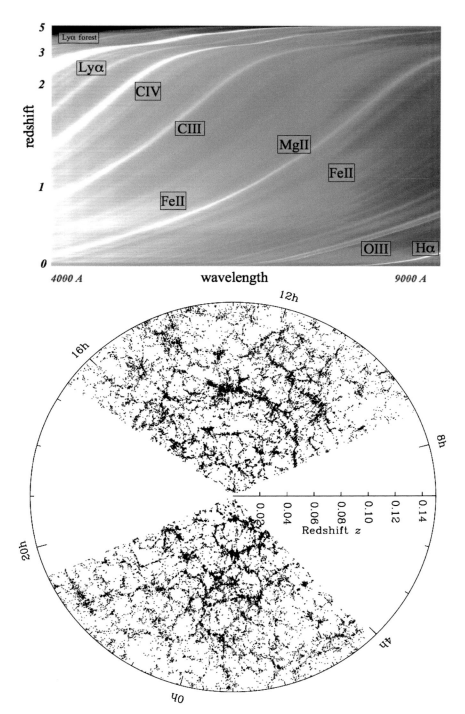

Fig. C.6 Two scientific results from large projects of the Sloan Digital Sky Survey. *Top:* Visualization of 46,420 quasar spectra ordered by distance showing emission line redshifted by the cosmological expansion. Credit X. Fan and the Sloan Digital Sky Survey. *Bottom:* Two-dimensional slice through a three-dimensional map of the distribution of SDSS galaxies showing the honeycomb structure of anisotropic clustering characteristic of large-scale structure in the Universe. (See SDSS press release, http://www.sdss.org/news/releases/20080817.vpf_final.html.)

and evolution of the Galaxy. For each star, the effective temperature, surface gravity, metallicity and approximate radial velocity are inferred. One important issue is to identify the star streams in the Galactic halo attributed to cannibalism of small galaxies in the past.

Sloan Supernova Survey Repeated imaging of 300 square degrees to discover supernovae in external galaxies and other variable objects. An important aspect is the identification of several hundred Type Ia supernovae that serve as standard candles to trace the expansion of the Universe.

Baryon Oscillation Spectroscopic Survey While galaxy locations in space show strong hierarchical clustering on scales ≤ 50 Mpc, Sloan studies of luminous red galaxies and quasars also found a faint signal on large scales around 150 Mpc attributed to baryon acoustic oscillations (Eisenstein *et al.* 2005). A detailed study of this effect will provide critical constraints on parameters in the consensus ΛCDM cosmological model.

APO Galactic Evolution Experiment The structure and metallicity of the bulge, bar, disk and halo of our Milky Way Galaxy will be studied using spectra of 100,000 red giants.

Multi-object APO Radial Velocity Exoplanet Large-area Survey The radial velocities of 11,000 bright stars will be monitored using a high-resolution spectrograph to detect Doppler wobbles caused by orbiting gas-giant planets (Section C.14).

Although data from the SDSS project are initially reserved for collaborating scientists, they are later released to the public at http://www.sdss.org. The Science Archive Server provides images and spectra, while the Catalog Archive Server provides sources and summary properties derived from a pipeline data analysis system. Data are accessed through interactive Web-based forms or SQL queries. The full data volume exceeds 100 terabytes.

Statistical exercises

The statistical analyses that can be performed using SDSS data are too varied to list. We note only that taking advantage of the full database will typically require advanced data mining tools with highly efficient algorithms and high-performance computing.

C.17 Fermi gamma-ray light curves

Statistical issues: time series analysis, heteroscedastic measurement errors, censoring

The field of gamma-ray astronomy involves the detection and characterization of sources emitting extremely energetic photons with energies in the MeV (mega electron volts), GeV and TeV range. Such emission cannot come from thermal processes that characterize bodies familiar at visible wavelengths like the surfaces of planets or stars. Gamma-rays can only

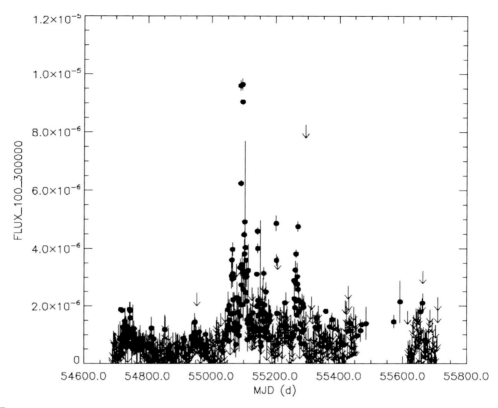

Time series of daily averaged gamma-ray fluxes from the jet of 3C273, a bright nearby quasar (accreting supermassive black hole), measured with the Large Area Telescope of NASA's Fermi Gamma-Ray Space Telescope. Plot provided by the Fermi Science Support Center. (http://fermi.gsfc.nasa.gov/ssc/data/access/lat/msl_lc.)

be produced by gas heated to billions of degrees while falling onto black holes or neutron stars, from nonthermal processes involving acceleration of particles in rapidly changing magnetic fields, or from the nonthermal scattering of lower energy photons by relativistic electrons. They provide a window into some of the most exotic phenomena in the Universe.

Gamma-ray astronomy is technically difficult because there is no surface that can reflect such energetic photons and produce images. Pseudo-imaging techniques are used, adopted from high-energy particle physics such as tracking second particles produced in a mutilayer detector. The most sophisticated telescope in the MeV band to date is the Large Area Telescope (LAT) on board the Fermi Gamma-Ray Space Telescope launched by NASA in 2008 (Atwood *et al.* 2009).

Time series of LAT brightness measurements (or light curves) for several dozen gamma-ray sources are provided graphically and in FITS binary table format by the Fermi Science Support Center (http://fermi.gsfc.nasa.gov/ssc/data/access/lat/msl_lc). An example is shown in Figure C.7. The signals are typically very weak leading to upper limits (i.e. left-censored data points) on many days and uncertain, low signal-to-noise detections on other days.

Statistical exercises

1. Ignoring measurement errors, estimate the distribution of daily intensities of 3C273 and other active galactic nuclei (AGN) using the Kaplan–Meier estimator. Compare the shapes of these distributions for three classes of AGN: quasars, broad-line radio galaxies, and blazars.
2. Estimate the distribution of daily intensities including treatment of heteroscedastic measurement errors. (†)

C.18   Swift gamma-ray bursts

Statistical issues: time series analysis, modeling

Gamma-ray bursts (GRBs) are sudden explosions occurring randomly in distant galaxies that are seen primarily in the gamma-ray band. They were discovered serendipitously in the 1960s by satellites designed to monitor treaties prohibiting nuclear tests in outer space. GRBs are now understood to arise from relativistically beamed jets produced as a massive star collapses into a black hole during a supernova explosion (Mészáros 2006). The prompt burst, lasting from a few seconds to minutes, is followed by a fading afterglow seen often in the X-ray band, and sometimes in the visual and radio bands. The prompt bursts and afterglows exhibit a wide range of characteristics, often not well-understood.

The most successful observatory characterizing GRBs is NASA's Swift explorer mission launched in 2004. When the wide-field Burst Alert Telescope detects an event, the satellite autonomously slews to the burst location so its narrow-field X-ray and Ultraviolet/Optical Telescopes can monitor the rapid variations during the early afterglow. Ground-based astronomers worldwide are informed of the burst location within one minute, and turn optical-band telescopes to study the burst and host galaxy. To date, Swift has detected nearly 700 bursts.

NASA's Swift Science Center (http://swift.gsfc.nasa.gov/docs/swift) provides data products including a summary catalog of GRBs, prompt gamma-ray light curves, X-ray light curves, optical light curves (when occasionally available) and links to the Gamma-ray Coordinates Network for ground-based findings. Figure C.8 shows the prompt and afterglow emission for a recent burst.

Statistical exercises

1. Smooth the gamma-ray and X-ray time series using density estimation techniques.
2. Smooth the time series including treatment of the heteroscedastic measurement errors. (†)

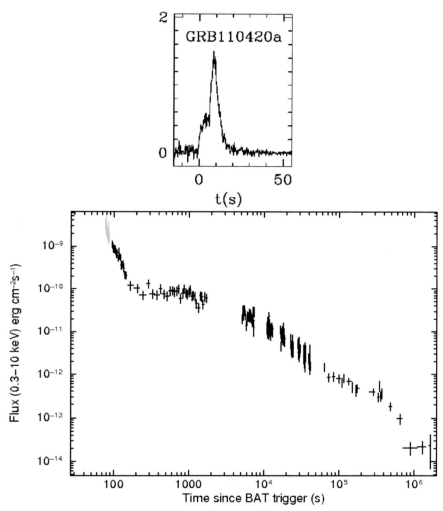

Fig. C.8   Time series for gamma-ray burst GRB 110420A obtained with NASA's Swift mission. *Top:* Prompt gamma-ray emission lasting seconds. *Bottom:* X-ray afterglow showing a million-fold decay from 1 minute to 2 weeks after the event. Plots from NASA's Swift archive, downloaded from http://swift.gsfc.nasa.gov/docs/swift.

3. Examine the full sample of Swift prompt gamma-ray time series and seek to classify them based on shape, multimodality, roughness and autocorrelation properties. Seek correlations of these classes with global properties of the GRBs and their host galaxies. (†)
4. Model the X-ray afterglows using astrophysical models of beamed relativistic fireballs. (†)

References

Abell, G. O. (1958) The distribution of rich clusters of galaxies, *Astrophys. J. Suppl.*, **3**, 211–288

Abraham, J. and 457 coauthors (2008) Observation of the suppression of the flux of cosmic rays above 4×10^{19} EV, *Phys. Rev. Lett.*, **101**, 061101

Adler, J. (2010) *R in a Nutshell*, O'Reilly, Sebastopol CA

Adler, D. & Murdoch, D. (2010) rgl: 3D visualization device system (OpenGL), R package version 0.92.794, http://CRAN.R-project.org/package=rgl

Agresti, A. (2002) *An Introduction to Categorical Data Analysis*, 2nd ed. Wiley-Interscience, New York

Aigner, D. J. & Goldberger, A. S. (1970) Estimation of Pareto's Law from grouped observations, *J. Amer. Stat. Assn.*, **65**, 712–723

Aigrain, S. & Irwin, M. (2004) Practical planet prospecting, *Mon. Not. Royal Astro. Soc.*, **350**, 331–345

Akaike, H. (1973) Information theory and an extension of the maximum likelihood principle, in *Second International Symposium on Information Theory* (B. N. Petrov & F. Csaki, eds.), Akademia Kiado, Budapest, pp. 267–281

Akritas, M. G. & Bershady, G. A. (1996) Linear regression for astronomical data with measurement errors and intrinsic scatter, *Astrophys. J.*, **470**, 706–714

Akritas, M. G. & Siebert, J. (1996) A test for partial correlation with censored astronomical data, *Mon. Not. Royal Astro. Soc.*, **278**, 919–924

Akritas, M. G., Murphy, S. A. & LaValley, M. P. (1995) The Theil–Sen estimator with doubly censored data and applications to astronomy, *J. Amer. Stat. Assn.*, **90**, 170–177

Albert, J. (2009) *Bayesian Computation with R*, Springer, Berlin

Aldrich, J. (1997) R. A. Fisher and the making of maximum likelihood 1912–1922, *Statist. Sci.*, **12**, 162–176

Alexander, T. (1997) Is AGN variability correlated with other AGN properties? ZDCF analysis of small samples of sparse light curves, in *Astronomical Time Series* (D. Maoz *et al.*, eds.), Reidel, Dordrecht, pp. 163–166

Andersson, J. (2002) An improvement of the GPH estimator, *Economics Lett.*, **77**, 137–146

Andersson, J. (2007) On the estimation of correlations for irregularly spaced time series, NHH Dept. of Finance & Management Science Discussion Paper No. 2007/19. Available at SSRN: http://ssrn.com/abstract=1020943

Andrews, F. (2010) latticist: A graphical user interface for exploratory visualisation, R package version 0.9-43, http://latticist.googlecode.com/

Ankerst, M., Breunig, M. M., Kriegel, H.-P. & Sander, J. (1999) OPTICS: Ordering Points To Identify the Clustering Structure, in *Proc. ACM SIGMOD99 Int. Conf. on Management of Data*, ACM Press, pp. 49–60

Anselin, L. (1995) Local indicators of spatial association – LISA, *Geographical Analysis*, **27**, 93–115

Anselin, L. & Rey, S. J. editors (2010) *Perspectives on Spatial Data Analysis*, Springer, Berlin

Apanasovich, T. V., Carroll, R. J. & Maity, A. (2009) Density estimation in the presence of heteroscedastic measurement error, *Electronic J. Statist.*, **3**, 318–348

Arnold, B. C. (1983) *Pareto Distributions*, International Co-operative Publishing House, Devon

Ashman, K. M. & Zepf, S. E. (1998) *Globular Cluster Systems*, Cambridge University Press

Atwood, W. B. *et al.* (2009) The Large Area Telescope on the Fermi Gamma-Ray Space Telescope Mission, *Astrophys. J.*, **697**, 1071–1102

Avni, Y. & Tananbaum, H. (1986) X-ray properties of optically selected QSOs, *Astrophys. J.*, **305**, 83–99

Avni, Y., Soltan, A., Tananbaum, H. & Zamorani, G. (1980) A method for determining luminosity functions incorporating both flux measurements and flux upper limits, with applications to the average X-ray to optical luminosity ratio for quasars, *Astrophys. J.*, **238**, 800–807

Babu, G. J. & Feigelson, E. D. (1996) *Astrostatistics*, Chapman & Hall, London

Babu, G. J. & Rao, C. R. (1993) Bootstrap methodology, in *Handbook of Statistics*, vol. 9, *Computational Statistics* (C. R. Rao, ed.), Elsevier, Amsterdam, pp. 627–659

Babu, G. J. & Rao, C. R. (2003) Confidence limits to the distance of the true distribution from a misspecified family by bootstrap, *J. Statist. Plann. Inference*, **115**, 471–478

Babu, G. J. & Rao, C. R. (2004) Goodness-of-fit tests when parameters are estimated, *Sankhyā*, **66**, 63–74

Babu, G. J. & Singh, K. (1983) Inference on means using the bootstrap, *Annals Statist.*, **11**, 999–1003

Baddeley, A. (2008) Analysing spatial point patterns in R, CSIRO, available on-line at http://www.csiro.au/files/files/pn0y.pdf

Baddeley, A. & Turner, R. (2005) Spatstat: An R package for analyzing spatial point patterns, *J. Statistical Software*, **12**, 1–42

Ball, N. M. & Brunner, R. J. (2010) Data mining and machine learning in astronomy, *Rev. Mod. Phys. D*, **19**, 1049–1106

Bally, J. & Reipurth, B. (2006) *The Birth of Stars and Planets*, Cambridge University Press

Baluev, R. V. (2008) Assessing the statistical significance of periodogram peaks, *Mon. Not. Royal Astro. Soc.*, **385**, 1279–1285

Bardeen, J. M., Bond, J. R., Kaiser, N. & Szalay, A. S. (1986) The statistics of peaks of Gaussian random fields, *Astrophys. J.*, **304**, 15–61

Barnett, V. (1999) *Comparative Statistical Inference*, John Wiley, New York

Basu, S. & Jones, C. E. (2004) On the power-law tail in the mass function of protostellar condensations and stars, *Mon. Not. R. Astron. Soc.*, 347, L47–L51

Baty, F. & Delignette-Muller, M. L. (2009) nlstools: tools for nonlinear regression diagnostics, R package version 0.0-9

Bauke, H. (2007) Parameter estimation for power-law distributions by maximum likelihood methods, *Eur. Phys. J. B*, **58**, 167–173

Behrend, R. *et al.* (2006) Four new binary minor planets: (854) Frostia, (1089) Tama, (1313) Berna, (4492) Debussy, *Astron. & Astrophys.*, **446**, 1177–1184

Bell, E. F. *et al.* (2004) Nearly 5000 distant early-type galaxies in COMBO-17: A red sequence and its evolution since $z \sim 1$, *Astrophys. J.*, **608**, 752–767

Belloni, T., Klein-Wolt, M., Méndez, M., van der Klis, M. & van Paradijs, J. (2000) A model-independent analysis of the variability of GRS 1915+105, *Astron. Astrophys.*, **355**, 271–290

Benedict, L. H., Nobach, H. & Tropea, C. (2000) Estimation of turbulent velocity spectra from laser Doppler data, *Meas. Sci. Technol.*, **11**, 1089–1104

Benjamini, Y. & Hochberg, Y. (1995) Controlling the false discovery rate: A practical and powerful approach to multiple testing, *J. Roy. Stat. Soc. B.*, **57**, 289–300

Beran, J. (1989) *Statistics for Long-Memory Processes*, Chapman & Hall, London

Berkson, J. (1980) Minimum chi-square, not maximum likelihood (with discussion), *Annals Statistics*, **8**, 457–487

Bevington, P. R. (1969) *Data Reduction and Error Analysis for the Physical Sciences*, McGraw-Hill, New York

Bevington, P. & Robinson, D. K. (2002) *Data Reduction and Error Analysis for the Physical Sciences*, 3rd ed., McGraw-Hill, New York

Bickel, P. & Rosenblatt, M. (1973) On some global measures of the deviations of density function estimators, *Ann. of Statistics*, **1**, 1071–1095

Billingsley, P. (1999) *Convergence of Probability Measures*, Wiley-Interscience, New York

Bivand, R. S., Pebesma, E. J. & Gómez-Rubio, V. (2008) *Applied Spatial Data Analysis with R*, Springer, Berlin

Bivand, R., with contributions by many others (2010) spdep: Spatial dependence: weighting schemes, statistics and models, R package version 0.5-26, http://CRAN.R-project.org/package=spdep

Blanton, M. R. (2004) The galaxy luminosity function and luminosity density at redshift $z = 0.1$, *Astrophys. J.*, **592**, 819–838

Blanton, M. R., Lupton, R. H., Schlegel, D. F., Strauss, M. A., Brinkmann, J., Fukugita, M. & Loveday, J. (2005), The properties and luminosity function of extremely low luminosity galaxies, *Astrophys. J.*, **631**, 208–230

Boggs, P. T., Byrd, R. H. & Schnabel, R. B. (1987) A stable and efficient algorithm for nonlinear orthogonal distance regression, *SIAM J. Sci. & Stat. Comput.*, **8**, 1052–1078

Bowman, A. W. & Azzalini, A. (1997) *Applied Smoothing Techniques for Data Analysis*, Clarendon Press, Oxford

Bowman, A. W. & Azzalini, A. (2010) R package "sm": nonparametric smoothing methods (version 2.2-4), http://www.stats.gla.ac.uk/~adrian/sm

Box, G. E. P. & Draper, N. R. (1987) *Empirical Model-Building and Response Surfaces*, John Wiley, New York

Box, G. E. P. & Pierce, D. A. (1970) Distribution of residual autocorrelations in autoregressive-integrated moving average time series models, *J. Amer. Stat. Assn.*, **65**, 1509–1526

Boyce, P. J. (2003) GammaFinder: a Java application to find galaxies in astronomical spectral line datacubes, M.S. thesis, http://users.cs.cf.ac.uk/Antonia J.Jones/GammaArchive/Theses/PeterBoyceGammaFinder.pdf

Brazdil, P., Giraud-Carrier, C., Soares, C. & Vilalta, R. (2009) *Metalearning: Applications to Data Mining*, Springer, Berlin

Breiman, L. (2001) Random forests, *Machine Learning*, **45**(1), 5–32. See also http://www.stat.berkeley.edu/b̃reiman/RandomForests

Breiman, L., Friedman, J. H., Olshen, R. A. & Stone, C. J. (1998) *Classification and Regression Trees*, CRC Press, Bora Raton

Breiman, L., Last, M. & Rice, J. (2003) Random forests: Finding quasars, in *Statistical Challenges in Astronomy* (E. D. Feigelson & G. J. Babu eds.), Springer, Berlin

Bretthorst, G. L. (2003) Frequency estimation and generalized Lomb–Scargle periodograms, in *Statistical Challenges in Astronomy* (E. D. Feigelson & G. J. Babu, eds.), Springer, Berlin, pp. 309–329

Breusch, T. S. & Pagan, A. R. (1979) Simple test for heteroscedasticity and random coefficient variation, *Econometrica*, **47**, 1287–1294

Brillinger, D. R. (1984) Statistical inference for irregularly observed processes, in *Time Series Analysis of Irregularly Observed Data* (E. Parzen, ed.), Springer, Berlin, pp. 38–57

Britt, D. T. & Consolmagno, G. J. (2003) Stony meteorite porosities and densities: A review of the data through 2001, *Meteor. Planet. Sci.*, **38**, 1161–1180

Britt, D. T., Yeomans, D., Housen, K. & Consolmagno, G. (2002) Asteroid density, porosity and structure, in *Asteroids III* (W. F. Bottke *et al.*, eds.), University of Arizona Press, pp. 485–500

Brodsky, B. E. & Darkhovsky, B. S. (1993) *Nonparametric Methods in Change Point Problems*, Springer, Berlin

Bronaugh, D. & Werner, W. for the Pacific Climate Impacts Consortium (2009) zyp: Zhang + Yue-Pilon trends package, R package version 0.9-1

Brönnimann, H. & Chazelle, B. (1998) Optimal slope selection via cuttings, *Comp. Geom. Theory & Appl.*, **10**, 23–29

Brooks, S., Gelman, A., Jones, G. & Meng, X.-L. (2011) *Handbook of Markov Chain Monte Carlo*, CRC Press, Bora Raton

Brown, A. (1950) On the determination of the convergent point of a moving cluster from proper motions, *Astrophys. J.*, **112**, 225–239

Brown, B. W., Hollander, M. & Korwar, R. M. (1974) Nonparametric tests of independence for censored data with applications to heart transplant studies, in *Reliability and Biometry* (F. Proschan & R. J. Serfling, eds.), SIAM, Philadelphia, pp. 327–354

Brown, L. D., Cai, T. T. & Das Gupta, A. (2001) Interval estimation for a binomial proportion, *Statist. Sci.*, **16**, 101–133

Buccheri, R. (1992) Search for periodicities in high-energy gamma-ray astronomical data, in *Statistical Challenges in Modern Astronomy* (E. D. Feigelson & G. J. Babu, eds.), Springer, Berlin

Buckley, J. & James, I. (1979) Linear regression with censored data, *Biometrika*, **66**, 429–36

Budavári, T., Wild, V., Szalay, A. S., Dobos, L. & Yip, C.-W. (2009) Reliable eigenspectra for new generation surveys, *Mon. Not. Royal Astro. Soc.*, **394**, 1496–1502

Buonaccorsi, J. P. (2010) *Measurement Error: Models, Methods, and Applications*, Chapman & Hall, London

Burnham, K. P. & Anderson, D. (2002) *Model Selection and Multi-Modal Inference*, 2nd ed., Springer, Berlin

Caimmi, R. (2011) Bivariate least squares linear regression: Towards a unified analytic formalism. I. Functional models, *New Astron.*, **16**, 337–356

Cameron, A. C. & Trivedi, P. K. (1998) *Regression Analysis of Count Data*, Cambridge University Press

Carroll, R. J., Ruppert, D., Stefanski, L. A. & Crainiceanu, C. M. (2006) *Measurement Error in Nonlinear Models: A Modern Perspective*, 2nd ed., Chapman & Hall, London

Cash, W. (1979) Parameter estimation in astronomy through application of the likelihood ratio, *Astrophys. J.*, **228**, 939–947

Chabrier, G. (2003) Galactic stellar and the substellar Initial Mass Function, *Pub. Astron. Soc. Pacific*, **115**, 763–795

Chambers, J. M. (2008) *Software for Data Analysis: Programming with R*, Springer, Berlin

Chatfield, C. (2004) *The Analysis of Time Series: An Introduction*, 6th ed., Chapman & Hall/CRC, Bota Raton

Chatterjee, A., Yarlagadda, S. & Chakrabarti, B. K., eds. (2005) *Econophysics of Wealth Distributions*, Springer-Verlag Italia

Chaudhuri, P., Ghosh, A. K. & Hannu Oja, H. (2009) Classification based on hybridization of parametric and nonparametric classifiers, *IEEE Trans. Pattern Anal. Machine Intell.*, **31**, 1153–1164

Cheeseman, P. & Stutz, J. (1995) Bayesian classification (AutoClass): Theory and results, in *Advances in Knowledge Discovery and Data Mining* (U. Fayyad *et al.*, eds.), AAAI Press, pp. 153–180

Chen, C.-H., Härdle, W. & Unwin, A., eds. (2008) *Handbook of Data Visualization*, Springer, Berlin

Chen, C.-H., Tsai, W.-Y., & Chao, W.-H. (1996) The product-moment correlation coefficient and linear regression for truncated data, *J. Amer. Stat. Assn.*, **91**, 1181–1186

Chen, J. (2006) Introducing hidden Markov models to LAMOST automatic data processing, *Proc. SPIE*, **6270**, 627022.

Chen, J. & Gupta, A. K. (2000) *Parametric Statistical Change Point Analysis*, Birkhäuser, Boston

Chen, S. X. (1996) Empirical likelihood confidence intervals for nonparametric density estimation, *Biometrika*, **83**, 329–341

Chernoff, H. & Lehmann, E. L. (1954) The use of maximum likelihood estimates in χ^2 tests for goodness of fit, *Ann. Math. Stat.*, **25**, 579–586

Chetwynd, A. & Diggle, P. (1998) On estimating the reduced second moment measure of a stationary spatial point process, *Australian J. Stat.*, **40**, 11–15

Chew, V. (1971) Point estimation of the parameter of the binomial distribution, *Amer. Statist.*, **15**, 47–50

Choi, Y.-Y., Park, C., Kim, J., Gott, J. R., Weinberg, D. H., Vogeley, M. S. & Kim, S. S. (2010) Galaxy clustering topology in the Sloan Digital Sky Survey main galaxy sample: A test for galaxy formation models, *Astrophys. J. Suppl.*, **190**, 181–202

Cleveland, W. S. & Loader, C. (1996) Smoothing by local regression: Principles and methods, in *Statistical Theory and Computational Aspects of Smoothing* (W. Härdle & M. G. Schimek, eds.), Springer, Berlin, pp. 10–49

Clyde, M. A., Berger, J. O., Bullard, F., Ford, E. B., Jefferys, W. H., Luo, R., Paulo, R. & Loredo, T. (2007) Current challenges in Bayesian model choice, in *Statistical Challenges in Modern Astronomy IV* (G. J. Babu & E. D. Feigelson, eds.), *Astron. Soc. Pacific*, **371**, 224–240

Cohen, A. C. (1991) *Truncated and Censored Samples: Theory and Applications*, Marcel Dekker, New York

Conover, W. J. (1999) *Practical Nonparametric Statistics*, 3rd ed., John Wiley, New York

Constantine, W. & Percival, D. (2010). fractal: Insightful fractal time series modeling and analysis, R package version1.0-3, http://CRAN.R-project.org/package=fractal

Cook, D. & Swayne, D. F. (2007) *Interactive and Dynamic Graphics for Data Analysis: With Examples Using R and GGobi*, Springer, Berlin

Cook, R. D. (1979) Influential observations in linear regression, *J. Amer. Stat. Assn.*, **74**, 169–174

Cousins, R. D. (1995) Why isn't every physicist a Bayesian?, *Amer. J. Physics*, **63**, 398–410

Cowan, G. (1998) The small-N problem in high energy physics, in *Statistical Challenges in Modern Astronomy IV* (G. J. Babu & E. D. Feigelson, eds.), *Astron. Soc. Pacific*, **371**, 75

Cowpertwait, P. S. P. & Metcalfe, A. V. (2009) *Introductory Time Series with R*, Springer, Berlin

Cox, D. R. (1958) *Planning of Experiments*, John Wiley, New York

Cox, D. R. (1972) Regression models and life-tables, *J. Royal Stat. Soc. Ser B.*, **34**, 187–220

Cox, D. R. (2006) *Principles of Statistical Inference*, Cambridge University Press

Crawford, D. F., Jauncey, D. L. & Murdoch, H. S. (1970) Maximum-likelihood estimation of the slope from number-flux counts of radio sources, *Astrophys. J.*, **162**, 405–410

Crawley, M. J. (2007) *The R Book*, John Wiley, New York

Cummins, D. J., Filloon, T. G. & Nychka, D. (2001) Confidence intervals for nonparametric curve estimates: Toward more uniform pointwise coverage, *J. Amer. Stat. Assoc.*, **96**, 233–246

D'Agostino, R. B. & Stephens, M. A., eds. (1986) *Goodness-of-Fit Techniques*, Marcel Dekker, New York

Dalgaard, P. (2008) *Introductory Statistics with R*, 2nd ed., Springer, Berlin

Davies, P. L. & Kovac, A. (2004) Densities, spectral densities and modality, *Ann. Statist.*, **32**, 1093–1136

Davis, A., Marshak, A., Wiscombe, W. & Cahalan, R. (1994) Multifractal characterizations of nonstationarity and intermittency in geophysical fields: Observed, retrieved, or simulated, *J. Geophys. Res.*, **99**, D4, 8055–8072

Deeming, T. J. (1968) The analysis of linear correlation in astronomy, *Vistas in Astron.*, **10**, 125–142

Deeming, T. J. (1975) Fourier analysis with unequally-spaced data, *Astrophys. Space Sci.*, **36**, 137–158

de Jager, O. C., Raubenheimer, B. C. & Swanepoel, J. W. H. (1989) A poweful test for weak periodic signals with unknown light curve shape in sparse data, *Astron. Astrophys.*, **221**, 180–190

Delaigle, A. & Meister, A. (2008) Density estimation with heteroscedastic error, *Bernoulli*, **14**, 562–579

Delaigle, A., Fan, J. & Carroll, R. J. (2009) A design-adaptive local polynomial estimator for the errors-in-variables problem, *J. Amer. Stat. Assoc.*, **104**, 348–359

Della Ceca, R. *et al.* (1994), The properties of X-ray~selected active galactic nuclei. III. The radio-quiet versus radio-loud samples, *Astrophys. J.*, **430**, 533–544

Dempster, A. P., Laird, N. M. & Rubin, D. B. (1977) Maximum likelihood from incomplete data via the EM algorithm (with discussion), *J. Roy. Statist. Soc.*, **39**, 1–38

Dennis, B. & Patil, G. P. (1984) The gamma distribution and the weight multimodal gamma distributions as models of population abundance, *Math. Biosci.*, **68**, 187–212

de Vaucouleurs, G. (1948) Recherches sur les nebuleuses extragalactiques, *Annales d'Astrophysique*, **11**, 247–287

Diggle, P. J. (2003) *Statistical Analysis of Spatial Point Patterns*, 2nd ed., Arnold, London

Diggle, P. J. & Ribeiro, P. J. (2007) *Model-based Geostatistics*, Springer, Berlin

Dimitriadou, E., Hornik, K., Leisch, F., Meyer, D. and Weingessel, A. (2010) e1071: Misc Functions of the Department of Statistics (e1071), TU Wien, R package version 1.5-24, http://CRAN.R-project.org/package=e1071

Djorgovski, S. & Davis. M. (1987) Fundamental properties of elliptical galaxies, *Astrophys. J.*, **313**, 59–68

Dobigeon, N., Tournert, J.-Y. & Scargle, J. D. (2007) Joint segmentation of multivariate astronomical time series: Bayesian sampling with a hierarchical model, *IEEE Trans. Signal Process.*, **55**, 414–423

Drinkwater, M. J., Parker, Q. A., Proust, D., Slezak, E. & Quintana, H. (2004) The large scale distribution of galaxies in the Shapley Supercluster, *Pub. Astron. Soc. Australia*, **21**, 89–96

Duda, R. O., Hart, P. E. & Stork, D. G. (2001) *Pattern Classification*, 2nd ed., John Wiley, New York

Durbin, J. & Koopman, S. J. (2001) *Time Series Analysis by State Space Methods*, Oxford University Press

Durbin, J. & Watson, G. S. (1950) Testing for serial correlation in least squares regression I, *Biometrika*, **37**, 409–428

Dworetsky, M. M. (1983) A period-finding method for sparse randomly spaced observations of "How long is a piece of string?", *Mon. Not. Royal Astro. Soc.*, **203**, 917–924

Ebeling, H. & Wiedenmann, G. (1993) Detecting structure in two dimensions combining Voronoi tessellation and percolation, *Phys. Rev. E*, **47**, 704–710

Edelson, R. A. & Krolik, J. H. (1988) The discrete correlation function – A new method for analyzing unevenly sampled variability data, *Astrophys. J.*, **333**, 646–659

Efron, B. (1967) The two sample problem with censored data, *Proc. Fifth Berkeley Symp.*, **4**, 831–853

Efron, B. (1982) The jackknife, the bootstrap and other resampling plans, in *CBMS-NSF Regional Conf. Ser. in Applied Mathematics*, **38**, SIAM, Philadelphia

Efron, B. (2004) Bayesians, frequentists, and physicists, in *Proc. PHYSTAT2003: Statistical Problems in Particle Physics, Astrophysics, and Cosmology*, SLAC, paper MOAT003

Efron, B. & Petrosian, V. (1992) A simple test for truncated data with applications to redshift surveys, *Astrophys. J.*, **399**, 345–352

Efron, B. & Tibshirani, R. J. (1993) *An Introduction to the Bootstrap*, Chapman & Hall, London

Efstathiou, G., Ellis, R. S. & Peterson, B. A. (1988) Analysis of a complete galaxy redshift survey. II – The field-galaxy luminosity function, *Mon. Not. Royal Astro. Soc.*, **232**, 431–461

Einasto, M and 12 coauthors (2010) The Sloan great wall: Rich clusters, *Astron. Astrophys.*, **522**, A92

Eisenstein, D. J. *et al.* (2005) Detection of the baryon acoustic peak in the large-scale correlation function of SDSS luminous red galaxies, *Astrophys. J.*, **633**, 560–574

Eisenstein, D. J., *et al.* (2006) A catalog of spectroscopically confirmed white dwarfs from the Sloan Digital Sky Survey Data Release 4, *Astrophys. J. Suppl.*, **167**, 40–58

Elsner, J. B. & Tsonis, A. A. (1996) *Singular Spectral Analysis: A New Tool in Time Series Analysis*, Plenum Press, New York

Emmanoulopoulos, D., McHardy, I. M. & Uttley, P. (2010) On the use of structure functions to study blazar variability: Caveats and problems, *Mon. Not. Royal Astro. Soc.*, **404**, 931–946

Engle, R. F. & Russell, J. R. (1998) Autoregressive conditional duration: A new model for irregularly spaced transaction data, *Econometrica*, **66**, 1127–62

EPA (2002) Calculating upper confidence limits for exposure point concentrations at hazardous waste sites, http://www.epa.gov/swerrims/riskassessment/pdf/ucl.pdf

Ester, M., Kriegel, H.-P., Sander, J. & Xu, X. (1996) A density-based algorithm for discovering clusters in large spatial databases with noise, in *Proc. 2nd Intl. Conf. Knowledge Discovery and Data Mining (KDD-96)* (E. Simoudis, J. Han & U. M. Fayyad, eds.), AAAI Press, pp. 226–231

Evans, M., Hastings, N. & Peacock, B. (2000) *Statistical Distributions*, 3rd ed., John Wiley, New York

Everitt, B. S., Landau, S. & Leese, M. (2001) *Cluster Analysis*, 4th ed., Arnold, London

Faber, S. M. *et al.* (2007) Galaxy luminosity functions to $z \sim 1$ from DEEP2 and COMBO-17: Implications for red galaxy formation, *Astrophys. J.*, **665**, 265–294

Fan, J. & Yao, Q. (2003) *Nonlinear Time Series: Nonparametric and Parametric Methods*, Springer, Berlin

Faraway, J. J. (2006) *Extending the Linear Model with R: Generalized Linear, Mixed Effects and Nonparametric Regression Models*, Chapman & Hall, London

Feigelson, E. D. (1997) Time series problems in astronomy: An introduction, in *Applications of Time Series Analysis in Astronomy and Meteorology* (T. Subba Rao *et al.*, eds.), Chapman & Hall, London

Feigelson, E. D. (2007) Discussion on "Nonparametric estimation of dark matter distributions" by Wang *et al.*, in *Statistical Challenges in Modern Astronomy IV* (G. J. Babu & E. D. Feigelson, eds.), *ASP Conf. Ser.*, **371**, 280–283

Feigelson, E. D. & Babu, G. J. (1992) Linear regression in astronomy. II, *Astrophys. J.*, **397**, 55–67

Feigelson, E. D. & Babu, G. J. (2012) *Statistical Challenges in Modern Astronomy V*, Springer, Berlin

Feigelson, E. D. & Nelson, P. I. (1985) Statistical methods for astronomical data with upper limits I: Univariate distributions, *Astrophys. J.*, **293**, 192–206

Fender, R. & Belloni, T. (2004) GRS 1915+105 and the disc-jet coupling in accreting black hole systems, *Ann. Rev. Astro. Astrophys.*, **42**, 317–364

Fermi, E. (1949) On the origin of the cosmic radiation, *Phys. Rev.*, **785**, 1169–1174

Feroz, F., Hobson, M. P., & Bridges, M. (2009) MULTINEST: An efficient and robust Bayesian inference tool for cosmology and particle physics, M.N.R.A.S., **398**, 1601–1614

Filzmoser, P. & Varmuza, K. (2010) chemometrics: Multivariate statistical analysis in chemometrics, R package version 1.2, http://CRAN.R-project.org/package= chemometrics

Finlay, A. O., Banerjee, S. & Carlin, B. P. (2007) spBayes: An R package for univariate and multivariate hierarchical point-referenced spatial models, *J. Statistical Software*, **19**(4), 1–24

Fisher, N. I., Lewis, T. & Embleton, B. J. J. (1987) *Statistical Analysis of Spherical Data*, Cambridge University Press

Fisher, R. A. (1922) On the mathematical foundations of theoretical statistics, *Phil. Trans. Royal Soc., A*, **222**, 309–368

Fisher, R. A. (1973) *Statistical Methods and Scientific Inference*, 3rd ed., Macmillan, London

Florek, K., Lukaszewicz, J., Prkal, J., Steinhaus, H. & Zubrzycki, S. (1951) Sur la liaison et la division des points d'un ensemble fini, *Colloq. Math.*, **2**, 282–285

Ford, E. B. & Gregory, P. C. (2007) Bayesian model selection and extrasolar planet detection, in *Statistical Challenges in Modern Astronomy IV* (G. J. Babu & E. D. Feigelson, eds.), ASP, Conf. Ser. **371**, 189–205

Fortin, M.-J. & Dale, M. (2005) *Spatial Analysis: A Guide for Ecologists*, Cambridge University Press

Foster, G. (1996) Wavelets for period analysis of unevenly sampled time series, *Astron. J.*, **112**, 1709–1729

Fotheringham, A. S. & Rogerson, P. A., eds. (2009) *The SAGE Handbook of Spatial Analysis*, Sage, London

Fotheringham, A. S., Brunsdon, C. & Charlton, M. (2002) *Geographically Weighted Regression: The Analysis of Spatially Varying Relationships*, John Wiley, New York

Fox, J. (2002) *An R and S-Plus Companion to Applied Regression*, Sage, London

Fraley, C. (**S** original by Chris Fraley, U. Washington, Seattle. R port by Fritz Leisch at TU-Wien; since 2003–12: Martin Maechler; fdGPH(), fdSperio(), etc. by Valderio Reisen and Artur Lemonte) (2009) fracdiff: Fractionally differenced ARIMA aka ARFIMA(p,d,q) models, R package version 1.3-2, http://CRAN.R-project.org/package=fracdiff

Fraley, C. & Raftery, A. E. (2002) Model-based clustering, discriminant analysis, and density estimation, *J. Amer. Stat. Assn.*, **97**, 611–631

Fraley, C. & Raftery, A. E. (2007) Bayesian regularization for normal mixture estimation and model-based clustering, *J. Classification*, **24**, 155–181

Fraser, A. M. (2009) *Hidden Markov Models and Dynamical Systems*, SIAM, Philadelphia

Freedman, D. & Diaconis, P. (1981) On the histogram as a density estimator: L_2 theory, *Zeitschrift Wahrscheinlichkeitstheorie verw. Gebiete*, **57**, 453–476

Friedman, J. H. (1984) A variable span scatterplot smoother, Laboratory for Computational Statistics, Stanford University Technical Report No. 5

Freedman, W. L. & Madore, B. F. (2010) The Hubble constant, *Ann. Rev. Astron. Astrophys.*, **48**, 673–710

Freund, Y. & Schapire, R. (1997) A decision-theoretical generalization of online learning and an application to boosting, *J. Comput. Syst. Sci.*, **55**, 119–139

Gehrels, N. (1986) Confidence limits for small numbers of events in astrophysical data, *Astrophys. J.*, **303**, 336–346

Geisser, S. (2006) *Modes of Parametric Statistical Inference*, Wiley-Interscience, New York

Gelman, A., Carlin, J. B., Stern, H. S. & Rubin, D. B. (2003) *Bayesian Data Analysis*, 2nd ed., Chapman & Hall, London

Gemerman, D. & Lopes, H. F. (2006) *Markov Chain Monte Carlo: Stochastic Simulation for Bayesian Inference*, 2nd ed., CRC Press, Boca Raton

Gencay, R., Selcuk, F. & Whitcher, B. (2001) *An Introduction to Wavelets and Other Filtering Methods in Finance and Economics*, Academic Press, New York

Gentleman, R. & Vandal, A. C. (2002) Nonparametric estimation of the bivariate CDF for arbitrarily censored data, *Canadian J. Stat.*, **30**, 557–571

Geweke, J. and Porter-Hudak, S. (1983) The estimation and application of long memory time series models, *J. Time Series Anal.*, **4**, 221–238

Ghosh, J. K., Delampady, M. & Samanta, T. (2006) *An Introduction to Bayesian Analysis: Theory and Methods*, Springer, Berlin

Ghysels, E., Sinko, A. & Valkanov, R. (2007) MIDAS regressions: Further results and new directions, *Econometric Reviews*, **26**, 53–90

Goldstein, M. L., Morris, S. A., & Yen, G. G. (2004) Problems with fitting to the power-law distribution, *Eur. Phys. J. B*, **41**, 255–258

Gorban, A., Kegl, B., Wunsch, D. & Zinovyev, A., eds. (2007) *Principal Manifolds for Data Visualisation and Dimension Reduction*, Springer, Berlin

Gower, J. C. & Hand, D. J. (1996), *Biplots*, Chapman & Hall, London

Graham, M. J., Fitzpatrick, M. J. & McGlynn, T. A., eds. (2008) *The National Virtual Observatory: Tools and Techniques for Astronomical Research*, ASP Conf. Ser **382**., Astronomical Society of the Pacific

Grandell, J. (1997) *Mixed Poisson Processes*, Chapman & Hall, London

Greene, W. H. (2003) *Econometric Analysis*, 6th ed., Prentice-Hall, Englewood Cliffs

Greenwood, P. E. & Nikulin, M. S. (1996) *A Guide to Chi-Squared Testing*, John Wiley, New York

Gregory, P. C. (2005) *Bayesian Logical Data Analysis for the Physical Sciences: A Comparative Approach with Mathematica Support*, Cambridge University Press

Gregory, P. C. & Loredo, T. J. (1992) A new method for the detection of a periodic signal of unknown shape and period, *Astrophys. J.*, **398**, 146–168

Griffin, E., Hanisch, R. & Seaman, R., eds. (2012) *New Horizons in Time Domain Astronomy*, IAU Symp. 285, ASP Conf. Ser. Vol. 387, Astronomical Society of the Pacific, San Francisco

Gross, J. (2010) nortest: Tests for normality, R package version 1.0.

Grothendieck, G. (2010) nls2: Non-linear regression with brute force. R package version 0.1-3, http://CRAN.R-project.org/package=nls2

Guis, V. & Barge, P. (2005) An image-processing method to detect planetary transits: The "gauging" filter, *Publ. Astron. Soc. Pacific*, **117**, 160–172

Hacking, I. (1990) *The Taming of Chance*, Cambridge University Press

Haight, F. A. (1967) *Handbook of the Poisson Distribution*, John Wiley, New York

Hald, A. (1998) *A History of Mathematical Statistics from 1750 to 1930*, Springer, Berlin

Hald, A. (2003) *A History of Probability and Statistics and Their Applications before 1750*, Springer, Berlin

Hall, P. (2008) Nonparametric methods for estimating periodic functions, with applications in astronomy, in *COMPSTAT 2008 Proceedings in Computational Statistics*, Springer, Berlin, pp. 3–18

Hall, P., Racine, J. S. & Li, Q. (2004) Cross-validation and the estimation of conditional probability densities, *J. Amer. Stat. Assn.*, **99**, 1015–1026

Hall, P., Watson, G. S. & Cabrera, J. (1987) Kernel density estimation with spherical data, *Biometrika*, **74**, 751–62

Halley, E. (1693) An estimate of the degrees of the mortality of mankind drawn from curious tables of the births and funerals at the city of Breslaw, *Phil. Trans. Roy. Soc. London*, **17**, 596–610

Hand, D. J. (2010) *Statistics: An Overview*, in International Encyclopedia of Statistical Science (M. Lovrik, ed.), Springer, New York

Hanisch, R. J., Farris, A., Greisen, E. W., Pence, W. D., Schlesinger, B. M., Teuben, P. J., Thompson, R. W. & Warnock, A. (2001) Definition of the Flexible Image Transport System (FITS), *Astron. Astrophys.*, **376**, 359–380

Hansen, C. & Johnson, C. R. (2004), *The Visualization Handbook*, Elsevier, Amsterdam

Hao, J. and 13 coauthors (2010) A GMBCG galaxy cluster catalog of 55,424 rich clusters from SDSS DR7, *Astrophys. J. Suppl.*, **191**, 254–274

Harikrishnan, K. P., Misra, R. & Ambika, G. (2011) Nonlinear time series analysis of the light curves from the black hole system GRS1915+105, *Res. Astron. Astrophys.*, **11**, 71–79

Hartigan, J. A. & Hartigan, P. M. (1985) The dip test of unimodality, *Ann. Statist.*, **13**, 70–84

Hastie, T. & Tibshirani, R. (1996) Discriminant adaptive nearest neighbor classification, *IEEE Trans. Pattern Anal. Machine Intell.*, **18**, 607–616

Hastie, T., Tibshirani, R. & Friedman, J. (2009) *The Elements of Statistical Learning: Data Mining, Inference, and Prediction*, 2nd ed., Springer, Berlin

Hayfield, T. & Racine, J. S. (2008) Nonparametric econometrics: The np package, *Journal Statist. Software*, **27**(5), http://www.jstatsoft.org/v27/i05/

Helsel, D. R. (2005) *Nondetects and Data Analysis: Statistics for Censored Environmental Data*, Wiley-Interscience, New York

Hennig, C. (2010) fpc: Flexible procedures for clustering, R package version 2.0-3, http://CRAN.R-project.org/package=fpc

Hernquist, L. (1990) An analytical model for spherical galaxies and bulges, *Astrophys. J.*, **356**, 359–364

Hertz, P. & Feigelson, E. D. (1995) A sample of astronomical time series, in *Applications of Time Series Analysis in Astronomy and Meteorology* (T. Subba Rao *et al.*, eds.), Chapman & Hall, London, pp. 340–356

Higdon, R. & Schafer, D. W. (2001) Maximum likelihood computations for regression with measurement error, *Comput. Statist. & Data Analysis*, **35**, 283–299

Higgins, J. J. (2004) *An Introduction to Modern Nonparametric Statistics*, Thomson/ Brooks-Cole, Belmont

Hilbe, J. M. (2011) *Negative Binomial Regression*, 2nd ed., Cambridge University Press

Hinneburg, A. & Keim, D. A. (1998) An efficient approach to clustering in large multimedia databases with noise, in *Proc. 1998 Int. Conf. Knowledge Discovery and Data Mining (KDD'98)*, pp. 58–65

Ho, L. C. *et al.* (2001) Detection of nuclear X-ray sources in nearby galaxies with Chandra, *Astrophys. J.*, **549**, L51–L54

Hobson, M. P., Parkinson, D., Jaffee, A. H., Liddle, A. J. & Mukherjee, P. (2010) *Bayesian Methods in Cosmology*, Cambridge University Press

Hogg, D. W., Myers, A. D. & Bovy, J. (2010) Inferring the eccentricity distribution, *Astrophys. J.*, **725**, 2166–2175

Hogg, R., McKean, J. & Craig, A. (2005) *Introduction to Mathematical Statistics*, 6th ed., Prentice-Hall, Englewood Cliffs

Hogg, R. V. & Tanis, E. (2009) *Probability and Statistical Inference*, 8th ed., Prentice-Hall, Englewood Cliffs

Holtzman, J. A., Watson, A. M., Baum, W. A., Grillmair, C. J., Groth, E. J., Light, R. M., Lynds, R. & O'Neil, E. J. (1998) The luminosity function and Initial Mass Function in the Galactic bulge, *Astron. J.*, **115**, 1946–1957

Hopkins, P. F., Richards, G. T. & Hernquist, L. (2007) An observational determination of the bolometric quasar luminosity function, *Astrophys. J.*, **654**, 731–753

Hörmann, W., Leydold, J. & Derflinger, G. (2004) *Automatic Nonuniform Random Variate Generation*, Springer, Berlin

Horne, J. H. & Baliunas, S. L. (1986) A prescription for period analysis of unevenly sampled time series, *Astrophys. J.*, **302**, 757–763

Hosmer, D. W., Lemeshow, S. & May, S. (2008) *Applied Survival Analysis*, Wiley-Interscience, New York

Hou, A., Parker, L. C., Harris, W. E. & Wilman, D. J. (2009) Statistical tools for classifying galaxy group dynamics, *Astrophys. J.*, **702**, 1199–1210

Howell, L. W. (2002) Maximum likelihood estimation of the broken power law spectral parameters with detector design applications, *Nucl. Instr. Methods Phys. Res. A*, **489**, 422–438

Hubble, E. P. (1926) Extragalactic nebulae, *Astrophys. J.*, **64**, 321–369

Hubble, E. P. (1930) Distribution of luminosity in elliptical nebulae, *Astrophys. J.*, **71**, 231–276

Hubble, E. & Humason, M. L. (1931) The velocity–distance relation among extra-galactic nebulae, *Astrophys. J.*, **74**, 43–80

Huber, P. J. (1967) The behavior of maximum likelihood estimates under nonstandard conditions, in *Proc. Fifth Berkeley Symposium in Mathematical Statistics and Probability*, University of California Press

Huber, P. J. & Ronchetti, E. M. (2009) *Robust Statistics*, 2nd ed., John Wiley, New York

Huber-Carol, C., Balakrishnan, N., Nikulin, M. S. & Mesbah, M., eds. (2002) *Goodness-of-Fit Tests and Model Validity*, Birkhäuser, Boston

Hubert, M., Rousseeuw, P. J. & Van Aelst, S. (2005) Multivariate outlier detection and robustness, in *Handbook of Statistics* (C. R. Rao *et al.*, eds.), vol. 24, Elsevier, Amsterdam, pp. 263–302

Huet, S., Bouvier, A., Poursat, M.-A. & Jolivet, E. (2004) *Statistical Tools for Nonlinear Regression: A Practical Guide with S-PLUS and R Examples*, 2nd ed., Springer, Berlin

Hufnagel, B. R. & Bregman, J. N. (1992) Optical and radio variability in blazars, *Astrophys. J.*, **386**, 473–484

Hurvich, C., Deo, R. & Brodsky, J. (1998) The mean squared error of Geweke and Porter-Hudak's estimator of the memory parameter of a long-memory time series, *J. Time Series Analysis*, **19**, 19–46

Hutchinson, T. P. & Lai, C. D. (1990) *Continuous Bivariate Distributions, Emphasising Applications*, Rumsby Scientific Publishing, Australia

Ihaka, R. & Gentleman, R. (1996) R: A language for data analysis and graphics, *J. Comput. Graph. Statist.*, **5**, 299–314

Ihaka, R. & Lang, D. T. (2008) Back to the future: Lisp as a base for a statistical computing system, in *COMPSTAT 2008: Proc. Computational Statistics* (P. Brito, ed.), pp. 21–34

Illian, J., Penttinen, A., Stoyan, H. & Stoyan, D. (2008) *Statistical Analysis and Modelling of Spatial Point Patterns*, John Wiley, New York

Ingram, A. & Done, C. (2010) A physical interpretation of the variability power spectral components in accreting neutron stars, *Mon. Not. Royal Astro. Soc.*, **405**, 2447–2452

Inselberg, A. (2009) *Parallel Coordinates: Visual Multidimensional Geometry and Its Applications*, Springer, Berlin

Irony, T. Z. (1992) Bayesian estimation for discrete distributions, *J. Appl. Stat.*, **19**, 533–549

Isobe, T., Feigelson, E. D. & Nelson, P. I. (1986) Statistical methods for astronomical data with upper limits II: Correlation and regression, *Astrophys. J.*, **306**, 490–507

Isobe, T., Feigelson, E. D., Akritas, M. G. & Babu, G. J. (1990) Linear regression in astronomy, *Astrophys. J.*, **364**, 104–113

Ivezić, Z. *et al.* (2002) Optical and radio properties of extragalactic sources observed by the FIRST survey and the Sloan Digital Sky Survey, *Astron. J.*, **124**, 2353–2400

Izenman, A. J. (2008) *Modern Multivariate Statistical Techniques: Regression, Classification, and Manifold Learning*, Springer, Berlin

James, F. (2006) *Statistical Methods in Experimental Physics*, 2nd ed., World Scientific, Singapore

Jammalamadaka, S. R. & SenGupta, A. (2001) *Topics in Circular Statistics*, World Scientific, Singapore

Johnson, N. L., Kemp, A. W. & Kotz, S. (2005) *Univariate Discrete Distributions*, 3rd ed., Wiley-Interscience, New York

Johnson, N. L., Kotz, S. & Balakrishnan, N. (1994) *Continuous Univariate Distributions*, 2nd ed., Vols 1 and 2, Wiley-Interscience, New York

Johnson, N. L., Kotz, S. & Balakrishnan, N. (1997) *Discrete Multivariate Distributions*, Wiley-Interscience, New York

Johnson, R. A. & Wichern, D. W. (2007) *Applied Multivariate Statistical Analysis*, 6th ed., Prentice-Hall, Englewood Cliffs

Johnstone, I. M. & Titterington, D. M. (2009) Statistical challenges of high-dimensional data, *Phil. Trans. Royal Soc.* A, **367**, 4237–4253

Jones, A. J. (2004) New tools in non-linear modelling and prediction, *Comput. Manag. Sci.*, **1**, 109–149

Jones, O., Maillardet, R. & Robinson, A. (2009) *Introduction to Scientific Programming and Simulation Using R*, CRC Press, Bora Raton

Jonker, P. G., van der Klis, M., Homan, J., Méndez, M., Lewin, W. H. G., Wijnands, R. & Zhang, W. (2002) Low- and high-frequency variability as a function of spectral properties in the bright X-ray binary GX 5–1, *Mon. Not. Royal Astro. Soc.*, **333**, 665–678

Jung, J. (1970) The derivation of absolute magnitudes from proper motions and radial velocities and the calibration of the H.R. diagram II, *Astron. Astrophys.*, **4**, 53–69

Jupp, P. E. & Mardia, K. V. (1989) Unified view of the theory of directional statistics, 1975–1985, *Intl. Statist. Review*, **57**, 261–294

Kaplan, E. L. & Meier, P. (1958) Nonparametric estimation from incomplete observations, *J. Amer. Stat. Assn.*, **53**, 457–481

Kapteyn, J. C. & van Rhijn, P. J. (1920) On the distribution of the stars in space especially in the high Galactic latitudes, *Astrophys. J.*, **52**, 23–38

Karypis, G., Han, E.-H. & Kumar, V. (1999) Chameleon: A hierarchical clustering algorithm using dynamic modeling, *IEEE Comput.*, **32**(8), 68–75

Kashyap, V. L., van Dyk, D. A., Connors, A., Freeman, P. E., Siemiginowska, A., Xu, J. & Zezas, A. (2010) On computing upper limits to source intensities, *Astrophys. J.*, **719**, 900–914

Kaufman, L. & Rousseeuw, P. J. (1990) *Finding Groups in Data: An Introduction to Cluster Analysis*, John Wiley, New York

Kazin, E. A. and 11 coauthors (2010) The baryonic acoustic feature and large-scale clustering in the Sloan Digital Sky Survey luminous red galaxy sample, *Astrophys. J.*, **710**, 1444–1461

Keiding, N. & Gill, R. D. (1990) Random truncation models and Markov processes, *Ann. Stat.*, **18**, 582–602

Kelly, B. C. (2007) Some aspects of measurement error in linear regression of astronomical data, *Astrophys. J.*, **665**, 1489–1506

Kingman, J. F. C. (1993) *Poisson Processes*, Clarendon Press, Oxford

Kitagawa, G. & Gersch, W. (1996) *Smoothness Priors Analysis of Time Series*, Springer, Berlin

Kitaura, F. S., Jasche, J., Li, C., Ensslin, T. A., Metcalf, R. B., Wandelt, B. D., Lemson, G. & White, S. D. M. (2009) Cosmic cartography of the large-scale structure with Sloan Digital Sky Survey data release 6, *Mon. Not. Royal Astro. Soc.*, **400**, 183–203

Klein, J. P. & Moeschberger, M. L. (2005) *Survival Analysis: Techniques for Censored and Truncated Data*, 2nd ed., Springer, Berlin

Koen, C. (1990) Significance testing of periodogram ordinates, *Astrophys. J.*, **348**, 700–702

Koen, C. & Kondlo, L. (2009) Fitting power-law distributions to data with measurement errors, *Mon. Not. Royal Astro. Soc.*, **397**, 495–505

Koen, C. & Lombard, F. (1993) The analysis of indexed astronomical time series. I. Basic methods, *Mon. Not. Royal Astro. Soc.*, **263**, 287–308

Koenker, R. (2005) *Quantile Regression*, Cambridge University Press

Koenker, R. (2010) quantreg: Quantile Regression, R package version 4.53, http://CRAN.R-project.org/package=quantreg

Kohavi, R. (1995) A study of cross-validation and bootstrap for accuracy estimation and model selection, *Proc. 12th Intl. Joint Conference on Artificial Intelligence*, 1137–1143

Komsta, L. & Novomestky, F. (2007) moments: Moments, cumulants, skewness, kurtosis and related tests, R package version 0.11, http://www.r-project.org, http://www.komsta.net/

Konishi, S. & Kitagawa, G. (2008) *Information Criteria and Statistical Modeling*, Springer, Berlin

Kormendy, J., Fisher, D. B., Cornell, M. E. & Bender, R. (2009) Structure and formation of elliptical and spheroidal galaxies, *Astrophys. J. Suppl.*, **182**, 216–309

Kotz, S., Balakrishnan, N. & Johnson, N. L. (2000) *Continuous Multivariate Distributions*, 2nd ed., Vols 1 and 2, Wiley-Interscience, New York

Kovács, G., Zucker, S. & Mazeh, T. (2002) A box-fitting algorithm in the search for periodic transits, *Astron. Astrophys.*, **391**, 369–377

Koyama, S. & Shinomoto, S. (2004) Histogram bin width selection for time-dependent Poisson processes, *J. Physics A*, **37**, 7255–7265

Kraus, D. (2008). surv2sample: Two-sample tests for survival analysis, R package version 0.1-2, http://www.davidkraus.net/surv2sample/

Krishnamoorthy, K. (2006) *Handbook of Statistical Distributions with Applications*, Chapman & Hall, London

Kroupa, P. (2001) On the variation of the initial mass function, *Mon. Not. Royal Astro. Soc.*, **322**, 231–246

Kruschke, J. K. (2011) *Doing Bayesian Data Analysis: A Tutorial with R and BUGS*, Academic Press, New York

Kutner, M. H., Nachtsheim, C. J., Neter, J. & Li, W. (2005) *Applied Linear Statistical Models*, 5th ed., McGraw-Hill, New Tork

Lafler, J. & Kinman, T. D. (1965) An RR Lyrae star survey with the Lick 20-inch astrograph II. The calculation of RR Lyrae periods by electronic computer, *Astrophys. J. Suppl.*, **11**, 216

Lahiri, P., ed. (2001) *Model Selection*, Institute of Mathematical Statisitcs

Lang, D. T., Swayne, D., Wickham, H. & Lawrence, M. (2010) rggobi: Interface between R and GGobi, R package version 2.1.16, http://CRAN.R-project.org/package=rggobi

LaValley, M., Isobe, T. & Feigelson, E. (1992) ASURV: Astronomy survival analysis package, in *Astronomical Data Analysis Software and Systems I* (D. Worrall *et al.*, eds.), *ASP Conf.* Ser. **25**, 245–247

Lawless, J. F. (2003) *Statistical Models and Methods for Lifetime Data*, 2nd ed., John Wiley, New York

Lawson, A. B. & Denison, D. G. T., eds. (2002) *Spatial Cluster Modelling*, Chapman & Hall/CRC, Boca Raton

Leahy, D. A., Elsner, R. F. & Weisskopf, M. C. (1983) On searches for periodic pulsed emission: The Rayleigh test compared to epoch folding, *Astrophys. J.*, **272**, 256–258

Lean, J. (1990) Evolution of the 155 day periodicity in sunspot areas during solar cycles 12 to 21, *Astrophys. J.*, **363**, 718–727

Lee, E. T. & Wang, J. W. (2003) *Statistical Methods for Survival Data Analysis*, 3rd ed., Wiley-Interscience, New York

Lee, J. A. & Verleysen, M. (2009) *Nonlinear Dimensionality Reduction*, Springer, Berlin

Lee, L. (2010) NADA: Nondetects And Data Analysis for environmental data, R package version 1.5-3, http://CRAN.R-project.org/package=NADA

Leggett, S. K. *et al.* (2000) The missing link: Early methane ("T") dwarfs in the Sloan Digital Sky Survey, *Astrophys. J.*, **536**, L35–L38

Lehmann, E. L. & Casella, G. (1998) *Theory of Point Estimation*, 2nd ed., Springer, Berlin

Lehmann, E. L. & Romano, J. P. (2005) *Testing Statistical Hypotheses*, 3rd ed., Springer, Berlin

Leisch, F. (2006) A toolbox for K-centroids cluster analysis, *Comput. Statist. Data Anal.*, **51**, 526–544

Leisch, F. (2010) Neighborhood graphs, stripes and shadow plots for cluster visualization, *Statist. Comput.*, **20**, 451–469

Levy, M. & Solomon, S. (1996) Power laws are logarithmic Boltzmann laws, *Intl. J. Mod. Phys. C*, **7**, 595–601

Lieshout, M. N. M. van & Baddeley, A. J. (1996) A nonparametric measure of spatial interaction in point patterns, *Statistica Neerlandica*, **3**, 344–361

Ligges, U. & Mächler, M. (2003) Scatterplot3d – an R package for visualizing multivariate data, *J. Statist. Software*, **8**(11), 1–20

Lilliefors, H. W. (1969) On the Kolmogorov–Smirnov test for the exponential distribution with mean unknown, *J. Amer. Statist. Assn.*, **64**, 387–389

Loader, C. (1999) *Local Regression and Likelihood*, Springer, New York

Loh, J. M. (2008) A valid and fast spatial bootstrap for correlation functions, *Astrophys. J.*, **681**, 726–734

Lomb, N. R. (1976) Least squares frequency analysis unequally spaced data, *Astrophys. Space Sci.*, **39**, 447–46

London, D. (1997) *Survival Models and Their Estimation*, 3rd. ed., ACTEX, Winsted, CT

Loredo, T. (1992) Promise of Bayesian inference for astrophysics, in *Statistical Challenges in Modern Astronomy* (Feigelson, E. D., & Babu, G. J., eds.), Springer, Berlin, pp. 275–306

Lucy, L. B. (1974) An iterative technique for the rectification of observed distributions, *Astron. J.*, **79**, 745–754

Luhman, K. L. *et al.* (2008) The disk population of the Chamaeleon I star-forming region, *Astrophys. J.*, **675**, 1375–1406

Luhman, K. L., Allen, P. R., Espaillat, C., Hartmann, L. & Calvet, N. (2010) The disk population of the Taurus star-forming region, *Astrophys. J. Suppl.*, **186**, 111–174

Lund, U. & Agostinelli, C. (2009) CircStats: Circular Statistics, from "Topics in Circular Statistics" 2001, R package version 0.2-4, http://CRAN.R-project.org/package=CircStats

Lupton, R. (1993) *Statistics in Theory and Practice*, Princeton University Press

Lynden-Bell, D. (1971) A method of allowing for known observational selection in small samples applied to 3CR quasars, *Mon. Not. Royal Astro. Soc.*, **155**, 95–118

Lynden-Bell, D., Faber, S. M., Burstein, D., Davies, R. L., Dressler, A., Terlevich, R. J. & Wegner, G. (1988) Spectroscopy and photometry of elliptical galaxies. V – Galaxy streaming toward the new supergalactic center, *Astrophys. J.*, **326**, 19–49

Maechler, M. (2010) diptest: Hartigan's dip test statistic for unimodality – corrected code (based on Fortran, S-plus from Dario Ringach and NYU.edu), R package version 0.25-3, http://CRAN.R-project.org/package=diptest

Maechler, M. and many others (2010) sfsmisc: Utilities from Seminar fuer Statistik ETH Zurich. R package version 1.0-11, http://CRAN.R-project.org/package=sfsmisc

Maíz Apellániz, J. & Úbeda, L. (2005) Numerical biases on Initial Mass Function determinations created by binning, *Astrophys. J.*, **629**, 873–880

Mallat, S. G. (2009) *A Wavelet Tour of Signal Processing*, 3rd ed., Academic Press, New York

Maller, A. H., McIntosh, D. H., Katz, N. & Weinberg, M. D. (2005) The galaxy angular correlation functions and power spectrum from the Two Micron All Sky Survey, *Astrophys. J.*, **619**, 147–160

Maloney, P. R. *et al.* (2005) A fluctuation analysis of the Bolocam 1.1 mm Lockman Hole survey, *Astrophys. J.*, **635**, 1044–1052

Mandel, K. S., Wood-Vasey, W. M., Friedman, A. S. & Kirshner, R. P. (2009) Type Ia supernova light-curve inference: Hierarchical Bayesian analysis in the near-infrared, *Astrophys. J.*, **704**, 629–651

Mandelkern, M. (2002) Setting confidence intervals for bounded parameters (with commentaries), *Statist. Sci.*, **17**, 149–172

Marchis, F. *et al.* (2006) A low density of 0.8 g cm^{-3} for the Trojan binary asteroid 617 Patroclus, *Nature*, **439**, 565–567

Mardia, K. V. (1962) Multivariate Pareto distributions, *Ann. Math. Statist.*, **33**, 1008–1015

Maronna, R. A. & Martin, R. D. (2006) *Robust Statistics: Theory and Methods*, John Wiley, Chichester

Marquardt, D. W. & Acuff, S. K. (1984) Direct quadratic spectrum estimation with irregularly spaced data, in *Time Series Analysis of Irregularly Observed Data* (E. Parzen, ed.), Lecture Notes in Statistics, **25**, Springer, Berlin

Marsland, S. (2009) *Machine Learning: An Algorithmic Perspective*, CRC Press, Boca Raton

Martínez, V. J. & Saar, E. (2001) *Statistics of the Galaxy Distribution*, Chapman & Hall/CRC, Boca Raton

Maschberger, T. & Kroupa, P. (2009) Estimators for the exponent and upper limit, and goodness-of-fit tests for (truncated) power-law distributions, *Mon. Not. Royal Astro. Soc.*, **395**, 931–942

Matsumoto, M. & Nishimura, T. (1998) Mersenne Twister: A 623-dimensionally equidistributed uniform pseudo-random number generator, *ACM Trans. Model. Comput. Simu.*, **8**, 3–30

McDermott, J. P., Babu, G. J., Liechty, J. C. & Lin, D. K. (2007) Data skeletons: simultaneous estimation of multiple quantiles for massive streaming datasets with applications to density estimation, *Statist. Comput.*, **17**, 311–321

McLachlan, G. J. & Krishnan, T. (2008) *The EM Algorithm and Extensions*, 2nd ed., John Wiley, New York

McLachlan, G. & Peel, D. (2000) *Finite Mixture Models*, Wiley-Interscience, New York

Mészáros, P. (2006) Gamma ray bursts, *Rept. Prog. Phys.*, **69**, 2259–2321

Meyer, D., Zeileis, A. & Hornik, K. (2006) The Strucplot framework: Visualizing multi-way contingency tables with vcd, *J. Statist. Software*, **17**(3), 1

Milborrow, S. derived from mda:mars by Trevor Hastie and Rob Tibshirani. (2010) earth: Multivariate Adaptive Regression Spline Models, R package version 2.4-5, http://CRAN.R-project.org/package=earth

Miller, C. J., Nichol, R. C., Genovese, C. & Wasserman, L. (2002), A nonparametric analysis of the cosmic microwave background power spectrum, *Astrophys. J.*, **565**, L67–L70

Minsky, M. L. & Papert, S. A. (1969) *Perceptrons*, MIT Press, Cambridge, MA

Mitzenmacher, M. (2004) A brief history of generative models for power law and lognormal distributions, *Internet Math.*, **1**, 226–251

Möller, J. & Waagepetersen, R. (2003) *Statistical Inference and Simulation for Spatial Point Processes*, CRC, Boca Raton

Morbey, C. L. & Griffin, R. F. (1987) On the reality of certain spectroscopic orbits, *Astrophys. J.*, **317**, 343–352

Moreira, C., de Uña-Álvarez, J. & Crujeiras, R. (2010) DTDA: An R package to analyze randomly truncated data, *J. Statist. Software*, **37**(7), 1–20, http://www.jstatsoft.org/v37/i07

Mukherjee, S., Feigelson, E. D., Babu, J. G., Murtagh, F., Fraley, C. & Raftery, A. (1998) Three types of gamma-ray bursts, *Astrophys. J.*, **508**, 314–327

Müller, D. W. & Sawitzki, G. (1991) Excess mass estimates and tests for multimodality, *J. Amer. Statist. Assoc.*, **86**, 738–746.

Müller, P. & Quintana, F. A. (2004) Nonparametric Bayesian data analysis, *Statistical Science*, **19**, 95–110

Nadarajah, S. & Kotz, S. (2006) R programs for computing truncated distributions, *J. Statist. Software*, **16**, Code Snippet 2

Nantais, J. B., Huchra, J. P., Barmby, P., Olsen K. A. & Jarrett, T. H. (2006) Nearby spiral globular cluster systems I: Luminosity functions, *Astron. J.*, **131**, 1416–1425

Narayan, R. & Nityananda, R. (1986) Maximum entropy image restoration in astronomy, *Ann. Rev. Astron. Astrophys.*, **24**, 127–170

Navarro, J. F., Frenk, C. S. & White, S. D. M. (1997) A universal density profile from hierarchical clustering, *Astrophys. J.*, **490**, 493–508

Nelder, J. & Wedderburn, R. (1972) Generalized linear models, *J. Royal Stat. Soc.*, Ser. A, **135**, 370–384

Neter, J., Wasserman, W. & Kutner, M. H. (1978) *Applied Linear Statistical Models*, McGraw-Hill, New York

Newman, M. E. J. (2005) Power laws, Pareto distributions and Zipf's law, *Contemp. Phys.*, **46**, 323–351

Nichols-Pagel, G. A., Percival, D. B., Reinhall, P. G. & Riley, J. J. (2008) Should structure functions be used to estimate power laws in turbulence? A comparative study, *Physica D: Nonlin. Phen.*, **237**, 665–677

Nocedal, J. & Wright, S. J. (2006) *Numerical Optimization*, 2nd ed., Springer, Berlin

Norris, J. P., Hertz, P., Wood, K. S., Vaughan, B. A., Michelson, P. F., Mitsuda, K. & Dotani, T. (1990) Independence of short time scale fluctuations of quasi-periodic oscillations and low frequency noise in GX5-1, *Astrophys. J.*, **361**, 514–526

Nousek, J. A. (2006) Evidence for a canonical gamma-ray burst afterglow light curve in the Swift XRT data, *Astrophys. J.*, **642**, 389–400

Okabe, A., Boots, B., Sugihara, K. & Chiu, S. N. (1999) *Spatial Tesselations: Concepts and Applications of Voronoi Diagrams*, John Wiley, New York

Ord, J. K. & Getis, A. (2001) Testing for local spatial autocorrelation in the present of global autocorrelation, *J. Regional Sci.*, **41**, 411–432

Osborne, C. (1991) Statistical calibration: A review, *Intl. Statist. Rev.*, **59**, 309–336

Palma, W. (2007) *Long-Memory Time Series: Theory and Methods*, Wiley-Interscience, New York

Palmer, D. M. (2009) A fast chi-squared technique for period search of irregularly sampled data, *Astrophys. J.*, **695**, 496–502

Paltani, S. (2004) Searching for periods in X-ray observations using Kuiper's test. Application to the ROSAT PSPC archive, *Astron. Astrophys.*, **420**, 789–797

Panik, M. J. (2005) *Advanced Statistics from an Elementary Point of View*, Academic Press, San Diego

Papadakis, I. E. & Lawrence, A. (1993) Improved methods for power spectrum modelling of red noise, *Mon. Not. Royal Astro. Soc.*, **261**, 612–624

Park, T., Kashyap, V. L., Siemiginowska, A., van Dyk, D. A., Zezas, A., Heinke, C. & Wargelin, B. J. (2006) Bayesian estimation of hardness ratios: Modeling and computations, *Astrophys. J.*, **652**, 610–628

Parzen, E., ed. (1984) *Time Series Analysis of Irregularly Spaced Data*, Lecture Notes in Statistics, vol. 25, Springer, Berlin

Peacock, J. A. (1983), Two-dimensional goodness-of-fit testing in astronomy, *Mon. Not. Royal Astron. Soc.*, **202**, 615–627

Pebesma, E. J. (2004) Multivariable geostatistics in S: The *gstat* package, *Comput. Geosci.*, **30**, 683–691

Peebles, P. J. E. (1973) Statistical analysis of catalogs of extragalactic objects. I. Theory, *Astrophys. J.*, **185**, 413–440

Pelt, J. (2009) High frequency limits in periodicity search from irregularly spaced data, *Baltic Astron.*, **18**, 83–92

Peng, C.-K., Buldyrev, S. V., Havlin, S., Simons, M., Stanley, H. E. & Goldberger, A. L. (1994) Mosaic organization of DNA nucleotides, *Phys. Rev. E*, **49**, 1685–1689

Percival, D. B. (1993) Three curious properties of the sample variance and autocovariance for stationary processes with unknown mean, *Amer. Statist.*, **47**, 274–276

Percival, D. B. & Walden, A. T. (1993) *Spectral Analysis for Physical Applications: Multi-taper and Conventional Univariate Techniques*, Cambridge University Press

Percival, D. B. & Walden, A. T. (2000) *Wavelet Methods for Time Series Analysis*, Cambridge University Press

Perryman, M. A. C. *et al.* (1998) The Hyades: distance, structure, dynamics, and age, *Astron. Astrophys.*, **331**, 81–120

Peterson, B. M., Wanders, I., Horne, K., Collier, S., Alexander, T., Kaspi, S. & Maoz, D. (1998) On uncertainties in cross-correlation lags and the reality of wavelength-dependent continuum lags in active galactic nuclei, *Publ. Astron. Soc. Pacific*, **110**, 660–670

Polonik, W. (1999) Concentration and goodness-of-fit in higher dimensions: (Asymptotically) distribution-free methods, *Ann. Statist.*, **27**, 1210–1229

Porter, T. M. (1986) *The Rise of Statistical Thinking: 1820–1900*, Princeton University Press

Press, W. H. (1978) Flicker noises in astronomy and elsewhere, *Comments Astrophys.*, **7**, 103–119

Press, W. H. & Schechter, P. (1974) Formation of galaxies and clusters of galaxies by self-similar gravitational condensation, *Astrophys. J.*, **187**, 425–438

Press, W. H. & Teukolsky, S. A. (1997) *Numerical Recipes*: Does this paradigm have a future?, *Computers in Physics*, **11**, 416

Press, W. H., Teukolsky, S. A. & Vetterling, W. T. (2007) *Numerical Recipes: The Art of Scientific Computing*, Cambridge University Press

Press, W. J. & Rybicki, G. B. (1989) Fast algorithm for spectral analysis of unevenly sampled data, *Astrophys. J.*, **338**, 277–280

Priestley, M. B. (1983) *Spectral Analysis and Time Series*, 2 vols., Academic Press, New York

Protassov, R., van Dyk, D. A., Connors, A., Kashyap, V. L., & Siemiginowska, A. (2002) Statistics, handle with care: Detecting multiple model components with the likelihood ratio test, *Astrophys. J.*, **571**, 545–559

Protheroe, R. J. (1987) Periodic analysis of gamma-ray data, *Proc. Astro. Soc. Australia*, **7**, 167–172

Quenouille, M. (1949) Approximate tests of correlation in time series, *J. Roy. Statist. Soc.*, Ser. B, **11**, 18–84

Quinlan, J. R. (1993) *C4.5: Programs for Machine Learning*, Morgan Kaufmann, New York

R Development Core Team (2010) R: A Language and Environment for Statistical Computing, R Foundation for Statistical Computing, Vienna, Austria

Rao, C. R. (1997) *Statistics and Truth: Putting Chance to Work*, 2nd ed., World Scientific, Singapore

Rao, C. R., Wegman, E. J. & Solka, J. L., eds. (2005), *Data Mining and Data Visualization*, Handbook of Statistics vol. 24, Elsevier, Amsterdam

Reed, W. J. & Hughes, B. D. (2002) From gene families and genera to incomes and internet file sizes: Why power laws are so common in nature, *Phys. Rev. E.*, **66**, 67103–67107

Reed, W. J. & Jorgensen, M. (2004) The double Pareto-lognormal distribution: A new parametric model for size distributions, *Commun. Stat. Theory Meth.*, **33**, 1733–1753

Reegen, P. (2007) SigSpec. I. Frequency- and phase-resolved significance in Fourier space, *Astron. Astrophys.*, **467**, 1353–1371

Reipurth, B., Rodríguez, L. F., Anglada, G. & Bally, J. (2004) Radio continuum jets from protostellar objects, *Astron. J.*, **127**, 1736–1746

Reisen, V. A. (1994) Estimation of the fractional difference parameter in the ARFIMA (p, d, q) model using the smoothed periodogram, *J. Time Series Anal.*, **15**, 335–350

Remy, N., Boucher, A. & Wu, J. (2009) *Applied Geostatistics with SGeMS*, Cambridge University Press

Renner, S., Rauer, H., Erikson, A., Hedelt, P., Kabath, P., Titz, R. & Voss, H. (2008) The BAST algorithm for transit detection, *Astron. Astrophys.*, **492**, 617–620

Ribeiro, P. J. & Diggle, P. J. (2001) geoR: a package for geostatistical analysis, *R-NEWS*, **1**(2), 15–18

Rice, J. A. (2007) *Mathematical Statistics and Data Analysis*, 3rd ed., Duxbury Press

Richardson, W. H. (1972) Bayesian-based iterative method for image restoration, *J. Optical Soc. Amer.*, **62**, 55–59

Rieger, E., Kanbach, G., Reppin, C., Share, G. H., Forrest, D. J. & Chupp, E. L. (1984) A 154-day periodicity in the occurrence of hard solar flares?, *Nature*, **312**, 623–625

Ripley, B. D. (1988) *Statistical Inference for Spatial Processes*, Cambridge University Press

Ritz, C. & Streibig, J. C. (2008) *Nonlinear Regression with R*, Springer, Berlin

Rivoira, A. & Fleury, G. A. (2004) A consistent nonparametric spectral estimator for randomly sampled signals, *IEEE Trans. Signal Proc.* **52**, 2383–2395

Roberts, D. H., Lehar, J. & Dreher, J. W. (1987) Time series analysis with CLEAN. I. Derivation of a spectrum. *Astron. J.*, **93**, 968–989

Robinson, P. D. (1995) Log-periodogram regression of time series with long range dependence, *Annals Stat.*, **23**, 1048–1072

Robinson, P. D., ed. (2003) *Time Series with Long Memory*, Oxford University Press

Ross, S. (2010) *Introduction to Probability Models*, 10th ed., Academic Press, New York

Rossini, A. J., Tierney, L. & Li, N. (2007) Simple parallel statistical computing in R, *J. Comput. Graph. Statist.*, **16**, 399–420

Rowlingson, B. & Diggle, P. (1993) Splancs: Spatial point pattern analysis code in S-Plus. *Comput. Geosci.*, **19**, 627–655

Santos, N. C., García López, R. J., Israelian, G., Mayor, M., Rebolo, R., García-Gil, A., Pérez de Taoro, M. R. & Randich, S. (2002) Beryllium abundances in stars hosting giant planets, *Astron. Astrophys.*, **386**, 1028–1038

Sarkar, D. (2008) *Lattice: Multivariate Data Visualization with R*, Springer, Berlin

Saslaw, W. C. (2000) *The Distribution of the Galaxies: Gravitational Clustering in Cosmology*, Cambridge University Press

Scargle, J. D. (1982) Studies in astronomical time series analysis. II – Statistical aspects of spectral analysis of unevenly spaced data, *Astrophys. J.*, **263**, 835–853

Scargle, J. D. (1992) Chaotic processes in astronomical data, in *Statistical Challenges in Modern Astronomy* (E. D. Feigelson & G. J. Babu, eds.), Springer, Berlin, pp. 411–436

Scargle, J. D. (1998) Studies in astronomical time series analysis. V. Bayesian blocks, a new method to analyze structure in photon counting data, *Astrophys. J.*, **504**, 405

Scargle, J. D. (2003) Advanced tools for astronomical time series and image analysis, in *Statistical Challenges in Modern Astronomy* (E. D. Feigelson & G. J. Babu, eds.), Springer, New York, pp. 293–308

Scargle, J. D., Steiman-Cameron, T., Young, K., Donoho, D. L., Crutchfield, J. P. & Imamura, J. (1993) The quasi-periodic oscillations and very low frequency noise of Scorpius X-1 as transient chaos – A dripping handrail?, *Astrophys. J.*, **411**, L91–L94

Schechter, P. (1976) An analytic expression for the luminosity function for galaxies, *Astrophys. J.*, **203**, 297–306

Schechtman, E. & Spiegelman, C. (2007) Mitigating the effect of measurement errors in quantile estimation, *Stat. Prob. Lett.*, **77**, 514–524

Schlesinger, F. (1916) The determination of the orbit of a spectroscopic binary by the method of least-squares, *Publ. Allegheny Obs.*, **1**(6), 33–44

Schmee, J. & Hahn, G. J. (1979) A simple method for regression analysis with censored data, *Technometrics*, **21**, 417–432

Schmidt, M. (1968) Space distribution and luminosity functions of quasi-stellar radio sources, *Astrophys. J.*, **151**, 393–409

Schmitt, J. H. M. M. (1985) Statistical analysis of astronomical data containing upper bounds: General methods and examples drawn from X-ray astronomy, *Astrophys. J.*, **293**, 178

Schneider, D. P. *et al.* (2007) The Sloan Digital Sky Survey Quasar Catalog. IV. Fifth Data Release, *Astron. J.*, **134**, 102–117

Schneider, D. P. *et al.* (2010) The Sloan Digital Sky Survey Quasar Catalog. V. Seventh Data Release, *Astron. J.*, **139**, 2360–2373

Schwarzenberg-Czerny, A. (1989) On the advantage of using analysis of variance for period search, *Mon. Not. Royal Astro. Soc.*, **241**, 153–165

Schwarzenberg-Czerny, A. (1999) Optimum period search: Quantitative analysis, *Astrophys. J.*, **516**, 315–323

Scott, D. W. (1979) On optimal and data-based histograms, *Biometrika* **66**, 605–610

Scott, D. W. (1992) *Multivariate Density Estimation: Theory, Practice and Visualization*, John Wiley, New York

Scott, D. W. (2009) (S original by David W. Scott, R port by Albrecht Gebhardt adopted to recent S-PLUS by Stephen Kaluzny) ash: David Scott's ASH routines, R package version 1.0-12, http://CRAN.R-project.org/package=ash

Scott, D. W. & Sain, S. R. (2005) Multidimensional density estimation, in *Data Mining and Data Visualization* (C. R. Rao *et al.*, eds.), *Handbook of Statistics*, vol. 24, Elsevier, Amsterdam

Secker, J. & Harris, W. E. (1993) A maximum likelihood analysis of globular cluster luminosity distributions in the Virgo ellipticals, *Astron. J.*, **105**, 1358–1368

Sen, P. K. 1968. Estimates of the regression coefficient based on Kendall's τ, *J. Amer. Stat. Assoc.*, **63**, 1379–1389

Sérsic, J. L. (1968) *Atlas de Galaxias Australes*, Obs. Astronómico de Córdoba

Shao, J. (1993) Linear model selection by cross-validation, *J. Amer. Stat. Assoc.*, **88**, 486–494

Shapiro, H. S. & Silverman, R. A. (1960) Alias-free sampling of random noise, *J. Soc. Indust. Appl. Math.*, **8**, 225–248

Sheather, S. J. (2009) *A Modern Approach to Regression with R*, Springer, Berlin

Sheynin, O. B. (1973) On the prehistory of the theory of probability, *Arch. Hist. Exact Sci.*, **12**, 97–141

Shkedy, Z., Decin, L., Molenberghs, G. & Aerts, C. (2007) Estimating stellar parameters from spectra using a hierarchical Bayesian approach, *Mon. Not. Royal Astro. Soc.*, **377**, 120–132

Shmueli, G., Minka, T., Kadane, J. B., Borle, S. & Boatwright, P. B. (2005) A useful distribution for fitting discrete data: revival of the Conway–Maxwell–Poisson distribution, *J. Royal Stat. Soc. Ser. C*, **54**, 127–142

Shumway, R. H. & Stoffer, D. S. (2006) *Time Series Analysis and Its Applications with R Examples*, 2nd ed., Springer, Berlin

Sicilia-Aguilar, A. *et al.* (2009) The long-lived disks in the η Chamaeleontis cluster, *Astrophys. J.*, **701**, 1188–1203

Silverman, B. W. (1981) Using kernel density estimates to investigate multimodality, *J. Amer. Stat. Assn.*, Ser B, **43**, 97–99

Silverman, B. W. (1986) *Density Estimation for Statistics and Data Analysis*, Chapman & Hall, London

Silverman, B. W. (1998) *Density Estimation*, Chapman & Hall/CRC, London

Simonetti, J. H., Cordes, J. M. & Heeschen, D. S. (1985) Flicker of extragalactic radio sources at two frequencies, *Astrophys. J.*, **296**, 46–59

Simpson, P. B. (1951) Note on the estimation of a bivariate distribution function, *Ann. Math. Stat.*, **22**, 476–478.

Simpson, T. (1756) A letter to the Right Honorable George Earl of Macclesfield, President of the Royal Society, on the advantage of taking the mean of a number of observations, in practical astronomy, *Phil. Trans. Royal Soc. London*, **49**, 82–93

Singh, K. (1981) On the asymptotic accuracy of Efron's bootstrap, *Ann. Statist.*, **9**, 1187–1195

Skellam, J. G. (1946) The frequency distribution of the difference between two Poisson variates belonging to different populations, *J. Royal Stat. Soc. Series A*, **109**, 296

Slanina, F. (2004) Inelastically scattering particles and wealth distribution in an open economy, *Phys. Rev. E*, **69**, 046102

Slutsky, E. (1913) On the criterion of goodness of fit of the regression lines and on the best method of fitting them to the data, *J. Royal Stat. Soc.*, **77**, 78–84

Sneath, P. H. A. (1957) The application of computers to taxonomy, *J. Gen. Microbiol.*, **17**, 201–226

Solanki, S. K., Krivova, N. A., Schüssler, M. & Fligge, M. (2002) Search for a relationship between solar cycle amplitude and length, *Astron. Astrophys.*, **396**, 1029–1035

Solomon, S. & Richmond, R. (2002) Stable power laws in variable economies; Lotka–Volterra implies Pareto–Zipf, *Euro. Phys J. B*, **27**, 257–261

Speed, T. (1992) Discussion: Statistics, mathematical statistics, and reality, in *Statistical Challenges in Modern Astronomy* (E. D. Feigelson & G. J. Babu, eds.), Springer, Berlin

Springel, V. *et al.* (2005) Simulations of the formation, evolution and clustering of galaxies and quasars, *Nature*, **435**, 629–636

Srivastava, S., Gupta, M. R. & Frigyik, B. A. (2007) Bayesian quadratic discriminant analysis, *J. Machine Learn. Res.*, **8**, 1277–1305

Starck, J.-L. & Murtagh, F. (2006) *Astronomical Image and Data Analysis*, 2nd ed., Springer, Berlin

Starck, J.-L., Murtagh, F. D. & Bijaoui. A. (1998) *Image Processing and Data Analysis: The Multiscale Approach*, Cambridge University Press

Starck, J.-L., Murtagh, F. & Fadili, J. (2010) *Sparse Image and Signal Processing: Wavelets, Curvelets, Morphological Diversity*, Cambridge University Press

Staudenmayer, J., Ruppert, D. & Buonaccorsi, J. P. (2008) Density estimation in the presence of heteroscedastic measurement error, *J. Amer. Stat. Assoc.*, **103**, 726–736

Stellingwerf, R. F. (1978) Period determination using phase dispersion minimization, *Astrophys. J.*, **224**, 953–960

Stephens, M. A. (1974) EDF statistics for goodness of fit and some comparisons, *J. Amer. Stat. Assn.*, **69**, 730–737

Stigler, S. M. (1986) *The History of Statistics: The Measurement of Uncertainty before 1900*, Harvard University Press

Stone, C. J. (1984) An asymptotically optimal window selection rule for kernel density estimators, *Annals Statist.*, **12**, 1285–1297

Stone, M. (1977) Asymptotics for and against cross-validation, *Biometrika* **64**, 29–35.

Stoyan, D. (2006) Fundamentals of point process statistics, in *Case Studies in Spatial Point Process Modeling* (A. Baddeley *et al.*, eds.), Springer, Berlin, pp. 3–22

Stoyan, D. & Stoyan, H. (1994) *Fractals, Random Shapes and Point Fields: Methods of Geometrical Statistics*, John Wiley, New York

Sturrock, P. A. & Scargle, J. D. (2009) A Bayesian assessment of *p*-values for significance estimation of power spectra and an alternative procedure, with application to solar neutrino data, *Astrophys. J.*, **706**, 393–398

Sutton, C. D. (2005), Classification and regression trees, bagging, and boosting, in *Data Mining and Data Visualization* (C. R. Rao *et al.*, eds.), Handbook for Statistics vol. 24, Elsevier, Amsterdam, p. 303

Swanepoel, J. W. H., de Beer, C. F. & Loots, H. (1996) Estimation of the strength of a periodic signal from photon arrival times, *Astrophys. J.*, **467**, 261–264

Tago, E., Saar, E., Tempel, D., Einasto, J., Einasto, M., Nurmi, P. & Heinämäki, P. (2010) Groups of galaxies in the SDSS Data Release 7: Flux- and volume-limited samples, *Astron. Astrophys.*, **514**, A102

Takeuchi, T. T. & Ishii, T. T. (2004) A general formulation of the source confusion statistics and application to infrared galaxy surveys, *Astrophys. J.*, **604**, 40–62

Takezawa, K. (2005) *Introduction to Nonparametric Regression*, Wiley-Interscience, New York

Tamuz, O., Mazeh, T. & Zucker, S. (2005) Correcting systematic effects in a large set of photometric light curves, *Mon. Not. Royal Astro. Soc.*, **356**, 1466–1470

Teerikorpi, P. (1997) Observational selection bias affecting the determination of the extragalactic distance scale, *Ann. Rev. Astro. Astrophys.*, **35**, 101–136

Telea, A. C. (2007) *Data Visualization: Principles and Practice*, A. K. Peters, Naticp, MA

Templ, M. (2010) clustTool: GUI for clustering data with spatial information, R package version 1.6.5, http://CRAN.R-project.org/package=clustTool

Theus, M. (2008) High-dimensional data visualization, in *Handbook of Data Visualization* (C. Chen, ed.) Springer, Berlin

Tibshirani, R. (1996) Regression shrinkage and selection via the lasso, *J. Royal. Statist. Soc. B*, **58**, 267–288

Totsuji, H. & Kihara, T. (1969) The correlation function for the distribution of galaxies, *Pub. Astron. Soc. Japan*, **21**, 221–229

Townshend, R. H. D. (2010) Fast calculation of the Lomb–Scargle periodogram using graphics processing units, *Astrophys. J. Suppl.*, **191**, 247–253

Tremaine, S. *et al.* (2002) The slope of the black hole mass versus velocity dispersion correlation, *Astrophys. J.*, **574**, 740–753

Tripodis, Y. & Buonaccorsi, J. P. (2009) Prediction and forecasting in linear models with measurement error, *J. Stat. Plan. Infer.*, **139**, 4039–4050

Trotta, R. (2008) Bayes in the sky: Bayesian inference and model selection in cosmology, *Contemp. Phys.*, **49**, 71–104

Trumpler, R. J. & Weaver, H. F. (1953) *Statistical Astronomy*, Dover, New York

Tsai, W.-Y. (1990) Testing the assumption of independence of truncation time and failure time, *Biometrika*, **77**, 169–77

Tukey, J. (1977) *Exploratory Data Analysis*, Addison-Wesley, Reading MA

Valdes, F. (1982) Resolution classifier, in *Instrumentation in Astronomy IV*, SPIE, **331**, 465–472

van der Klis, M. (1989) Quasi-periodic oscillations and noise in low mass X-ray binaries, *Ann. Rev. Astron. Astrophys.*, **27**, 517–553

van der Klis, M. (2000) Millisecond oscillations in X-ray binaries, *Ann. Rev. Astron. Astrophys.*, **38**, 717–760

van Dyk, D. A., Connors, A., Kashyap, V. L. & Siemiginowska, A. (2001) Analysis of energy spectra with low photon counts via Bayesian posterior simulation, *Astrophys. J.*, **548**, 224–243

van Leeuwen, F. (2007) Validation of the new Hipparcos reduction, *Astron. Astrophys.*, **474**, 653–664

Vapnik, V. (2000) *The Nature of Statistical Learning Theory*, Springer, Berlin

Varmuza, K. & Filzmoser P. (2009) *Introduction to Multivariate Statistical Analysis in Chemometrics*, CRC Press, Boca Raton

Veilleux, S. & Osterbrock, D. E. (1987) Spectral classification of emission-line galaxies, *Astrophys. J. Suppl.*, **63**, 295–310

Venables, W. N. & Ripley, B. D. (2002) *Modern Applied Statistics with S*, 4th ed., Springer, Berlin

Vio, R., Andreani, P. & Biggs, A. (2010) Unevenly-sampled signals: a general formalism for the Lomb–Scargle periodogram, *Astron. Astrophys.*, **519**, id.A85

Vio, R., Strohmer, T. & Wamsteker, W. (2000), On the reconstruction of irregularly sampled time series, *Pub. Astron. Soc. Pacific*, **112**, 74–90

Vio, R., Kristensen, N. R., Madsen, H. & Wamsteker, W. (2005) Time series analysis in astronomy: Limits and potentialities, *Astron. Astrophys.*, **435**, 773–780

Vogt, S. S., Butler, R. P., Rivera, E. J., Haghighipour, N., Henry, G. W. & Williamson, M. H. (2010) The Lick–Carnegie Exoplanet Survey: A 3.1 M ? planet in the habitable zone of the nearby M3V star Gliese 581, *Astrophys. J.*, **723**, 954–965

Volk, P. B., Rosenthal, F., Hahmann, M., Habich, D. & Lehner, W. (2009) Clustering uncertain data with possible worlds, in *IEEE 25th Conf. Data Engineering (ICDE '09)*, pp. 1625–1632

von Neumann, J. (1941) Distribution of the ratio of the mean square successive difference to the variance, *Ann. Math. Stat.*, **12**, 367–395

Wackernagel, H. (1995) *Multivariate Geostatistics: An Introduction with Applications*, Springer, Berlin

Wall, J. V. & Jenkins, C. R. (2003) *Practical Statistics for Astronomers*, Cambridge University Press

Wand, M. P. & Jones, M. C. (1995) *Kernel Smoothing*, Chapman and Hall, London

Wang, J., Feigelson, E. D., Townsley, L. K., Román-Zúñiga, C. G., Lada, E. & Garmire, G. (2009) A Chandra study of the Rosette star-forming complex. II. Clusters in the Rosette Molecular Cloud, *Astrophys. J.*, **696**, 47–65

Wang, L. (1998) Estimation of censored linear errors-in-variables models, *J. Econometrics*, **84**, 383–400

Wang, X., Woodroofe, M., Walker, M. G., Mateo, M. & Olszewski, E. (2005) Estimating dark matter distributions, *Astrophys. J.*, **626**, 145–158

Wang, X.-F. & Wang, B. (2010) decon: Deconvolution estimation in measurement error models, R package version 1.2-1, http://CRAN.R-project.org/package=decon

Warner, R. M. (1998) *Spectral Analysis of Time-Series Data*, Guilford Press, New York

Wasserman, L. (2004) *All of Statistics: A Concise Course in Statistical Inference*, Springer, Berlin

Way, M. J., Scargle, J. D., Ali, K. & Srivastava, A. N., eds. (2011) *Advances in Machine Learning and Data Mining for Astronomy*, Chapman & Hall, London

Webbink, R. F. (1985) Structure parameters of galactic globular clusters, in *Dynamics of Star Clusters*, Reidel, Dordrecht, pp. 541–577

Wei, W. W. S. (2006) *Time Series Analysis: Univariate and Multivariate Method*, 2nd ed., Pearson, Harlow

Weinstein, M. A. *et al.* (2004) An empirical algorithm for broadband photometric redshifts of quasars from the Sloan Digital Sky Survey, *Astrophys. J. Suppl.*, **155**, 243–256

Wells, D. C., Greisen, E. W. & Harten, R. H. (1981) FITS – a Flexible Image Transport System, *Astron. Astrophys. Suppl.*, **44**, 363–370

Wheatland, M. S. (2000) Do solar flares exhibit an interval-size relationship?, *Solar Phys.*, **191**, 381–399

Whitcher, B. (2010) waveslim: Basic wavelet routines for one-, two- and three-dimensional signal processing, R package version 1.6.4, http://CRAN.R-project.org/package=waveslim

White, H. (1982) Maximum likelihood estimation of misspecified models, *Econometrica*, **50**, 1–25

Whitmore, B. C. (1984) An objective classification system for spiral galaxies. I. The two dominant dimensions, *Astrophys. J.*, **278**, 61–80

Whittle, P. (1962) Gaussian estimation in stationary time series, *Bull. Inst. Internat. Statist.*, **39**, 105–129

Wickham, H. (2009) *ggplot2: Elegant Graphics for Data Analysis*, Springer, Berlin

Wielen, R., Jahreiss, J. & Krüser, R. (1983) The determination of the luminosity function of nearby stars, in *The Nearby Stars and the Stellar Luminosity Function*, L. Davis Press, Schenectady, NY, pp. 163–170

Williams, G. J. (2009) Rattle: A data mining GUI for R, *The R Journal*, **1**(2), 45–55, http://journal.r-project.org/archive/2009–2/RJournal_2009-2_Williams.pdf.

Wilson, R. E. & Devinney, E. J. (1971) Realization of accurate close-binary light curves: Application to MR Cygni, *Astrophys. J.*, **166**, 605–619

Winkler, G. (2002) *Image Analysis, Random Fields and Markov Chain Monte Carlo: A Mathematical Introduction*, 2nd ed., Springer, Berlin

Winston, E. *et al.* (2007) A combined Spitzer and Chandra survey of young stellar objects in the Serpens cloud core, *Astrophys. J.*, **669**, 493–518

Witten, I. H., Frank, E. & Hall, M. A. (2011) *Data Mining: Practical Machine Learning Tools and Techniques*, 3rd ed., Morgan Kaufman, New York

Wolf, C., Meisenheimer, K., Rix, H.-W., Borch, A., Dye, S. & Kleinheinrich, M. (2003) The COMBO-17 survey: Evolution of the galaxy luminosity function from 25 000 galaxies with $0.2 < z < 1.2$, *Astron. Astrophys.*, **401**, 73–98

Wolpert, D. H. & Macready, W. G. (1997) No free lunch theorems for optimization, *IEEE Trans. Evol. Comput.*, **1**, 67–82

Wolynetz, M. S. (1979) Algorithm AS 138: Maximum likelihood estimation from confined and censored normal data, *Applied Stat.*, **28**, 185–195

Woodroofe, M. (1985) Estimating a distribution function with truncated data, *Annals Statist.*, **13**, 163–177

Wu, C. F. J. (1983) On the convergence properties of the EM Algorithm, *Annals Statist.*, **11**, 95–103

Wu, C. F. J. (1986) Jackknife, bootstrap and other resampling methods in regression analysis, *Annals Statist.*, **14**, 1261–1295

Wu, X.-B. & Jia, Z. (2010) Quasar candidate selection and photometric redshift estimation based on SDSS and UKIDSS data, *Mon. Not. Royal Astron. Soc.*, **406**, 1583–1594

Yèche, C. *et al.* (2010) Artificial neural networks for quasar selection and photometric redshift determination, *Astron. Astrophys.*, **523**, id.A14

Yee, T. W. (2010) The VGAM package for categorical data analysis, *J. Statist. Software*, **32**(10), 1–34, http://www.jstatsoft.org/v32/i10/

York, D. (1966) Least-squares fitting of a straight line. *Canad. J. Phys.*, **44**, 1079–1086

York, D. *et al.* (2000) The Sloan Digital Sky Survey: Technical summary, *Astron. J.*, **20**, 1579–1587

Young, G. A. & Smith, R. L. (2005) *Essentials of Statistical Inference*, Cambridge University Press

Zamorani, G. *et al.* (1981) X-ray properties of quasars with the Einstein Observatory. II, *Astrophys. J.*, **245**, 357–374

Zechmeister, M. & Kürster, M. (2009), The generalised Lomb–Scargle periodogram. A new formalism for the floating-mean and Keplerian periodograms, *Astron. Astrophys.*, **496**, 577–584

Zeileis, A. & Grothendieck, G. (2005) zoo: S3 infrastructure for regular and irregular time series, *J. Statist. Software*, **14**, 1–27

Zhang, T., Ramakrishnan, R. & Livny, M. (1997) BIRCH: A new data clustering algorithm and its applications, *Data Mining & Knowledge Discovery*, **1**, 141–182

Zoubir, A. M. & Iskander, D. R. (2004), *Bootstrap Techniques for Signal Processing*, Cambridge University Press

Zwicky, F. (1937) On the masses of nebulae and of clusters of nebulae, *Astrophys. J.*, **86**, 217–246

Subject index

R and CRAN commands

Printed in the United States
By Bookmasters